# Lecture Notes in Computer Science 8988

Commenced Publication in 1973
Founding and Former Series Editors:
Gerhard Goos, Juris Hartmanis, and Jan van Leeuwen

More information about this series at http://www.springer.com/series/7151

James F. Peters · Andrzej Skowron
Dominik Ślęzak · Hung Son Nguyen
Jan G. Bazan (Eds.)

# Transactions on Rough Sets XIX

Springer

*Editors-in-Chief*

James F. Peters
Electrical and Computer Engineering
University of Manitoba
Winnipeg, MB
Canada

Andrzej Skowron
University of Warsaw
Warsaw
Poland

*Guest Editors*

Dominik Ślęzak
University of Warsaw
Warsaw
Poland

Jan G. Bazan
University of Rzeszów
Rzeszów
Poland

Hung Son Nguyen
University of Warsaw
Warsaw
Poland

ISSN 0302-9743          ISSN 1611-3349  (electronic)
Lecture Notes in Computer Science
ISBN 978-3-662-47814-1          ISBN 978-3-662-47815-8    (eBook)
DOI 10.1007/978-3-662-47815-8

Library of Congress Control Number: 2015943053

Springer Heidelberg New York Dordrecht London

Printed on acid-free paper

Springer-Verlag GmbH Berlin Heidelberg is part of Springer Science+Business Media
(www.springer.com)

# Preface

Volume XIX of the *Transactions on Rough Sets* was inspired by the 2012 Workshop on Rough Set Applications (RSA 2012) organized within the 7th International Symposium on Advances in Artificial Intelligence and Applications (AAIA 2012), held as a part of the 2012 Federated Conference on Computer Science and Information Systems (FedCSIS 2012) in Wrocław, Poland, during September 9–12, 2012. The volume consists of two parts. The first part gathers seven regular contributions, which are extended and re-reviewed versions of the papers originally presented at the afore mentioned symposium. The second part contains three contributions in the category of short surveys and monographs. We believe that this way we achieved a well-balanced content reflecting the current trends and advances in both the foundations and practical applications of rough sets.

The paper co-authored by Tuan-Fang Fan, Churn-Jung Liau, and Duen-Ren Liu introduces a new uniform theoretical framework for rough approximations based on generalized quantifiers. The paper co-authored by Ivo Düntsch and Günther Gediga proposes a parameter-free and monotonic alternative to the parametric variable precision model developed within the theory of rough sets. The paper co-authored by Mohammad Azad, Igor Chikalov, Mikhail Moshkov, and Beata Zielosko compares three approaches to define and search for superreducts in inconsistent decision tables. The paper co-authored by Long Giang Nguyen and Hung Son Nguyen outlines an alternative method of eliminating the attributes that do not occur in any reducts of a given decision table. The paper by Beata Zielosko utilizes the tools of dynamic programming to extract inexact decision rules optimized with respect to their length, coverage, and confidence. The paper by Krzysztof Pancerz discusses how to incorporate information about semantic relations between decision classes to better model the accuracy of rough approximations. The paper co-authored by Jan G. Bazan, Sylwia Buregwa-Czuma, Przemysław W. Pardel, Stanisława Bazan-Socha, Barbara Sokołowska, and Sylwia Dziedzina describes the rough-set-based classification method for predicting coronary stenosis demanding revascularization for patients diagnosed with stable coronary heart disease. The survey co-authored by Pulak Samanta and Mihir K. Chakraborty summarizes correspondences between different extensions of the theory of rough sets and modal logic systems. The monograph co-authored by Gloria Virginia and Hung Son Nguyen illustrates how to use the tolerance rough set model to conduct semantic text retrieval for the Indonesian language. The monograph authored by Pradip Kundu shows how to formulate and solve transportation problems in uncertain environments modeled by means of fuzzy sets and rough sets.

The editors would like to acknowledge the authors of all submitted papers and anonymous reviewers for their excellent work and insightful comments. The guest editors are grateful to James F. Peters and Andrzej Skowron for agreeing to include extended post-conference materials in the series of the *Transactions on Rough Sets*. The editors and the authors extend their gratitude to Alfred Hofmann and the whole

LNCS team at Springer for their support in making this volume possible. Special thanks go to Sheela Ramanna and Marcin Szczuka for their help in the process of collecting, tracking, and editing the articles.

The Editors-in-Chief were supported by the Polish National Science Centre (NCN) grants DEC-2012/05/B/ST6/03215 and DEC-2013/09/B/ST6/01568, the Polish National Centre for Research and Development (NCBiR) grants O ROB/0010/03/001 and PBS2/B9/20/2013, as well as the Natural Sciences and Engineering Research Council of Canada (NSERC) discovery grant 185986.

April 2015                                                      Dominik Ślęzak
                                                               Jan G. Bazan
                                                          Hung Son Nguyen
                                                           James F. Peters
                                                          Andrzej Skowron

# LNCS Transactions on Rough Sets

The *Transactions on Rough Sets* series has as its principal aim the fostering of professional exchanges between scientists and practitioners who are interested in the foundations and applications of rough sets. Topics include foundations and applications of rough sets as well as foundations and applications of hybrid methods combining rough sets with other approaches important for the development of intelligent systems. The journal includes high-quality research articles accepted for publication on the basis of thorough peer reviews. Dissertations and monographs of up to 250 pages that include new research results can also be considered as regular papers. Extended and revised versions of selected papers from conferences can also be included in regular or special issues of the journal.

**Editors-in-Chief:** James F. Peters, Andrzej Skowron
**Managing Editor:** Sheela Ramanna
**Technical Editor:** Marcin Szczuka

## Editorial Board

# Contents

# A Uniform Framework for Rough Approximations Based on Generalized Quantifiers

Tuan-Fang Fan[1], Churn-Jung Liau[2]([⊠]), and Duen-Ren Liu[3]

[1] Department of Computer Science and Information Engineering,
National Penghu University of Science and Technology, Penghu 880, Taiwan
dffan@npu.edu.tw
[2] Institute of Information Science, Academia Sinica, Taipei 115, Taiwan
liaucj@iis.sinica.edu.tw
[3] Institute of Information Management, National Chiao-Tung University,
Hsinchu 300, Taiwan
dliu@iim.nctu.edu.tw

**Abstract.** The rough set theory provides an effective tool for decision analysis in the way of extracting decision rules from information systems. The rule induction process is based on the definitions of lower and upper approximations of the decision class. The condition attributes of the information system constitute an indiscernibility relation on the universe of objects. An object is in the lower approximation of the decision class if all objects indiscernible with it are in the decision class and it is in the upper approximation of the decision class if some objects indiscernible with it are in the decision class. Various generalizations of rough set theory have been proposed to enhance the capability of the theory. For example, variable precision rough set theory is used to improve the robustness of rough set analysis and fuzzy rough set approach is proposed to deal with vague information. In this paper, we present a uniform framework for different variants of rough set theory by using generalized quantifiers. In the framework, the lower and upper approximations of classical rough set theory are defined with universal and existential quantifiers respectively, whereas variable precision rough approximations correspond to probability quantifiers. Moreover, fuzzy rough set approximations can be defined by using different fuzzy quantifiers. We show that the framework can enhance the expressive power of the decision rules induced by rough set-based decision analysis.

**Keywords:** Fuzzy set · Rough set · Variable precision rough set · Fuzzy cardinality

## 1 Introduction

The rough set theory proposed by [2] provides an effective tool for extracting knowledge from information systems. The rule induction process is based on the

---

A preliminary version of this paper was published in [1].

© Springer-Verlag Berlin Heidelberg 2015
J. Peters et al. (Eds.): TRS XIX, LNCS 8988, pp. 1–16, 2015.
DOI: 10.1007/978-3-662-47815-8_1

definitions of lower and upper approximations of the decision class. The condition attributes of the information system constitute an indiscernibility relation on the universe of objects. An object is in the lower approximation of the decision class if all objects indiscernible with it are in the decision class and it is in the upper approximation of the decision class if some objects indiscernible with it are in the decision class.

A strong assumption about information systems is that each object takes exactly one value with respect to an attribute. However, in practice, we may only have incomplete information about the values of an object's attributes. Thus, more general information systems have been introduced to represent incomplete information ([3–7]), whereas fuzzy rough set theory proposed in [8] has been considered as an important mathematical tool to deal with such information systems.

Another limitation of rough set analysis is its sensitivity to noisy information, because a mistakenly labeled sample may deteriorate the quality of approximations significantly. The variable precision rough set (VPRS) theory introduced in [9,10] is a main approach to improve the robustness of rough set analysis. In VPRS, classification rules can be induced even though not satisfied by all objects. It is only required that the proportion of objects satisfying the rules must be above a threshold called a precision level. The idea of error-tolerance by variable precision has been also applied to fuzzy rough set theory [11–17]. In particular, Zadeh's fuzzy quantifiers [18] is used to define lower and upper approximations in the vaguely quantified rough set (VQRS) theory proposed in [11].

In this paper, we extend VQRS to a uniform framework for rough approximations based on generalized quantifiers. The theory of generalized quantifiers (TGQ) has been extensively studied in linguistics and logic [19–24] and different kinds of fuzzy quantifiers have been also formalized by using TGQ [18,25,26]. In addition to the well-known results on the correspondence between classical quantifiers and rough approximations, we also propose new models of variable precision fuzzy rough set (VPFRS) as instances of the uniform framework. The new models are developed by using the notion of fuzzy cardinality introduced in [27]. The proportion of objects satisfying a rule is modeled as a relative fuzzy cardinality in our approach. Because a fuzzy cardinality may be a scalar, a fuzzy number, or a random variable, we can induce three types of models depending on what kinds of fuzzy cardinalities are taken as the precision levels.

The remainder of the paper is organized as follows. In Sect. 2, we review several variants of rough set theory and present a uniform framework for rough approximations based on generalized quantifiers. In Sect. 3, we review the notion of fuzzy cardinality. In Sect. 4, we show that the framework can accommodate different existing variants of rough approximations and introduce three types of VPFRS models based on the relative cardinalities of fuzzy sets as instances of our uniform framework. Section 5 contains some concluding remarks.

## 2 A Uniform Framework for Rough Set Theory

### 2.1 Classical Rough Set

The basic construct of rough set theory is an *approximation space*, which is defined as a pair $(U, R)$, where $U$ is a finite universe and $R \subseteq U \times U$ is an equivalence relation on $U$. We write an equivalence class of $R$ as $[x]_R$ if it contains the element $x$. For any subset $X$ of the universe, the lower approximation and upper approximation of $X$ are defined as follows:

$$\underline{R}X = \{x \in U \mid \forall y((x, y) \in R \to y \in X)\}, \tag{1}$$

$$\overline{R}X = \{x \in U \mid \exists y((x, y) \in R \wedge y \in X)\}. \tag{2}$$

This definition of rough set is called the *logic-based* definition [11].

An alternative way to define rough sets is to use the *rough membership function* [28]. Given an approximation space $(U, R)$ and a subset $X \subseteq U$, the rough membership function $\nu_X^R : U \to [0, 1]$ is defined as

$$\nu_X^R(x) = \frac{|[x]_R \cap X|}{|[x]_R|}. \tag{3}$$

The value $\nu_X^R(x)$ is interpreted as the degree that $x$ belongs to $X$ in view of knowledge about $x$ expressed by the indiscernibility relation $R$ or the degree to which the $R$-equivalence class $[x]_R$ is included in the set $X$. Then, the lower approximation and upper approximation of $X$ are defined as follows:

$$\underline{R}X = \{x \in U \mid \nu_X^R(x) = 1\}, \tag{4}$$

$$\overline{R}X = \{x \in U \mid \nu_X^R(x) > 0\}. \tag{5}$$

This definition of rough set is called the *frequency-based* definition [11].

Although an approximation space is an abstract structure used to represent classification knowledge, it can easily be derived from a concrete information system. Pawlak ([29]) defined an information system[1] as a tuple $T = (U, A, \{V_i \mid i \in A\}, \{f_i \mid i \in A\})$, where $U$ is a nonempty finite set, called the universe; $A$ is a nonempty finite set of primitive attributes; for each $i \in A$, $V_i$ is the domain of values of $i$; and for each $i \in A$, $f_i : U \to V_i$ is a total function. In decision analysis, we assume the set of attributes is partitioned into $\{d\} \cup (A - \{d\})$, where $d$ is called the *decision attribute*, and the remaining attributes in $C = A - \{d\}$ are called *condition attributes*. Given a subset of attributes $B$, the *indiscernibility relation* with respect to $B$ is defined as $ind(B) = \{(x, y) \mid x, y \in U, f_i(x) = f_i(y) \forall i \in B\}$. Obviously, for each $B \subseteq A$, $(U, ind(B))$ is an approximation space.

---

[1] Also called knowledge representation systems, data tables, or attribute-value systems.

## 2.2   Fuzzy Rough Set

In [8], it is shown that fuzzy sets and rough sets are essentially different but complementary for the modeling of uncertainty. Despite the essential difference between fuzzy sets and rough sets, there are approaches to incorporate the notion of fuzzy sets into rough set models [8]. One approach is to consider the lower and upper approximations of a fuzzy concept in an approximation space, which results in the *rough fuzzy set*. The other approach is to consider the approximations of a crisp or fuzzy concept in a *fuzzy approximation space*, which is defined as a pair $(U, R)$, where $R$ is a fuzzy binary relation on $U$, i.e., $R : U \times U \to [0, 1]$. This leads to the *fuzzy rough set*. Let $(U, R)$ be a fuzzy approximation space and let $X$ be a fuzzy subset of $U$ with membership function $\mu_X : U \to [0, 1]$. Then, $\underline{R}X, \overline{R}X : U \to [0, 1]$ are defined by

$$\underline{R}X(x) = \inf_{y \in U} R(x, y) \to_{\otimes} \mu_X(y), \tag{6}$$

$$\overline{R}X(x) = \sup_{y \in U} R(x, y) \otimes \mu_X(y), \tag{7}$$

where $\otimes : [0, 1] \times [0, 1] \to [0, 1]$ is a t-norm and $\to_{\otimes}: [0, 1] \times [0, 1] \to [0, 1]$ is an implication with respect to $\otimes$. There are several different definitions of implication functions $\to_{\otimes}$ for a given t-norm, including the S-implication defined by $a \to_{\otimes} b = 1 - (a \otimes (1 - b))$ and the R-implication defined by $a \to_{\otimes} b = \sup\{c \mid a \otimes c \leq b\}$.

## 2.3   Variable Precision Rough Set

By using the frequency-based definition of the rough set, an object $x$ belongs to the lower approximation of a set $X$ if its rough membership value is 1, i.e., $x \in \underline{R}X$ iff $\nu_X^R(x) = 1$. However, the requirement seems overly strict since in the noisy environment, it may be difficult to require an $R$-equivalence class $[x]_R$ is totally included in a set $X$. The purpose of variable precision rough set (VPRS) theory is to address the issue by relaxing the strict requirement of total inclusion to partial inclusion. Technically, this is achieved by two parameters called precision levels. Let $l$ and $u$ be real numbers such that $0 \leq l < u \leq 1$. Then, the $u$-lower approximation and the $l$-upper approximation of $X$ are defined as follows:

$$\underline{R}_u X = \{x \in U \mid \nu_X^R(x) \geq u\}, \tag{8}$$

$$\overline{R}_l X = \{x \in U \mid \nu_X^R(x) > l\}. \tag{9}$$

## 2.4   Bayesian Rough Set

Although VPRS can alleviate the effect of noisy data, it is not clear how to choose the parameter $u$ and $l$ appropriately. In [30,31], the Bayesian rough set (BRS) model is proposed to overcome the difficulty. Instead of using a fixed but arbitrarily given parameter, BRS sets the precision level as the prior probability of the occurrence of the decision class in the general population. Specifically, the positive, negative and boundary regions of $X$ in BRS are defined as follows:

$$POS_R^*(X) = \{x \in U \mid \nu_X^R(x) > \frac{|X|}{|U|}\}, \tag{10}$$

$$NEG_R^*(X) = \{x \in U \mid \nu_X^R(x) < \frac{|X|}{|U|}\}, \tag{11}$$

$$BND_R^*(X) = \{x \in U \mid \nu_X^R(x) = \frac{|X|}{|U|}\}, \tag{12}$$

where $\frac{|X|}{|U|}$ can be seen as the prior probability of an arbitrarily chosen object in the universe belonging to $X$, whereas the rough membership value $\nu_X^R(x)$ is used to denote the conditional probabilities of an object indiscernible with $x$ belonging to $X$. Thus, the BRS positive region defines an area of the universe where the probability of $X$ is higher than the prior probability. It is an area of certainty improvement or gain with respect to predicting the occurrence of $X$. On the other hand, the BRS negative region defines an area of the universe where the probability of $X$ is lower than the prior probability. It is an area of certainty loss with respect to predicting the occurrence of $X$. As usual, the lower approximation $\underline{R}^* X$ is identified with the positive region and the upper approximation $\overline{R}^* X$ is defined as the complement of the negative region (or the union of the positive and the boundary regions).

## 2.5    Vaguely Quantified Rough Set

The VPRS definitions for upper and lower approximation can be softened by introducing vague quantifiers, to express that $x$ belongs to the upper approximation of $X$ to the extent that some elements of $[x]_R$ are in $X$, and $x$ belongs to the lower approximation of $X$ to the extent that most elements of $[x]_R$ are in $X$. The resultant model is called the vaguely quantified rough set (VQRS) [11]. In VQRS, the lower and upper approximations of a subset $X$ are defined as fuzzy subsets with the following membership functions:

$$\underline{R}_{q_u} X(x) = q_u(\nu_X^R(x)), \tag{13}$$

$$\overline{R}_{q_l} X(x) = q_l(\nu_X^R(x)), \tag{14}$$

where $q_l$ and $q_u$ are two vague quantifiers which are defined as mappings from $[0, 1]$ to $[0, 1]$ according to Zadeh's concept [18].

## 2.6    Generalized Quantifier-Based Rough Set

The notion of VQRS can be further extended to a uniform framework for rough approximations if we consider the vague quantifiers as simply instances of generalized quantifiers. The theory of generalized quantifiers (TGQ) were originally introduced as generalizations of the standard quantifiers of modern logic, $\forall$ and $\exists$ [19–24]. Since the invention of logic by Aristotle, the study of quantification has been the main part of the discipline. The syllogistics can be seen as a formal study of the meaning of the four basic quantifier expressions "all", "no", "some", "not all", and of their properties. In 1870s, Frege introduced the language of predicate logic, with sentential connectives, identity, and the variable-binding operators $\forall$ and $\exists$, which then become the standard quantifiers of modern logic. His formulation of a quantifier as a second-order relation, or, as he called it, a second level concept (Begriff zweiter Stufe) constitutes the foundation of the TGQ except that Frege did not have the idea of an interpretation or model, which we now see as a universe that the quantifiers range over [24].

The modern TGQ was first formulated by Mostowski [21] and then further generalized by Lindström [20]. Formally, a generalized quantifier $Q$ of the type $\langle n_1, n_2, \cdots, n_k \rangle$ assigns to each non-empty universe $U$ a $k$-ary relation $Q_U$ between subsets of $U^{n_1}, \ldots, U^{n_k}$. For a fixed universe $U$, we usually abuse the notation $Q$ to denote the relation $Q_U$. The generalized quantifier $Q$ is *monadic* if all $n_i = 1$ and *polyadic* otherwise. For our purpose, we need to consider only quantifiers of type $\langle 1, 1 \rangle$. Thus, hereafter, we will not specify the types of the generalized quantifiers and assume that their types are all $\langle 1, 1 \rangle$. However, to take fuzzy sets into consideration, we have to further generalize the notion of generalized quantifiers. Therefore, in our context of applications, we define a (fuzzy) generalized quantifier $Q$ on the universe $U$ as a mapping $Q : \widetilde{\mathcal{P}}(U) \times \widetilde{\mathcal{P}}(U) \to [0, 1]$, where $\widetilde{\mathcal{P}}(U)$ denote the class of fuzzy subsets of $U$.

Based on the notion of generalized quantifier, we can now define a uniform framework for rough set theory. We call the framework *generalized quantifier-based rough set theory*(GQRS). By considering the Pawlak's approximation space as a special case of the fuzzy approximation space, our framework is defined simply for the general case of a fuzzy approximation space. Let $(U, R)$ be a fuzzy approximation space and $Q$ be a generalized quantifier on $U$. Then, in GQRS, the $Q$-approximation of a fuzzy subset $X \in \widetilde{\mathcal{P}}(U)$ is a fuzzy subset $R_Q X$ with the following membership function:

$$R_Q X(x) = Q(R_x, X) \tag{15}$$

for $x \in U$, where $R_x$ is a fuzzy subset with membership function $\mu_{R_x}(u) = R(x, u)$ for $u \in U$. In GQRS, we do not have to distinguish lower and upper approximations any more because they are simply different instances of the $Q$-approximations.

Both the original theories of Mostowski and Lindström impose the condition of isomorphism closure on generalized quantifiers, which means that a generalized quantifier should not distinguish isomorphic models. Although not all natural language quantifiers satisfy the condition, logic is supposed to be topic-neutral, so the condition is almost always imposed for quantifiers in logical languages. An interesting implication of the condition for a monadic quantifier $Q$ is that the value $Q(X_1, \cdots, X_k)$ depends only on the cardinalities of the arguments $X_i$ and their Boolean combinations. Monadic quantifiers satisfying isomorphism closure are called *quantitative* quantifiers and play an important role in the evaluation of quantified statements in [25]. Since most variants of rough set theory can be seen as instances of GQRS with quantitative quantifiers, we will restrict our attention to this class of quantifiers. However, because our quantifiers are applied to fuzzy sets, we have to review the notions of fuzzy cardinalities before we can consider real instances of GQRS.

## 3   Fuzzy Cardinality

The notion of fuzzy cardinality is closely related with generalized quantifiers because it is normally used to evaluate quantified sentences in fuzzy logic. For example, to evaluate the truth degree of the sentence "Most students are young," we have to determine if the cardinality of the set of young students satisfies the interpretation of the fuzzy quantifier "most." In applications, two kinds of cardinality are considered: *absolute cardinality*, which measures the number of elements in a set; and *relative cardinality*, which measures the percentage of elements of one set (called the referential set) that are also present in another set ([27]).

Several approaches for measuring the cardinality of a fuzzy set have been proposed in the literature. The approaches, which extend the classic approach in different ways, can be classified into two categories: *scalar cardinality* approaches and *fuzzy cardinality* approaches. The former measure the cardinality of a fuzzy set by means of a scalar value, either an integer or a real value; whereas the latter assume that the cardinality of a fuzzy set is just another fuzzy set over the non-negative numbers ([27]). The most simple scalar cardinality of a fuzzy set is its *power* (also called the $\Sigma$-count), which is defined as the summation of the membership degrees of all elements ([32]). Formally, for a given fuzzy subset $F$ on the universe $U$, the $\Sigma$-count of $F$ is defined as

$$\Sigma\sharp(F) = \sum_{x \in U} \mu_F(x). \tag{16}$$

The relative cardinality of a fuzzy set $G$ with respect to another fuzzy set $F$ is then defined as ([33]):

$$\Sigma\sharp(G/F) = \frac{\Sigma\sharp(F \cap G)}{\Sigma\sharp(F)}. \tag{17}$$

Subsequently, [34] proposed a fuzzy subset $Z(F)$ of $\mathbb{N}$ as the measure of the absolute cardinality of a fuzzy set $F$ such that the membership degree of a natural number $k \in \mathbb{N}$ in $Z(F)$ is defined as

$$Z(F, k) = \sup\{\alpha \mid |F_\alpha| = k\}, \tag{18}$$

where $F_\alpha = \{x \in U \mid \mu_F(x) \geq \alpha\}$ is the $\alpha$-cut of $F$. In addition, a fuzzy multiset $Z(G/F)$ over $[0, 1]$ is introduced in ([18]) to measure the fuzzy relative cardinality of $G$ with respect to $F$. The membership function of $Z(G/F)$, written in the standard integral notation, is defined as

$$Z(G/F) = \sum_{\alpha \in \Lambda(F) \cup \Lambda(G)} \alpha / \frac{|F_\alpha \cap G_\alpha|}{|F_\alpha|}, \tag{19}$$

where $\Lambda(F)$ and $\Lambda(G)$ are the level sets of $F$ and $G$ respectively, i.e., $\Lambda(F) = \{\mu_F(x) \mid x \in U\}$. Delgado et al. ([27]) proposed a more compact representation of $Z(G/F)$ by transforming the fuzzy multiset into a fuzzy subset of rational numbers in $[0,1]$. The representation is formulated as follows:

$$ES(G/F, q) = \sup\{\alpha \in \Lambda(G/F) \mid \frac{|(F \cap G)_\alpha|}{|F_\alpha|} = q\} \tag{20}$$

for any $q \in \mathbb{Q} \cap [0,1]$, where $\Lambda(G/F) = \Lambda(F \cap G) \cup \Lambda(F)$.

In the context of a finite universe $U$, Delgado et al. ([27]) proposed a family of fuzzy measures $\mathcal{E}$ for absolute cardinalities based on the evaluation of fuzzy logic sentences. To define the measures, the possibility of a fuzzy set $F$ containing at least $k$ elements is identified with the truth degree of the fuzzy sentence $\exists X \subseteq U(|X| = k \wedge X \subseteq F)$, which can be formally defined as

$$L(F, k) = \begin{cases} 1, & \text{if } k = 0, \\ 0, & \text{if } k > |U|, \\ \bigoplus_{X \subseteq_k U} \bigotimes_{x \in X} \mu_F(x), & \text{if } 1 \leq k \leq |U|, \end{cases} \tag{21}$$

where $X \subseteq_k U$ denotes that $X$ is any $k$-element subset of $U$ and $\oplus$ is the corresponding s-norm of the t-norm $\otimes$. Then, the possibility that $F$ contains exactly $k$ elements is formulated as follows:

$$E(F, k) = L(F, k) \otimes \neg L(F, k + 1), \tag{22}$$

where $\otimes$ is any t-norm (not necessarily the same as that used in the definition of $L(F, k)$), and $\neg$ stands for a fuzzy negation. Each member of the family $\mathcal{E}$ is determined by the choice of s-norm, t-norms and negation in (21) and (22). Using max and min in (21) and standard negation as well as Łukasiewicz's t-norm $\max(0, a + b - 1)$ in (22), a probabilistic measure of absolute cardinality $ED$ defined as

$$ED(F, k) = \alpha_k - \alpha_{k+1} \tag{23}$$

is shown to be a member of the family $\mathcal{E}$, where $\alpha_k$ is the $k$th largest value of the multiset $\{\mu_F(x) \mid x \in U\}$ for $1 \leq k \leq |U|$, $\alpha_0 = 1$, and $\alpha_k = 0$ when $k > |U|$. The relative version of $ED$ is also defined as

$$ER(G/F, q) = \sum_{i: \frac{|(F \cap G)_{\alpha_i}|}{|F_{\alpha_i}|} = q} (\alpha_i - \alpha_{i+1}) \tag{24}$$

for any $q \in \mathbb{Q} \cap [0,1]$, where $\alpha_i$ is the $i$th largest value of $\Lambda(G/F)$.

# 4    Instances of the Uniform Framework

In this section, we consider several instances of GQRS by instantiating $Q$ to particular quantifiers. We show that the variants of rough set theory reviewed above are all instances of GQRS and present three variable precision fuzzy rough set models as new instances of GQRS.

## 4.1    Existing Instances: Variants of Rough Set Theory

A straightforward instance of $Q$-approximation is the instantiation of $Q$ as the universal and existential quantifiers in the classical predicate logic. When applied to fuzzy sets, these quantifiers correspond to the degree of inclusion and the degree of intersection between sets. Formally, the universal and the existential quantifiers $\forall$ and $\exists : \widetilde{\mathcal{P}}(U) \times \widetilde{\mathcal{P}}(U) \rightarrow [0, 1]$ are defined by

$$\forall(F, G) = \inf_{x \in U} \mu_F(x) \rightarrow_\otimes \mu_G(x) \tag{25}$$

and

$$\exists(F, G) = \sup_{x \in U} \mu_F(x) \otimes \mu_G(x) \tag{26}$$

respectively. Obviously, the lower and upper approximations in fuzzy rough set theories are instances of $Q$-approximation for $Q = \forall$ and $\exists$ respectively.

**Proposition 1.** *Let $(U, R)$ be a fuzzy approximation space and $X$ be a fuzzy subset of $U$. Then,*

1. $R_\forall X = \underline{R}X$
2. $R_\exists X = \overline{R}X$

As a corollary of the proposition, the result holds in the special case when $(U, R)$ is a Pawlak's approximation space and $X$ is a crisp subset of $U$.

It is also quite obvious that instantiating GQRS with *proportional* quantifiers can lead to VPRS. The proportional quantifiers include "most", "at least", "more than", "many", "almost all", etc. and constitute an important subclass of quantitative quantifiers [25]. Formally, a quantifier $Q : \widetilde{\mathcal{P}}(U) \times \widetilde{\mathcal{P}}(U) \rightarrow [0, 1]$ is called proportional if there exists a function $f : [0, 1] \rightarrow [0, 1]$ and $c \in [0, 1]$ such that

$$Q(F, G) = \begin{cases} f(\Sigma\sharp(G/F)), & \text{if } \Sigma\sharp(F) \neq 0, \\ c, & \text{otherwise} \end{cases} \tag{27}$$

Let us define the proportional quantifiers $(\geq u)$ (at least $u * 100\,\%$) and $(> l)$ (more than $l * 100\,\%$) with $f$ in ([27]) being the two-valued functions

$$f_{\geq u}(r) = \begin{cases} 1, & \text{if } r \geq u, \\ 0, & \text{otherwise} \end{cases}$$

and

$$f_{>l}(r) = \begin{cases} 1, & \text{if } r > l, \\ 0, & \text{otherwise} \end{cases}$$

respectively. Then, we have the following proposition.

**Proposition 2.** *Let $(U, R)$ be an approximation space and $X$ be a subset of $U$. Then,*

1. $R_{(\geq u)}X = \underline{R}_u X$
2. $R_{(>l)}X = \overline{R}_l X$

In an analogous way, the VQRS approximations are reformulated as instances of Q-approximations. Specifically, we have the following proposition.

**Proposition 3.** *Let $q_u$ and $q_l$ be two vague quantifiers in VQRS and let us define proportional quantifiers $Q_u$ and $Q_l$ with $f = q_u$ and $q_l$ respectively. Then,*

1. $R_{Q_u}X = \underline{R}_{q_u}X$
2. $R_{Q_l}X = \overline{R}_{q_l}X$

To formulate BRS as an instance of GQRS, we have to introduce *cardinal comparatives*, which are quantifiers of type $\langle 1, 1, 1 \rangle$ that express a comparison of two cardinalities sampled from two restriction arguments $F, G$ and a common scope argument $H$ [25]. A typical example is "more than" in "More students than teachers went to the ballgame" where $F = $ "students", $G = $ "teachers", and $H = $"persons who went to the ballgame" [23]. The formal definition of the quantifier $Q_{mt} : \widetilde{\mathcal{P}}(U)^3 \to \{0, 1\}$ is as follows:

$$Q_{mt}(F, G, H) = (\Sigma\sharp(F \cap H) > \Sigma\sharp(G \cap H)).$$

More generally, a quantifier $Q : \widetilde{\mathcal{P}}(U)^3 \to [0, 1]$ is called cardinal comparative if there exists $q : [0, |U|]^2 \to [0, 1]$ such that

$$Q(F, G, H) = q(\Sigma\sharp(F \cap H), \Sigma\sharp(G \cap H)).$$

Although a cardinal comparative is of type $\langle 1, 1, 1 \rangle$, it is definable in terms of type $\langle 1, 1 \rangle$ quantifiers [23]. In fact, most monadic quantifiers are definable in terms of $\langle 1, 1 \rangle$ quantifiers except the one like "proportionately more than", which typically occurs in sentence like "Proportionately more students than teachers went to the ballgame" [23]. Formally, we can define the quantifier $Q_{pmt} : \widetilde{\mathcal{P}}(U)^3 \to \{0, 1\}$ as follows:

$$Q_{pmt}(F, G, H) = ((\Sigma\sharp(F) > 0) \wedge (\Sigma\sharp(G) > 0) \wedge (\Sigma\sharp(H/F) > \Sigma\sharp(H/G)).$$

Analogously, we can define $Q_{plt}$ ("proportionately less than") and $Q_{pama}$ ("proportionately as many as") by replacing $\Sigma\sharp(H/F) > \Sigma\sharp(H/G)$ in the definition of $Q_{pmt}$ with $\Sigma\sharp(H/F) < \Sigma\sharp(H/G)$ and $\Sigma\sharp(H/F) = \Sigma\sharp(H/G)$ respectively. While $Q_{pmt}, Q_{plt}$, and $Q_{pama}$ are not definable in terms of type $\langle 1, 1 \rangle$ quantifiers, they can be reduced to type $\langle 1, 1 \rangle$ quantifiers if one argument of these quantifiers is fixed. In particular, we can define a type $\langle 1, 1 \rangle$ quantifiers $Q^*$ from a type $\langle 1, 1, 1 \rangle$ quantifier $Q$ by $Q^*(F, G) = Q(F, U, G)$, where $U$ is the universe. Then, we have the following proposition.

**Proposition 4.** *Let $(U, R)$ be an approximation space and $X$ be a subset of $U$. Then,*

1. $R_{Q^*_{pmt}} X = POS^*_R(X)$,
2. $R_{Q^*_{plt}} X = NEG^*_R(X)$,
3. $R_{Q^*_{pama}} X = BND^*_R(X)$

## 4.2  New Instances: Variable Precision Fuzzy Rough Set

The main feature of VPRS is to allow objects that partially violate the indiscernibility principle. For example, if an object belongs to the lower approximation of a set, it is not necessary that all objects that are indiscernible with the object belong to the target set. Instead, the only requirement for an object to be included in the lower approximation of a target set is that a sufficiently large portion of the object's indiscernibility class belongs to the target set. From the viewpoint of the logic-based definition, this amounts to relax the universal quantifier in (1) and strengthen the existential quantifier in (2) to a proportional quantifier determined by $u$ and $l$ respectively. On the other hand, in the frequency-based definition, the lower approximation and the upper approximation of $X$ correspond to the 1-cut and strict 0-cut of the rough membership function respectively[2]. Therefore, from the viewpoint of the frequency-based definition, VPRS simply decreases the cutting point from 1 to $u$ for the lower approximation and increases the cutting point from 0 to $l$ for the upper approximation.

Since the logic-based definition and frequency-based definition are equivalent for classical rough set, the changes of quantifiers or cutting points lead to the same VPRS model. However, for the fuzzy rough set, the situation is quite different. The definition of fuzzy rough approximations in (6) and (7) is essentially a generalization of the logic-based definition of the classical rough set. By using the notion of relative cardinality, we can now present the frequency-based definition of fuzzy rough set.

There have been several VPFRS models that are derived from modifying the logic-based definition of the fuzzy rough set [12–17]. However, the frequency-based approach to VPFRS remains largely unexplored except the vaguely quantified fuzzy rough set (VQFRS) approach [11]. As in the case of VQRS, the main idea of VQFRS is that the membership degree of an object in the lower and upper approximations of a target set is determined by applying fuzzy quantifiers, such as "most" and "some", to the object's fuzzy-rough membership degree in the target set. Nevertheless, the fuzzy-rough membership function used in VQFRS is simply denoted by the $\Sigma^\sharp$ relative cardinality. Therefore, the purpose of this paper is to investigate frequency-based VPFRS models by using different relative cardinalities. In this regard, we can consider three kinds of generalizations of VPRS to VPFRS.

---

[2] Recall that the $\alpha$-cut and the strict $\alpha$-cut of a membership function $\nu : U \to [0,1]$ are defined as $\{x \in U \mid \nu(x) \geq \alpha\}$ and $\{x \in U \mid \nu(x) > \alpha\}$ respectively.

First, if the scalar precision levels $l \in [0, 0.5]$ and $u \in (0.5, 1]$ are given, then we use the relative $\Sigma$-count to measure if an object satisfies the partial precision requirement. Thus, the $u$-lower approximation and the $l$-upper approximation of $X$ are defined as crisp subsets of $U$ as follows:

$$\underline{R}_u X = \{x \in U \mid \Sigma\sharp(X/R_x) \geq u\}, \tag{28}$$

$$\overline{R}_l X = \{x \in U \mid \Sigma\sharp(X/R_x) > l\}. \tag{29}$$

Note that we only apply the quantifiers $(\geq u)$ and $(> l)$ to crisp sets in Proposition 2. Hence, the relative $\Sigma$-count $\Sigma\sharp(X/R_x)$ is reduced to the rough membership value $\nu_X^R(x)$. However, Proposition 2 still holds for fuzzy approximation space according to the definition of VPFRS given here. In other words, it is straightforward to view the new definition as an instance of the GQRS. In particular, this is a special case of VQFRS, which is also an instance of the GQRS.

Second, if the precision levels are fuzzy numbers, then we use the relative cardinality $ES$ to measure an object's fuzzy-rough membership degree. Let $\tilde{l}$ : $[0, 0.5] \cap \mathbb{Q} \to [0, 1]$ and $\tilde{u}$ : $(0.5, 1] \cap \mathbb{Q} \to [0, 1]$ be two fuzzy numbers. Then, $\tilde{u}$-lower approximation and the $\tilde{l}$-upper approximation of $X$ are defined as follows:

$$\underline{R}_{\tilde{u}} X(x) = \pi(ES(X/R_x) \geq \tilde{u}), \tag{30}$$

$$\overline{R}_{\tilde{l}} X(x) = \pi(ES(X/R_x) > \tilde{l}), \tag{31}$$

In the above definition, by slightly abusing the notation defined in (20), the relative cardinality $ES$ is regarded as a fuzzy subset of $\mathbb{Q} \cap [0, 1]$ (i.e., a fuzzy number) with membership function $\mu_{ES(G/F)}(q) = ES(G/F, q)$, and $\pi(\cdot)$ returns the possibility of the comparison statement between two fuzzy numbers based on the extension principle. For example, the possibility of a fuzzy number $\tilde{l}_1$ being greater than another fuzzy number $\tilde{l}_2$ is defined as $\pi(\tilde{l}_1 > \tilde{l}_2) = \sup_{x_1 > x_2} \min(\mu_{\tilde{l}_1}(x_1), \mu_{\tilde{l}_2}(x_2))$.

To formulate $\tilde{u}$-lower approximation and the $\tilde{l}$-upper approximation as instance of $Q$-approximation, we can define two quantifiers $(\geq \tilde{u})$ and $(> \tilde{l})$ as follows:

$$(\geq \tilde{u})(F, G) = \pi(ES(G/F) \geq \tilde{u}), \tag{32}$$

$$(> \tilde{l})(F, G) = \pi(ES(G/F) > \tilde{l}). \tag{33}$$

sNote that these two quantifiers are no longer proportional quantifiers since they do not depend on the relative $\Sigma$-count between $F$ and $G$. Then, straightforwardly, we have the following proposition.

**Proposition 5.** *Let $(U, R)$ be a fuzzy approximation space and $X$ be a fuzzy subset of $U$. Then,*

1. $R_{(\geq \tilde{u})} X = \underline{R}_{\tilde{u}} X$
2. $R_{(> \tilde{l})} X = \overline{R}_{\tilde{l}} X$

Third, if the precision levels are random variables, we can use the relative cardinality $ER$ to represent an object's fuzzy-rough membership degree. Let us overload the notation $ER(G/F)$ to denote a $[0,1] \cap \mathbb{Q}$-valued random variable whose probability mass function is defined as $Pr(ER(G/F) = q) = ER(G/F, q)$ for any fuzzy sets $F$ and $G$, and let $\hat{l}$ and $\hat{u}$ be $[0, 0.5] \cap \mathbb{Q}$-valued and $(0.5, 1] \cap \mathbb{Q}$-valued random variables respectively. Then, the $\hat{u}$-lower approximation and the $\hat{l}$-upper approximation of $X$ are defined as follows:

$$\underline{R}_{\hat{u}}X(x) = Pr(ER(X/R_x) \geq \hat{u}), \tag{34}$$

$$\overline{R}_{\hat{l}}X(x) = Pr(ER(X/R_x) > \hat{l}). \tag{35}$$

As above, we can define two quantifiers $(\geq \hat{u})$ and $(> \hat{l})$ as follows:

$$(\geq \hat{u})(F, G) = Pr(ER(G/F) \geq \hat{u}), \tag{36}$$

$$(> \hat{l})(F, G) = Pr(ER(G/F) > \hat{l}), \tag{37}$$

and derive the following proposition.

**Proposition 6.** *Let $(U, R)$ be a fuzzy approximation space and $X$ be a fuzzy subset of $U$. Then,*

1. $R_{(\geq \hat{u})}X = \underline{R}_{\hat{u}}X$
2. $R_{(> \hat{l})}X = \overline{R}_{\hat{l}}X$

Note that the precision levels $\hat{u}$ and $\hat{l}$ are independent of the data under consideration. In fact, they play the same role as linguistic terms do in (30) and (31). For example, if we set the precision level $\tilde{u}$ in (30) as a linguistic term "about 0.8", then the fuzzy number corresponding to $\tilde{u}$ depends only on the interpretation of the linguistic term and does not have any dependence on $ER(X/R_x)$. Analogously, because the precision levels $\hat{u}$ and $\hat{l}$ are specified externally, they are completely independent of the approximation space $(U, R)$ and the approximated fuzzy set $X$. Therefore, in calculating the probabilities $Pr(ER(X/R_x) \geq \hat{u})$ and $Pr(ER(X/R_x) > \hat{l})$, we can simply assume that the random variable $ER(X/R_x)$ is independent of $\hat{u}$ and $\hat{l}$.

The three types of VPFRS models discussed above are called VPFRS1, VPFRS2, and VPFRS3 respectively. As a scalar can be regarded as a degenerate (possibility or probability) distribution concentrated on a single point, all three types of models can be applied when the precision level is a scalar. However, a typical application of VPFRS2 is in the case of using linguistic terms as precision levels. For example, it may be required that the precision level is moderately high. On the other hand, VPFRS3 can be applied when the precision level is set as a sub-interval of $(0.5, 1] \cap \mathbb{Q}$. In this case, the precision level is regarded as a uniform distribution on the sub-interval, so it is actually a random variable.

# 5   Concluding Remarks

In this paper, we present a uniform framework for rough set theory from the perspective of generalized quantifiers. The advantages of the framework are twofold.

First, the framework can accommodate existing variants of rough set theory as well as motivate new versions of rough set theory. To show this advantage, we instantiate the framework to classical rough set, fuzzy rough set, VPRS, VQRS, and BRS. We also propose three VPFRS models as instances of the framework. These VPFRS models can improve the robustness of the fuzzy rough set. As rough set analysis is sensitive to noisy samples, VPRS can avoid the problem by tolerating partially inconsistency data. The transition from classical rough set theory to VPRS can be achieved by generalizing the quantifiers in the logic-based definition or by changing the cutting points of the rough membership function. Although these two approaches result in the same VPRS model due to the equivalence between logic-based and frequency-based definitions of classical rough set, the situation become radically different in the case of fuzzy rough set. Since the logic-based definition of fuzzy rough set can not be derived from the fuzzy-rough membership function, the VPFRS models obtained from modifying the former should be significantly different from those based on the latter. While most existing VPFRS models adopt the modification of the logic-based definition, the VQRS based on the fuzzy-rough membership function is also shown to outperform the original fuzzy rough set in a benchmark data set [11]. However, since VQRS only consider proportional quantifiers based on the relative $\Sigma$-count, it does not utilize the full potential of generalized quantifiers. It has been also suggested that fuzzy numbers are more appropriate than scalars as cardinalities of fuzzy sets [27]. Thus, our work extends the VQRS model by using different types of relative cardinalities to define the fuzzy-rough membership function. This approach provides a greater flexibility to specify the precision levels when the parameters of the VPFRS models can not be determined precisely, since we allow the precision levels to be scalars, linguistic terms, or random variables.

Second, an advantage of using the uniform framework is to enhance the expressive power of the decision rules induced by rough set analysis. It is well-known that the certain rules and possible rules induced from the lower and upper approximations in classical rough set correspond to universally and existentially quantified statements. By allowing a lot of generalized quantifiers in the rough approximations, we can discover different kinds of rules that contain natural language quantifiers or logical quantifiers. For example, we can induce rules like "Many $A$'s are $B$'s", "Most $A$'s are $B$'s", and "Almost all $A$'s are $B$'s" by using appropriate interpretations of the quantifiers "Many", "Most", and "Almost all". Although we do not explore the impact of this aspect in much detail, it is quite reasonable to expect a very expressive and flexible extension of the Pawlak's decision logic [29] by including generalized quantifiers in the language.

Furthermore, as the main objective of the paper is to develop a theoretical framework that can unify different kinds of rough approximations, the computational aspect is largely ignored. However, there exist efficient procedures to compute (fuzzy) generalized quantifiers. For example, practical implementations of efficient methods for computing a broad class of generalized quantifiers are presented by using the determiner fuzzification scheme in [25]. The complexity analysis demonstrates the computational feasibility of that approach. Our framework can benefit from these existing techniques straightforwardly because the $Q$-approximation of any fuzzy subset $X$ under a fuzzy approximation space $(U, R)$ can be computed simply by evaluating $Q(R_x, X)$ for any $x \in U$.

# References

1. Fan, T., Liau, C., Liu, D.: Variable precision fuzzy rough set based on relative cardinality. In: Proceedings of the Federated Conference on Computer Science and Information Systems (FedCSIS 2012), pp. 43–47 (2012)
2. Pawlak, Z.: Rough sets. Int. J. Comput. Inf. Sci. **11**(15), 341–356 (1982)
3. Kryszkiewicz, M.: Properties of incomplete information systems in the framework of rough sets. In: Polkowski, L., Skowron, A. (eds.) Rough Sets in Knowledge Discovery, pp. 422–450. Physica-Verlag, Heidelberg (1998)
4. Lipski, W.: On databases with incomplete information. J. ACM **28**(1), 41–70 (1981)
5. Myszkorowski, K.: Multiargument relationships in fuzzy databases with attributes represented by interval-valued possibility distributions. In: Kuznetsov, S.O., Slezak, D., Hepting, D.H., Mirkin, B.G. (eds.) RSFDGrC 2011. LNCS, vol. 6743, pp. 199–206. Springer, Heidelberg (2011)
6. Sakai, H., Nakata, M., Slezak, D.: A prototype system for rule generation in Lipski's incomplete information databases. In: Kuznetsov, S.O., Slezak, D., Hepting, D.H., Mirkin, B.G. (eds.) RSFDGrC 2011. LNCS, vol. 6743, pp. 175–182. Springer, Heidelberg (2011)
7. Yao, Y.Y., Liu, Q.: A generalized decision logic in interval-set-valued information tables. In: Zhong, N., Skowron, A., Ohsuga, S. (eds.) RSFDGrC 1999. LNCS (LNAI), vol. 1711, pp. 285–293. Springer, Heidelberg (1999)
8. Dubois, D., Prade, H.: Rough fuzzy sets and fuzzy rough sets. Int. J. Gen. Syst. **17**(2–3), 191–209 (1990)
9. Katzberg, J., Ziarko, W.: Variable precision extension of rough sets. Fundam. Inf. **27**(2/3), 155–168 (1996)
10. Ziarko, W.: Variable precision rough set model. J. Comput. Syst. Sci. **46**(1), 39–59 (1993)
11. Cornelis, C., De Cock, M., Radzikowska, A.M.: Vaguely quantified rough sets. In: An, A., Stefanowski, J., Ramanna, S., Butz, C.J., Pedrycz, W., Wang, G. (eds.) RSFDGrC 2007. LNCS (LNAI), vol. 4482, pp. 87–94. Springer, Heidelberg (2007)
12. Cornelis, C., Verbiest, N., Jensen, R.: Ordered weighted average based fuzzy rough sets. In: Yu, J., Greco, S., Lingras, P., Wang, G., Skowron, A. (eds.) RSKT 2010. LNCS, vol. 6401, pp. 78–85. Springer, Heidelberg (2010)
13. Mieszkowicz-Rolka, A., Rolka, L.: Fuzzy implication operators in variable precision fuzzy rough sets model. In: Rutkowski, L., Siekmann, J.H., Tadeusiewicz, R., Zadeh, L.A. (eds.) ICAISC 2004. LNCS (LNAI), vol. 3070, pp. 498–503. Springer, Heidelberg (2004)
14. Mieszkowicz-Rolka, A., Rolka, L.: Remarks on approximation quality in variable precision fuzzy rough sets model. In: Tsumoto, S., Słowiński, R., Komorowski, J., Grzymała-Busse, J.W. (eds.) RSCTC 2004. LNCS (LNAI), vol. 3066, pp. 402–411. Springer, Heidelberg (2004)

15. Mieszkowicz-Rolka, A., Rolka, L.: Variable precision fuzzy rough sets. In: Peters, J.F., Skowron, A., Grzymała-Busse, J.W., Kostek, B., Swiniarski, R.W., Szczuka, M.S. (eds.) TRS I. LNCS, vol. 3100, pp. 144–160. Springer, Heidelberg (2004)
16. Mieszkowicz-Rolka, A., Rolka, L.: An approach to parameterized approximation of crisp and fuzzy sets. In: Greco, S., Hata, Y., Hirano, S., Inuiguchi, M., Miyamoto, S., Nguyen, H.S., Słowiński, R. (eds.) RSCTC 2006. LNCS (LNAI), vol. 4259, pp. 127–136. Springer, Heidelberg (2006)
17. Salido, J.F., Murakami, S.: Rough set analysis of a general type of fuzzy data using transitive aggregations of fuzzy similarity relations. Fuzzy Sets Syst. **139**(3), 635–660 (2003)
18. Zadeh, L.: A computational approach to fuzzy quantifiers in natural languages. Comput. Math. Appl. **9**(1), 149–184 (1983)
19. Keenan, D., Westerståhl, D.: Generalized quantifiers in linguistics and logic. In: Benthem, J.V., ter Meulen, A. (eds.) Handbook of Logic and Language, pp. 837–893. Elsevier, Beijing (2011)
20. Lindström, P.: First order predicate logic with generalized quantifiers. Theoria **32**, 186–195 (1966)
21. Mostowski, A.: On a generalization of quantifiers. Fundam. Math. **44**(1), 12–36 (1957)
22. Peters, S., Westerståhl, D.: Quantifiers in Language and Logic. Clarendon Press, Wotton-under-Edge, UK (2006)
23. Westerståhl, D.: Quantifiers in natural language: a survey of some recent work. In: Krynicki, M., Mostowski, M., Szczerba, L. (eds.) Quantifiers: Logic, Models and Computation, vol. 1, pp. 359–408. Kluwer Academic Publishers, Dordrecht (1995)
24. Westerståhl, D.: Generalized quantifiers. In: Zalta, E. (ed.) The Stanford Encyclopedia of Philosophy (2011)
25. Glöckner, I.: Evaluation of quantified propositions in generalized models of fuzzy quantification. Int. J. Approx. Reason. **37**(2), 93–126 (2004)
26. Thiele, H.: On T-quantifiers and S-quantifiers. In: Proceedings of the 24th IEEE International Symposium on Multiple-Valued Logic, pp. 264–269 (1994)
27. Delgado, M., Sánchez, D., Martín-Bautista, M., Vila, M.: A probabilistic definition of a nonconvex fuzzy cardinality. Fuzzy Sets Syst. **126**(2), 177–190 (2002)
28. Pawlak, Z., Skowron, A.: Rough membership functions. In: Yager, R., Fedrizzi, M., Kacprzyk, J. (eds.) Advances in the Dempster-Shafer Theory of Evidence, pp. 251–271. John Wiley & Sons, New York (1994)
29. Pawlak, Z.: Rough Sets-Theoretical Aspects of Reasoning about Data. Kluwer Academic Publishers, Dordrecht (1991)
30. Slezak, D.: Rough sets and bayes factor. In: Peters, J.F., Skowron, A. (eds.) TRS III. LNCS, vol. 3400, pp. 202–229. Springer, Heidelberg (2005)
31. Slezak, D., Ziarko, W.: The investigation of the Bayesian rough set model. Int. J. Approx. Reason. **40**(1–2), 81–91 (2005)
32. de Luca, A., Termini, S.: A definition of a nonprobabilistic entropy in the setting of fuzzy sets theory. Inf. Control **20**(4), 301–312 (1972)
33. Zadeh, L.: The concept of a linguistic variable and its applications in approximate reasoning. Inf. Sci. **8**, 199–251 (1975)
34. Zadeh, L.: A theory of approximate reasoning. In: Hayes, J., Mitchie, D., Mikulich, L. (eds.) Machine Intelligence, vol. 9, pp. 149–194 (1979)

# PRE and Variable Precision Models in Rough Set Data Analysis

Ivo Düntsch[1]([⊠]) and Günther Gediga[2]

[1] Brock University, St. Catharines, Ontario L2S 3A1, Canada
duentsch@brocku.ca
[2] Department of Psychology, Institut IV, Universität Münster,
Fliednerstr. 21, Münster, Germany
gediga@uni-muenster.de

**Abstract.** We present a parameter free and monotonic alternative to the parametric variable precision model of rough set data analysis. The proposed model is based on the well known PRE index $\lambda$ of Goodman and Kruskal. Using a weighted $\lambda$ model it is possible to define a two dimensional space based on (Rough) sensitivity and (Rough) specificity, for which the monotonicity of sensitivity in a chain of sets is a nice feature of the model. As specificity is often monotone as well, the results of a rough set analysis can be displayed like a receiver operation curve (ROC) in statistics. Another aspect deals with the precision of the prediction of categories – normally measured by an index $\alpha$ in classical rough set data analysis. We offer a statistical theory for $\alpha$ and a modification of $\alpha$ which fits the needs of our proposed model. Furthermore, we show how expert knowledge can be integrated without losing the monotonic property of the index. Based on a weighted $\lambda$, we present a polynomial algorithm to determine an approximately optimal set of predicting attributes. Finally, we exhibit a connection to Bayesian analysis. We present several simulation studies for the presented concepts. The current paper is an extended version of [1].

## 1 Introduction

Rough sets were introduced by Z. Pawlak in the early 1980s [2] and have since become an established tool in information analysis and decision making. Given a finite set $U$ and an equivalence relation $\theta$ on $U$ the idea behind rough sets is that we know the world only up to the equivalence classes of $\theta$. This leads to the following definition: Suppose that $X \subseteq U$. Then, the *lower approximation of $X$* is the set $\text{Low}(X) = \{x \in U : \theta(x) \subseteq U\}$, and the *upper approximation of $X$* is the set $\text{Upp}(X) = \{x \in U : \theta(x) \cap X \neq \emptyset\}$. Here, $\theta(x)$ is the equivalence class of $x$, i.e. $\theta(x) = \{y \in U : x\theta y\}$. A *rough set* now is a pair $\langle \text{Low}(X), \text{Upp}(X) \rangle$ for

---

Ordering of authors is alphabetical, and equal authorship is implied

Ivo Düntsch– gratefully acknowledges support by the Natural Sciences and Engineering Research Council of Canada.

J. Peters et al. (Eds.): TRS XIX, LNCS 8988, pp. 17–37, 2015.
DOI: 10.1007/978-3-662-47815-8_2

each $X \subseteq U$. A subset $X$ of $U$ is called *definable*, if $\text{Low}(X) = \text{Upp}(X)$. In this case, $X$ is a union of classes of $\theta$.

Rough set data analysis (RSDA) is an important tool in reasoning with uncertain information. Its basic data type is as follows: A *decision system* in the sense of rough sets is a tuple $\langle U, \Omega, (D_a)_{a \in \Omega}, (f_a)_{a \in \Omega}, d, D_d, f_d \rangle$, where

- $U, \Omega, D_a, D_d$ are nonempty finite sets. $U$ is the set of objects, $\Omega$ is the set of (independent) attributes, and $D_a$ is the domain of attribute $a$. The decision attribute is $d$, and $D_d$ is its domain.
- For each $a \in \Omega$, $f_a : U \to D_a$ is a mapping; furthermore $f_d : U \to D_d$ is a mapping, called the *decision function*.

Since all sets under consideration are finite, an information system can be visualized as a matrix where the columns are labeled by the attributes and the rows correspond to feature vectors. An example from [3] is shown in Table 1.

**Table 1.** A decision system from [3]

| U | A | b | c | d | U | a | b | c | d |
|---|---|---|---|---|----|---|---|---|---|
| 1 | 1 | 0 | 0 | 1 | 12 | 0 | 1 | 1 | 1 |
| 2 | 1 | 0 | 0 | 1 | 13 | 0 | 1 | 1 | 2 |
| 3 | 1 | 1 | 1 | 1 | 14 | 1 | 1 | 0 | 2 |
| 4 | 0 | 1 | 1 | 1 | 15 | 1 | 1 | 0 | 2 |
| 5 | 0 | 1 | 1 | 1 | 16 | 1 | 1 | 0 | 2 |
| 6 | 0 | 1 | 1 | 1 | 17 | 1 | 1 | 0 | 2 |
| 7 | 0 | 1 | 1 | 1 | 18 | 1 | 1 | 0 | 3 |
| 8 | 0 | 1 | 1 | 1 | 19 | 1 | 0 | 0 | 3 |
| 9 | 0 | 1 | 1 | 1 | 20 | 1 | 0 | 0 | 3 |
| 10 | 0 | 1 | 1 | 1 | 21 | 1 | 0 | 0 | 3 |
| 11 | 0 | 1 | 1 | 1 | | | | | |

There, $U = \{1, \ldots, 21\}$ and $\Omega = \{a, b, c\}$. Each nonempty set $Q$ of attributes leads to an equivalence relation $\equiv_Q$ on $U$ in the following way: For all $x, y \in U$,

$$x \equiv_Q y \iff (\forall a \in Q)[f_a(x) = f_a(y)]. \tag{1.1}$$

According to the philosophy of rough sets, given a set $Q$ of attributes, the elements of the universe $U$ can only be distinguished up to the classes of $\equiv_Q$. A similar assumption holds for the decision classes of $\theta_d$.

To continue the example of Table 1, the classes of $\theta_\Omega$ are

$$
\begin{aligned}
&X_1 = \{1, 2, 19, 20, 21\}, && X_2 = \{3\}, &&& (1.2)\\
&X_3 = \{4, \ldots, 13\}, && X_4 = \{14, \ldots, 18\},
\end{aligned}
$$

and the decision classes are

$$Y_1 = \{1, \ldots, 12\}, \quad Y_2 = \{13, \ldots, 17\}, \quad Y_3 = \{18, \ldots, 21\}.$$

A class $X$ of $\theta_Q$ is called *deterministic (with respect to $d$)* if there is a class $Y$ of $\theta_d$ such that $X \subseteq Y$. In this case, all members of $X$ have the same decision value. The set of all deterministic classes is denoted by $\mathrm{Pos}(Q, d)$.

The basic statistic used in RSDA is as follows:

$$\gamma(Q, d) = \frac{|\bigcup \mathrm{Pos}(Q, d)|}{|U|}. \tag{1.3}$$

$\gamma(Q, d)$ is called the *approximation quality of $Q$ with respect to $d$*. If $\gamma(Q, d) = 1$, then each element of $U$ can be correctly classified with the granularity given by $Q$. In the example, the only deterministic class is $\{3\}$, and thus, $\gamma(\Omega, d) = \frac{1}{21}$.

An important property of $\gamma$ is monotony: If $Q \subseteq Q'$ then, $\gamma(Q, d) \leq \gamma(Q', d)$. In other words, increasing the granularity does not reduce the quality of classification.

In the sequel we exclude trivial cases and suppose that $\theta_Q$ and $\theta_d$ have more than one class.

## 2   The Variable Precision Model

One problem of decision making using $\gamma$ is the assumption of error free measurements, i.e. that the attribute functions $f_a$ are exact, and even one error may reduce the approximation quality dramatically [4]. Therefore, it would be advantageous to have a procedure which allows some errors in order to result in a more stable prediction success.

A well established model which is less strict in terms of classification errors is the *variable precision rough set model* (VP – model) [3] with the following basic constructions: Let $U$ be a finite universe, $X, Y \subseteq U$, and first define

$$c(X, Y) = \begin{cases} 1 - \frac{|X \cap Y|}{|X|}, & \text{if } |X| \neq 0, \\ 0, & \text{if } |X| = 0. \end{cases}$$

Clearly, $c(X, Y) = 0$ if and only if $X = 0$ or $X \subseteq Y$, and $c(X, Y) = 1$ if and only if $X \neq \emptyset$ and $X \cap Y = \emptyset$. The *majority requirement* of the VP – model says that more than $50\%$ of the elements in $X$ should be in $Y$; this can be specified by an additional parameter $\beta$ which is interpreted as an admissible classification error, where $0 \leq \beta < 0.5$. The *majority inclusion relation* $\overset{\beta}{\subseteq}$ (with respect to $\beta$) is now defined as

$$X \overset{\beta}{\subseteq} Y \Longleftrightarrow c(X, Y) \leq \beta. \tag{2.1}$$

Given a family of nonempty subsets $\mathscr{X} = \{X_1, \ldots, X_k\}$ of $U$ and $Y \subseteq U$, the *lower approximation* $\underline{Y}_\beta$ of $Y$ *given $\mathscr{X}$ and $\beta$* is defined as the union of all those

$X_i$, which are in relation $X_i \overset{\beta}{\subseteq} Y$, in other words,

$$\underline{Y}_\beta = \bigcup \{X \in \mathscr{X} : c(X,Y) \leq \beta\} \tag{2.2}$$

The classical approximation quality $\gamma(Q,d)$ is now replaced by a three-parametric version which includes the external parameter $\beta$, namely,

$$\gamma(Q,d,\beta) = \frac{|\operatorname{Pos}(Q,d,\beta)|}{|U|}, \tag{2.3}$$

where $\operatorname{Pos}(Q,d,\beta)$ is the union of those equivalence classes $X$ of $\theta_Q$ for which $X \overset{\beta}{\subseteq} Y$ for some decision class $Y$. Note that $\gamma(Q,d,0) = \gamma(Q,d)$. Continuing the example from the original paper ([3], p. 55), we obtain

$$\gamma(\Omega,d,0) = \frac{|X_2|}{|U|} \qquad\qquad = 1/21$$

$$\gamma(\Omega,d,0.1) = \frac{|X_2 \cup X_3|}{|U|} \qquad\qquad = 11/21$$

$$\gamma(\Omega,d,0.2) = \frac{|X_2 \cup X_3 \cup X_4|}{|U|} \qquad\qquad = 16/21$$

$$\gamma(\Omega,d,0.4) = \frac{|X_2 \cup X_3 \cup X_4 \cup X_1|}{|U|} \qquad\qquad = 21/21$$

Although the approach shows some nice properties, we think that care must be taken in at least three situations:

1. If we have a closer look at $\gamma(\Omega,d,0.1)$, we observe that, according to the table, object 13 is classified as being in class in $Y_2$, but with $\beta = 0.1$ it is assigned to the lower bound of $Y_1$. Intuitively, this assignment can be supported when the classification of the dependent attribute is assumed to be erroneous, and therefore, the observation is "moved" to a more plausible equivalence class due to approximation of the predicting variables. However, this may be problematic: Assume the decision classes arise from a medical diagnosis - why should an automatic device overrule the given diagnosis? Furthermore, the class changes are dependent on the actual predicting attributes in use, which may be problematic as well. This is evident if we assume for a moment that we want to predict $d$ with only one class $X = U$. If we set $\beta = \frac{9}{21} < 0.5$, we observe that $U \overset{\frac{9}{21}}{\subseteq} Y_1$, resulting in $\gamma(\{U\},d,\frac{9}{21}) = 1$.

2. Classical reduct search is based on the monotone relation

$$P \subseteq Q \quad \text{implies} \quad \gamma(P,d) \leq \gamma(Q,d).$$

Unfortunately, the generalized $\gamma(Q,d,\beta)$ is not necessarily monotone [5]. As a counterexample, consider the information system shown in Table 2 which adds an additional independent attribute $e$ to the system of Table 1.

**Table 2.** An enhanced decision system

| U | a | b | c | e | d | U | a | b | c | e | d |
|---|---|---|---|---|---|----|---|---|---|---|---|
| 1 | 1 | 0 | 0 | 0 | 1 | 12 | 0 | 1 | 1 | 1 | 1 |
| 2 | 1 | 0 | 0 | 0 | 1 | 13 | 0 | 1 | 1 | 1 | 2 |
| 3 | 1 | 1 | 1 | 0 | 1 | 14 | 1 | 1 | 0 | 0 | 2 |
| 4 | 0 | 1 | 1 | 0 | 1 | 15 | 1 | 1 | 0 | 0 | 2 |
| 5 | 0 | 1 | 1 | 0 | 1 | 16 | 1 | 1 | 0 | 0 | 2 |
| 6 | 0 | 1 | 1 | 0 | 1 | 17 | 1 | 1 | 0 | 0 | 2 |
| 7 | 0 | 1 | 1 | 0 | 1 | 18 | 1 | 1 | 0 | 0 | 3 |
| 8 | 0 | 1 | 1 | 0 | 1 | 19 | 1 | 0 | 0 | 0 | 3 |
| 9 | 0 | 1 | 1 | 1 | 1 | 20 | 1 | 0 | 0 | 0 | 3 |
| 10 | 0 | 1 | 1 | 1 | 1 | 21 | 1 | 0 | 0 | 0 | 3 |
| 11 | 0 | 1 | 1 | 1 | 1 |    |   |   |   |   |   |

Setting $P = \{a, b, c\}$ and $Q = \{a, b, c, e\}$, we observe that $Q$ generates five classes for prediction. The three classes $X_1$, $X_2$, and $X_4$ are identical to those of the first example – given in (1.2) –, here given by $P$, but $Q$ splits the class $X_3$ into the new classes $X_{3,0} = \{4...8\}$ and $X_{3,1} = \{9...13\}$. We now have

$$\gamma(Q, d, 0.1) = \frac{|X_2 \cup X_{3,0}|}{|U|} = \frac{6}{21} < \gamma(P, d, 0.1) = \frac{11}{21}.$$

The reason for this behavior is that $c(X_{3,1}, Y) > 0.1$.
3. A third – perhaps minor – problem is the choice of $|U|$ as the denominator in $\gamma(Q, d, \beta)$. Using $|U|$ makes sense, when a no-knowledge-model cannot predict anything of $d$, and therefore any prediction success of $\Omega$ can be attributed to the predicting variables in $\Omega$. But, as we have shown in the current section, there are situations in which a simple guessing model serves as a "perfect" model in terms of approximation quality.

*Simulation 1.* We conducted a simulation study based on an information system $\mathscr{I}$ with binary attributes $A_1, A_2, A_3$, and a decision attribute $d$ with classes $C_1, C_2, C_3$ whose relative frequency is 0.6, 0.35, 0.05, respectively; there are 300 objects. Initially, for each $x \in U$ and $1 \le i \le 3$ we set

$$f_{A_i}(x) = \begin{cases} 1, & \text{if } f_d(x) \in C_i, \\ 0, & \text{otherwise.} \end{cases}$$

The information system $\mathscr{I}$ is assumed to be error free, i.e. its reliability is 100 %. For each simulation, a certain percentage $p$ of attribute values is changed to their opposite value, i.e. $f'_{A_i}(x) = 1 - f_{A_i}$ to obtain a different reliability $1 - p$. The expectation values of the prediction success for various values of $\beta$ and reliabilities are shown in Table 3, based on 1000 simulations each. We observe

**Table 3.** Simulation for the variable precision model

| $\beta$ | Reliability | | |
|---|---|---|---|
| | .95 | .90 | 0.85 |
| 0.00 | 0.5740 | 0.1553 | 0.0341 |
| 0.05 | 0.7315 | 0.4945 | 0.3754 |
| 0.10 | 0.7284 | 0.5045 | 0.3774 |
| 0.15 | 0.7356 | 0.4821 | 0.3734 |
| 0.20 | 0.7349 | 0.5146 | 0.3838 |
| 0.25 | 0.7349 | 0.4855 | 0.3731 |

that the prediction values are quite low for $\beta = 0$ (no error) and are maximal already for small values of $\beta$. $\qquad\qquad\qquad\qquad\qquad\qquad\qquad\square$

## 3   Contingency Tables and Information Systems

In this and the following sections we describe a formal connection of statistical and rough set data analysis. First of all, we need data structures which can be used for both types of analysis. It is helpful to observe that rough set data analysis is concept free because of its nominal scale assumption; in other words, only cardinalities of classes and intersection of classes are recorded. As $Q \subseteq \Omega$ and $d$ induce partitions on $U$, say, $\mathscr{X}$ with classes $X_j$, $1 \leq j \leq J$, respectively, $\mathscr{Y}$ with classes $Y_i$, $1 \leq i \leq I$, it is straightforward to cross–classify the classes and list the cardinalities of the intersections $Y_i \cap X_j$ in a contingency table (see also [6]). As an example, the information system of Table 1 is shown as a contingency array in Table 4.

**Table 4.** Contingency table of the decision system of Table 1

| | $X_1$ | $X_2$ | $X_3$ | $X_4$ | $n_{i\bullet}$ |
|---|---|---|---|---|---|
| $Y_1$ | 2 | 1 | 9 | 0 | 12 |
| $Y_2$ | 0 | 0 | 1 | 4 | 5 |
| $Y_3$ | 3 | 0 | 0 | 1 | 4 |
| $n_{\bullet j}$ | 5 | 1 | 10 | 5 | 21 |

The actual frequency of the occurrence, i.e. the cardinality of $Y_i \cap X_j$, is denoted by $n_{ij}$ and the row and column sums by $n_{i\bullet}$ and $n_{\bullet j}$ respectively. The maximum of each column is shown in bold.

If a column $X_j$ consists of only one non-zero entry, the corresponding set $X_j$ is a deterministic class, i.e. it is totally contained in a decision class. In terms of classical rough set analysis, any column $X_j$ which has at least two non-zero entries is not deterministic. The approximation quality $\gamma(Q, d)$ can now easily be derived by adding the frequencies $n_{ij}$ in the columns with exactly one non-zero entry and dividing the sum by $|U|$. In the example we see that $X_2$ is the only column with exactly one nonzero entry, and $\gamma = \frac{1}{21}$. To be consistent with statistical notation, we will frequently speak of the classes of $\theta_Q$ as categories of the variable $X$ and of the classes of $\theta_d$ as categories of the variable $Y$.

## 4    PRE Measures and the Goodman-Kruskal $\lambda$

Statistical measures of prediction success – such as $R^2$ in multiple regression or $\eta^2$ in the analysis of variance – are often based on the comparison of the prediction success of a chosen model with the success of a simple zero model. In categorical data analysis the idea behind the *Proportional Reduction of Errors* (PRE) approach is to count the number of errors, i.e. events which should not be observed in terms of an assumed theory, and to compare the result with an "expected number of errors", given a zero ("baseline") model [4,7,8]. If the number of expected errors is not zero, then

$$\text{PRE} = 1 - \frac{\text{number of observed errors}}{\text{number of expected errors}}$$

More formally, starting with a measure of error $\epsilon_0$, the relative success of the model is defined by its proportional reduction of error in comparison to the baseline model,

$$\text{PRE} = 1 - \frac{\epsilon_1}{\epsilon_0}.$$

A very simple strategy in the analysis of categorical data is betting on the highest frequency; this strategy is normally used as the zero model benchmark ("baseline accuracy") in machine learning.

A simple modification which fits the contingency table was proposed by Goodman and Kruskal in the 1950s [9]. When no other information is given, it is reasonable to choose a decision category with highest frequency (such as $Y_1$ in Table 4). If the categories of $\mathscr{X}$ and the distribution of $\mathscr{Y}$ in each $X_j$ are known, it makes sense to guess within each $X_j$ some $Y_i$ which shows the highest frequency, see also [10]. The PRE of knowing $\mathscr{X}$ instead of (uninformed) guessing is given by

$$\lambda = 1 - \frac{n - \sum_{j=1}^{J} \max_{i=1}^{I} n_{ij}}{n - \max_{i=1}^{I} n_{i\bullet}}. \tag{4.1}$$

Here, $n = |U|$. Note that $n - \max_{i=1}^{I} n_{i\bullet} \neq 0$, since we have assumed that $\theta_d$ has at least two classes. For our example we obtain

$$\lambda = 1 - \frac{21 - (3 + 1 + 9 + 4)}{21 - 12} = 1 - \frac{5}{9} = 0.444$$

We conclude that knowing $\mathscr{X}$ reduces the error of the pure guessing procedure by 44.4 % in comparison to the baseline accuracy.

The $\lambda$-index is one of the most effective methods in ID3 [11], and a slightly modified approach in [10] – known as the *1R learning procedure* – was shown to be a quite effective tool as well [12].

## 5   Weighted $\lambda$

If we compare the set of classes $C(\beta)$ of $\theta_Q$ used to determine $\text{Pos}(Q, d, \beta)$ in the VP-model, and the set of classes $C$ used in the computation of $\lambda$, we observe that $C(\beta) \subseteq C$ for any value of $0 \leq \beta < 0.5$. The proof is simple: For every $j$ more than 50 % of the observations must be collected in one $n_{ij}$, and so these frequencies are the maximal frequency in column $j$.

The connection of $\lambda$ and the approximation quality $\gamma$ is straightforward: Whereas $\lambda$ counts the maximum of each column $j$, $\gamma$ counts this maximum only in the deterministic case if $n_{ij} = n_{\bullet j}$, i.e. if exactly one entry in column $j$ is nonzero.

Assume that we want to predict the decision attribute by one class only. In case that there is one attribute value of the decision attribute for which $n_{i\bullet} = |U| = n$ holds, we result in a situation in which the expected error is 0. Since this situation is of no interest for prediction, we should exclude it – the decision attribute is deterministic itself.

In all other cases, the decision attribute is indeterministic in the sense that there is no deterministic rule for prediction given no attributes; hence, in this case, the expected error is $n$. We observe that $|U| = n$ is a suitable denominator for $\gamma$.

As $\gamma$ is a special case by filtering maximal categories by an additional condition, we define a *weighted* $\lambda$ by

$$\lambda(w) = 1 - \frac{n - \sum_{j=1}^{J}(\max_{i=1}^{I} n_{ij}) \cdot w(j)}{n - (\max_{i=1}^{I} n_{i\bullet}) \cdot w(U)}. \tag{5.1}$$

where $w : \{1, \ldots, J\} \cup \{U\} \to [0, 1]$ is a function weighting the maxima of the columns of the contingency table. In the cases we consider, $w$ will be an indicator function taking its values from $\{0, 1\}$.

Now we set

$$X_j \subseteq_w Y_i \Longleftrightarrow n_{ij} = \max_{k=1}^{I} n_{kj} \text{ and } w(j) > 0,$$

and define the lower approximation of $Y_i$ by $\mathscr{X}$ with respect to $w$ by

$$\text{Low}_w(\mathscr{X}, Y_i) = Y_i \cap \bigcup \{X_j : X_j \subseteq_w Y_i\}.$$

Observe that $\text{Low}_w(Y_i) \subseteq Y_i$, unlike in the lower approximation of the VP – model. For the upper approximation we choose the "classical" definition

$$\text{Upp}(\mathscr{X}, Y_i) = \bigcup \{X_j : X_j \cap Y_i \neq \emptyset\}.$$

The *w-boundary* now is the set

$$\mathrm{Bnd}_w(\mathscr{X}, Y_i) = \mathrm{Upp}(\mathscr{X}, Y_i) \setminus \mathrm{Low}_w(\mathscr{X}, Y_i).$$

Unlike in the VP – model, elements of non–deterministic classes are not re–classified with respect to the decision attribute but are left in the boundary region.

We can now specify the error of the lower bound classification by

$$\mathrm{Err}_w(\mathscr{X}, Y_i) = \bigcup \{X_j \setminus Y_i : X_j \subseteq_w Y_i\}.$$

If we assume that errors are proportional to the number of entries in the contingency table – but independent of the joint distribution – it makes sense to count the absolute error $c_j = n_{\bullet j} - \max_{i=1}^I n_{ij}$ for every column $j$ and compare it to some cutpoint $C$. This leads to the following definition:

$$w_{\mathrm{eq}}^C(j) = \begin{cases} 1, & \text{if } n_{\bullet j} - \max_{i=1}^I n_{ij} \leq C, \\ 0, & \text{otherwise} \end{cases}$$

and

$$w_{\mathrm{eq}}^C(U) = \begin{cases} 1, & \text{if } n - \max_{i=1}^I n_{i\bullet} \leq C, \\ 0, & \text{otherwise.} \end{cases}$$

respectively.

It is easy to see that $\lambda_{\mathrm{eq}} = \gamma$ if $C = 0$, and $\lambda_{\mathrm{eq}} = \lambda$ if $C = \infty$, i.e. if $\lambda_{\mathrm{eq}} \equiv 1$. Furthermore, if $C \leq \max_{j=1}^J(n_{\bullet j} - \max_{i=1}^I n_{ij})$, then the denominator of $\lambda(w_{\mathrm{eq}})$ is $|U|$.

In classical rough set theory, adding an independent attribute while keeping the same decision attribute will not decrease the approximation quality $\gamma$. The same holds for $\gamma_{w_{\mathrm{eq}}}$:

**Proposition 1.** *Let* $Q_a = Q \cup \{a\}$ *and* $\mathscr{X}_a$ *be its associated partition. Then,* $\gamma_{w_{\mathrm{eq}}}^C(\mathscr{X}, \mathscr{Y}) \leq \gamma_{w_{\mathrm{eq}}}^C(\mathscr{X}_a, \mathscr{Y})$.

*Proof.* We assume w.l.o.g. that $a$ takes only the two values $0, 1$ (see e.g. [13] for the binarization of attributes). Let $Z_0, Z_1$ be the classes of $\theta_a$. The classes of $\theta_{Q_a}$ are the non–empty elements of $\{X_i \cap Z_0 : 1 \leq i \leq I\} \cup \{X_i \cap Z_1 : 1 \leq i \leq I\}$. Each $n_{ij}$ is split into $n_{ij}^0 = |X_i \cap Y_j \cap Z_0|$ and $n_{ij}^1 = |X_i \cap Y_j \cap Z_1|$ with respective columns $j0$ and $j1$, and sums $n_{\bullet j}^0$ and $n_{\bullet j}^1$. Then, $n_{ij}^0 + n_{ij}^1 = n_{ij}$, $n_{\bullet j}^0 + n_{\bullet j}^1 = n_{\bullet j}$, and $\max_{i=1}^I n_{ij}^0 + \max_{i=1}^I n_{ij}^1 \geq \max_{i=1}^I n_{ij}$ by the triangle inequality. Thus, if $n_{\bullet j} - \max_{i=1}^I n_{ij} \leq C$, then

$$n_{\bullet j}^0 - \max_{i=1}^I n_{ij}^0 \leq n_{\bullet j}^0 - \max_{i=1}^I n_{ij}^0 + n_{\bullet j}^1 - \max_{i=1}^I n_{ij}^1$$
$$= n_{\bullet j}^0 + n_{\bullet j}^1 - (\max_{i=1}^I n_{ij}^0 + \max_{i=1}^I n_{ij}^1)$$
$$= n_{\bullet j} - (\max_{i=1}^I n_{ij}^0 + \max_{i=1}^I n_{ij}^1)$$
$$\leq n_{\bullet j} - \max_{i=1}^I n_{ij}$$
$$\leq C.$$

**Table 5.** Simulation for the $\lambda$ model

| | Reliability | | |
|---|---|---|---|
| C | 0.95 | 0.90 | 0.85 |
| 0 | 0.5740 | 0.1553 | 0.0341 |
| 1 | 0.7985 | 0.4798 | 0.1339 |
| 2 | 0.8508 | 0.5916 | 0.2813 |
| 3 | 0.8615 | 0.6707 | 0.4603 |

Similarly,

$$n^1_{\bullet j} - \max^I_{i=1} n^1_{ij} \le C.$$

Therefore, if $w_{\text{eq}}(j) = 1$, then $w_{\text{eq}}(j0) = w_{\text{eq}}(j1) = 1$.

Again by the triangle inequality, the sum of errors in the two $j0$ and $j1$ columns is no more than the error in the original column $j$. As the overall error is simply the sum of the errors per column, the proof is complete. □

*Simulation 2.* In order to compare the new model with the variable precision model, we assume the same setup as in Simulation 1, but use the equal weights $\lambda$ PRE model instead. The results can be seen in Table 5.

We note that the prediction quality increases with the cutpoint $C$, and from a certain $C$ is larger than the asymptotic value of the variable precision model. The value depends on the chosen reliability. As a rule, the percentage of successful predictions is higher than that of the variable precision model. However, choosing $C = 1$ will not increase the prediction quality compared to the variable precision model when the reliability is low.

We may consider the $\lambda$ PRE model as a "variable precision model" when we assume that the $\beta$-boundaries vary in dependence of the group sizes. Table 6 shows the dependencies. □

**Table 6.** Variable $\beta$-values to mimic the $\lambda$ model in the example given a sample size of $n = 300$

| Group sizes | C | | | |
|---|---|---|---|---|
| | 0 | 1 | 2 | 3 |
| 0.60 | 0.000 | 0.006 | 0.011 | 0.017 |
| 0.35 | 0.000 | 0.010 | 0.019 | 0.029 |
| 0.05 | 0.000 | 0.067 | 0.133 | 0.200 |

## 6    Rough–sensitivity and Rough–specificity

Various other indices may be defined: Let $\mathscr{X}$ be the partition associated with $\theta_Q$ and $Y_i$ be a decision class. In a slightly different meaning than in machine learning, we will use the terms *Rough-sensitivity* and *Rough-specificity* for the results of our analysis: If $\mathscr{Y}$ is the partition induced by the decision attribute, we consider

1. The *Rough-sensitivity* of the partition $\mathscr{X}$ with respect to the partition $\mathscr{Y}$

$$\gamma_w(\mathscr{X},\mathscr{Y}) = \frac{\sum_{Y_i \in \mathscr{Y}} |\operatorname{Low}_w(\mathscr{X},Y_i)|}{|U|}$$

2. The *Rough-specificity* of the partition $\mathscr{X}$ with respect to the partition $\mathscr{Y}$ is based on $\zeta_w(\mathscr{X},\mathscr{Y})$, which is defined by

$$\zeta_w(\mathscr{X},\mathscr{Y}) = \begin{cases} \frac{\sum_{Y_i \in \mathscr{Y}} |\operatorname{Err}_w(\mathscr{X},Y_i)|}{\sum_{Y_i \in \mathscr{Y}} |\operatorname{Bnd}_w(\mathscr{X},Y_i)|}, & \text{if } \sum_{Y_i \in \mathscr{Y}} |\operatorname{Bnd}_w(\mathscr{X},Y_i)| > 0 \\ 1, & \text{otherwise.} \end{cases}$$

The *Rough-specificity* is defined by $1 - \zeta_w(\mathscr{X},\mathscr{Y})$.

If $\mathscr{X}$ and $\mathscr{Y}$ are understood, we will just write $\gamma_w$ and $\zeta_w$ or just $\zeta$. The Rough-sensitivity tells us about the approximation of the set or partition, whereas $\zeta$ is an index which expresses the relative error of the classification procedure. Both indices are bounded by 0 and 1, and in most cases monotonically related (a counter example is discussed below). Rough-sensitivity reflects the relative precision of deterministic rules, which are true up to some specified error. It captures the rough set approximation quality $\gamma$ in case $w$ is defined as

$$w(j) = \begin{cases} 1, & \text{if } n_{\bullet j} = \max_{i=1}^I n_{ij} \\ 0, & \text{otherwise,} \end{cases}$$

and $w(U) = 0$.

Rough-specificity is a new concept: Whereas errors are addressed to the lower bound in the classic variable precision model, in our model an error is an instance of the boundary – it addresses those elements which are errors of the prediction rules in contrast to indeterministic elements, which cannot be predicted by prediction rules. The value of $\zeta$ tells us the relative magnitude of the "hard boundary" within the boundary. In other words, $1 - \zeta$ (the Rough-specificity) is the relative number of elements of the boundary which may become deterministic, if we consider more attributes.

Using the data of Table 1 and the chain $(\{\}, \{a\}, \{a,b\}, \{a,b,c\}, \{a,b,c,d\})$ of attributes and $C = 1$ we obtain the results shown in Table 7.

A diagram of our results – which we may call a *rough receiver operation curve* (Rough–ROC) is depicted in Fig. 1.

Apart from the boundary values 0 and 1, we find that the sensitivity $\gamma_w$ is much higher than $\zeta_w$. We call the difference $\gamma_w - \zeta_w$ the *Rough-Youden-index* (RY). In ROC analysis – the statistical counterpart to analyse sensitivity and specificity – the Youden index is a good heuristic to capture a good cut–point for prediction [14,15]. In rough set data analysis, the largest RY within a chain

**Table 7.** Rough–Sensitivity and Rough–Specificity given the data of Table 1

| Attribute Sets | {} | {a} | {a,b} | {a,b,c} | {a,b,c,d} |
|---|---|---|---|---|---|
| Lower Bound | {} | {4–12} | {4–12} | {3–12,14–17} | $U$ |
| Error | {} | {13} | {13} | {13,18} | {} |
| Upper Bound | $U$ | {1,2,3,13–21} | {1,2,3,13–21} | {1,2,13,18–21} | {} |
| Sensitivity | 0.000 | 0.429 | 0.429 | 0.714 | 1.000 |
| $\zeta = 1 -$ Specificity | 0.000 | 0.083 | 0.083 | 0.286 | 1.000 |
| Difference | 0.000 | 0.345 | 0.345 | 0.429 | 1.000 |

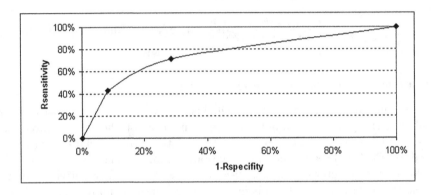

**Fig. 1.** A Rough–sensitivity/$\zeta = 1$-Rough–specificity diagram

of attributes tells us a promising set for prediction – in case of the example the set $\{a, b, c\}$ seems to be good choice for prediction.

Note that if the decision attribute contains more than 2 classes, $\zeta$ – unlike $\gamma$ – need not be monotonically increasing in case an error class changes to a deterministic class when adding a new independent attribute. As long as we do not split a deterministic class by adopting a further attribute, any new granule will not decrease the lower bound, and will not increase the number of elements in the boundary; therefore, the error will not decrease. Hence, $\zeta$ will not decrease, when we add a further attribute for prediction, and the deterministic classes are unchanged.

**Table 8.** A non–monotonic $\zeta$

| X | $X_0$ | $X_1$ |
|---|---|---|
| $Y_0$ | 0 | 0 | 0 |
| $Y_1$ | 3 | 3 | 0 |
| $Y_2$ | 2 | 0 | 2 |

As an example of a non monotonic $\zeta$, consider an information system with a granule $G = \langle \langle X, Y_0, 0 \rangle, \langle X, Y_1, 3 \rangle, \langle X, Y_2, 2 \rangle \rangle$, see Table 8. $G$ is deterministic, if we choose $C = 2$. Let $n_e$ be the number of errors outside this granule, and $n_b > 0$ be the number of elements of the boundary outside this granule; then, $n_e < n_b$ and $\zeta = \frac{n_e + 2}{n_b + 2}$.

Suppose that a new attribute splits exactly $G$ into $G_0$ and $G_1$ according to Table 8. Then, we obtain $\zeta = \frac{n_e}{n_b}$ as two elements are moved from the boundary to the lower bound. Since $n_b > 0$ and $n_e < n_b$, it follows that $\frac{n_e}{n_b} < \frac{n_e + 2}{n_b + 2}$.

There are various ways how to deal with the non-monotonicity of $\zeta$:

1. Use a rule in the algorithm that prevent the split of deterministic classes.
2. Require that any deterministic class has to consist of more than $C$ elements. Hence, using

$$\tilde{w}_{eq}^C(j) = \begin{cases} 1, & \text{if } n_{\bullet j} - \max_{i=1}^I n_{ij} \leq C \text{ and } \max_{i=1}^I n_{ij} > C, \\ 0, & \text{otherwise} \end{cases}$$

is a weighting function for which $\zeta$ is monotone when classes are split. It is straightforward to show that $\gamma$ is monotone as well when using $\tilde{w}$ as the weight function.
3. We leave everything unchanged. If $\zeta$ decreases when adding an attribute, we assume that this behaviour is due to spurious deterministic rules, and consider this as another stopping rule for adopting more attributes.

## 7    Precision and Its Confidence Bounds

In order to avoid technical problems, we assume that $\text{Low}(Y_i) > 0$ for each class $Y_i$ of the decision attribute, so that there is at least one rule which predicts membership in $Y_i$.

In classical rough set data analysis the accuracy of approximation $\alpha$ of $Y_i$ is defined as

$$\alpha(Y_i) = \frac{|\text{Low}(Y_i)|}{|\text{Upp}(Y_i)|} = \frac{|\text{Low}(Y_i)|}{|\text{Low}(Y_i)| + |\text{Bnd}(Y_i)|} = \frac{1}{1 + |\text{Bnd}(Y_i)|/|\text{Low}(Y_i)|}.$$

To approximate the standard error of the accuracy we use the Delta method. Broadly speaking, in a first step we linearize the fractions by taking the logarithms, secondly, we approximate the logarithm by the first order Taylor expansion, see e.g. [16] for details.

Letting $\pi_1 = |\text{Low}(Y_i)|/|U|$ and $\pi = |\text{Bnd}(Y_i)|/|U|$, we obtain

$$Var(\ln(\alpha(Y_i))) = \pi_1/|U| \cdot \left(\frac{\pi}{\pi_1^2 + \pi_1 \pi}\right)^2 + \pi/|U| \cdot \left(\frac{1}{\pi_1 + \pi}\right)^2.$$

*Example 1.* Suppose that in a company with 500 employees there are 80 employees who are involved in accidents per year $(Y_1)$. Of these, 70 can be predicted correctly, while of the other 420 cases $(Y_0)$ 300 can be predicted correctly. As the decision attribute consists of two categories only, the number of subjects in the boundary is determined by $500 - 70 - 300 = 130$.

For the category $Y_1$ ("had accidents") we obtain the precision

$$\alpha(Y_1) = \frac{70}{70 + 130} = 0.35$$

The standard error of $\ln(\alpha(Y_1))$ is given by $SE(\ln \alpha(Y_1))) = 0.120$ resulting in a 95 % confidence interval $[0.277, 0.442]$ for $\alpha(Y_1)$.

For the category $Y_0$ ("no accidents") the precision is given by

$$\alpha(Y_0) = \frac{300}{300 + 130} = 0.698$$

The standard error of $\ln(\alpha(Y_0))$ is $SE(\ln \alpha(Y_0))) = 0.103$ resulting in a 95 % confidence interval $[0.570, 0.854]$ for $\alpha(Y_0)$. As both confidence intervals do not intersect, we conclude, that the precision of $Y_0$ is higher than the precision of $Y_1$.    □

Assume now that there are some errors in the prediction. It makes no sense to count the errors for the prediction of the other categories as possible indeterministic rules for the category under study. Therefore we eliminate the errors from the other categories from the boundary by

$$|\,\text{Bnd}_{\text{corrected};Y_i}\,| = |\,\text{Bnd}(Y_i)| - \sum_{k \neq i} |\,\text{Err}(Y_k)|$$

and use the corrected boundary instead in the computation of $\alpha_c$ (a corrected $\alpha$).

*Example 2.* We use the data from the preceding example, but assume additionally, that there were 30 errors to predict $Y_1$ and 100 errors to predict $Y_0$ due to application of our PRE model.

$$\alpha_c(Y_1) = \frac{70}{70 + (130 - 100)} = 0.7 \qquad (95\,\% \; CI = [0.485, 1.000])$$

$$\alpha_c(Y_0) = \frac{300}{300 + (130 - 30)} = 0.75 \qquad (95\,\% \; CI = [0.601, 0.936]).$$

Obviously, the estimated precision of $Y_1$ is enhanced dramatically. Note that $\alpha_c(Y_1)$ and $\alpha_c(Y_0)$ cannot be improved, as the boundary consists of error elements only. $\alpha_c(Y_1) = 0.7$ means in this case that 70 % of the rules are deterministic and lead to the correct result $Y_1$, but 30 % of the rules for $Y_1$ cannot be described in this way.    □

In the variable precision model the error is moved to the lower bound. It is interesting to see how $\alpha_c$ looks like in this case. We select $\beta$ large enough that the same errors occur as in our example using the PRE model. In this case we observe.

*Example 3.*

$$\alpha_c(Y_1; \text{VPRM}) = \frac{70 + 30}{70 + 30 + (130 - 100 - 30)} = 1$$

and

$$\alpha_c(Y_0; \text{VPRM}) = \frac{300 + 100}{300 + 100 + (130 - 100 - 30)} = 1$$

In case of the VPRM, the $\alpha_c$ values signal a "perfect" precision of the model. $\square$

## 8   Using Additional Expert Knowledge

Weights given by experts or a priori probabilities of the outcomes $Y_i$ $(1 \le i \le I)$ are one of the simplest assumptions of additional knowledge which can be applied to a given situation: We let $\pi_i$ $(1 \le i \le I)$ be weights of the outcomes and w.l.o.g. we assume that $\sum_i \pi_i = 1$. Now, we obtain a weighted contingency table simply by defining $n_{ij}^* = n_{ij} \cdot \pi_i$ and use $n_{ij}^*$ instead of $n_{ij}$ of the original table.

**Table 9.** Weighted contingency table of the decision system of Table 1 using $\pi = \langle 0.5, 0.3, 0.2 \rangle$

|       | $X_1$ | $X_2$ | $X_3$ | $X_4$ | $n_{i\bullet}^*$ |
|-------|-------|-------|-------|-------|------------------|
| $Y_1$ | 1     | 0.5   | 4.5   | 0     | 6                |
| $Y_2$ | 0     | 0     | 0.3   | 1.2   | 1.5              |
| $Y_3$ | 0.6   | 0     | 0     | 0.2   | 0.8              |
| $n_{\bullet j}^*$ | 1.6 | .5 | 4.8 | 1.4 | 8.3          |

Using Table 9 and applying the bounds $E = 0, 0.2, 0.3, 0.6$ to compute $w_{\text{eq}}^E(j)$, we observe the approximation qualities shown in Table 10.

**Table 10.** $\lambda$ given various bounds

| $E$ | Formula (4.1) | Weighted $\lambda$ |
|-----|---------------|--------------------|
| 0.0 | $1 - \frac{8.3 - 0.5}{8.3}$ | 0.060 |
| 0.2 | $1 - \frac{8.3 - 0.5 - 1.2}{8.3}$ | 0.250 |
| 0.3 | $1 - \frac{8.3 - 0.5 - 1.2 - 4.5}{8.3}$ | 0.747 |
| 0.6 | $1 - \frac{8.3 - 0.5 - 1.2 - 4.5 - 1}{8.3}$ | 0.867 |

We see that $\lambda$ increases here as well as in case of the unweighted $\lambda$, but if we consider the weighted $\lambda$, the approximation qualities differ from those in the unweighted case. Furthermore, even the (approximate) deterministic class may change, if the weights differ largely: Note, that in case $E = 0.6$ we choose class $X_1$ as the (approximate) deterministic class, whereas $X_3$ would be chosen, if we use equal (or no) weights.

The algorithm given below and the monotonicity of $\lambda$ given a split (or using an additional attribute) stay valid in case of introducing weights for the decision category as in the unweighted case. This holds because we have changed the entries of the table only – the structure of the table remains unchanged.

## 9   A Simple Decision Tree Algorithm Based on Rough Sets

In order to find an algorithm for optimization, not only the Rough-sensitivity but also the Rough-specificity must be taken into account, and we have to find a function which reflects the status of the partitions in a suitable way. Numerical experiments show that neither the difference $\gamma_w - \zeta_w$ (the RY-index) nor the odds $\frac{\gamma_w}{\zeta_w}$ are appropriate for the evaluation of the partitions. The reason for this seems to be that the amount of deterministic classification, which is a function of $|U| \cdot \gamma_w$, as well as the amount of the probabilistic part of $\zeta_w$ are not taken into account.

Therefore we define an objective function based on entropy measures, which computes the fitness of the partition $\mathscr{X}$ on the basis of the difference of the coding complexity of the approximate deterministic and indeterministic classes, which is an instance of a mutual entropy [17]:

$$\mathbf{O}(\mathscr{Y}|\mathscr{X}) = -\gamma_w \ln(|U| \cdot \gamma_w) + \zeta_w \ln( \sum_{Y_i \in \mathscr{Y}} |\operatorname{Bnd}_w(\mathscr{X}, Y_i)| \cdot \zeta_w)$$

The algorithm proceeds as follows:

1. Set a cutpoint $C$ for the algorithm.
2. Start with $Q = \emptyset$.
3. Add any attribute from $\Omega \setminus Q$ to $Q$. Compute $\mathbf{O}$ for the chosen cutpoint $C$
4. Choose a new attribute which shows the maximum in $\mathbf{O}$.
5. If the new maximum is less than or equal to the maximum of the preceding step, then stop. Otherwise add the new attribute to $Q$ and proceed with step 2.

The time complexity of the algorithm is bounded by $\mathcal{O}(J^2)$ and it will find a partition $\mathscr{X}$ which shows a good approximation of $Y$ with an error less than $C$.

Applying the algorithm to the decision system given in Table 1 and using $C = 1$ (we allow 1 error per column), results in the following steps:

Step 1     $C = 1$
Step 2.0   $Q = \emptyset$
Step 3.0.a Test attribute $a$

|        | $X_1$ (a=0) | $X_2(a = 1)$ |
| ------ | ----------- | ------------ |
| $Y_1$  | **9**       | 3            |
| $Y_2$  | 1           | 4            |
| $Y_3$  | 0           | 4            |
| $n_{\bullet j}$ | 10 | 11           |
| **O**  | 0.942       |              |

Step 3.0.b Test attribute $b$

|        | $X_1$ (b=0) | $X_2(b = 1)$ |
| ------ | ----------- | ------------ |
| $Y_1$  | 2           | 10           |
| $Y_2$  | 0           | 5            |
| $Y_3$  | 3           | 1            |
| $n_{\bullet j}$ | 6  | 16           |
| **O**  | 0.000       |              |

Step 3.0.c Test attribute $c$

|        | $X_1$ (b=0) | $X_2(b = 1)$ |
| ------ | ----------- | ------------ |
| $Y_1$  | 2           | **10**       |
| $Y_2$  | 4           | 1            |
| $Y_3$  | 4           | 0            |
| $n_{\bullet j}$ | 10 | 11           |
| **O**  | 1.096       |              |

Step 4.0     Choose attribute $c$, because it is maximal
in terms of **O**

Step 5.0     Iterate step 2.1

Step 2.1    $Q = \{c\}$

Step 3.1.a Test attribute $a$.

|        | $X_1$ | $X_2$ | $X_3$ |
|        | $(c = 0, a = 1)$ | $(c = 1, a = 0)$ | $(c = 1, a = 1)$ |
| ------ | ---------------- | ---------------- | ---------------- |
| $Y_1$  | 2                | 9                | 1                |
| $Y_2$  | 4                | 1                | 0                |
| $Y_3$  | 4                | 0                | 0                |
| $n_{\bullet j}$ | 10      | 10               | 1                |
| **O**  | 1.096            |                  |                  |

Step 3.1.b Test attribute $b$

|        | $X_1$ | $X_2$ | $X_3$ |
|        | $(c = 0, b = 0)$ | $(c = 0, b = 1)$ | $(c = 1, b = 1)$ |
| ------ | ---------------- | ---------------- | ---------------- |
| $Y_1$  | 2                | 0                | **10**           |
| $Y_2$  | 0                | 4                | 1                |
| $Y_3$  | 3                | 1                | 0                |
| $n_{\bullet j}$ | 5       | 5                | 11               |
| **O**  | 1.561            |                  |                  |

**Step 4.1**   Choose attribute $b$, because it is maximal
in terms of **O**

**Step 5.2**   Iterate step 2.2

**Step 2.2**   $Q = \{b, c\}$

**Step 3.2.a** Test attribute $a$.

|       | $X_1$ | $X_2$ | $X_3$ | $X_4$ |
|-------|-------|-------|-------|-------|
| $Y_1$ | 2     | 1     | 9     | 0     |
| $Y_2$ | 0     | 0     | 1     | 4     |
| $Y_3$ | 3     | 0     | 0     | 1     |
| $n_{\bullet j}$ | 5 | 1 | 10 | 5 |
| **O** | 1.561 |       |       |       |

**Step 4.2** Stop, because **O** does not increase.

The attributes.$Q = \{b, c\}$ show the best behaviour in terms of **O**.

## 10   Bayesian Considerations

As we introduced weights for the decision attribute, and since the weights may
be interpreted as prior probabilities, it is worthwhile to find a connection to
Bayesian posterior probabilities[1]. Choose some cutpoint $C$; we shall define a two
dimensional strength function $s_C(i, j)$ ($1 \leq i \leq I, 1 \leq j \leq J$), which reflects the
knowledge given in column $X_i$ to predict the category $Y_j$. As we use approximate deterministic classes as basis of our knowledge, the strength function is
dependent on $C$ as well.

First consider the case that the column $X_j$ satisfies the condition

$$n_{\bullet j} - \max_{i=1}^{I} n_{ij} \leq C. \tag{10.1}$$

In that case there is one class with frequency $\max_{i=1}^{I} n_{ij}$ which is interpreted
as the approximate deterministic class; all other frequencies are assumed as error.
In this case we define $s_C(i, j) := \frac{n(i,j)}{n}$. This is simply the joint relative frequency
$p(i, j)$ of the occurrence of $Y = Y_i$ and $X = X_j$. If the column $X_j$ does not fulfill
condition (10.1), we conclude that $X_j$ cannot be used for approximation.

In this case no entry of column $X_j$ contains (approximate) rough information
about the decision attribute. Therefore we define $s_C(i, j) := 0$ for $1 \leq i \leq I$.

Now we define a conditional strength $s_C(X = X_j | Y = Y_i)$: If there is a
least one $1 \leq j \leq J$ with $s_C(i, j) > 0$, then there is at least one (approximate)
deterministic class $X_j$, which predicts $Y_i$. In this case we set

$$s_C(X = X_j | Y = Y_i) = \frac{s_C(i, j)}{\sum_{k=1}^{I} s_C(k, j)}. \tag{10.2}$$

---

[1] For other views of Bayes' Theorem and its connection to rough sets see e.g. [18–20].

Obviously, $s_C(X = X_j | Y = Y_i)$ reflects the relative strength of a rule predicting $Y = Y_i$.

If there is no (approximate) deterministic attribute $X = X_j$, which predicts $Y = Y_i$, the fraction $s_C(X = X_j | Y = Y_i)$ of (10.2) is undefined, since its denominator is 0. In this case – as we do not know the result –, we use $\underline{s}_C(X = X_j | Y = Y_i) = 0$ as the lower bound, and $\overline{s}_C(X = j | Y = Y_i) = 1$ as the upper bound.

Now we are able to define lower and upper posterior strength values by setting

$$\overline{s}_C(Y = Y_i | X = X_j) = \frac{\underline{s}_C(X = X_j | Y = Y_i)\pi_i}{\sum_r \underline{s}_C(X = X_j | Y = Y_r)\pi_r}$$

and

$$\underline{s}_C(Y = Y_i | X = X_j) = \frac{\overline{s}_C(X = X_j | Y = Y_i)\pi_i}{\sum_r \overline{s}_C(X = X_j | Y = Y_r)\pi_r}$$

If $C \geq n$, i.e. if the cutpoint is not less than the number of objects, then (10.1) is true for every $X_j$, and we observe that $s_C(Y = Y_i | X = X_j) = \frac{n(i,j)}{n} = p(i,j)$ for any $i, j$. Hence,

$$\overline{s}_n(Y = Y_i | X = X_j) = \underline{s}_n(Y = Y_i | X = X_j) = p(Y = Y_i | X = X_j)$$

and we result in the ordinary posterior probability of $Y = Y_i$ given $X = X_j$. Note, that although $\overline{s}_C \geq \underline{s}_C$ holds, the probability estimators $p(Y = Y_i | X = X_j)$ may be greater than $\overline{s}_C$ or smaller than $\underline{s}_C$. This is due the fact that the strength tables for different cutpoints $C$ may looks quite different.

## 11    Summary and Outlook

Whereas the variable precision model uses a parameter $\beta$ to relax the strict inclusion requirement of the classical rough set model and to compute an approximation quality, a parameter free $\lambda$ model based on proportional reduction of errors can be adapted to the rough set approach to data analysis. This index has the additional property that it is monotone in terms of attributes, i.e. if our knowledge of the world increases, so does the approximation quality. Weighted $\lambda$ measures can be used to include expert or other context knowledge into the model, and an algorithm was given which approximates optimal sets of independent attributes and that is polynomial in the number of attributes. In the final section we showed how to explain Bayesian reasoning into this model. In future work we shall compare our algorithm with other machine learning procedures and extend our approach to unsupervised learning.

Furthermore, we would like to point out that the approach can be characterized as a task to "generate deterministic structures which allow $C$ errors within a substructure", and that this approach can be generalized for other structures as well. For example, finding deterministic orders of objects may be quite unsatisfactory, because given a linear order and adding one error could result in a much larger deterministic structure.

As an example note that the data table

| Object | Attr 1 | Attr 2 | Attr 3 | Attr 4 | Attr 5 |
|--------|--------|--------|--------|--------|--------|
| Obj 1  | 1      | 0      | 0      | 0      | 0      |
| Obj 2  | 1      | 1      | 0      | 0      | 0      |
| Obj 3  | 1      | 1      | 1      | 0      | 0      |
| Obj 4  | 1      | 1      | 1      | 1      | 0      |
| Obj 5  | 1      | 1      | 1      | 1      | 1      |

Produces a linear order as a concept lattice [21]. Now consider the following table with one erroneous observation:

| Object | Attr 1 | Attr 2 | Attr 3 | Attr 4 | Attr 5 |
|--------|--------|--------|--------|--------|--------|
| Obj 1  | 1      | 0      | 0      | 0      | 0      |
| Obj 2  | 1      | 1      | 0      | 0      | 0      |
| Obj 3  | 1      | 1      | 1      | 0      | 0      |
| Obj 4  | 1      | 1      | 1      | 1      | 0      |
| Obj 5  | **0**  | 1      | 1      | 1      | 1      |

This system results in a concept lattice consisting of $|U| - 2$ more nodes than the simple order structure, see Fig. 2.

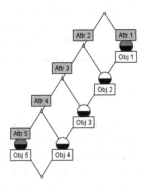

**Fig. 2.** Concept lattice resulting from one error

Hence, leaving out some erroneous observation may lead to a smaller, stronger and mutually more stable structure. We will investigate this in future work.

# References

1. Düntsch, I., Gediga, G.: Weighted λ precision models in rough set data analysis. In: Proceedings of the Federated Conference on Computer Science and Information Systems, pp. 309–316. IEEE, Wrocław, Poland (2012)
2. Pawlak, Z.: Rough sets. Int. J. Comput. Inform. Sci. **11**, 341–356 (1982)
3. Ziarko, W.: Variable precision rough set model. J. Comput. Syst. Sci. **46**, 39–59 (1993)

4. Gediga, G., Düntsch, I.: Rough approximation quality revisited. Artif. Intell. **132**, 219–234 (2001)
5. Beynon, M.: Reducts within the variable precision rough sets model: a further investigation. Eur. J. Oper. Res. **134**, 592–605 (2001)
6. Zytkow, J.M.: Granularity refined by knowledge: contingency tables and rough sets as tools of discovery. In: Dasarathy, B. (ed.) Proceedings of SPIE 4057, Data Mining and Knowledge Discovery: Theory, Tools, and Technology II., pp. 82–91 (2000)
7. Hildebrand, D., Laing, J., Rosenthal, H.: Prediction logic and quasi-independence in empirical evaluation of formal theory. J. Math. Sociol. **3**, 197–209 (1974)
8. Hildebrand, D., Laing, J., Rosenthal, H.: Prediction Analysis of Cross Classification. Wiley, New York (1977)
9. Goodman, L.A., Kruskal, W.H.: Measures of association for cross classification. J. Am. Stat. Assoc. **49**, 732–764 (1954)
10. Holte, R.C.: Very simple classification rules perform well on most commonly used datasets. Mach. Learn. **11**, 63–90 (1993)
11. Wu, S., Flach, P.A.: Feature selection with labelled and unlabelled data. In: Bohanec, M., Kasek, B., Lavrac, N., Mladenic, D. (eds.) ECML/PKDD 2002 workshop on Integration and Collaboration Aspects of Data Mining, pp. 156–167. University of Helsinki (August, Decision Support and Meta-Learning (2002)
12. Nevill-Manning, C.G., Holmes, G., Witten, I.H.: The development of Holte's 1R classifier. In: Proceedings of the 2nd New Zealand Two-Stream International Conference on Artificial Neural Networks and Expert Systems. ANNES 1995, pp. 239–246. IEEE Computer Society, Washington, DC, USA(1995)
13. Düntsch, I., Gediga, G.: Simple data filtering in rough set systems. Int. J. Approx. Reason. **18**(1–2), 93–106 (1998)
14. Youden, W.: Index for rating diagnostic tests. Cancer **3**, 32–35 (1950)
15. Böhning, D., Böhning, W., Holling, H.: Revisiting youden's index as a useful measure of the misclassification error in meta-analysis of diagnostic studies. Stat. Methods Med. Res. **17**, 543–554 (2008)
16. Oehlert, G.: A note on the Delta method. Am. Stat. **46**, 27–29 (1992)
17. Chen, C.B., Wang, L.Y.: Rough set based clustering with refinement using Shannon's entropy theory. Comput. Math. Appl. **52**, 1563–1576 (2006)
18. Pawlak, Z.: A rough set view on Bayes' theorem. Int. J. Intell. Syst. **18**, 487–498 (2003)
19. Ślęzak, D.: Rough sets and Bayes factor. In: Peters, J.F., Skowron, A. (eds.) Transactions on Rough Sets III. LNCS, vol. 3400, pp. 202–229. Springer, Heidelberg (2005)
20. Yao, Y.: Probabilistic rough set approximations. Int. J. Approx. Reason. **49**(2), 255–271 (2008)
21. Wille, R.: Restructuring lattice theory: an approach based on hierarchies of concepts. In: Rival, I. (ed.) Ordered Sets. NATO Advanced Studies Institute, vol. 83, pp. 445–470. Springer, Reidel, Dordrecht (1982)

# Three Approaches to Deal with Tests for Inconsistent Decision Tables – Comparative Study

Mohammad Azad[1], Igor Chikalov[1], Mikhail Moshkov[1]([⊠]), and Beata Zielosko[2]

[1] Computer, Electrical and Mathematical Sciences and Engineering Division,
King Abdullah University of Science and Technology,
Thuwal 23955-6900, Saudi Arabia
{mohammad.azad,igor.chikalov,mikhail.moshkov}@kaust.edu.sa
[2] Institute of Computer Science, University of Silesia,
39, Będzińska St., 41-200 Sosnowiec, Poland
beata.zielosko@us.edu.pl

**Abstract.** We present three approaches to deal with tests (super-reducts) for inconsistent decision tables. In such tables, we have groups of rows with equal values of conditional attributes and different decisions (values of the decision attribute). Instead of a group of equal rows, we consider one row given by values of conditional attributes and we attach to this row: (i) the set of all decisions for rows from the group (many-valued decisions approach); (ii) the most common decision for rows from the group (the most common decision approach); and (iii) unique code of the set of all decisions for rows from the group (generalized decision approach). For many-valued decisions approach, we consider the problem of finding an arbitrary decision from the set of decisions. For the most common decision approach, we consider the problem of finding the most common decision from the set of decisions. For generalized decision approach, we consider the problem of finding all decisions from the set of decisions. We present experimental results connected with the cardinality of tests and comparative study for the considered approaches.

## 1 Introduction

Tests (super-reducts) has been used for single valued decision tables extensively in literature. But in this paper, we are interested to construct tests for inconsistent decision tables. Here, we have multiple rows (objects) with equal values of conditional attributes but with different decisions (values of the decision attribute). We present three approaches to deal with tests for inconsistent decision tables.

For approach called many-valued decisions – $MVD$, we transform inconsistent decision table into a decision table with many-valued decisions. Instead of a group of equal rows with different decisions we consider one row given by values of conditional attributes and we attach to this row the set of all decisions for

© Springer-Verlag Berlin Heidelberg 2015
J. Peters et al. (Eds.): TRS XIX, LNCS 8988, pp. 38–50, 2015.
DOI: 10.1007/978-3-662-47815-8_3

rows from the group [7]. The second approach is called the most common decision – *MCD*. We transform inconsistent decision table into consistent decision table with one-valued decision. Instead of a group of equal rows with different decisions, we consider one row given by values of conditional attributes and we attach to this row the most common decision for rows from the group. The third approach is well known in the rough set theory [10, 14] and is called generalized decision – *GD*. In this case, we transform inconsistent decision table into the table with many-valued decisions and after that encode each set of decisions by a number (decision) such that equal sets are encoded by equal numbers and different sets by different numbers.

In *MVD* approach, we consider the problem of finding arbitrary decision from the set of decisions. In *MCD* approach, our aim is to find the most common decision from the set of decisions. In *GD* approach, we should find the whole set of decisions attached to rows from a group of equal rows.

In the rough set theory, reducts usually are considered as minimal subsets of attributes (with respect to inclusion) which discern all pairs of objects with different decisions that are discernible by the whole set of attributes. However, there exist different modifications of the notion of reduct and different approaches for reduct construction [4–6, 9, 11–13, 15–19]. Exact and approximate reducts and tests (super-reducts) are used for feature selection, knowledge representation and for construction of classifiers. Problems of minimization of exact and approximate tests are NP-hard [6, 9, 17].

We define the notions of reduct and test for decision tables with many-valued decisions. For the case when each set of decisions attached to the rows of table has only one decision, the considered notion of reduct coincides with the usual one. In [2, 8] we studied a greedy algorithm for construction of tests for decision tables with many-valued decisions. This algorithm can be used also in the cases of *MCD* and *GD* approaches: we can consider decision tables with one-valued decision as decision tables with many-valued decisions where sets of decisions attached to rows have one element.

This paper is an extension of the conference publication [3]. It is devoted to the comparison of the cardinality of tests constructed by the greedy algorithm for *MVD*, *MCD* and *GD* approaches. In this paper, we present experimental results for data sets from UCI Machine Learning Repository [1] that were converted to inconsistent decision tables by removal of some conditional attributes. These results show that the use of *MCD* and, especially, *MVD* approaches can reduce the number of attributes in tests in comparison with *GD* approach. It means that *MVD* and *MCD* approaches can be useful from the point of view of knowledge representation.

This paper consists of six sections. Section 2 contains main notions. In Sect. 3, we consider decision tables which have at most $t$ decisions in each set of decisions attached to rows. In Sect. 4, we present the greedy algorithm for construction of approximate tests. Section 5 contains results of experiments and Sect. 6 – conclusions.

## 2  Main Notions

A *decision table with one-valued decision* is a rectangular table $T$ filled by non-negative integers. Columns of this table are labeled with conditional attributes $f_1, \ldots, f_n$. Each row is labeled with a natural number (decision) which is interpreted as a value of the decision attribute. It is possible that $T$ is inconsistent, i.e., contains equal rows with different decisions.

A *Decision table with many-valued decisions*, $T$ is a rectangular table filled by nonnegative integers. Columns of this table are labeled with conditional attributes $f_1, \ldots, f_n$. If we have strings as values of attributes, we have to encode the values as nonnegative integers. We do not have any duplicate rows, and each row is labeled with a nonempty finite set of natural numbers (set of decisions). We have $N(T)$ number of rows. We denote row $i$ by $r_i$ where $i = 1, \ldots, N(T)$. For example, $r_1$ means the first row, $r_2$ means the second rows and so on. Note that each consistent decision table with one-valued decision can be interpreted also as a decision table with many-valued decisions. In such table, each row is labeled with a set of decision which has one element.

The most frequent decision attached to rows from a group of rows in a decision table $T$ is called the *most common decision* for this group of rows. If we have more than one such decision we choose the minimum one.

To work with inconsistent decision tables we consider three approaches:

1. many-valued decisions – *MVD*,
2. the most common decision – *MCD*,
3. generalized decision – *GD*.

For approach called *many-valued decisions* – *MVD*, we transform an inconsistent decision table $T$ into a decision table $T_{MVD}$ with many-valued decisions. Instead of a group of equal rows with different decisions, we consider one row from the group and we attach to this row the set of all decisions for rows from the group [7]. Figure 1 presents transformation of an inconsistent decision table $T^0$ for *MVD* approach.

$$
T^0 = \begin{array}{c|ccc|c}
 & f_1 & f_2 & f_3 & \\
\hline
r_1 & 1 & 1 & 1 & 1 \\
r_2 & 0 & 1 & 0 & 1 \\
r_3 & 0 & 1 & 0 & 3 \\
r_4 & 1 & 1 & 0 & 2 \\
r_5 & 0 & 0 & 1 & 2 \\
r_6 & 0 & 0 & 1 & 3 \\
r_7 & 1 & 0 & 0 & 1 \\
r_8 & 1 & 0 & 0 & 2 \\
\end{array}
\implies
T^0_{MVD} = \begin{array}{c|ccc|c}
 & f_1 & f_2 & f_3 & \\
\hline
r_1 & 1 & 1 & 1 & \{1\} \\
r_2 & 0 & 1 & 0 & \{1,3\} \\
r_3 & 1 & 1 & 0 & \{2\} \\
r_4 & 0 & 0 & 1 & \{2,3\} \\
r_5 & 1 & 0 & 0 & \{1,2\} \\
\end{array}
$$

**Fig. 1.** Transformation of inconsistent decision table $T^0$ into decision table $T^0_{MVD}$

$$T^0 = \begin{array}{c|ccc|c} & f_1 & f_2 & f_3 & \\ \hline r_1 & 1 & 1 & 1 & 1 \\ r_2 & 0 & 1 & 0 & 1 \\ r_3 & 0 & 1 & 0 & 3 \\ r_4 & 1 & 1 & 0 & 2 \\ r_5 & 0 & 0 & 1 & 2 \\ r_6 & 0 & 0 & 1 & 3 \\ r_7 & 1 & 0 & 0 & 1 \\ r_8 & 1 & 0 & 0 & 2 \end{array} \implies T^0_{MCD} = \begin{array}{c|ccc|c} & f_1 & f_2 & f_3 & \\ \hline r_1 & 1 & 1 & 1 & 1 \\ r_2 & 0 & 1 & 0 & 1 \\ r_3 & 1 & 1 & 0 & 2 \\ r_4 & 0 & 0 & 1 & 2 \\ r_5 & 1 & 0 & 0 & 1 \end{array}$$

**Fig. 2.** Transformation of inconsistent decision table $T^0$ into decision table $T^0_{MCD}$

$$T^0 = \begin{array}{c|ccc|c} & f_1 & f_2 & f_3 & \\ \hline r_1 & 1 & 1 & 1 & 1 \\ r_2 & 0 & 1 & 0 & 1 \\ r_3 & 0 & 1 & 0 & 3 \\ r_4 & 1 & 1 & 0 & 2 \\ r_5 & 0 & 0 & 1 & 2 \\ r_6 & 0 & 0 & 1 & 3 \\ r_7 & 1 & 0 & 0 & 1 \\ r_8 & 1 & 0 & 0 & 2 \end{array} \implies T^0_{GD} = \begin{array}{c|ccc|c} & f_1 & f_2 & f_3 & \\ \hline r_1 & 1 & 1 & 1 & 1 \\ r_2 & 0 & 1 & 0 & 2 \\ r_3 & 1 & 1 & 0 & 3 \\ r_4 & 0 & 0 & 1 & 4 \\ r_5 & 1 & 0 & 0 & 5 \end{array}$$

**Fig. 3.** Transformation of inconsistent decision table $T^0$ into decision table $T^0_{GD}$

For approach called *the most common decision – MCD*, we transform inconsistent decision table $T$ into consistent decision table $T_{MCD}$ with one-valued decision. Instead of a group of equal rows with different decisions, we consider one row from the group and we attach to this row the most common decision for the considered group of rows. Figure 2 presents transformation of the inconsistent decision table $T^0$ for *MCD* approach.

For approach called *generalized decision – GD*, we transform inconsistent decision table $T$ into consistent decision table $T_{GD}$ with one-valued decision. Instead of a group of equal rows with different decisions, we consider one row from the group and we attach to this row the set of all decisions for rows from the group. Then instead of a set of decisions we attach to each row a code of this set – a natural number such that the codes of equal sets are equal and the codes of different sets are different. Figure 3 presents transformation of the inconsistent decision table $T^0$ for *GD* approach.

To unify some notions for decision tables with one-valued and many-valued decisions, we will interpret decision table with one-valued decision as a decision table with many-valued decisions where each row is labeled with a set of decision that has one element.

We will say that $T$ is a *degenerate* table if either $T$ is empty (has no rows), or the intersection of sets of decisions attached to rows of $T$ is nonempty.

A table obtained from $T$ by removal of some rows is called a *subtable* of $T$. A subtable $T'$ of $T$ is called *boundary* subtable if $T'$ is not degenerate but each

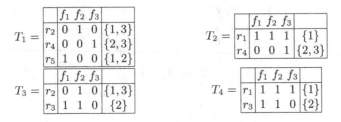

**Fig. 4.** All boundary subtables of the decision table $T^0_{MVD}$ (see Fig. 1)

proper subtable of $T'$ is degenerate. We denote by $B(T)$ the number of boundary subtables of the table $T$. It is clear that $T$ is a degenerate table if and only if $B(T) = 0$. The value $B(T)$ will be interpreted as *uncertainty* of $T$.

Figure 4 presents all four boundary subtables of the decision table $T^0_{MVD}$ depicted in Fig. 1. The number of boundary subtables of the decision table $T^0_{MCD}$ depicted in Fig. 2 is equal to six. The number of boundary subtables of the decision table $T^0_{GD}$ depicted in Fig. 3 is equal to 10. Each boundary subtable of tables $T^0_{MCD}$ and $T^0_{GD}$ has two rows (see Corollary 1).

We will say that an attribute $f_i$ *divides* a boundary subtable $\Theta$ of the table $T$ if and only if this attribute is not constant on the rows of $\Theta$ (for example, for a binary decision table at the intersection with the column $f_i$ we can find some rows which contain 1 and some rows which contain 0).

Let us define the notion of an $\alpha$-test for the table $T$. Let $\alpha$ be a real number such that $0 \le \alpha < 1$.

An $\alpha$-*test for the table* $T$ is a subset of attributes $\{f_{i_1}, \dots, f_{i_m}\}$ such that these attributes divide at least $(1 - \alpha)B(T)$ boundary subtables of $T$. Empty set is an $\alpha$-test for $T$ if and only if $T$ is a degenerate table. An $\alpha$-*reduct for the table* $T$ is an $\alpha$-test for $T$ for which each proper subset is not an $\alpha$-test. We denote by $R_{\min}(\alpha, T)$ the minimum cardinality of an $\alpha$-test for the table $T$. It is clear that each $\alpha$-test has an $\alpha$-reduct as a subset. Therefore $R_{\min}(\alpha, T)$ is the minimum cardinality of an $\alpha$-reduct for the table $T$.

## 3   Set $Tab(t)$ of Decision Tables

We denote by $Tab(t)$, where $t$ is a natural number, the set of decision tables with many-valued decisions such that each row in the table has at most $t$ decisions (is labeled with a set of decisions which cardinality is at most $t$).

The next statement was proved in [7]. For the completeness, we will give it with the proof.

**Lemma 1.** *Let $T'$ be a boundary subtable with $m$ rows. Then each row of $T'$ is labeled with a set of decisions whose cardinality is at least $m - 1$.*

*Proof.* Let rows of $T'$ be labeled with sets of decisions $D_1, \dots, D_m$ respectively. Then $D_1 \cap \dots \cap D_m = \emptyset$ and, for any $i \in \{1, \dots, m\}$, the set $D_1 \cap \dots \cap D_{i-1} \cap$

$D_{i+1} \cap \ldots \cap D_m$ contains a number $d_i$. Assume that $i \neq j$ and $d_i = d_j$. Then $D_1 \cap \ldots \cap D_m \neq \emptyset$ which is impossible. Therefore $d_1, \ldots, d_m$ are pairwise different numbers. It is clear that, for $i = 1, \ldots, m$, the set $\{d_1, \ldots, d_m\} \setminus \{d_i\}$ is a subset of the set $D_i$.

**Corollary 1.** *Each boundary subtable of a table* $T \in Tab(t)$ *has at most* $t + 1$ *rows.*

Therefore, for tables from $Tab(t)$, there exists a polynomial time algorithm for the finding of all boundary subtables and the computation of parameter $B(T)$. For example, for any decision table $T$ with one-valued decision (in fact, for any table from $Tab(1)$), the equality $B(T) = P(T)$ holds, where $P(T)$ is the number of unordered pairs of rows of $T$ with different decisions.

# 4    Greedy Algorithm for $\alpha$-Test Construction

Note one more time that each consistent decision table with one-valued decision can be interpreted as a decision table with many-valued decisions.

Now we present a greedy algorithm for $\alpha$-test construction (if $\alpha = 0$ we are working with exact tests). Let $T$ be a table with many-valued decisions containing $n$ columns labeled with attributes $f_1, \ldots, f_n$. Let $\alpha$ be a real number such that $0 \leq \alpha < 1$, and $B(T)$ be the number of boundary subtables of the table $T$. Greedy algorithm at each iteration chooses an attribute which divides the maximum number of not divided boundary subtables. This algorithm stops if attributes from the constructed set divide at least $(1 - \alpha)B(T)$ boundary subtables (see Algorithm 1). For example, if $\alpha = 0.1$ then attributes from $\alpha$-test divide at least $90\%$ of boundary subtables. By $R_{\text{greedy}}(\alpha, T)$ we denote the cardinality of $\alpha$-test constructed by the greedy algorithm. Remind that $R_{\min}(\alpha, T)$ is the minimum cardinality of $\alpha$-test.

---

**Algorithm 1.** Greedy algorithm for $\alpha$-test construction

---

**Require:** decision table $T$ with attributes $f_1, \ldots, f_n$, $\alpha \in \mathbb{R}$, $0 \leq \alpha < 1$,
**Ensure:** an $\alpha$-test for $T$
   $Q \leftarrow \emptyset$;
   **while** attributes from $Q$ divide less than $(1 - \alpha)B(T)$ boundary subtables of $T$ **do**
      select $f_i \in \{f_1, \ldots, f_n\}$ with the minimum index $i$ such that $f_i$ divides the maximum number of boundary subtables not divided by attributes from $Q$
      $Q \leftarrow Q \cup \{f_i\}$;
   **end while**

---

If, for a fixed natural $t$, we apply the considered algorithm to decision tables from $Tab(t)$ then the time complexity of this algorithm (including construction of all boundary subtables) will be bounded from above by a polynomial depending on the length of decision table description.

Results presented in [2] show that the problem of minimization of $\alpha$-test cardinality, $0 \leq \alpha < 1$, is NP-hard. We know (see [8]) that under the assumption $NP \nsubseteq DTIME(n^{O(\log \log n)})$ the greedy algorithm is close to the best (from the point of view of precision) approximate polynomial algorithms for minimization of $\alpha$-test cardinality.

**Theorem 1.** *Reference [8] Let $T$ be a nondegenerate decision table with many-valued decisions and $\alpha$ be a real number such that $0 < \alpha < 1$. Then*

$$R_{\text{greedy}}(\alpha, T) \leq R_{\min}(0, T) \ln(1/\alpha) + 1.$$

## 5   Experimental Results

We consider a number of decision tables from UCI Machine Learning Repository [1]. In some tables there were missing values. Each such value was replaced with the most common value of the corresponding attribute. Some decision tables contain conditional attributes that take unique value for each row. Such attributes were removed.

**Table 1.** Characteristics of decision tables $T_{MVD}$, $T_{MCD}$ and $T_{GD}$

| Decision table $T$ | Rows | Attr | Spectrum | | | | | |
|---|---|---|---|---|---|---|---|---|
| | | | #1 | #2 | #3 | #4 | #5 | #6 |
| balance-scale-1 | 125 | 3 | 45 | 50 | 30 | | | |
| breast-cancer-1 | 193 | 8 | 169 | 24 | | | | |
| breast-cancer-5 | 98 | 4 | 58 | 40 | | | | |
| cars-1 | 432 | 5 | 258 | 161 | 13 | | | |
| flags-5 | 171 | 21 | 159 | 12 | | | | |
| hayes-roth-data-1 | 39 | 3 | 22 | 13 | 4 | | | |
| kr-vs-kp-5 | 1987 | 31 | 1564 | 423 | | | | |
| kr-vs-kp-4 | 2061 | 32 | 1652 | 409 | | | | |
| lymphography-5 | 122 | 13 | 113 | 9 | | | | |
| mushroom-5 | 4078 | 17 | 4048 | 30 | | | | |
| nursery-1 | 4320 | 7 | 2858 | 1460 | 2 | | | |
| nursery-4 | 240 | 4 | 97 | 96 | 47 | | | |
| spect-test-1 | 164 | 21 | 161 | 3 | | | | |
| teeth-1 | 22 | 7 | 12 | 10 | | | | |
| teeth-5 | 14 | 3 | 6 | 3 | 0 | 5 | 0 | 2 |
| tic-tac-toe-4 | 231 | 5 | 102 | 129 | | | | |
| tic-tac-toe-3 | 449 | 6 | 300 | 149 | | | | |
| zoo-data-5 | 42 | 11 | 36 | 6 | | | | |

**Table 2.** Number of boundary subtables and cardinality of $\alpha$-tests for tables $T_{MVD}$

| Decision table $T$ | Number of boundary subtables $B(T_{MVD})$ | cardinality of $\alpha$-tests for $\alpha =$ | | | | | |
|---|---|---|---|---|---|---|---|
| | | 0.0 | 0.001 | 0.01 | 0.1 | 0.2 | 0.3 |
| balance-scale-1 | 3699 | 2 | 2 | 2 | 1 | 1 | 1 |
| breast-cancer-1 | 6328 | 8 | 6 | 4 | 2 | 2 | 1 |
| breast-cancer-5 | 480 | 4 | 4 | 3 | 1 | 1 | 1 |
| cars-1 | 18576 | 5 | 4 | 3 | 2 | 2 | 1 |
| flags-5 | 11424 | 13 | 9 | 4 | 2 | 1 | 1 |
| hayes-roth-data-1 | 297 | 2 | 2 | 2 | 2 | 1 | 1 |
| kr-vs-kp-5 | 609843 | 26 | 12 | 7 | 4 | 3 | 2 |
| kr-vs-kp-4 | 681120 | 27 | 12 | 7 | 4 | 3 | 2 |
| lymphography-5 | 3434 | 11 | 9 | 5 | 2 | 2 | 1 |
| mushroom-5 | 3068380 | 8 | 4 | 3 | 2 | 1 | 1 |
| nursery-1 | 4797564 | 7 | 3 | 2 | 1 | 1 | 1 |
| nursery-4 | 12860 | 2 | 2 | 1 | 1 | 1 | 1 |
| spect-test-1 | 930 | 10 | 10 | 6 | 3 | 3 | 2 |
| teeth-1 | 231 | 5 | 5 | 4 | 2 | 2 | 1 |
| teeth-5 | 91 | 3 | 3 | 3 | 2 | 2 | 1 |
| tic-tac-toe-4 | 1976 | 5 | 5 | 4 | 2 | 2 | 1 |
| tic-tac-toe-3 | 18404 | 6 | 5 | 4 | 2 | 2 | 2 |
| zoo-data-5 | 686 | 9 | 9 | 7 | 3 | 2 | 2 |
| **Average** | 513129.06 | 8.50 | 5.89 | 3.94 | 2.11 | 1.78 | 1.28 |

We removed from these tables some conditional attributes. As a result, we obtain inconsistent decision tables. After that we transform each such table $T$ into tables $T_{MVD}$, $T_{MCD}$ and $T_{GD}$ as it was described in Sect. 2. The information about these decision tables can be found in Table 1. This table contains name of inconsistent table $T$ in the form "name of initial table from [1]"-"number of removed conditional attributes", number of rows in $T_{MVD}$, $T_{MCD}$, $T_{GD}$ (column "Rows"), number of attributes in $T_{MVD}$, $T_{MCD}$, $T_{GD}$ (column "Attr"), and spectrum of the table $T_{MVD}$ (column "Spectrum"). Spectrum of a decision table with many-valued decisions is a sequence #1, #2,..., where #$i$, $i = 1, 2, \ldots$, is the number of rows labeled with sets of decision with the cardinality equal to $i$.

For decision tables described in Table 1 and $\alpha \in \{0.0, 0.001, 0.01, 0.1, 0.2, 0.3\}$, we constructed $\alpha$-tests by the greedy algorithm. Table 2 presents the number of boundary subtables (column "$B(T_{MVD})$") and the cardinality of $\alpha$-tests for tables $T_{MVD}$. Table 3 presents the number of boundary subtables (column "$B(T_{MCD})$") and the cardinality of $\alpha$-tests for tables $T_{MCD}$. Table 4 presents the number of boundary subtables (column "$B(T_{GD})$") and the cardinality of

**Table 3.** Number of boundary subtables and cardinality of $\alpha$-tests for tables $T_{MCD}$

| Decision table $T$ | Number of boundary subtables $B(T_{MCD})$ | cardinality of $\alpha$-tests for $\alpha =$ | | | | | |
|---|---|---|---|---|---|---|---|
| | | 0.0 | 0.001 | 0.01 | 0.1 | 0.2 | 0.3 |
| balance-scale-1 | 3906 | 3 | 3 | 3 | 2 | 1 | 1 |
| breast-cancer-1 | 7752 | 8 | 6 | 4 | 2 | 2 | 1 |
| breast-cancer-5 | 1377 | 4 | 4 | 3 | 2 | 1 | 1 |
| cars-1 | 34992 | 5 | 4 | 3 | 2 | 2 | 1 |
| flags-5 | 11747 | 13 | 9 | 4 | 2 | 1 | 1 |
| hayes-roth-data-1 | 494 | 3 | 3 | 3 | 2 | 1 | 1 |
| kr-vs-kp-5 | 956242 | 27 | 12 | 7 | 4 | 3 | 2 |
| kr-vs-kp-4 | 1030424 | 28 | 13 | 7 | 4 | 3 | 2 |
| lymphography-5 | 4019 | 11 | 9 | 5 | 2 | 2 | 1 |
| mushroom-5 | 3110992 | 9 | 4 | 3 | 2 | 1 | 1 |
| nursery-1 | 6385972 | 7 | 4 | 3 | 2 | 1 | 1 |
| nursery-4 | 19079 | 4 | 3 | 2 | 2 | 1 | 1 |
| spect-test-1 | 1099 | 12 | 11 | 7 | 3 | 2 | 2 |
| teeth-1 | 231 | 5 | 5 | 4 | 2 | 2 | 1 |
| teeth-5 | 91 | 3 | 3 | 3 | 2 | 2 | 1 |
| tic-tac-toe-4 | 9800 | 5 | 5 | 4 | 2 | 2 | 2 |
| tic-tac-toe-3 | 39688 | 6 | 6 | 4 | 3 | 2 | 2 |
| zoo-data-5 | 704 | 9 | 9 | 7 | 3 | 2 | 2 |
| **Average** | 645478.28 | 9.00 | 6.28 | 4.22 | 2.39 | 1.72 | 1.33 |

$\alpha$-tests for tables $T_{GD}$. The last row in Tables 2, 3 and 4 contains the average number of boundary subtables and the average cardinality of $\alpha$-tests.

Based on results presented in Table 2 we can see that the cardinality of $\alpha$-test is decreasing or nonincreasing when the value of $\alpha$ is increasing. For example, for data sets "kr-vs-kp-5" and "kr-vs-kp-4", the exact tests contain 26 and 27 attributes respectively but 0.01-tests contain only 7 attributes. Similar situation can be observed in Tables 3 and 4. We can also notice that the average number of boundary subtables is the smallest for $MVD$ approach and the biggest for $GD$ approach.

Table 5, based on results from Tables 2 and 3, presents comparison of the number of boundary subtables and $\alpha$-test cardinality for $MVD$ and $MCD$ approaches. Each input of this table is equal to the value from Table 3 divided by the corresponding value from Table 2. Presented results show that, greedy algorithm sometimes constructs two times shorter tests for $MVD$ approach than $MCD$ approach: "balance-scale-1" ($\alpha = 0.1$), "breast-cancer-5" ($\alpha = 0.1$), "nursery-1" ($\alpha = 0.1$), "nursery-4" ($\alpha = 0.0$, $\alpha = 0.01$, and $\alpha = 0.1$), and "tic-tac-toe-4" ($\alpha = 0.3$). Only for "spect-test-1" the cardinality of 0.2-test is smaller

**Table 4.** Number of boundary subtables and cardinality of $\alpha$-tests for tables $T_{GD}$

| Decision table $T$ | Number of boundary subtables $B(T_{GD})$ | cardinality of $\alpha$-tests for $\alpha =$ | | | | | |
|---|---|---|---|---|---|---|---|
| | | 0.0 | 0.001 | 0.01 | 0.1 | 0.2 | 0.3 |
| balance-scale-1 | 6255 | 3 | 3 | 3 | 2 | 1 | 1 |
| breast-cancer-1 | 10384 | 8 | 6 | 4 | 2 | 2 | 1 |
| breast-cancer-5 | 2800 | 4 | 4 | 3 | 1 | 1 | 1 |
| cars-1 | 54589 | 5 | 5 | 3 | 2 | 2 | 1 |
| flags-5 | 12105 | 14 | 9 | 4 | 2 | 1 | 1 |
| hayes-roth-data-1 | 571 | 3 | 3 | 3 | 2 | 1 | 1 |
| kr-vs-kp-5 | 1271415 | 27 | 13 | 8 | 4 | 3 | 2 |
| kr-vs-kp-4 | 1356788 | 28 | 13 | 8 | 4 | 3 | 2 |
| lymphography-5 | 4397 | 11 | 10 | 6 | 2 | 2 | 1 |
| mushroom-5 | 3189820 | 9 | 4 | 3 | 2 | 1 | 1 |
| nursery-1 | 6971260 | 7 | 5 | 3 | 2 | 1 | 1 |
| nursery-4 | 19914 | 4 | 3 | 2 | 2 | 1 | 1 |
| spect-test-1 | 1413 | 11 | 10 | 7 | 3 | 3 | 2 |
| teeth-1 | 231 | 5 | 5 | 4 | 2 | 2 | 1 |
| teeth-5 | 91 | 3 | 3 | 3 | 2 | 2 | 1 |
| tic-tac-toe-4 | 15134 | 5 | 5 | 4 | 2 | 2 | 2 |
| tic-tac-toe-3 | 63104 | 6 | 6 | 4 | 3 | 2 | 2 |
| zoo-data-5 | 750 | 9 | 9 | 7 | 3 | 2 | 2 |
| **Average** | 721167.83 | 9.00 | 6.44 | 4.39 | 2.33 | 1.78 | 1.33 |

for $MCD$ approach. The number of boundary subtables is usually greater for $MCD$ approach than for $MVD$ approach, for "tic-tac-toe-4" – almost five times.

Table 6, based on results from Tables 2 and 4, presents comparison of the number of boundary subtables and $\alpha$-test cardinality for $MVD$ and $GD$ approaches. Each input of this table is equal to the value from Table 4 divided by the corresponding value from Table 2. Presented results show that, the cardinality of $\alpha$-tests is sometimes two times smaller for $MVD$ approach than for $GD$ approach: "balance-scale-1" ($\alpha = 0.1$), "nursery-1" ($\alpha = 0.1$), "nursery-4" ($\alpha = 0.0$, $\alpha = 0.01$, and $\alpha = 0.1$), and "tic-tac-toe-4" ($\alpha = 0.3$). The number of boundary subtables is usually greater for $GD$ approach than for $MVD$ approach, for "breast-cancer-5" – almost six times, for "tic-tac-toe-4" – more than seven times.

Table 7, based on results from Tables 3 and 4, presents comparison of the number of boundary subtables and $\alpha$-test cardinality for $MCD$ and $GD$ approaches. Each input of this table is equal to the value from Table 4 divided by the corresponding value from Table 3. Presented results show that the cardinality of $\alpha$-tests is comparable for $MCD$ and $GD$ approaches. Only for "breast-cancer-5" ($\alpha = 0.1$) the greedy algorithm for $GD$ approach constructs 50 %

**Table 5.** Comparison of number of boundary subtables and $\alpha$-test cardinality for tables $T_{MVD}$ and $T_{MCD}$ ($\frac{T_{MCD}}{T_{MVD}}$)

| Decision table $T$ | Ratio of number of boundary subtables | Ratio of cardinality of $\alpha$-tests for $\alpha =$ | | | | | |
|---|---|---|---|---|---|---|---|
| | | 0.0 | 0.001 | 0.01 | 0.1 | 0.2 | 0.3 |
| balance-scale-1 | 1.06 | 1.50 | 1.50 | 1.50 | 2.00 | 1.00 | 1.00 |
| breast-cancer-1 | 1.23 | 1.00 | 1.00 | 1.00 | 1.00 | 1.00 | 1.00 |
| breast-cancer-5 | 2.87 | 1.00 | 1.00 | 1.00 | 2.00 | 1.00 | 1.00 |
| cars-1 | 1.88 | 1.00 | 1.00 | 1.00 | 1.00 | 1.00 | 1.00 |
| flags-5 | 1.03 | 1.00 | 1.00 | 1.00 | 1.00 | 1.00 | 1.00 |
| hayes-roth-data-1 | 1.66 | 1.50 | 1.50 | 1.50 | 1.00 | 1.00 | 1.00 |
| kr-vs-kp-5 | 1.57 | 1.04 | 1.00 | 1.00 | 1.00 | 1.00 | 1.00 |
| kr-vs-kp-4 | 1.51 | 1.04 | 1.08 | 1.00 | 1.00 | 1.00 | 1.00 |
| lymphography-5 | 1.17 | 1.00 | 1.00 | 1.00 | 1.00 | 1.00 | 1.00 |
| mushroom-5 | 1.01 | 1.13 | 1.00 | 1.00 | 1.00 | 1.00 | 1.00 |
| nursery-1 | 1.33 | 1.00 | 1.33 | 1.50 | 2.00 | 1.00 | 1.00 |
| nursery-4 | 1.48 | 2.00 | 1.50 | 2.00 | 2.00 | 1.00 | 1.00 |
| spect-test-1 | 1.18 | 1.20 | 1.10 | 1.17 | 1.00 | 0.67 | 1.00 |
| teeth-1 | 1.00 | 1.00 | 1.00 | 1.00 | 1.00 | 1.00 | 1.00 |
| teeth-5 | 1.00 | 1.00 | 1.00 | 1.00 | 1.00 | 1.00 | 1.00 |
| tic-tac-toe-4 | 4.96 | 1.00 | 1.00 | 1.00 | 1.00 | 1.00 | 2.00 |
| tic-tac-toe-3 | 2.16 | 1.00 | 1.20 | 1.00 | 1.50 | 1.00 | 1.00 |
| zoo-data-5 | 1.03 | 1.00 | 1.00 | 1.00 | 1.00 | 1.00 | 1.00 |

shorter test than for $MCD$ approach. The number of boundary subtables is usually greater for $GD$ approach than for $MCD$ approach, however the difference is not significant, only for "breast-cancer-5" – two times.

**Table 6.** Comparison of number of boundary subtables and $\alpha$-test cardinality for tables $T_{MVD}$ and $T_{GD}$ ($\frac{T_{GD}}{T_{MVD}}$)

| Decision table $T$ | Ratio of number of boundary subtables | Ratio of cardinality of $\alpha$-tests for $\alpha =$ | | | | | |
|---|---|---|---|---|---|---|---|
| | | 0.0 | 0.001 | 0.01 | 0.1 | 0.2 | 0.3 |
| balance-scale-1 | 1.69 | 1.50 | 1.50 | 1.50 | 2.00 | 1.00 | 1.00 |
| breast-cancer-1 | 1.64 | 1.00 | 1.00 | 1.00 | 1.00 | 1.00 | 1.00 |
| breast-cancer-5 | 5.83 | 1.00 | 1.00 | 1.00 | 1.00 | 1.00 | 1.00 |
| cars-1 | 2.94 | 1.00 | 1.25 | 1.00 | 1.00 | 1.00 | 1.00 |
| flags-5 | 1.06 | 1.08 | 1.00 | 1.00 | 1.00 | 1.00 | 1.00 |
| hayes-roth-data-1 | 1.92 | 1.50 | 1.50 | 1.50 | 1.00 | 1.00 | 1.00 |
| kr-vs-kp-5 | 2.08 | 1.04 | 1.08 | 1.14 | 1.00 | 1.00 | 1.00 |
| kr-vs-kp-4 | 1.99 | 1.04 | 1.08 | 1.14 | 1.00 | 1.00 | 1.00 |
| lymphography-5 | 1.28 | 1.00 | 1.11 | 1.20 | 1.00 | 1.00 | 1.00 |
| mushroom-5 | 1.04 | 1.13 | 1.00 | 1.00 | 1.00 | 1.00 | 1.00 |
| nursery-1 | 1.45 | 1.00 | 1.67 | 1.50 | 2.00 | 1.00 | 1.00 |
| nursery-4 | 1.55 | 2.00 | 1.50 | 2.00 | 2.00 | 1.00 | 1.00 |
| spect-test-1 | 1.52 | 1.10 | 1.00 | 1.17 | 1.00 | 1.00 | 1.00 |
| teeth-1 | 1.00 | 1.00 | 1.00 | 1.00 | 1.00 | 1.00 | 1.00 |
| teeth-5 | 1.00 | 1.00 | 1.00 | 1.00 | 1.00 | 1.00 | 1.00 |
| tic-tac-toe-4 | 7.66 | 1.00 | 1.00 | 1.00 | 1.00 | 1.00 | 2.00 |
| tic-tac-toe-3 | 3.43 | 1.00 | 1.20 | 1.00 | 1.50 | 1.00 | 1.00 |
| zoo-data-5 | 1.09 | 1.00 | 1.00 | 1.00 | 1.00 | 1.00 | 1.00 |

**Table 7.** Comparison of number of boundary subtables and $\alpha$-test cardinality for tables $T_{MCD}$ and $T_{GD}$ ($\frac{T_{GD}}{T_{MCD}}$)

| Decision table $T$ | Ratio of number of boundary subtables | Ratio of cardinality of $\alpha$-tests for $\alpha =$ | | | | | |
|---|---|---|---|---|---|---|---|
| | | 0.0 | 0.001 | 0.01 | 0.1 | 0.2 | 0.3 |
| balance-scale-1 | 1.60 | 1.00 | 1.00 | 1.00 | 1.00 | 1.00 | 1.00 |
| breast-cancer-1 | 1.34 | 1.00 | 1.00 | 1.00 | 1.00 | 1.00 | 1.00 |
| breast-cancer-5 | 2.03 | 1.00 | 1.00 | 1.00 | 0.50 | 1.00 | 1.00 |
| cars-1 | 1.56 | 1.00 | 1.25 | 1.00 | 1.00 | 1.00 | 1.00 |
| flags-5 | 1.03 | 1.08 | 1.00 | 1.00 | 1.00 | 1.00 | 1.00 |
| hayes-roth-data-1 | 1.16 | 1.00 | 1.00 | 1.00 | 1.00 | 1.00 | 1.00 |
| kr-vs-kp-5 | 1.33 | 1.00 | 1.08 | 1.14 | 1.00 | 1.00 | 1.00 |
| kr-vs-kp-4 | 1.32 | 1.00 | 1.00 | 1.14 | 1.00 | 1.00 | 1.00 |
| lymphography-5 | 1.09 | 1.00 | 1.11 | 1.20 | 1.00 | 1.00 | 1.00 |
| mushroom-5 | 1.03 | 1.00 | 1.00 | 1.00 | 1.00 | 1.00 | 1.00 |
| nursery-1 | 1.09 | 1.00 | 1.25 | 1.00 | 1.00 | 1.00 | 1.00 |
| nursery-4 | 1.04 | 1.00 | 1.00 | 1.00 | 1.00 | 1.00 | 1.00 |
| spect-test-1 | 1.29 | 0.92 | 0.91 | 1.00 | 1.00 | 1.50 | 1.00 |
| teeth-1 | 1.00 | 1.00 | 1.00 | 1.00 | 1.00 | 1.00 | 1.00 |
| teeth-5 | 1.00 | 1.00 | 1.00 | 1.00 | 1.00 | 1.00 | 1.00 |
| tic-tac-toe-4 | 1.54 | 1.00 | 1.00 | 1.00 | 1.00 | 1.00 | 1.00 |
| tic-tac-toe-3 | 1.59 | 1.00 | 1.00 | 1.00 | 1.00 | 1.00 | 1.00 |
| zoo-data-5 | 1.07 | 1.00 | 1.00 | 1.00 | 1.00 | 1.00 | 1.00 |

# 6  Conclusions

We studied the greedy algorithm for construction of $\alpha$-tests. This algorithm requires the construction of all boundary subtables of the initial decision table. We proved that for an arbitrary natural $t$, the considered algorithm has polynomial time complexity on tables which have at most $t$ decisions in each set of decisions attached to rows.

We considered the cardinality of $\alpha$-tests and the number of boundary subtables for *MVD*, *MCD* and *GD* approaches. We removed some conditional attributes from a number of data sets from UCI ML Repository to obtain inconsistent decision tables. Experimental results show that, the greedy algorithm constructs relatively short $\alpha$-tests for all considered approaches. We can observe also that the cardinality of $\alpha$-tests for *MVD* approach is sometimes two times smaller than for *MCD* and *GD* approaches. Differences in the cardinality of $\alpha$-tests for *MCD* and *GD* approaches are not significant. We can also notice that the average number of boundary subtables is the smallest for *MVD* approach and the biggest for *GD* approach.

# References

1. Asuncion, A., Newman, D.J.: UCI Machine Learning Repository (http://www.ics. uci.edu/mlearn/)
2. Azad, M., Chikalov, I., Moshkov, M., Zielosko, B.: Greedy algorithms for construction of approximate tests. Fundam. Inform. **120**(3–4), 231–242 (2012)
3. Azad, M., Chikalov, I., Moshkov, M., Zielosko, B.: Tests for decision tables with many-valued decisions - comparative study. In: Ganzha, M., Maciaszek, L.A., Paprzycki, M. (eds.) FedCSIS, pp. 271–277 (2012)

4. Chikalov, I., Lozin, V., Lozina, I., Moshkov, M., Nguyen, H.S., Skowron, A., Zielosko, B.: Three Approaches to Data Analysis. ISRL, vol. 41. Springer, Heidelberg (2013)
5. Kryszkiewicz, M.: Comparative study of alternative types of knowledge reduction in inconsistent systems. Int. J. Intell. Syst. **16**(1), 105–120 (2001)
6. Moshkov, M., Piliszczuk, M., Zielosko, B.: Partial Covers, Reducts and Decision Rules in Rough Sets - Theory and Applications. Studies in Computational Intelligence. Springer, Heidelberg (2008)
7. Moshkov, M., Zielosko, B.: Combinatorial Machine Learning - A Rough Set Approach. Studies in Computational Intelligence. Springer, Heidelberg (2011)
8. Moshkov, M., Zielosko, B.: Construction of tests for tables with many-valued decisions. In: Szczuka, M., Czaja, L., Skowron, A., Kacprzak, M. (eds.) CS and P, pp. 376–384, Białystok University of Technology (2011)
9. Nguyen, H.S., Ślęzak, D.: Approximate reducts and association rules. In: Zhong, N., Skowron, A., Ohsuga, S. (eds.) RSFDGrC 1999. LNCS (LNAI), vol. 1711, pp. 137–145. Springer, Heidelberg (1999)
10. Pawlak, Z.: Rough Sets - Theoretical Aspects of Reasoning about Data. Kluwer Academic Publishers, Dordrecht (1991)
11. Pawlak, Z.: Rough set elements. In: Polkowski, L., Skowron, A. (eds.) Rough Sets in Knowledge Discovery, pp. 10–30. Physica-Verlag, Heidelberg (1998)
12. Pawlak, Z., Skowron, A.: Rough sets and boolean reasoning. Inf. Sci. **177**(1), 41–73 (2007)
13. Qian, Y., Liang, J., Li, D., Wang, F., Ma, N.: Approximation reduction in inconsistent incomplete decision tables. Knowl.-Based Syst. **23**(5), 427–433 (2010)
14. Skowron, A., Rauszer, C.: The discernibility matrices and functions in information systems. In: Slowinski, R. (ed.) Intelligent Decision Support. Handbook of Applications and Advances of the Rough Set Theory, pp. 331–362. Kluwer Academic Publishers, Dordrecht (1992)
15. Ślęzak, D.: Searching for dynamic reducts in inconsistent decision tables. In: Proceedings of IPMU 1998, pp. 1362–1369, vol. 2, Paris, France, 6–10 July 1998
16. Ślęzak, D.: Normalized decision functions and measures for inconsistent decision tables analysis. Fundam. Inform. **44**, 291–319 (2000)
17. Ślęzak, D.: Approximate entropy reducts. Fundam. Inform. **53**, 365–390 (2002)
18. Zhang, W.X., Mi, J.S., Wu, W.Z.: Approaches to knowledge reductions in inconsistent systems. Int. J. Intell. Syst. **18**(9), 989–1000 (2003)
19. Ziarko, W.: Variable precision rough set model. J. Comput. Syst. Sci. **46**(1), 39–59 (1993)

# Searching for Reductive Attributes in Decision Tables

Long Giang Nguyen[1] and Hung Son Nguyen[2(✉)]

[1] Institute of Information Technology,
Vietnamese Academy of Science and Technology,
18 Hoang Quoc Viet Road, Cau Giay District, Hanoi, Vietnam
nlgiang@ioit.ac.vn
[2] Institute of Mathematics, The University of Warsaw,
Banacha 2, 02-097 Warsaw, Poland
son@mimuw.edu.pl

**Abstract.** Most decision support systems based on rough set theory are related to the minimal reduct calculation problem, which is NP-hard. This paper investigates the problem of searching for the set of useful attributes that occur in at least one reduct. By complement, this problem is equivalent to searching for the set of redundant attributes, i.e. the attributes that do not occur in any reducts of the given decision table. We show that the considered problem is equivalent to a Sperner system for relational data base system and prove that it can be solved in polynomial time. On the base of these theoretical results, we also propose two different algorithms for elimination of redundant attributes in decision tables.

**Keywords:** Rough sets · Reducts · Relational database · Minimal keys · Sperner system

## 1 Introduction

Feature selection is one of the crucial problems in machine learning and data mining. The accuracy of many classification algorithms depends on the quality of selected attributes. Rough set approach to feature selection problem is based on reducts, which are in fact the minimal (with respect to inclusion) sets of attributes that preserve some necessary amount of information. Unfortunately, the number of all reducts for a given decision table can be exponential with respect to the number of attributes. Therefore we are forced to search either for minimal length reducts or for core attributes, i.e. the attributes that occur in all reducts. The minimal reduct problem is NP-hard whilst the searching for core attribute problem can be solved in polynomial time.

L.G. Nguyen—This work is partially supported by the Vietnam's National Foundation for Science and Technology Development (NAFOSTED) via a research grant for fundamental sciences, grant number 102.01-2010.09 and Polish National Science Centre grant DEC-2012/05/B/ST6/03215.

© Springer-Verlag Berlin Heidelberg 2015
J. Peters et al. (Eds.): TRS XIX, LNCS 8988, pp. 51–64, 2015.
DOI: 10.1007/978-3-662-47815-8_4

This paper investigates the problem of identifying the set of attributes, that are present in at least one reduct. Such attributes are called the *reductive attributes*. The non reductive attributes are called *redundant attributes* because they do not play any role in object classification. For a given decision table, the problem of searching for all reductive attributes becomes the problem of determining the union of all reducts of the given decision table, or determining the set of all redundant attributes [1] of a decision table.

In this paper we present two approaches to the investigated problem. Firstly, we present the fundamental analysis of the problem of searching for reductive attributes. Using Boolean reasoning approach we prove that the problem can be solved completely in polynomial time. Moreover, we can consider the decision table as the relation over the set of attributes and apply some results in relational database theory to solve the mentioned problems. We propose an algorithm to determine the set of all reductive attributes of consistent decision tables based on the methods of searching for keys, antikeys and prime attributes in decision table (see [2,3]).

This paper is the extended version of [4].

The structure of this paper is as follows. Sections 2 and 3 presents some basic concepts in rough set theory as well as the computational complexity of the reduct calculation problems. Section 4 presents the concept of reducts in decision table from the view point of relational database theory. We also propose an algorithm to determine the set of all reductive attributes of a consistent decision table. In Sect. 5, we perform some experiments of the proposed algorithm. The conclusions and future remarks are presented in the last section.

## 2    Basic Concepts

An *information system* is a pair $\mathbb{A} = (U, A)$, where the set $U$ denotes the *universe of objects* and $A$ is the set of *attributes*, i.e. the mappings of the form: $a : U \rightarrow V_a$. The set $V_a$ is called the *domain* or *the value set* of attribute $a$.

A decision system is an information system $\mathbb{D} = (U, A \cup \{dec\})$ where $dec$ is a distinguished attribute called the *decision attribute* or briefly *decision*. The remaining attributes are called *conditional attributes* or briefly *conditions*. For convenience, we assume that the domain of decision attribute consists of two or very few values. For any $k \in V_{dec}$ the set

$$CLASS_k = \{u \in U : dec(u) = k\}$$

is called the *decision class* of $\mathbb{D}$.

As an example, let us consider the decision system below (Table 1). Attributes *Diploma, Experience, French* and *Reference* are the *condition attributes*, whereas *Decision* is the decision attribute. We will refer to decision attribute *Decision* as $dec$, and to conditional attributes *Diploma, Experience, French* and *Reference* as to $a_1, \ldots, a_4$ in this order. In this example there are two decision classes related to the values *Accept* and *Reject* of the decision attribute domain. These decision classes are as follow:

$$CLASS_{Accept} = \{x_1, x_4, x_6, x_7\}$$

$$CLASS_{Reject} = \{x_2, x_3, x_5, x_8\}$$

**Table 1.** An example decision system represented as a table.

|  | $a_1$ | $a_2$ | $a_3$ | $a_4$ | $dec$ |
|---|---|---|---|---|---|
|  | Diploma | Experience | French | Reference | Decision |
| $x_1$ | MBA | Medium | Yes | Excellent | Accept |
| $x_2$ | MBA | Low | Yes | Neutral | Reject |
| $x_3$ | MCE | Low | Yes | Good | Reject |
| $x_4$ | MSc | High | Yes | Neutral | Accept |
| $x_5$ | MSc | Medium | Yes | Neutral | Reject |
| $x_6$ | MSc | High | Yes | Excellent | Accept |
| $x_7$ | MBA | High | No | Good | Accept |
| $x_8$ | MCE | Low | No | Excellent | Reject |

Rough set theory has been introduced by Professor Z. Pawlak [5] as a tool for concept approximation under uncertainty. The idea is to approximate an unknown concept by two descriptive sets called the *lower and upper approximations*. One of the assumptions in rough set theory that differs it from other methods in soft computing and concept approximation is that the lower and upper approximations must be extracted from the information that is available in training data.

One of the simplest ways to define the lower and upper approximations has been proposed by Prof. Z. Pawlak in [1]. This approach to concept approximation is based on the indiscernibility relation.

For a subset of attributes $B \subseteq A$ we define *B-indiscernibility relation* $IND(B)$ and *decision-relative indiscernibility relation* $IND_{dec}(B)$ (both defined on $U \times U$) as following:

$$(x, y) \in IND(B) \iff \forall_{a \in A} a(x) = a(y)$$
$$(x, y) \in IND_{dec}(B) \iff dec(x) = dec(y) \vee \forall_{a \in A} a(x) = a(y)$$

The relation $IND(B)$ is an equivalence relation and it defines a partitioning of $U$ into equivalence classes which we denote by $[x]_B$ $(x \in U)$. The complement of $IND(B)$ in $U \times U$ is called *discernibility relation*, denoted by $DISC(B)$. The lower and upper approximations of a concept $X$ (using attributes from $B$) are defined by:

$$\mathbf{L}_B(X) = \{x \in U : [x]_{IND(B)} \subseteq X\} \quad \text{and}$$
$$\mathbf{U}_B(X) = \{x \in U : [x]_{IND(B)} \cap X \neq \varnothing\}.$$

The main philosophy of rough set approach to concept approximation problem is based on minimizing the difference between upper and lower approximations (also called the *boundary region*). This simple, but brilliant idea, leads to many efficient applications of rough sets in machine learning and data mining like feature extraction and selection, rule induction, discretization or classifier construction [6].

It has been shown that all of the problems mentioned above are related to one of the crucial concepts in rough set theory, called *reducts* or *decision reducts* (see [1,7]). In general, reducts are minimal subsets (with respect to the set inclusion relation) of attributes which contain a necessary portion of information about the set of all attributes [6,8].

There are several ways to define reducts in Rough set theory. We will further focus on the following one.

A *decision-relative reduct* is a minimal set of attributes $R \subseteq A$ such that

$$IND_{dec}(R) = IND_{dec}(A).$$

This condition guarantees that $R$ contains all information necessary to discern objects belonging to different classes. The set of all decision reducts of a given decision table $\mathbb{D} = (U, A \cup \{dec\})$ is denoted by

$$\mathcal{RED}(\mathbb{D}) = \{R \subseteq A : R \text{ is a reduct of } \mathbb{D}\}$$

The attribute $a \in A$ called *core attribute* iff $a$ presents in all reducts of $A$. The set of all core attributes is denoted by

$$CORE(\mathbb{D}) = \bigcap_{R \in \mathcal{RED}(\mathbb{D})} R$$

The attribute $a \in A$ is called *reductive attribute* if and only if $a$ belongs to at least one reduct of $A$. The set of all reductive attributes is denoted by

$$REAT(\mathbb{D}) = \bigcup_{R \in \mathcal{RED}(\mathbb{D})} R$$

It is obvious that
$$CORE(\mathbb{D}) \subseteq R \subseteq REAT(\mathbb{D})$$

for any reduct $R \in \mathcal{RED}(\mathbb{D})$.

The attribute is called *redundant attribute* if it is not a reductive attribute. In other words, redundant attribute is not presented in any reduct of $A$.

For example, the set of all reducts of the decision table in Table 1 is

$$\mathcal{RED}(\mathbb{D}) = \{\{a_1, a_2\}, \{a_2, a_4\}\}.$$

Thus
$$CORE(\mathbb{D}) = \{a_2\}; \quad REAT(\mathbb{D}) = \{a_1, a_2, a_4\}$$

In this example, $a_3$ is the redundant attribute.

From historical point of view, the classification of attributes into three groups: core, reductive and redundant attributes was introduced by Professor Pawlak in [1]. This topic has been also studied by some other authors using different names, e.g., in the paper [9] the authors used the three types of attributes: *absolutely necessary attributes relatively necessary attributes*, and *absolutely unnecessary attributes*, while in the paper [10] the author used the names *useful attributes* and *useless attributes* to label the different types of attributes.

For any fixed natural $k$, there exists a polynomial in time algorithm which, for a given decision table T and given k conditional attributes, recognizes if there exists a decision reduct of T containing these $k$ attributes

# 3   Complexity Results

The concept of decision reducts using discernibility matrix has been explained in [8]. This simple and nice idea is also a tool for showing that the reduct calculation problem is equivalent to the problem of searching for prime implicants of boolean functions.

In fact, discernibility matrix for a given decision table $\mathbb{D} = (U, A \cup \{dec\})$, denoted by $\mathbb{M}_{\mathbb{D}}(A) = [c_{ij}]$, is a $n \times n$ table, where $n$ is the number of object, and the entry $c_{ij}$ is referring to the pair of objects $(x_i, x_j)$ that belong to different decision classes. The entry $c_{ij}$ is the set of all conditional attributes which discern the objects $x_i$ and $x_j$, i.e.

$$c_{ij} = \{a \in A : a(x_i) \neq a(x_j)\}$$

In Table 2 we present a compact form of *decision-relative discernibility matrix* corresponding to the decision system from Table 1, where the objects corresponding to class *Accept* are listed as columns and the objects corresponding to class *Reject* are listed as rows.

**Table 2.** The compact form of decision-relative discernibility matrix corresponding to decision system in Table 1.

|       | $x_1$          | $x_4$                | $x_6$           | $x_7$                |
|-------|----------------|----------------------|-----------------|----------------------|
| $x_2$ | $a_2, a_4$     | $a_1, a_2$           | $a_1, a_2, a_4$ | $a_2, a_3, a_4$      |
| $x_3$ | $a_1, a_2, a_4$ | $a_1, a_2, a_4$     | $a_1, a_2, a_4$ | $a_1, a_2, a_3$      |
| $x_5$ | $a_1, a_4$     | $a_2$                | $a_2, a_4$      | $a_1, a_2, a_3, a_4$ |
| $x_8$ | $a_1, a_2, a_3$ | $a_1, a_2, a_3, a_4$ | $a_1, a_2, a_3$ | $a_1, a_2, a_4$      |

The Boolean reasoning approach to reduct calculation problem is based on encoding it by the boolean *discernibility function* defined as follows:

$$\Delta_{\mathbb{D}}(a_1, \ldots, a_k) = \prod_{i,j:d(x_i) \neq d(x_j)} \sum_{a \in C_{ij}} a$$

where $a_1, \ldots, a_k$ are the boolean variables related to attributes from $A$, and $\prod, \sum$ denote the Boolean conjunction and Boolean disjunction operators. Thus, for the discernibility matrix in Table 2, the discernibility function can be written as follows:

$$
\begin{aligned}
\Delta_{\mathbb{D}}(a_1, \ldots, a_4) =& (a_2 + a_4)(a_1 + a_2)(a_1 + a_2 + a_4) \\
& (a_2 + a_3 + a_4)(a_1 + a_2 + a_4)(a_1 + a_2 + a_4) \\
& (a_1 + a_2 + a_4)(a_1 + a_2 + a_3)(a_1 + a_4)(a_2)(a_2 + a_4) \\
& (a_1 + a_2 + a_3 + a_4)(a_1 + a_2 + a_3)(a_1 + a_2 + a_3 + a_4) \\
& (a_1 + a_2 + a_3)(a_1 + a_2 + a_4)
\end{aligned}
\tag{1}
$$

It has been shown in [6,8] that the set of attributes $R = \{a_{i_1}, \ldots, a_{i_j}\}$ is a reduct in $\mathbb{D}$ if and only if the monomial

$$
m_R = a_{i_1} \cdot \ldots \cdot a_{i_j}
$$

is a prime implicant of $\Delta_{\mathbb{D}}(a_1, \ldots, a_k)$. As a consequence of this fact, both the problem of searching for minimal length reducts as well as the problem of searching for all reducts of a given decision table are NP-hard.

In terms of decision-relative discernibility matrix, a decision reduct $R$ is a minimal subset of attributes so that for each non-empty entry $C_{ij}$ of $M(\mathbb{A})$, $C_{ij} \cap R \neq \emptyset$.

Discernibility matrix and discernibility function are very important tools for calculation and analysis of reducts. Let us recall the following well known fact (see [6,8]).

**Theorem 1.** For any attribute $a \in A$, $a$ is a core attribute if and only if $a$ occurs in discernibility matrix as a singleton. As a consequence, the problem of searching for core attributes can be solved in polynomial time.

For the example from Table 1, according to the Theorem 1, attribute $a_2$ (Experience) is the core attribute, because this is the only attribute that discerns $x_4$ and $x_5$ (see also Table 2). And we can determine it without calculation of all reducts of this table.

The question is related to computational complexity of the problems of reductive attributes. We will use the discernibility matrix and discernibility function to prove that this problem can be solved in polynomial time. Therefore, once again, the Boolean reasoning approach shows to be a useful tool for reduct calculation problem.

The main idea is based on application of the *absorption law* in Boolean algebra, which states that

$$
x + (x \cdot y) = x \qquad x \cdot (x + y) = x
$$

where $x, y$ are the arbitrary Boolean functions. In other words, in Boolean algebra, the longer expressions are absorbed by the shorter ones. For the Boolean function in Eq. 1, $(a_1 + a_2)$ absorbs $(a_1 + a_2 + a_4)$ but, at the same time, it is absorbed by $(a_2)$.

The Boolean expression is called the irreducible CNF if it is in CNF (conjunctive normal form) and it is not possible to apply the absorption law on its clauses.

As an example, the irreducible CNF of the discernibility function in Eq. 1 is as follows:

$$\Delta_{\mathbb{D}}(a_1, \ldots, a_4) = a_2 \cdot (a_1 + a_4)$$

We have the following theorem

**Theorem 2.** For any decision table $\mathbb{D} = (U, A \cup \{dec\})$, if

$$\Delta_{\mathbb{D}}(a_1, \ldots, a_k) = \left( \sum_{a \in C_1} a \right) \cdot \left( \sum_{a \in C_2} a \right) \cdots \left( \sum_{a \in C_m} a \right) \tag{2}$$

is the irreducible CNF of discernibility function $\Delta_{\mathbb{D}}(a_1, \ldots, a_k)$, then

$$REAT(\mathbb{D}) = \bigcup_{i=1}^{m} C_i \tag{3}$$

*Proof (Proof of Theorem 2).* As (2) is the irreducible CNF of discernibility function, the family $\{C_1, \ldots, C_m\}$ should satisfy the following properties:

- It is an antichain, i.e. $C_i \nsubseteq C_j$ and $C_j \nsubseteq C_i$ for any $i, j \in \{1, \ldots, m\}$
- If $R$ is a reduct, i.e. $R \in \mathcal{RED}(\mathbb{D})$, then $R \cap C_i \neq \emptyset$ for any $i \in \{1, \ldots, m\}$.

We will prove that the inclusions in both directions of the Eq. (3) hold:

1. The proof of $REAT(\mathbb{D}) \subseteq \bigcup_{i=1}^{m} C_i$:

   Let $a \in REAT(\mathbb{D})$. From the definition, there exists a reduct $R \in \mathcal{RED}(\mathbb{D})$ such that $a \in R$. This means that $R \cap C_i \neq \emptyset$ for $i = 1, \ldots, m$. If $a \notin \bigcup_{i=1}^{m} C_i$ then for any $i \in \{1, \ldots, m\}$ we have $a \notin C_i$, which implies that

   $$(R - \{a\}) \cap C_i = R \cap C_i \neq \emptyset.$$

   Thus there exists a subset of $R - \{a\}$ which is also a reduct of $\mathbb{D}$, and this is a contradiction.
   Hence we have $a \in \bigcup_{i=1}^{m} C_i$.

2. The proof of $\bigcup_{i=1}^{m} C_i \subseteq REAT(\mathbb{D})$:

   We can use the fact that the irreducible CNF of monotone Boolean function is unique to prove this inverse inclusion.
   Indeed, if $a \in \bigcup_{i=1}^{m} C_i$ and $a$ is a redundant attribute, then $R \cap C_i - \{a\} \neq \emptyset$ for each reduct $R \in \mathcal{RED}(\mathbb{D})$. Thus

   $$\Delta_{\mathbb{D}}^{(1)}(a_1, \ldots, a_k) = \prod_{i=1}^{m} \left( \sum_{a_j \in C_i} a_j \right)$$

and

$$\Delta_{\mathbb{D}}^{(2)}(a_1,\ldots,a_k) = \prod_{i=1}^{m}\left(\sum_{a_j \in C_i - \{a\}} a_j\right)$$

are the two different irreducible CNF form of the discernibility function $\Delta_{\mathbb{D}}(a_1,\ldots,a_k)$, what is the contradiction.

The following algorithm is the straightforward application of the presented above theorem:

---

**Algorithm 1.** Determining the set of all reductive attributes of a given decision table.

---

**Data**: a consistent decision table $\mathbb{D} = (U, A \cup \{dec\})$;
**Result**: $REAT(A)$ – the set of all reductive attributes of $\mathbb{D}$;
1 Calculate the discernibility matrix $\mathbb{M}_{\mathbb{D}}(A)$;
2 Reduce $\mathbb{M}_{\mathbb{D}}(A)$ using absorption law;
3 Let $\{C_1,\ldots,C_m\}$ be the set of nonempty entries of $\mathbb{M}_{\mathbb{D}}(A)$ after reduction;
4 Return $REAT(A) = \bigcup_{i=1}^{m} C_i$ as the set of all reductive attributes of $\mathbb{D}$.

---

If $|A| = k$ and $|U| = n$ then the time complexity of construction of discernibility matrix (step 1) is $O(n^2 k)$. Since there are $O(n^2)$ subsets of $A$ in the discernibility matrix, the reducing phase using absorbtion law requires $O(n^4)$ set comparison operations, thus the time complexity of Step 2 in Algorithm 1 is $O(n^4 k)$. Therefore the time complexity of reductive attributes calculation problem is at most $O(n^4 k)$.

The polynomial time complexity of reductive attribute calculation problem can be also derived from another fact presented in [11]. In this paper, it has been shown that for any fixed natural $k$, there exists a polynomial time algorithm which, for a given decision table $\mathbb{D}$ and given $k$ conditional attributes, recognizes if there exists a decision reduct of $\mathbb{D}$ containing these $k$ attributes. If we chose $k = 1$ and for each attribute $a \in A$, we apply the mentioned algorithm to check the existence of reduct that contains $a$, we can also construct the set of all reductive attributes of $\mathbb{D}$.

## 4    Decision Tables in Terms of Relational Databases

Let us give some necessary definitions and results of the theory of relation database that can be found in [2,3,12].

### 4.1    Relational Database Theory

Let $A = \{a_1,\ldots,a_k\}$ be a finite set of attributes and let $D(a_i) \subseteq V_{a_i}$ be the set of all possible values of attribute $a_i$, for $i = 1,\ldots,k$. Any subset of the Cartesian product

$$\mathcal{R} \subseteq D(a_1) \times D(a_2) \times \ldots \times D(a_k)$$

is called *the relation over A*. In other words, relation over $A$ is the set of tuples $\{h_1, \ldots, h_n\}$ where

$$h_j : A \to \bigcup_{a_i \in A} D(a_i),$$

is a function that $h_j(a_i) \in D(a_i))$ for $1 \le j \le n$.

Let $\mathcal{R} = \{h_1, \ldots, h_n\}$ be a given relation over the set of attributes $A = \{a_1, \ldots, a_k\}$. Any pair of attribute sets $B, C \subseteq A$ is called the functional dependency (FD for short) over $A$, and denoted by $B \to C$, if and only if for any pair of tuples $h_i, h_j \in \mathcal{R}$:

$$\forall_{a \in B}(h_i(a) = h_j(a)) \implies \forall_{a \in C}(h_i(a) = h_j(a))$$

The set $\mathcal{F}_{\mathcal{R}} = \{(B, C) : B, C \subset A; B \to C\}$ is called the *full family of functional dependencies in* $\mathcal{R}$.

Let $\mathbb{P}(A)$ be the power set of attribute set $A$. A family $\mathcal{R} \subset \mathbb{P}(A) \times \mathbb{P}(A)$ is called *an f-family over A* if and only if for all subsets of attributes $P, Q, S, T \subseteq A$ the following properties hold:

R1.  $(P, P) \in \mathcal{R}$                                                                  (4)

R2.  $(P, Q) \in \mathcal{R}, (Q, S) \in \mathcal{R} \implies (P, S) \in \mathcal{R}$                (5)

R3.  $(P, Q) \in \mathcal{R}, P \subseteq S, T \subseteq Q \implies (S, T) \in \mathcal{R}$             (6)

R4.  $(P, Q) \in \mathcal{R}, (R, S) \in \mathcal{R} \implies (P \cup R, Q \cup T) \in \mathcal{R}$      (7)

Clearly $\mathcal{F}_{\mathcal{R}}$ is an $f$-family over $A$. It is also known that if $\mathcal{F}$ is an $f$-family over $A$ then there is a relation $\mathcal{R}$ such that $\mathcal{F}_{\mathcal{R}} = \mathcal{F}$.

A pair $\mathbb{S} = (A, \mathcal{F})$, where $A$ is a set of attributes and $\mathcal{F}$ is a set of functional dependency on $A$, is called the *relation scheme*. Let us denote by $\mathcal{F}^+$ the set of all functional dependecies, which can be derived from $\mathcal{F}$ by using the rules $R1 - R4$.

For any subset of attributes $B \subseteq A$, the set

$$B^+ = \{a \in A : B \to a \in \mathcal{F}^+\}$$

is called *the closure of B* on $\mathbb{S}$. It is clear that $B \to C \in \mathcal{F}^+$ if and only if $C \subseteq B^+$.

A set of attributes $B \subset A$ is called the *key* of $\mathbb{S} = (A, \mathcal{F})$ iff $B \to A \in \mathcal{F}^+$. The set $B$ is the *minimal key* of $\mathbb{S} = (A, \mathcal{F})$ if $B$ is a key of $\mathbb{S}$ and any proper subset of $B$ is not a key of $\mathbb{S}$. Let us denote by $\mathcal{K}(\mathbb{S})$ the set of all minimal keys of the given relation scheme $\mathbb{S}$.

Recall that a family $\mathcal{K} \subseteq \mathbb{P}(A)$ is a Sperner system if for any $K_1, K_2 \in \mathcal{K}$ implies $K_1 \not\subseteq K_2$. Clearly $\mathcal{K}(\mathbb{S})$ is a Sperner system.

Let $\mathcal{K} = \mathcal{K}_{\mathbb{S}}$ be a Sperner system over $A$ containing all minimal keys of $\mathbb{S}$. We defined the set of antikeys of $\mathcal{K}$, denoted by $\mathcal{K}^{-1}$, as follows:

$$\mathcal{K}^{-1} = \{B \subseteq A : \forall_{C \subset A}(C \in \mathcal{K} \implies C \not\subseteq B) \wedge$$
$$\forall_{D \subset A}(B \subseteq D \implies \exists_{C \in \mathcal{K}} C \subseteq D)\}$$

It is easy to see that $\mathcal{K}^{-1}$ is the set of subsets of $A$, which does not contain the elements of $\mathcal{K}$ and which is maximal for this property. They are the maximal non-keys. Clearly, $\mathcal{K}^{-1}$ is also a Sperner system.

Let us now define the concept of equality system for a given relation. For $\mathcal{R} = \{h_1, \ldots, h_n\}$ over $A = \{a_1, \ldots, a_k\}$, we define

$$\mathbb{E}(\mathcal{R}) = \{E_{ij} : 1 \leq i < j \leq n\},$$
$$\text{where } E_{ij} = \{a \in A : h_i(a) = h_j(a)\} \tag{8}$$

The family $\mathbb{E}(\mathcal{R})$ is called *the equality system of* the relation $\mathcal{R}$. It is easy to notice that in the worse case, $\mathbb{E}(\mathcal{R})$ contains $O(n^2)$ subsets of attributes, where $n$ is the number of records in the relation $\mathcal{R}$.

It is known (see [2]) that for each subset of attributes $B \subseteq A$, the following property holds:

$$B^+ = \begin{cases} E_{ij} & \text{if } B \subseteq E_{ij} \text{ for some } E_{ij} \in \mathbb{E}(\mathcal{R}) \\ A & \text{otherwise} \end{cases}$$

Let $\mathbb{S} = (A, \mathcal{F})$ be a relation scheme over attribute set $A$. For any attribute $a \in A$, the set

$$\mathcal{K}_{\mathbb{S}}(a) = \{B \subseteq A : B \rightarrow a \wedge \nexists_{C \subset B} C \rightarrow a\} \tag{9}$$

is called the *family of minimal sets of the attribute a over* $\mathbb{S}$. It is known that $\{a\} \in \mathcal{K}_{\mathbb{S}}(a)$, $A \notin \mathcal{K}_{\mathbb{S}}(a)$ and $\mathcal{K}_{\mathbb{S}}(a)$ is a Sperner system over $A$.

## 4.2    Relational Database Theory and Reducts

In rough set theory the minimal sets from $\mathcal{K}_{\mathbb{S}}(a)$ are strongly related to the concept of decision reducts. Any decision table $\mathbb{D} = (U, A \cup \{dec\})$ can be treated as a relation $U = \{u_1, \ldots, u_n\}$ over the set of all attributes $A \cup \{dec\} = \{a_1, \ldots, a_k, dec\}$.

Moreover, the following facts state that in some cases the concept of decision reduct in rough set theory and the concept of functional dependency are equivalent.

**Theorem 3.** *For any decision table* $\mathbb{D} = (U, A \cup \{dec\})$, *the following equation holds*

$$\mathcal{K}_{\mathbb{S}}(dec) = \mathcal{RED}(\mathbb{D}) \cup \{dec\}$$

*where* $\mathbb{S}$ *is the relation scheme induced from the decision table* $\mathbb{D}$ *as it was defined in Eq. (9).*

In relational database theory, the following important facts has been proven (see e.g. in [3]).

**Lemma 1.** The following equality holds for any Sperner system $\mathcal{K}$ over the set of attribute $A$:

$$\bigcup_{K \in \mathcal{K}} K = A - \bigcap_{K \in \mathcal{K}^{-1}} K$$

As the consequence we have the following theorem

**Theorem 4.** Let $\mathbb{D} = (U, A \cup \{dec\})$ be a consistent decision table, the set of reductive attributes can be determined by:

$$REAT(A) = \bigcup_{K \in \mathcal{K}_\mathbb{S}(dec)} (K - \{dec\})$$

$$= A - \bigcap_{K \in (\mathcal{K}_\mathbb{S}(dec))^{-1}} K \qquad (10)$$

Therefore, the main problem is to calculate the family $(\mathcal{K}_\mathbb{S}(dec))^{-1}$. According to the theory of relational database in previous Section we have

$$(\mathcal{K}_\mathbb{S}(dec))^{-1} = \{B \subseteq A : (B \to dec \notin \mathcal{F}^+) \wedge$$
$$(B \subsetneq C \Rightarrow C \to dec \in \mathcal{F}^+)\}$$

It is clear that for any set of attributes $B \subseteq A$ we have $B \in (\mathcal{K}_\mathbb{S}(dec))^{-1}$ if and only if

$$B \in \mathbb{E}(\mathcal{R}) \wedge \nexists_{C \in \mathbb{E}(\mathcal{R})}(dec \in C \text{ and } (B \subsetneq C - \{dec\})$$

The method of determining the set of reductive attributes using the equality set $\mathbb{E}(\mathcal{R})$ is presented in Algorithm 2. Similar to Algorithm 1, the size of $E_{ij}$ is $O(n^2 k)$, where $k$ and $n$ are the number of attributes and number of objects. Thus, in the worse case, the construction of $\mathcal{M}(dec)$ requires at most $O(n^4 k)$ steps. Therefore the problem of calculation of all reductive attributes can be solved by a straightforward algorithm, which runs in $O(n^4 k)$ time and uses $O(n^2 k)$ space.

---

**Algorithm 2.** Determining the set of all reductive attributes of a given decision table.

---

**Data**: a consistent decision table $\mathbb{D} = (U, A \cup \{dec\})$;
**Result**: $REAT(A)$ – the set of all reductive attributes of $\mathbb{D}$;
1 Calculate the equality system

$$\mathbb{E}(\mathcal{R}) = \{E_{ij} : 1 \leq i < j \leq n\}$$

where $E_{ij}$ is the set of attributes that have the same values for $u_i$ and $u_j$;
2 Let

$$\mathbb{E}_d = \{B \in \mathbb{E}(\mathcal{R}) : dec \in B\}$$
$$\mathbb{E}_0 = \{B \in \mathbb{E}(\mathcal{R}) : dec \notin B\}$$

3 Construct the family of subsets of $A$:

$$\mathcal{M}(dec) = \{B \in \mathbb{E}_0 : \forall_{C \in \mathbb{E}_0}(B \cap C \neq B)\}$$

4 Construct the set $V = \bigcap_{K \in \mathcal{M}(dec)} K$
5 Return $REAT(A) = A - V$ as the set of all reductive attributes of $\mathbb{D}$.

---

### 4.3   Example

Let us consider the exemplary decision table in Table 1. The equality set $\mathbb{E}(\mathcal{R})$ of this table is presented in Table 3.

In fact, $\mathbb{E}(\mathcal{R})$ consists of different subsets of the attribute set

$$A \cup \{dec\} = \{a_1, a_2, a_3, a_4, dec\}.$$

However, for the clear representation, we divided $\mathbb{E}(\mathcal{R})$ into two parts, i.e., $\mathbb{E}_0$ – the family of subsets that do not contain the decision $dec$, and $\mathbb{E}_d$ – the family of subsets that contain the decision $dec$. The family $\mathbb{E}_0$ is placed in the left column, while the family $\mathbb{E}_d$ is placed in the right column of Table 3.

**Table 3.** The equality set of the decision table from Table 1

| $\mathbb{E}(\mathcal{R})$: | $\mathbb{E}_0$ | $\mathbb{E}_d$ |
|---|---|---|
| | without $dec$ | with $dec$ |
| | $\{a_1, a_3\}$ | $\{a_3, dec\}$ |
| | $\{a_3, a_4\}$ | $\{a_3, a_4, dec\}$ |
| | $\{a_3\}$ | $\{a_1, dec\}$ |
| | $\{a_1\}$ | $\{a_1, a_2, a_3, dec\}$ |
| | $\{a_4\}$ | $\{a_2, dec\}$ |
| | $\{a_2, a_3\}$ | $\{a_2, a_3, dec\}$ |
| | $\{a_1, a_3, a_4\}$ | $\{a_1, a_2, dec\}$ |
| | | $\{dec\}$ |

One can see that the left column can be calculated by taking the complements of all entries of the discernibility matrix in Table 2.

The next step is to calculate $\mathcal{M}(dec)$, which is the family maximal subsets among the subsets of $\mathbb{E}(\mathcal{R})$ that do not contain $dec$. In this example

$$\mathcal{M}(dec) = \{\{a_2, a_3\}, \{a_1, a_3, a_4\}\}$$

Thus $V = \{a_3\}$ and $REAT(A) = \{a_1, a_2, a_4\}$.

## 5   Experiments

The experiments are performed on 4 data sets obtained from UCI Machine Learning Repository[1]. Due to the high time and space complexity, we demonstrate the presented algorithms on the data sets containing a small number of objects and relatively large number of attributes. The selected data sets are: a part of *Adult*, a small data set of *Soybean*, *Sponge.data* and *Zoo.data*.

---

[1] The UCI machine learning repository, http://archive.ics.uci.edu/ml/.

We present the results of calculation the set of all reductive attributes and the set of all redundant attributes in Table 4. In this Table $|U|$, $|A|$ are the numbers of objects and condition attributes, and $t$ is the time of operation (calculated by second) calculated on PC (Pentium Dual Core 2.13 GHz, 1 GB RAM, WINXP). The conditional attributes are denoted by $1, 2, \ldots |A|$.

**Table 4.** The results of experiment on some benchmark data sets using the proposed algorithm

| Data sets | $|U|$ | $|A|$ | $t$ | The reductive attributes | The redundant attributes |
|---|---|---|---|---|---|
| Adult stretch | 20 | 4 | 0.93 | 3, 4 | 1, 2 |
| Soybean small data | 47 | 35 | 2.74 | 1, 2, 3, 4, 5, 6, 7, 8, 9, 10, 12, 20, 21, 22, 23, 24, 25, 26, 27, 28, 35 | 11, 13, 14, 15, 16, 17, 18, 19, 29, 30, 31, 32, 33, 34 |
| Sponge.data | 76 | 45 | 2.1 | 1, ..., 11, 13, ..., 34, 36, ..., 45 | 12, 35 |
| Zoo.data | 101 | 17 | 3.19 | 1, 2, 4, 5, 7, 8, 9, 10, 11, 12, 13, 14, 15, 17 | 3, 6, 16 |

# 6  Conclusions

In this paper, we presented two alternative approaches to the problem of determining the set of all reductive attributes for a decision table. The first approach is based on discernibility matrix and Boolean reasoning methodology.

The second approach is based on Sperner system using the equality set. We defined the family of all minimal sets of an attribute over a relation based on the definition of the family of minimal sets of an attribute over a relation scheme [3], so the concept of reduct in decision tables is equivalent to that of minimal set of an attribute in a relation. As a result, an algorithm for determining all reduced attributes of a consistent decision table was proposed based on some results proposed in [2]. We also proved that the time complexity of proposed algorithm is polynomial in the number of rows and columns of the decision table. This results play an important role in rejecting redundant attributes in decision tables before attribute reduction and rule extraction.

The positive result is related to the fact that the set of reductive attributes can be calculated in polynomial time. However both proposed methods seem to have quite a high complexity. In the worst case, the proposed solutions may need $O(n^4 k)$ steps, where $n$ in the number of objects and $k$ is the number of attributes in the decision table.

We are planing to work on the more efficient methods to reduce the time complexity of the proposed solutions. The idea may be based on the attempt to realize the same algorithms without implementation of discernibility matrix.

# References

1. Pawlak, Z.: Rough Sets-Theoretical Aspect of Reasoning About Data. Kluwer Academic Publishers, Norwell (1991)
2. Demetrovics, J., Thi, V.D.: Keys, antikeys and prime attributes. Ann. Univ. Sci. Bp. Sect. Comp. **8**, 35–52 (1987)
3. Demetrovics, J., Thi, V.D.: Describing candidate keys by hypergraphs. Comput. Artif. Intell. **18**(2), 191–207 (1999)
4. Nguyen, L.G., Nguyen, H.S.: On elimination of redundant attributes in decision tables. In: 2012 Federated Conference on Computer Science and Information Systems (FedCSIS). IEEE, pp. 317–322 (2012)
5. Pawlak, Z.: Rough sets. Int. J. Comput. Inf. Sci. **11**(5), 341–356 (1982)
6. Nguyen, H.S.: Approximate boolean reasoning: foundations and applications in data mining. In: Peters, J.F., Skowron, A. (eds.) Transactions on Rough Sets V. LNCS, vol. 4100, pp. 334–506. Springer, Heidelberg (2006)
7. Pawlak, Z.: Rough sets and intelligent data analysis. Inf. Sci. **147**(1), 1–12 (2002)
8. Skowron, A., Rauszer, C.: The discernibility matrices and functions in information systems. In: Słoński, R. (ed.) Intelligent Decision Support, vol. 11, pp. 331–362. Springer, Heidelberg (1992)
9. Wei, L., Li, H.R., Zhang, W.X.: Knowledge reduction based on the equivalence relations defined on attribute set and its power set. Inf. Sci. **177**(15), 3178–3185 (2007)
10. Yao, Y.: Duality in rough set theory based on the square of opposition. Fundam. Inf. **127**(1–4), 49–64 (2013)
11. Moshkov, M.J., Skowron, A., Suraj, Z.: On covering attribute sets by reducts. In: Kryszkiewicz, M., Peters, J.F., Rybiński, H., Skowron, A. (eds.) RSEISP 2007. LNCS (LNAI), vol. 4585, pp. 175–180. Springer, Heidelberg (2007)
12. Thi, V.D., Son, N.H.: On armstrong relations for strong dependencies. Acta Cybernetica **17**(3), 521–531 (2006)

# Sequential Optimization of $\gamma$-Decision Rules Relative to Length, Coverage and Number of Misclassifications

Beata Zielosko$^{(\boxtimes)}$

Institute of Computer Science, University of Silesia,
39, Będzińska St., 41-200 Sosnowiec, Poland
beata.zielosko@us.edu.pl

**Abstract.** The paper is devoted to the study of an extension of dynamic programming approach which allows sequential optimization of approximate decision rules relative to length, coverage and number of misclassifications. Presented algorithm constructs a directed acyclic graph $\Delta_\gamma(T)$ which nodes are subtables of the decision table $T$. Based on the graph $\Delta_\gamma(T)$ we can describe all irredundant $\gamma$-decision rules with the minimum length, after that among these rules describe all rules with the maximum coverage, and among such rules describe all rules with the minimum number of misclassifications. We can also change the set of cost functions and order of optimization. Sequential optimization can be considered as a tool that helps to construct simpler rules for understanding and interpreting by experts.

## 1 Introduction

Decision rules are one of popular ways for data representation used in machine learning and knowledge discovery. Exact decision rules can be overfitted, i.e., dependent essentially on the noise or adjusted too much to the existing examples. If decision rules are considered as a way of knowledge representation then instead of exact decision rules with many attributes, it is more appropriate to work with approximate decision rules which contain smaller number of attributes and have relatively good accuracy. Moreover, classifiers based on approximate decision rules have often better accuracy than the classifiers based on exact decision rules. Therefore, approximate decision rules and also closely connected with them approximate reducts are studied intensively last years [8,9,12,19,20,22, 23,25,28,30].

There are many approaches to the construction of decision rules: brute-force approach which is applicable to tables with relatively small number of attributes, Boolean reasoning [21,24,29], separate-and-conquer approach (algorithms based on a sequential covering procedure) [6,10–12,14–16], algorithms based on decision tree construction [13,17,20,26], different kinds of greedy algorithms [19,21]. Each method has different modifications, e.g., in the case of decision trees, we can use greedy algorithms based on different uncertainty measures (Gini index, entropy, etc.) for construction of decision rules.

© Springer-Verlag Berlin Heidelberg 2015
J. Peters et al. (Eds.): TRS XIX, LNCS 8988, pp. 65–82, 2015.
DOI: 10.1007/978-3-662-47815-8_5

The paper, extending a conference publication [31], presents, based on dynamic programming algorithm, one more approach that allows sequential optimization of approximate decision rules. We introduce an uncertainty measure that is the difference between number of rows in a given decision table and number of rows labeled with the most common decision for this table. We fix a nonnegative threshold $\gamma$, and study so-called $\gamma$-decision rules that localize rows in subtables which uncertainty is at most $\gamma$. For each of such rules the number of misclassifications is at most $\gamma$.

We consider three cost functions: length, coverage and number of misclassifications. The choice of length is connected with the Minimum Description Length principle [27]. The rule coverage is important to discover major patterns in the data. Number of misclassifications is important from the viewpoint of accuracy of classification. Considered approach allows sequential optimization of $\gamma$-decision rules relative to the mentioned cost functions.

Sequential optimization can be considered as a tool that helps to construct rules which are simpler for understanding and interpreting by experts, e.g., among rules with the maximum coverage we can find rules with the minimum length. Such rules can be considered as part of knowledge and experts can easier analyze them. Moreover, sequential optimization allows to find optimal rules relative to the considered cost functions, e.g., rules with the minimum length and the maximum coverage. In this case, results of sequential optimization do not depend on the order of optimization. Besides, sequential optimization of system of decision rules also can help to discover some regularities or anomalies in the data.

First results for decision rules based on dynamic programming approach were obtained in [32]. The aim of this study was to find one decision rule with the minimum length for each row. In [4] we studied dynamic programming approach for exact decision rule optimization. In [5] we studied dynamic programming approach for approximate decision rule optimization and we used another uncertainty measure that is the number of unordered pairs of rows with different decisions in decision table $T$. In [2] we presented procedures of optimization of $\gamma$-decision rules relative to the length and coverage, and in [3] – relative to the number of misclassifications.

In this paper, we concentrate on sequential optimization of $\gamma$-decision rules relative to the length, coverage and number of misclassifications, and totally optimal rules relative to these cost functions.

We present results of experiments with some decision tables from UCI Machine Learning Repository [7] based on Dagger software system [1] created in King Abdullah University of Science and Technology (KAUST).

This paper consists of seven sections. Section 2 contains definitions of main notions. In Sect. 3, we study a directed acyclic graph which allows to describe the whole set of irredundant $\gamma$-decision rules. In Sect. 4, we describe procedures of optimization of irredundant $\gamma$-decision rules relative to the length, coverage and number of misclassifications. In Sect. 5, we discuss possibilities of sequential optimization of rules relative to a number of cost functions. Section 6 contains results of experiments with decision tables from UCI Machine Learning Repository. Section 7 contains conclusions.

## 2    Main Notions

In this section, we consider definitions of notions corresponding to decision table and decision rules.

A *decision table* $T$ is a rectangular table with $n$ columns labeled with conditional attributes $f_1, \ldots, f_n$. Rows of this table are filled by nonnegative integers which are interpreted as values of conditional attributes. Rows of $T$ are pairwise different and each row is labeled with a nonnegative integer which is interpreted as a value of the decision attribute. It is possible that $T$ is empty, i.e., has no rows.

A minimum decision value which is attached to the maximum number of rows in $T$ will be called the *most common decision for $T$*. The most common decision for empty table is equal to 0.

We denote by $N(T)$ the number of rows in the table $T$ and by $N_{mcd}(T)$ we denote the number of rows in the table $T$ labeled with the most common decision for $T$. We will interpret the value $J(T) = N(T) - N_{mcd}(T)$ as *uncertainty* of the table $T$.

The table $T$ is called *degenerate* if $T$ is empty or all rows of $T$ are labeled with the same decision. It is clear that $J(T) = 0$ if and only if $T$ is a degenerate table.

A table obtained from $T$ by the removal of some rows is called a *subtable* of the table $T$. Let $T$ be a nonempty, $f_{i_1}, \ldots, f_{i_k} \in \{f_1, \ldots, f_n\}$ and $a_1, \ldots, a_k$ be nonnegative integers. We denote by $T(f_{i_1}, a_1) \ldots (f_{i_k}, a_k)$ the subtable of the table $T$ which contains only rows that have numbers $a_1, \ldots, a_k$ at the intersection with columns $f_{i_1}, \ldots, f_{i_k}$. Such nonempty subtables (including the table $T$) are called *separable subtables* of $T$.

We denote by $E(T)$ the set of attributes from $\{f_1, \ldots, f_n\}$ which are not constant on $T$. For any $f_i \in E(T)$, we denote by $E(T, f_i)$ the set of values of the attribute $f_i$ in $T$.

The expression

$$f_{i_1} = a_1 \wedge \ldots \wedge f_{i_k} = a_k \to d \qquad (1)$$

is called a *decision rule over $T$* if $f_{i_1}, \ldots, f_{i_k} \in \{f_1, \ldots, f_n\}$, and $a_1, \ldots a_k, d$ are nonnegative integers. It is possible that $k = 0$. In this case (1) is equal to the rule

$$\to d. \qquad (2)$$

Let $r = (b_1, \ldots, b_n)$ be a row of $T$. We will say that the rule (1) is *realizable for $r$*, if $a_1 = b_{i_1}, \ldots, a_k = b_{i_k}$. If $k = 0$ then the rule (2) is realizable for any row from $T$.

Let $\gamma$ be a nonnegative integer. We will say that the rule (1) is a $\gamma$-*true for $T$* if $d$ is the most common decision for $T' = T(f_{i_1}, a_1) \ldots (f_{i_k}, a_k)$ and $J(T') \leq \gamma$.

If $k = 0$ then the rule (2) is a $\gamma$-true for $T$ if $d$ is the most common decision for $T$ and $J(T) \leq \gamma$.

If the rule (1) is a $\gamma$-true for $T$ and realizable for $r$, we will say that (1) is a $\gamma$-*decision rule for $T$ and $r$*. Note that if $\gamma = 0$ we have an exact decision rule for $T$ and $r$.

We will say that the rule (1) with $k > 0$ is an *irredundant* $\gamma$-decision rule for $T$ and $r$ if (1) is a $\gamma$-decision rule for $T$ and $r$ and the following conditions hold:

(i) $f_{i_1} \in E(T)$, and if $k > 1$ then $f_{i_j} \in E(T(f_{i_1}, a_1) \ldots (f_{i_{j-1}}, a_{j-1}))$ for $j = 2, \ldots, k$;

(ii) $J(T) > \gamma$, and if $k > 1$ then $J(T(f_{i_1}, a_1) \ldots (f_{i_j}, a_j)) > \gamma$ for $j = 1, \ldots, k - 1$.

If $k = 0$ then the rule (2) is an *irredundant* $\gamma$-decision rule for $T$ and $r$ if (2) is a $\gamma$-decision rule for $T$ and $r$, i.e., if $d$ is the most common decision for $T$ and $J(T) \leq \gamma$.

Let $\tau$ be a decision rule over $T$ and $\tau$ be equal to (1).

The number $k$ of descriptors (pairs "attribute=value") on the left-hand side of $\tau$ is called the *length* of this rule and is denoted by $l(\tau)$. The length of decision rule (2) is equal to 0.

The *coverage* of $\tau$ is the number of rows in $T$ for which $\tau$ is realizable and which are labeled with the decision $d$. We denote it by $c(\tau)$. The coverage of decision rule (2) is equal to the number of rows in $T$ which are labeled with the decision $d$.

The *number of misclassifications* of $\tau$ is the number of rows in $T$ for which $\tau$ is realizable and which are labeled with decisions different from $d$. We denote it by $\mu(\tau)$. The number of misclassifications of the decision rule (2) is equal to the number of rows in $T$ which are labeled with decisions different from $d$.

**Proposition 1.** *[2] Let $T$ be a nonempty decision table, $r$ be a row of $T$ and $\tau$ be a $\gamma$-decision rule for $T$ and $r$ which is not irredundant. Then by removal of some descriptors from the left-hand side of $\tau$ and by changing the decision on the right-hand side of $\tau$ we can obtain an irredundant $\gamma$-decision rule $irr(\tau)$ for $T$ and $r$ such that $l(irr(\tau)) \leq l(\tau)$ and $c(irr(\tau)) \geq c(\tau)$.*

Unfortunately, it is impossible to prove similar result for the number of misclassifications.

## 3   Directed Acyclic Graph $\Delta_\gamma(T)$

In this section, we present an algorithm that constructs a directed acyclic graph $\Delta_\gamma(T)$. Based on this graph we can describe the set of irredundant $\gamma$-decision rules for $T$ and for each row $r$ of $T$. Nodes of the graph are separable subtables of the table $T$. During each step, the algorithm processes one node and marks it with the symbol *. At the first step, the algorithm constructs a graph containing a single node $T$ which is not marked with the symbol *.

Let the algorithm have already performed $p$ steps. Let us describe the step $(p + 1)$. If all nodes are marked with the symbol * as processed, the algorithm finishes its work and presents the resulting graph as $\Delta_\gamma(T)$. Otherwise, choose a node (table) $\Theta$, which has not been processed yet. Let $d$ be the most common decision for $\Theta$. If $J(\Theta) \leq \gamma$ label the considered node with the decision $d$, mark it with symbol * and proceed to the step $(p+2)$. If $J(\Theta) > \gamma$, for each $f_i \in E(\Theta)$, draw a bundle of edges from the node $\Theta$. Let $E(\Theta, f_i) = \{b_1, \ldots, b_t\}$. Then draw $t$ edges from $\Theta$ and label these edges with pairs $(f_i, b_1), \ldots, (f_i, b_t)$ respectively. These edges enter to nodes $\Theta(f_i, b_1), \ldots, \Theta(f_i, b_t)$. If some of nodes $\Theta(f_i, b_1), \ldots, \Theta(f_i, b_t)$ are absent in the graph then add these nodes to the graph. We label each row $r$ of $\Theta$ with the set of attributes $E_{\Delta_\gamma(T)}(\Theta, r) = E(\Theta)$. Mark the node $\Theta$ with the symbol * and proceed to the step $(p+2)$. The graph $\Delta_\gamma(T)$ is a directed acyclic graph. A node of such graph will be called *terminal* if there are no edges leaving this node. Note that a node $\Theta$ of $\Delta_\gamma(T)$ is terminal if and only if $J(\Theta) \leq \gamma$.

Later, we will describe the procedures of optimization of the graph $\Delta_\gamma(T)$. As a result we will obtain a graph $G$ with the same sets of nodes and edges as in $\Delta_\gamma(T)$. The only difference is that any row $r$ of each nonterminal node $\Theta$ of $G$ is labeled with a nonempty set of attributes $E_G(\Theta, r) \subseteq E(\Theta)$. It is possible also that $G = \Delta_\gamma(T)$.

Now, for each node $\Theta$ of $G$ and for each row $r$ of $\Theta$, we describe the set of $\gamma$-decision rules $Rul_G(\Theta, r)$. We will move from terminal nodes of $G$ to the node $T$.

Let $\Theta$ be a terminal node of $G$ labeled with the most common decision $d$ for $\Theta$. Then

$$Rul_G(\Theta, r) = \{\rightarrow d\}.$$

Let now $\Theta$ be a nonterminal node of $G$ such that for each child $\Theta'$ of $\Theta$ and for each row $r'$ of $\Theta'$, the set of rules $Rul_G(\Theta', r')$ is already defined. Let $r = (b_1, \ldots, b_n)$ be a row of $\Theta$. For any $f_i \in E_G(\Theta, r)$, we define the set of rules $Rul_G(\Theta, r, f_i)$ as follows: $Rul_G(\Theta, r, f_i) = \{f_i = b_i \wedge \sigma \rightarrow s : \sigma \rightarrow s \in Rul_G(\Theta(f_i, b_i), r)\}$. Then

$$Rul_G(\Theta, r) = \bigcup_{f_i \in E_G(\Theta, r)} Rul_G(\Theta, r, f_i)$$

**Theorem 1.** [2] *For any node $\Theta$ of $\Delta_\gamma(T)$ and for any row $r$ of $\Theta$, the set $Rul_{\Delta_\gamma(T)}(\Theta, r)$ is equal to the set of all irredundant $\gamma$-decision rules for $\Theta$ and $r$.*

*Example 1.* To illustrate the work of the presented algorithm we consider an example based on decision table $T_0$ (see Table 1).

In example, we set $\gamma = 1$, so during the construction of the graph $\Delta_1(T_0)$ we stop the partitioning of a subtable $\Theta$ of $T_0$ when $J(\Theta) \leq 1$. Figure 1 presents the obtained directed acyclic graph for $T_0$. We denote $G = \Delta_1(T_0)$.

For each node $\Theta$ of the graph $G$ and for each row $r$ of $\Theta$ we describe the set $Rul_G(\Theta, r)$. We will move from terminal nodes of $G$ to the node $T_0$. Terminal nodes of the graph $G$ are $\Theta_1, \Theta_2, \Theta_3, \Theta_4, \Theta_6, \Theta_7, \Theta_8$. For these nodes,

**Table 1.** Decision table $T_0$

|        | $f_1$ | $f_2$ | $f_3$ |   |
|--------|-------|-------|-------|---|
| $r_1$  | 1     | 1     | 1     | 2 |
| $r_2$  | 1     | 0     | 0     | 2 |
| $r_3$  | 1     | 1     | 0     | 3 |
| $r_4$  | 0     | 1     | 0     | 3 |
| $r_5$  | 0     | 0     | 0     | 1 |

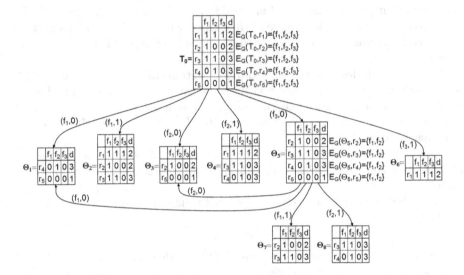

**Fig. 1.** Directed acyclic graph $G = \Delta_1(T_0)$

$Rul_G(\Theta_1, r_4) = Rul_G(\Theta_1, r_5) = \{\to 1\};$
$Rul_G(\Theta_2, r_1) = Rul_G(\Theta_2, r_2) = Rul_G(\Theta_2, r_3) = \{\to 2\};$
$Rul_G(\Theta_3, r_2) = Rul_G(\Theta_3, r_5) = \{\to 1\};$
$Rul_G(\Theta_4, r_1) = Rul_G(\Theta_4, r_3) = Rul_G(\Theta_4, r_4) = \{\to 3\};$
$Rul_G(\Theta_6, r_1) = \{\to 2\};$
$Rul_G(\Theta_7, r_2) = Rul_G(\Theta_7, r_3) = \{\to 2\};$
$Rul_G(\Theta_8, r_3) = Rul_G(\Theta_8, r_4) = \{\to 3\}.$

Now, we can describe the sets of rules attached to rows of $\Theta_5$. This is a nonterminal node of $G$ for which all children $\Theta_1$, $\Theta_3$, $\Theta_7$, and $\Theta_8$ are already treated. We have:

$Rul_G(\Theta_5, r_2) = \{f_2 = 0 \to 1, f_1 = 1 \to 2\};$
$Rul_G(\Theta_5, r_3) = \{f_1 = 1 \to 2, f_2 = 1 \to 3\};$
$Rul_G(\Theta_5, r_4) = \{f_1 = 0 \to 1, f_2 = 1 \to 3\};$
$Rul_G(\Theta_5, r_5) = \{f_1 = 0 \to 1, f_2 = 0 \to 1\}.$

Finally, we can describe the sets of rules attached to rows of $T_0$:

$Rul_G(T_0, r_1) = \{f_1 = 1 \to 2, f_2 = 1 \to 3, f_3 = 1 \to 2\};$
$Rul_G(T_0, r_2) = \{f_1 = 1 \to 2, f_2 = 0 \to 1, f_3 = 0 \wedge f_2 = 0 \to 1,$
$f_3 = 0 \wedge f_1 = 1 \to 2\};$
$Rul_G(T_0, r_3) = \{f_1 = 1 \to 2, f_2 = 1 \to 3, f_3 = 0 \wedge f_1 = 1 \to 2$
$f_3 = 0 \wedge f_2 = 1 \to 3\};$
$Rul_G(T_0, r_4) = \{f_1 = 0 \to 1, f_2 = 1 \to 3, f_3 = 0 \wedge f_1 = 0 \to 1,$
$f_3 = 0 \wedge f_2 = 1 \to 3\};$
$Rul_G(T_0, r_5) = \{f_1 = 0 \to 1, f_2 = 0 \to 1, f_3 = 0 \wedge f_1 = 0 \to 1,$
$f_3 = 0 \wedge f_2 = 0 \to 1\};$

# 4 Procedures of Optimization Relative to Length, Coverage and Number of Misclassifications

In this section, we describe procedures of optimization of the graph $G$ relative to the length, coverage and number of misclassifications.

First, we describe the procedure of optimization of the graph $G$ relative to the length $l$. For each node $\Theta$ in the graph $G$, this procedure assigns to each row $r$ of $\Theta$ the set $Rul_G^l(\Theta, r)$ of $\gamma$-decision rules with the minimum length from $Rul_G(\Theta, r)$ and the number $Opt_G^l(\Theta, r)$ – the minimum length of a $\gamma$-decision rule from $Rul_G(\Theta, r)$.

We will move from the terminal nodes of the graph $G$ to the node $T$. We will assign to each row $r$ of each table $\Theta$ the number $Opt_G^l(\Theta, r)$ and we will change the set $E_G(\Theta, r)$ attached to the row $r$ in $\Theta$ if $\Theta$ is a nonterminal node of $G$. We denote the obtained graph by $G^l$.

Let $\Theta$ be a terminal node of $G$ and $d$ be the most common decision for $\Theta$. Then we assign the number

$$Opt_G^l(\Theta, r) = 0$$

to each row $r$ of $\Theta$.

Let $\Theta$ be a nonterminal node of $G$ and all children of $\Theta$ have already been treated. Let $r = (b_1, \ldots, b_n)$ be a row of $\Theta$. We assign the number

$$Opt_G^l(\Theta, r) = \min\{Opt_G^l(\Theta(f_i, b_i), r) + 1 : f_i \in E_G(\Theta, r)\}$$

to the row $r$ in the table $\Theta$ and we set

$$E_{G^l}(\Theta, r) = \{f_i : f_i \in E_G(\Theta, r), Opt_G^l(\Theta(f_i, b_i), r) + 1 = Opt_G^l(\Theta, r)\}.$$

**Theorem 2.** [2] *For each node $\Theta$ of the graph $G^l$ and for each row $r$ of $\Theta$, the set $Rul_{G^l}(\Theta, r)$ is equal to the set $Rul_G^l(\Theta, r)$ of all $\gamma$-decision rules with the minimum length from the set $Rul_G(\Theta, r)$.*

Now, we describe the procedure of optimization of the graph $G$ relative to the coverage $c$. For each node $\Theta$ in the graph $G$, this procedure assigns to each row $r$ of $\Theta$ the set $Rul_G^c(\Theta, r)$ of $\gamma$-decision rules with the maximum coverage from $Rul_G(\Theta, r)$ and the number $Opt_G^c(\Theta, r)$ – the maximum coverage of a $\gamma$-decision rule from $Rul_G(\Theta, r)$.

We will move from the terminal nodes of the graph $G$ to the node $T$. We will assign to each row $r$ of each table $\Theta$ the number $Opt_G^c(\Theta, r)$ and we will change the set $E_G(\Theta, r)$ attached to the row $r$ in $\Theta$ if $\Theta$ is a nonterminal node of $G$. We denote the obtained graph by $G^c$.

Let $\Theta$ be a terminal node of $G$ and $d$ be the most common decision for $\Theta$. Then we assign to each row $r$ of $\Theta$ the number $Opt_G^c(\Theta, r)$ that is equal to the number of rows in $\Theta$ which are labeled with the decision $d$.

Let $\Theta$ be a nonterminal node of $G$ and all children of $\Theta$ have already been treated. Let $r = (b_1, \ldots, b_n)$ be a row of $\Theta$. We assign the number

$$Opt_G^c(\Theta, r) = \max\{Opt_G^c(\Theta(f_i, b_i), r) : f_i \in E_G(\Theta, r)\}$$

to the row $r$ in the table $\Theta$ and we set

$$E_{G^c}(\Theta, r) = \{f_i : f_i \in E_G(\Theta, r), Opt_G^c(\Theta(f_i, b_i), r) = Opt_G^c(\Theta, r)\}.$$

**Theorem 3.** [2] *For each node $\Theta$ of the graph $G^c$ and for each row $r$ of $\Theta$, the set $Rul_{G^c}(\Theta, r)$ is equal to the set $Rul_G^c(\Theta, r)$ of all $\gamma$-decision rules with the maximum coverage from the set $Rul_G(\Theta, r)$.*

Detailed descriptions of the procedures of optimization of irredundant $\gamma$-decision rules relative to the length and the coverage with examples, can be found in [2].

Now, we describe the procedure of optimization of the graph $G$ relative to the number of misclassifications $\mu$. For each node $\Theta$ in the graph $G$, this procedure assigns to each row $r$ of $\Theta$ the set $Rul_G^\mu(\Theta, r)$ of $\gamma$-decision rules with the minimum number of misclassifications from $Rul_G(\Theta, r)$ and the number $Opt_G^\mu(\Theta, r)$ – the minimum number of misclassifications of a $\gamma$-decision rule from $Rul_G(\Theta, r)$.

We will move from the terminal nodes of the graph $G$ to the node $T$. We will assign to each row $r$ of each table $\Theta$ the number $Opt_G^\mu(\Theta, r)$ and we will change the set $E_G(\Theta, r)$ attached to the row $r$ in $\Theta$ if $\Theta$ is a nonterminal node of $G$. We denote the obtained graph by $G^\mu$.

Let $\Theta$ be a terminal node of $G$ and $d$ be the most common decision for $\Theta$. Then we assign to each row $r$ of $\Theta$ the number $Opt_G^\mu(\Theta, r)$ which is equal to the number of rows in $\Theta$ which are labeled with decisions different from $d$.

Let $\Theta$ be a nonterminal node of $G$ and all children of $\Theta$ have already been treated. Let $r = (b_1, \ldots, b_n)$ be a row of $\Theta$. We assign the number

$$Opt_G^\mu(\Theta, r) = \min\{Opt_G^\mu(\Theta(f_i, b_i), r) : f_i \in E_G(\Theta, r)\}$$

to the row $r$ in the table $\Theta$ and we set

$$E_{G^\mu}(\Theta, r) = \{f_i : f_i \in E_G(\Theta, r), Opt_G^\mu(\Theta(f_i, b_i), r) = Opt_G^\mu(\Theta, r)\}.$$

**Theorem 4.** [3] *For each node $\Theta$ of the graph $G^\mu$ and for each row $r$ of $\Theta$, the set $Rul_{G^\mu}(\Theta, r)$ is equal to the set $Rul_G^\mu(\Theta, r)$ of all $\gamma$-decision rules with the minimum number of misclassifications from the set $Rul_G(\Theta, r)$.*

Detailed description of the procedure of optimization of irredundant $\gamma$-decision rules relative to the number of misclassifications with example can be found in [3].

## 5  Sequential Optimization

Theorems 2, 3 and 4 show that for a given decision table $T$ and row $r$ of $T$, we can make sequential optimization of rules relative to the length, coverage and number of misclassifications. We can find all irredundant $\gamma$-decision rules for $T$ and $r$ with the minimum length, after that among these rules find all rules with the maximum coverage, and finally among the obtained rules find all rules with the minimum number of misclassifications. We can use an arbitrary set of cost functions and an arbitrary order of optimization.

We have three cost functions: length $l$, coverage $c$, and number of misclassifications $\mu$. Let $F$ be one of sets $\{l, c, \mu\}$, $\{l, c, \}$, $\{l, \mu\}$, and $\{c, \mu\}$. An irredundant $\gamma$-decision rule $\tau$ for $T$ and $r$ is called *totally optimal relative to the cost functions from $F$* if, for each cost function $f \in F$, the value $f(\tau)$ is minimum if $f \in \{l, \mu\}$ or maximum if $f = c$ among all irredundant $\gamma$-decision rules for $T$ and $r$. In particular, we will say that an irredundant $\gamma$-decision rule for $T$ and $r$ is *totally optimal relative to the length and coverage* if it has the minimum length and the maximum coverage among all irredundant $\gamma$-decision rules for $T$ and $r$. We can describe all totally optimal rules relative to the cost functions from $F$ using the procedures of optimization relative to these cost functions.

To describe process of sequential optimization we set $G = \Delta_\gamma(T)$ and first, we consider the case when $F$ contains two cost functions. Without the loss of generality we can assume that $F = \{l, c\}$.

We apply the procedure of optimization relative to the coverage to the graph $G$. As a result we obtain the graph $G^c$ and, for each row $r$ of $T$, the value $Opt_G^c(T, r)$ which is equal to the maximum coverage of an irredundant $\gamma$-decision rule for $T$ and $r$.

After that, we apply the procedure of optimization relative to the length to the graph $G$. As a result we obtain the graph $G^l$. Finally, we apply the procedure of optimization relative to the coverage to the graph $G^l$. As a result we obtain the graph $G^{lc}$ and, for each row $r$ of $T$, the value $Opt_{G^l}^c(T, r)$ which is equal to the maximum coverage of an irredundant $\gamma$-decision rule for $T$ and $r$ among all irredundant $\gamma$-decision rules for $T$ and $r$ with the minimum length.

One can show that a totally optimal relative to the length and coverage irredundant $\gamma$-decision rule for $T$ and $r$ exists if and only if $Opt_G^c(T, r) = Opt_{G^l}^c(T, r)$. If the last equality holds then the set $Rul_{G^{lc}}(T, r)$ is equal to the set of all totally optimal relative to the length and coverage irredundant $\gamma$-decision rules for $T$ and $r$.

It is clear that the results of sequential optimization of irredundant decision rules for $T$ and $r$ relative to the length and coverage does not depend on the

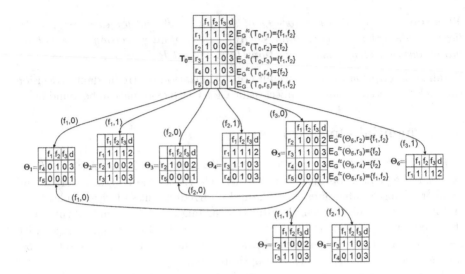

**Fig. 2.** Graph $G^{lc}$

order of optimization (length+coverage or coverage+length) if and only if there exists a totally optimal relative to the length and coverage irredundant $\gamma$-decision rule for $T$ and $r$. We can find all irredundant $\gamma$-decision rules for $T$ and $r$ with the minimum length and after that among these rules find all rules with the maximum coverage. We can use an arbitrary order of optimization, e.g., coverage and length.

*Example 2.* Figure 2 presents the directed acyclic graph $G^{lc}$ obtained from the graph $G$, depicted in Fig. 1, after sequential optimization relative to the length and coverage. Using the graph $G^{lc}$ we can describe for each row $r_i$, $i = 1, \ldots, 5$, of the table $T_0$ the set $Rul_{G^{lc}}(T_0, r_i)$ of all irredundant 1-decision rules for $T_0$ and $r_i$ which have the maximum coverage among all irredundant 1-decision rules for $T_0$ and $r_i$ with the minimum length. We will give also the value $Opt^c_{G^l}(T_0, r_i)$ which is equal to the maximum coverage of a 1-decision rule for $T_0$ and $r_i$ among all irredundant 1-decision rules for $T_0$ and $r_i$ with the minimum length. This value was obtained during the procedure of optimization of the graph $G^l$ relative to the coverage. We have:

$$Rul^{lc}_G(T_0, r_1) = \{f_1 = 1 \rightarrow 2, f_2 = 1 \rightarrow 3\}, Opt^c_{G^l}(T_0, r_1) = 2;$$
$$Rul^{lc}_G(T_0, r_2) = \{f_1 = 1 \rightarrow 2\}, Opt^c_{G^l}(T_0, r_2) = 2;$$
$$Rul^{lc}_G(T_0, r_3) = \{f_1 = 1 \rightarrow 2, f_2 = 1 \rightarrow 3\}, Opt^c_{G^l}(T_0, r_3) = 2;$$
$$Rul^{lc}_G(T_0, r_4) = \{f_2 = 1 \rightarrow 3\}, Opt^c_{G^l}(T_0, r_4) = 2;$$
$$Rul^{lc}_G(T_0, r_5) = \{f_1 = 0 \rightarrow 1, f_2 = 0 \rightarrow 1\}, Opt^c_{G^l}(T_0, r_5) = 1.$$

Values $Opt^c_G(T_0, r_i)$, obtained by the procedure of optimization the graph $G$ relative to the coverage, are the same as $Opt^c_{G^l}(T_0, r_i)$, for $i = 1, \ldots, 5$. Therefore, for $i = 1, \ldots, 5$, $Rul_{G^{lc}}(T_0, r_i)$ is the set of all totally optimal relative to the length and coverage irredundant 1-decision rules for $T_0$ and $r_i$.

Now we consider the case when $F = \{l, c, \mu\}$. At the beginning, we will make the same steps as in the case $F = \{l, c\}$. As a result we obtain the graph $G^{lc}$ and values $Opt_G^c(T, r)$, $Opt_{G^l}^c(T, r)$ for any row $r$ of $T$.

After that, we apply the procedure of optimization relative to the number of misclassifications to the graph $G$. As a result we obtain the graph $G^\mu$ and, for each row $r$ of $T$, the value $Opt_G^\mu(T, r)$ which is equal to the minimum number of misclassifications of an irredundant $\gamma$-decision rule for $T$ and $r$.

Finally, we apply the procedure of optimization relative to the number of misclassifications to the graph $G^{lc}$. As a result we obtain the graph $G^{lc\mu}$ and, for each row $r$ of $T$, the value $Opt_{G^{lc}}^\mu(T, r)$ which is equal to the minimum number of misclassifications of an irredundant $\gamma$-decision rule for $T$ and $r$ among all irredundant $\gamma$-decision rules for $T$ and $r$ with the maximum coverage, and among all irredundant $\gamma$-decision rules for $T$ and $r$ with the minimum length.

One can show that a totally optimal relative to $l$, $c$ and $\mu$ irredundant $\gamma$-decision rule for $T$ and $r$ exists if and only if $Opt_G^c(T, r) = Opt_{G^l}^c(T, r)$ and $Opt_G^\mu(T, r) = Opt_{G^{lc}}^\mu(T, r)$. If these equalities hold then the set $Rul_{G^{lc\mu}}(T, r)$ is equal to the set of all totally optimal relative to $l$, $c$ and $\mu$ irredundant $\gamma$-decision rules for $T$ and $r$.

It is clear that the results of sequential optimization of irredundant decision rules for $T$ and $r$ relative to the length, coverage and number of misclassifications do not depend on the order of optimization ($l+c+\mu$, $l+\mu+c$, $c+l+\mu$, $c+\mu+l$, $\mu+l+c$, or $\mu+c+l$) if and only if there exists a totally optimal relative to $l$, $c$ and $\mu$ irredundant $\gamma$-decision rule for $T$ and $r$.

For decision table $T_0$ (see Table 1), the graph $G^{lc\mu}$ is the same as the graph $G^{lc}$ presented in Fig. 2: $E_{G^{lc}}(\Theta_5, r_i) = E_{G^{lc\mu}}(\Theta_5, r_i)$ for $i = 2, \ldots, 5$, and $E_{G^{lc}}(T_0, r_i) = E_{G^{lc\mu}}(T_0, r_i)$ for $i = 1, \ldots, 5$. However, totally optimal relative to $l$, $c$ and $\mu$ irredundant 1-decision rules exist only for rows $r_2$ and $r_5$ of the table $T_0$.

Considering complexities of the presented algorithms, it is possible to show (see analysis of similar algorithms in [20], page 64) that the time complexities of algorithms which construct the graph $\Delta_\gamma(T)$ and make sequential optimization of $\gamma$-decision rules relative to the length, coverage and number of misclassifications, are bounded from above by polynomials on the number of separable subtables of $T$, and the number of attributes in $T$. In [18] it was shown that the number of separable subtables for decision tables with attributes from a restricted infinite information systems is bounded from above by a polynomial on the number of attributes in the table. Examples of restricted infinite information system were considered, in particular, in [20].

# 6    Experimental Results

We studied a number of decision tables from UCI Machine Learning Repository [7]. Some decision tables contain conditional attributes that take unique

**Table 2.** Parameters of decision tables and values of $\gamma \in \Gamma(T)$

| Name of decision table | Alias | Rows | Attr | $J(T)$ | $\gamma \in \Gamma(T)$ | | |
|---|---|---|---|---|---|---|---|
| | | | | | $\lfloor J(T) \times 0.01 \rfloor$ | $\lfloor J(T) \times 0.2 \rfloor$ | $\lfloor J(T) \times 0.3 \rfloor$ |
| Adult-stretch | Ad | 16 | 4 | 4 | 0 | 0 | 1 |
| Agaricus-lepiota | Ag | 8124 | 22 | 3916 | 39 | 783 | 1174 |
| Balance-scale | Ba | 625 | 4 | 337 | 3 | 67 | 101 |
| Breast-cancer | Br | 266 | 9 | 76 | 0 | 15 | 22 |
| Cars | Ca | 1728 | 6 | 518 | 5 | 103 | 155 |
| Hayes-roth-data | Ha | 69 | 4 | 39 | 0 | 7 | 11 |
| Lymphography | Ly | 148 | 18 | 67 | 0 | 13 | 20 |
| Nursery | Nu | 12960 | 8 | 8640 | 86 | 1728 | 2592 |
| Shuttle-landing-control | Sh | 15 | 6 | 6 | 0 | 1 | 1 |
| Soybean-small | So | 47 | 35 | 30 | 0 | 6 | 9 |
| Teeth | Te | 32 | 8 | 22 | 0 | 4 | 6 |
| Zoo | Zo | 59 | 16 | 40 | 0 | 8 | 12 |

value for each row. Such attributes were removed. In some tables there were equal rows with, possibly, different decisions. In this case each group of identical rows was replaced with a single row from the group with the most common decision for this group. In some tables there were missing values. Each such value was replaced with the most common value of the corresponding attribute.

Let $T$ be one of these decision tables. We consider for this table the value of $J(T)$ and values of $\gamma$ from the set $\Gamma(T) = \{\lfloor J(T) \times 0.01 \rfloor, \lfloor J(T) \times 0.2 \rfloor, \lfloor J(T) \times 0.3 \rfloor\}$. These parameters can be found in Table 2, where column "Alias" denotes a shortcut of the name of decision table, column "Rows" contains number of rows, column "Attr" contains number of conditional attributes, column "$J(T)$" contains difference between number of rows in decision table and number of rows with the most common decision for this decision table, column "$\gamma \in \Gamma(T)$" contains values from $\Gamma(T)$.

Table 3 presents some results about totally optimal rules relative to the length, coverage, and number of misclassifications, i.e., irredundant $\gamma$-decision rules with the minimum length, the maximum coverage, and the minimum number of misclassifications among all irredundant $\gamma$-decision rules for $T$ and $r$. Column "Dt" denotes decision table and contains names of decision tables, column "Rows" contains number of rows in $T$. For the values $\gamma$ from the set $\Gamma(T)$, for each of the considered decision tables $T$, we count the number of rows $r$ such that there exists a totally optimal irredundant $\gamma$-decision rule for $T$ and $r$ (column "Rows with t-opt rules"). We also find for rows of $T$, the minimum (column "min"), the average (column "avg") and the maximum (column "max") number of totally optimal irredundant $\gamma$-decision rules for $T$ and $r$ among all rows $r$ of $T$. Note that two irredundant rules with different order of the same descriptors are considered as two different rules.

Values in bold denotes that the number of rows with totally optimal rules relative to $l$, $c$, $\mu$ in table $T$ is equal to the number of rows in this table, so each

**Table 3.** Totally optimal rules relative to length, coverage and number of misclassifications

| Dt | Rows | $\gamma = \lfloor J(T) \times 0.01 \rfloor$ | | | | $\gamma = \lfloor J(T) \times 0.2 \rfloor$ | | | | $\gamma = \lfloor J(T) \times 0.3 \rfloor$ | | | |
|----|------|----------------|----------|-----|-----|----------------|----------|-----|-----|----------------|----------|-----|-----|
| | | Rows with | t-opt rules | | | Rows with | t-opt rules | | | Rows with | t-opt rules | | |
| | | t-opt rules | min | avg | max | t-opt rules | min | avg | max | t-opt rules | min | avg | max |
| Ad | 16 | **16** | 1 | 1.5 | 2 | **16** | 1 | 1.5 | 2 | **16** | 1 | 1.5 | 2 |
| Ag | 8124 | 288 | 0 | 0.0 | 1 | 0 | 0 | 0.0 | 0 | 0 | 0 | 0.0 | 0 |
| Ba | 625 | 217 | 0 | 3.0 | 24 | **625** | 1 | 1.4 | 4 | **625** | 1 | 1.4 | 4 |
| Br | 266 | 92 | 0 | 3.3 | 96 | 0 | 0 | 0.0 | 0 | 0 | 0 | 0.0 | 0 |
| Ca | 1728 | 1188 | 0 | 1.2 | 24 | 1002 | 0 | 0.7 | 2 | 960 | 0 | 0.7 | 2 |
| Ha | 69 | **69** | 1 | 4.5 | 24 | 30 | 0 | 0.5 | 2 | 30 | 0 | 0.5 | 2 |
| Ly | 148 | 49 | 0 | 1.0 | 24 | 0 | 0 | 0.0 | 0 | 0 | 0 | 0.0 | 0 |
| Nu | 12960 | 5184 | 0 | 0.7 | 6 | 4320 | 0 | 0.3 | 1 | 4320 | 0 | 0.3 | 1 |
| Sh | 15 | 13 | 0 | 2.6 | 24 | 7 | 0 | 0.6 | 2 | 4 | 0 | 0.4 | 2 |
| So | 47 | 37 | 0 | 1.9 | 5 | 24 | 0 | 1.0 | 5 | 20 | 0 | 0.7 | 5 |
| Te | 23 | **23** | 2 | 13.5 | 72 | 4 | 0 | 0.3 | 2 | 2 | 0 | 0.1 | 2 |
| Zo | 59 | 44 | 0 | 8.4 | 408 | 24 | 0 | 0.4 | 1 | 21 | 0 | 0.4 | 1 |

row has at least one totally optimal rule relative to $l$, $c$, $\mu$ (see column "min"). Presented results show that data sets for which exists, for each row, at least one totally optimal relative to $l$, $c$, $\mu$ irredundant γ-decision rule are different (with the exception "Adult-stretch") for $\gamma = \lfloor J(T) \times 0.01 \rfloor$ and $\gamma = \lfloor J(T) \times 0.2 \rfloor$.

Tables 4, 5 and 6 present results of sequential optimization of irredundant γ-decision rules relative to the length, coverage and number of misclassifications for corresponding values of γ. Columns "$l$", "$c$", "$\mu$" contain respectively the average length, the average coverage and the average number of misclassifications after three steps of corresponding order of optimization: $c+l+\mu$, $c+\mu+l$, $l+\mu+c$, $l+c+\mu$, $\mu+c+l$, and $\mu+l+c$. For example, for the order $c+l+\mu$ we make three steps of optimization: relative to the the coverage, then relative to the length, and then relative to the number of misclassifications. After that, we find the average length, the average coverage and the average number of misclassifications of rules after three steps of optimization.

For data sets which are in bold (Tables 4, 5 and 6), for all combinations of order of optimization relative to $l$, $c$, $\mu$, we can see the same values of the average length, the average coverage and the average number of misclassifications after three steps of optimization. It confirms some results from Table 3 that for such data sets each row has at least one totally optimal rule relative to the length, coverage and number of misclassifications.

Sequential optimization can be considered as a problem of multi-criteria optimization with hierarchically dependent criteria. For example, if the length of rules is the most important criterium and we would like to construct short rules, length should be considered as the first cost function in order of optimization.

Based on results presented in Tables 4, 5 and 6 we can find, for considered γ, the minimum values of the average length (column "$l$" in order "$l+c+\mu$" and "$l+\mu+c$"), the maximum values of the average coverage (column "$c$" in order

**Table 4.** Sequential optimization for $\gamma = \lfloor J(T) \times 0.01 \rfloor$

| Dt | $\gamma = \lfloor J(T) \times 0.01 \rfloor$ | | | | | | | | | | | | | | | | |
|---|---|---|---|---|---|---|---|---|---|---|---|---|---|---|---|---|---|
| | $c+l+\mu$ | | | $c+\mu+l$ | | | $l+c+\mu$ | | | $l+\mu+c$ | | | $\mu+c+l$ | | | $\mu+l+c$ | | |
| | $l$ | $c$ | $\mu$ | $l$ | $c$ | $\mu$ | $l$ | $c$ | $\mu$ | $l$ | $c$ | $\mu$ | $l$ | $c$ | $\mu$ | $l$ | $c$ | $\mu$ |
| **Ad** | 1.3 | 7.0 | 0.0 | 1.3 | 7.0 | 0.0 | 1.3 | 7.0 | 0.0 | 1.3 | 7.0 | 0.0 | 1.3 | 7.0 | 0.0 | 1.3 | 7.0 | 0.0 |
| **Ag** | 3.4 | 2278.9 | 9.8 | 3.4 | 2278.9 | 9.8 | 1.2 | 1380.6 | 0.7 | 1.2 | 1366.5 | 0.1 | 2.5 | 2135.5 | 0.0 | 1.2 | 1370.1 | 0.0 |
| **Bae** | 2.3 | 17.8 | 1.4 | 2.3 | 17.8 | 1.4 | 2.3 | 17.8 | 1.4 | 2.3 | 17.8 | 1.4 | 2.9 | 5.9 | 0.3 | 2.9 | 5.9 | 0.3 |
| **Br** | 3.3 | 9.9 | 0.2 | 3.4 | 9.9 | 0.2 | 2.5 | 7.3 | 0.2 | 2.5 | 7.0 | 0.1 | 3.4 | 9.5 | 0.0 | 2.7 | 7.0 | 0.0 |
| **Ca** | 2.0 | 336.9 | 2.4 | 2.0 | 336.9 | 2.4 | 2.0 | 336.9 | 2.4 | 2.0 | 336.5 | 2.2 | 2.3 | 333.1 | 0.1 | 2.3 | 333.1 | 0.1 |
| **Ha** | 2.1 | 6.5 | 0.0 | 2.1 | 6.5 | 0.0 | 2.1 | 6.5 | 0.0 | 2.1 | 6.5 | 0.0 | 2.1 | 6.5 | 0.0 | 2.1 | 6.5 | 0.0 |
| **Ly** | 3.0 | 21.6 | 0.0 | 3.0 | 21.6 | 0.0 | 2.0 | 15.2 | 0.0 | 2.0 | 15.1 | 0.0 | 2.9 | 21.5 | 0.0 | 2.0 | 15.2 | 0.0 |
| **Nu** | 2.3 | 1695.1 | 34.0 | 2.3 | 1695.1 | 34.0 | 2.3 | 1694.9 | 34.1 | 2.3 | 1642.2 | 18.2 | 3.0 | 1537.7 | 1.2 | 3.0 | 1537.7 | 1.2 |
| **Sh** | 1.7 | 2.1 | 0.0 | 1.7 | 2.1 | 0.0 | 1.4 | 1.9 | 0.0 | 1.4 | 1.9 | 0.0 | 1.7 | 2.1 | 0.0 | 1.4 | 1.9 | 0.0 |
| **So** | 1.2 | 12.5 | 0.0 | 1.2 | 12.5 | 0.0 | 1.0 | 12.2 | 0.0 | 1.0 | 12.2 | 0.0 | 1.2 | 12.5 | 0.0 | 1.0 | 12.2 | 0.0 |
| **Te** | 2.3 | 1.0 | 0.0 | 2.3 | 1.0 | 0.0 | 2.3 | 1.0 | 0.0 | 2.3 | 1.0 | 0.0 | 2.3 | 1.0 | 0.0 | 2.3 | 1.0 | 0.0 |
| **Zo** | 1.9 | 11.1 | 0.0 | 1.9 | 11.1 | 0.0 | 1.6 | 10.5 | 0.0 | 1.6 | 10.5 | 0.0 | 1.9 | 11.1 | 0.0 | 1.6 | 10.5 | 0.0 |

**Table 5.** Sequential optimization for $\gamma = \lfloor J(T) \times 0.2 \rfloor$

| Dt table | $\gamma = \lfloor J(T) \times 0.2 \rfloor$ | | | | | | | | | | | | | | | | |
|---|---|---|---|---|---|---|---|---|---|---|---|---|---|---|---|---|---|---|
| | $c+l+\mu$ | | | $c+\mu+l$ | | | $l+c+\mu$ | | | $l+\mu+c$ | | | $\mu+c+l$ | | | $\mu+l+c$ | | |
| | $l$ | $c$ | $\mu$ | $l$ | $c$ | $\mu$ | $l$ | $c$ | $\mu$ | $l$ | $c$ | $\mu$ | $l$ | $c$ | $\mu$ | $l$ | $c$ | $\mu$ |
| **Ad** | 1.3 | 7.0 | 0.0 | 1.3 | 7.0 | 0.0 | 1.3 | 7.0 | 0.0 | 1.3 | 7.0 | 0.0 | 1.3 | 7.0 | 0.0 | 1.3 | 7.0 | 0.0 |
| **Ag** | 1.6 | 3316.4 | 344.8 | 2.4 | 3316.4 | 147.6 | 1.0 | 2826.6 | 231.7 | 1.0 | 1229.8 | 16.6 | 2.4 | 2116.1 | 0.0 | 1.2 | 1367.0 | 0.0 |
| **Ba** | 1.0 | 92.3 | 32.7 | 1.0 | 92.3 | 32.7 | 1.0 | 92.3 | 32.7 | 1.0 | 92.3 | 32.7 | 1.0 | 92.3 | 32.7 | 1.0 | 92.3 | 32.7 |
| **Br** | 2.3 | 70.4 | 14.5 | 2.3 | 70.4 | 14.5 | 1.1 | 35.3 | 12.9 | 1.1 | 20.7 | 7.1 | 2.6 | 10.7 | 0.7 | 2.5 | 9.9 | 0.7 |
| **Ca** | 1.3 | 419.4 | 28.4 | 1.3 | 419.4 | 28.4 | 1.3 | 419.4 | 28.4 | 1.3 | 412.8 | 21.8 | 1.6 | 355.6 | 8.8 | 1.6 | 355.6 | 8.8 |
| **Ha** | 1.3 | 8.0 | 3.6 | 1.4 | 8.0 | 2.9 | 1.3 | 8.0 | 3.6 | 1.3 | 7.8 | 2.7 | 1.6 | 7.2 | 0.7 | 1.6 | 7.2 | 0.7 |
| **Ly** | 1.7 | 55.1 | 12.6 | 1.7 | 55.1 | 12.6 | 1.0 | 46.7 | 10.0 | 1.0 | 18.7 | 5.0 | 2.6 | 16.9 | 0.0 | 2.1 | 14.1 | 0.0 |
| **Nu** | 1.2 | 2334.2 | 753.8 | 1.2 | 2334.2 | 753.8 | 1.0 | 2289.9 | 878.1 | 1.0 | 2289.9 | 878.1 | 1.7 | 1877.9 | 137.5 | 1.7 | 1877.9 | 137.5 |
| **Sh** | 2.4 | 3.0 | 0.6 | 2.5 | 3.0 | 0.5 | 1.1 | 2.0 | 0.3 | 1.1 | 2.0 | 0.3 | 1.7 | 2.2 | 0.1 | 1.3 | 1.9 | 0.1 |
| **So** | 1.6 | 14.6 | 2.3 | 1.7 | 14.6 | 1.9 | 1.0 | 12.7 | 1.2 | 1.0 | 12.2 | 0.0 | 1.2 | 12.5 | 0.0 | 1.0 | 12.2 | 0.0 |
| **Te** | 1.2 | 1.0 | 2.0 | 2.0 | 1.0 | 0.3 | 1.2 | 1.0 | 2.0 | 1.2 | 1.0 | 2.0 | 2.0 | 1.0 | 0.3 | 2.0 | 1.0 | 0.3 |
| **Zo** | 1.8 | 16.1 | 3.9 | 2.2 | 16.1 | 3.1 | 1.0 | 13.3 | 2.7 | 1.0 | 11.7 | 1.7 | 1.8 | 11.0 | 0.0 | 1.6 | 10.6 | 0.0 |

"$c+l+\mu$" and "$c+\mu+l$"), and the minimum values of the average number of misclassifications (column "$\mu$" in order "$\mu+c+l$" and "$\mu+l+c$").

For Tables 4, 5 and 6 we can observe also that the average length is of irredundant $\gamma$-decision rules is nonincreasing when $\gamma$ in increasing (column "$l$" in order "$l+c+\mu$" and "$l+\mu+c$"), the average coverage is nondecreasing when $\gamma$ in increasing (column "$c$" in order "$c+l+\mu$" and "$c+\mu+l$"), and the average number of misclassifications of irredundant $\gamma$-decision rules is nondecreasing when the $\gamma$ is increasing (column "$\mu$" in order "$\mu+c+l$" and "$\mu+l+c$").

We can consider $\gamma$ as an upper bound on the number of misclassifications of irredundant $\gamma$-decision rules (see column $\gamma \in \Gamma(T)$ in Table 2). Results in Tables 4, 5 and 6 show that average values of the minimum number of misclassifications are often less than upper bound on the number of misclassifications given by $\gamma$.

**Table 6.** Sequential optimization for $\gamma = \lfloor J(T) \times 0.3 \rfloor$

| Dt | $\gamma = \lfloor J(T) \times 0.3 \rfloor$ | | | | | | | | | | | | | | | | | |
|---|---|---|---|---|---|---|---|---|---|---|---|---|---|---|---|---|---|---|
| | $c+l+\mu$ | | | $c+\mu+l$ | | | $l+c+\mu$ | | | $l+\mu+c$ | | | $\mu+c+l$ | | | $\mu+l+c$ | | |
| | $l$ | $c$ | $\mu$ | $l$ | $c$ | $\mu$ | $l$ | $c$ | $\mu$ | $l$ | $c$ | $\mu$ | $l$ | $c$ | $\mu$ | $l$ | $c$ | $\mu$ |
| Ad | 1.3 | 7.0 | 0.0 | 1.3 | 7.0 | 0.0 | 1.3 | 7.0 | 0.0 | 1.3 | 7.0 | 0.0 | 1.3 | 7.0 | 0.0 | 1.3 | 7.0 | 0.0 |
| Ag | 1.5 | 3328.2 | 576.8 | 1.9 | 3328.2 | 499.9 | 1.0 | 3096.5 | 567.1 | 1.0 | 1229.8 | 16.6 | 2.4 | 2096.0 | 0.0 | 1.2 | 1365.0 | 0.0 |
| Ba | 1.0 | 92.3 | 32.7 | 1.0 | 92.3 | 32.7 | 1.0 | 92.3 | 32.7 | 1.0 | 92.3 | 32.7 | 1.0 | 92.3 | 32.7 | 1.0 | 92.3 | 32.7 |
| Br | 2.0 | 79.7 | 19.7 | 2.0 | 79.7 | 19.7 | 1.0 | 48.7 | 20.9 | 1.0 | 22.2 | 7.8 | 2.3 | 11.3 | 1.3 | 2.2 | 10.8 | 1.3 |
| Ca | 1.0 | 485.5 | 47.8 | 1.0 | 485.5 | 47.8 | 1.0 | 485.5 | 47.8 | 1.0 | 472.6 | 43.4 | 1.4 | 364.2 | 15.2 | 1.4 | 364.2 | 15.2 |
| Ha | 1.0 | 11.3 | 6.3 | 1.0 | 11.3 | 6.3 | 1.0 | 11.3 | 6.3 | 1.0 | 9.6 | 5.3 | 1.6 | 7.1 | 1.5 | 1.6 | 7.1 | 1.5 |
| Ly | 1.3 | 57.9 | 15.3 | 1.4 | 57.9 | 14.4 | 1.0 | 56.8 | 15.8 | 1.0 | 18.8 | 5.1 | 2.7 | 11.5 | 0.3 | 2.3 | 9.5 | 0.3 |
| Nu | 1.0 | 3066.1 | 1254 | 1.0 | 3066.1 | 1254 | 1.0 | 3066.1 | 1254 | 1.0 | 2289.9 | 878.1 | 1.7 | 1757.6 | 271.2 | 1.7 | 1757.6 | 271.2 |
| Sh | 2.7 | 3.6 | 1.2 | 2.9 | 3.6 | 1.1 | 1.1 | 2.1 | 0.3 | 1.1 | 2.0 | 0.3 | 1.7 | 2.2 | 0.1 | 1.3 | 1.9 | 0.1 |
| So | 1.7 | 15.7 | 3.7 | 1.9 | 15.7 | 3.3 | 1.0 | 13.8 | 2.1 | 1.0 | 12.2 | 0.0 | 1.2 | 12.5 | 0.0 | 1.0 | 12.2 | 0.0 |
| Te | 1.0 | 1.0 | 2.7 | 1.9 | 1.0 | 0.5 | 1.0 | 1.0 | 2.7 | 1.0 | 1.0 | 2.7 | 1.9 | 1.0 | 0.5 | 1.9 | 1.0 | 0.5 |
| Zo | 1.9 | 17.4 | 6.2 | 2.3 | 17.4 | 5.4 | 1.0 | 13.7 | 4.0 | 1.0 | 11.7 | 1.7 | 1.8 | 11.2 | 0.0 | 1.6 | 10.8 | 0.0 |

**Table 7.** Size of the directed acyclic graph for $\gamma \in \Gamma(T)$

| Dt | $\gamma = \lfloor J(T) \times 0.01 \rfloor$ | | $\gamma = \lfloor J(T) \times 0.2 \rfloor$ | | $\gamma = \lfloor J(T) \times 0.3 \rfloor$ | |
|---|---|---|---|---|---|---|
| | nodes | edges | nodes | edges | nodes | edges |
| Ad | 72 | 108 | 72 | 108 | 52 | 76 |
| Ag | 83959 | 947650 | 6483 | 26469 | 2894 | 10438 |
| Ba | 647 | 1460 | 21 | 20 | 21 | 20 |
| Br | 6001 | 60387 | 1132 | 1786 | 633 | 883 |
| Ca | 4669 | 11379 | 270 | 392 | 158 | 197 |
| Ha | 236 | 572 | 97 | 150 | 73 | 84 |
| Ly | 40928 | 814815 | 9029 | 34787 | 4223 | 13397 |
| Nu | 16952 | 42001 | 339 | 525 | 217 | 245 |
| Sh | 85 | 513 | 82 | 458 | 82 | 458 |
| So | 3592 | 103520 | 2438 | 11386 | 1473 | 4789 |
| Te | 135 | 1075 | 122 | 512 | 99 | 288 |
| Zo | 4568 | 83043 | 3702 | 23619 | 2533 | 10760 |

Table 7 presents a size of the directed acyclic graph, i.e., number of nonterminal nodes (column "nodes") and number of edges (column "edges") in the graph constructed by the dynamic programming algorithm, for $\gamma \in \Gamma(T)$.

The obtained results show that the number of nodes and the number of edges for $\Delta_\gamma(T)$ decrease with the growth of $\gamma$. It means that the parameter $\gamma$ can help to control algorithm complexity. They show also that the structure of graph $\Delta_\gamma(T)$ is usually far from a tree: the number of edges is larger than the number of nodes.

Experiments were done using software system Dagger [1]. It is implemented in C++ and uses Pthreads and MPI libraries for managing threads and processes respectively. It runs on a single-processor computer or multiprocessor system with shared memory. Parameters of computer which was used for experiments are following: desktop with 2 Xeon x5550 processors running at 2.66 GHz (each with 4 cores and 8 threads) all sharing 16 GB of RAM. Time of the preformed experiments (for $\gamma \in \Gamma(T)$) for the biggest data sets was 36 min. for "Agaricus-lepiota", and 3 min. for "Nursery".

## 7    Conclusions

We studied an extension of dynamic programming approach for the sequential optimization of $\gamma$-decision rules relative to the length, coverage and number of misclassifications. The considered approach allows to describe the whole set of irredundant $\gamma$-decision rules and optimize these rules sequentially relative to arbitrary subset and order of cost functions.

Results of sequential optimization of irredundant $\gamma$-decision rules for $T$ and $r$ depend on the order of optimization if there are no

totally optimal relative to $l$, $c$, $\mu$ irredundant $\gamma$-decision rules for $T$ and $r$.

Sequential optimization of irredundant $\gamma$-decision rules can be considered as a tool that supports design of classifiers. To predict a value of the decision attribute for a new object we can use in a classifier sequentially optimized rules, e.g., among all irredundant $\gamma$ decision rules choose rules with the maximum coverage, and among them – rules with the minimum length. Short rules which cover many objects can be useful also from the point of view of knowledge representation. In this case, rules with smaller number of descriptors are more understandable.

**Acknowledgment.** The author would like to thank you Prof. Mikhail Moshkov, Dr. Igor Chikalov and Talha Amin for possibility to use Dagger software system.

## References

1. Alkhalid, A., Amin, T., Chikalov, I., Hussain, S., Moshkov, M., Zielosko, B.: Dagger: a tool for analysis and optimization of decision trees and rules. In: Ficarra, F.V.C. (ed.) Computational Informatics, Social Factors and New Information technologies: Hypermedia Perspectives and Avant-Garde Experienes in the Era of Communicability Expansion, pp. 29–39. Blue Herons, Bergamo (2011)
2. Amin, T., Chikalov, I., Moshkov, M., Zielosko, B.: Dynamic programming approach for partial decision rule optimization. Fundam. Inform. **119**(3–4), 233–248 (2012)
3. Amin, T., Chikalov, I., Moshkov, M., Zielosko, B.: Optimization of approximate decision rules relative to number of misclassifications. In: Graña, M., Toro, C., Posada, J., Howlett, R.J., Jain, L.C. (eds.) KES. Frontiers in Artificial Intelligence and Applications, vol. 243, pp. 674–683. IOS Press, Amsterdam (2012)
4. Amin, T., Chikalov, I., Moshkov, M., Zielosko, B.: Dynamic programming approach for exact decision rule optimization. In: Skowron, A., Suraj, Z. (eds.) Rough Sets and Intelligent Systems - Professor Zdzisław Pawlak in Memoriam. ISRL, vol. 42, pp. 211–228. Springer, Heidelberg (2013)

5. Amin, T., Chikalov, I., Moshkov, M., Zielosko, B.: Dynamic programming approach to optimization of approximate decision rules. Inf. Sci. **221**, 403–418 (2013)
6. An, A., Cercone, N.: ELEM2: a learning system for more accurate classifications. In: Mercer, R.E. (ed.) Canadian AI 1998. LNCS, vol. 1418. Springer, Heidelberg (1998)
7. Asuncion, A., Newman, D.J.: UCI Machine Learning Repository (2007)
8. Bazan, J.G., Nguyen, H.S., Nguyen, T.T., Skowron, A., Stepaniuk, J.: Synthesis of decision rules for object classification. In: Orłowska, E. (ed.) Incomplete Information: Rough Set Analysis, pp. 23–57. Physica-Verlag, Heidelberg (1998)
9. Błaszczyński, J., Słowiński, R., Susmaga, R.: Rule-based estimation of attribute relevance. In: Yao, J.T., Ramanna, S., Wang, G., Suraj, Z. (eds.) RSKT 2011. LNCS, vol. 6954, pp. 36–44. Springer, Heidelberg (2011)
10. Błaszczyński, J., Słowiński, R., Szeląg, M.: Sequential covering rule induction algorithm for variable consistency rough set approaches. Inf. Sci. **181**(5), 987–1002 (2011)
11. Clark, P., Niblett, T.: The CN2 induction algorithm. Mach. Learn. **3**(4), 261–283 (1989)
12. Dembczyński, K., Kotłowski, W., Słowiński, R.: Ender: a statistical framework for boosting decision rules. Data Min. Knowl. Discov. **21**(1), 52–90 (2010)
13. Frank, E., Witten, I.H.: Generating accurate rule sets without global optimization. In: Proceedings of the Fifteenth International Conference on Machine Learning, ICML 1998, pp. 144–151. Morgan Kaufmann Publishers Inc., Burlington (1998)
14. Fürnkranz, J., Flach, P.A.: ROC 'n' rule learning-towards a better understanding of covering algorithms. Mach. Learn. **58**(1), 39–77 (2005)
15. Grzymala-Busse, J.W.: LERS-a system for learning from examples based on rough sets. In: Slowański, R. (ed.) Intelligent Decision Support. Handbook of Applications and Advances of the Rough Sets Theory, pp. 3–18. Springer, The Netherlands (1992)
16. Michalski, R.S.: A theory and methodology of inductive learning. Artif. Intell. **20**(2), 111–161 (1983)
17. Michalski, S., Pietrzykowski, J.: iAQ: A program that discovers rules. AAAI-07 AI Video Competition (2007)
18. Moshkov, M., Chikalov, I.: On algorithm for constructing of decision trees with minimal depth. Fundam. Inform. **41**(3), 295–299 (2000)
19. Moshkov, M., Piliszczuk, M., Zielosko, B.: Partial Covers, Reducts and Decision Rules in Rough Sets Theory and Applications. Springer, Heidelberg (2008)
20. Moshkov, M., Zielosko, B.: Combinatorial Machine Learning - A Rough Set Approach. Springer, Heidelberg (2011)
21. Nguyen, H.S.: Approximate boolean reasoning: foundations and applications in data mining. In: Peters, J.F., Skowron, A. (eds.) Transactions on Rough Sets V. LNCS, vol. 4100, pp. 334–506. Springer, Heidelberg (2006)
22. Nguyen, Hung Son, Ślęzak, Dominik: Approximate Reducts and Association Rules. In: Zhong, Ning, Skowron, Andrzej, Ohsuga, Setsuo (eds.) RSFDGrC 1999. LNCS (LNAI), vol. 1711, pp. 137–145. Springer, Heidelberg (1999)
23. Pawlak, Z.: Rough set elements. In: Polkowski, L., Skowron, A. (eds.) Rough Sets in Knowledge Discovery, pp. 10–30. Physica-Verlag, Heidelberg (1998)
24. Pawlak, Z., Skowron, A.: Rough sets and boolean reasoning. Inf. Sci. **177**(1), 41–73 (2007)
25. Pawlak, Z., Skowron, A.: Rudiments of rough sets. Inf. Sci. **177**(1), 3–27 (2007)
26. Quinlan, J.R.: C4.5: Programs for Machine Learning. Morgan Kaufmann Publishers Inc., San Francisco (1993)

27. Rissanen, J.: Modeling by shortest data description. Automatica **14**(5), 465–471 (1978)
28. Sikora, M.: Decision rule-based data models using TRS and NetTRS – methods and algorithms. In: Peters, J.F., Skowron, A. (eds.) Transactions on Rough Sets XI. LNCS, vol. 5946, pp. 130–160. Springer, Heidelberg (2010)
29. Skowron, A., Rauszer, C.: The discernibility matrices and functions in information systems. In: Slowański, R. (ed.) Intelligent Decision Support. Handbook of Applications and Advances of the Rough Set Theory., pp. 331–362. Kluwer Academic Publishers, Dordrecht (1992)
30. Ślęzak, D., Wróblewski, J.: Order based genetic algorithms for the search of approximate entropy reducts. In: Wang, G., Liu, Q., Yao, Y., Skowron, A. (eds.) RSFD-GrC 2003. LNCS, vol. 2639, pp. 308–311. Springer, Heidelberg (2003)
31. Zielosko, B.: Sequential optimization of $\gamma$-decision rules. In: Ganzha, M., Maciaszek, L.A., Paprzycki, M., eds.: FedCSIS, pp. 339–346 (2012)
32. Zielosko, B., Moshkov, M., Chikalov, I.: Optimization of decision rules based on methods of dynamic programming. Vestnik of Lobachevsky State University of Nizhny Novgorod **6**, 195–200 (2010). (in Russian)

# Toward Qualitative Assessment of Rough Sets in Terms of Decision Attribute Values in Simple Decision Systems over Ontological Graphs

Krzysztof Pancerz[1,2]([envelope])

[1] University of Management and Administration,
Akademicka Str. 4, 22-400 Zamość, Poland
[2] University of Information Technology and Management,
Sucharskiego Str. 2, 35-225 Rzeszów, Poland
kpancerz@wszia.edu.pl

**Abstract.** Approximation of sets is a fundamental notion of rough set theory (RST) proposed by Z. Pawlak. Each rough set can be characterized numerically by the coefficient called the accuracy of approximation. This coefficient determines quantitatively a degree of roughness. Such an approach does not take into consideration semantics of data. In the paper, we show that adding information on semantic relations between decision attribute values in the form of ontological graphs enables us to determine qualitatively the accuracy of approximation. The qualitative assessment of approximation should be treated as some additional characteristic of rough sets. The proposed approach enriches application of rough sets if decision attribute values classifying objects are symbolical (e.g., words, terms, linguistic concepts, etc.). The presented approach refers to a general trend in computations proposed by L. Zadeh and called "computing with words".

**Keywords:** Approximations of sets · Accuracy of approximation · Ontological graphs · Rough sets · Semantic relations

## 1 Introduction

In [16], information (decision) systems were proposed as the knowledge representation systems. In simple case, they consist of vectors of numbers or symbols (attribute values) describing objects from a given universe of discourse. In [14] and [15], ontologies were incorporated into information (decision) systems, i.e., attribute values were considered in the ontological (semantic) spaces. Similar approaches have been considered in the literature, e.g., DAG-Decision Systems [10], Dominance-Based Rough Set Approach (DRSA) [5], Rough Ontology [7], Attribute Value Ontology (AVO) [9], etc. In our approach, we replace, in a classic definition of information (decision) systems, simple sets of attribute values with ontological graphs, which deliver us some new knowledge about meanings of attribute values. This knowledge enables us to assess qualitatively

© Springer-Verlag Berlin Heidelberg 2015
J. Peters et al. (Eds.): TRS XIX, LNCS 8988, pp. 83–94, 2015.
DOI: 10.1007/978-3-662-47815-8_6

the accuracy of approximation used in rough set theory. In the paper, we consider one of the possible cases, where the accuracy is assessed in terms of decision attribute values. The second case worth considering in the future covers the problem of the assessment of the accuracy of approximation in terms of condition attribute values placed in semantic spaces of ontological graphs. Some remarks related to that problem were given in [15], where we showed how ontological graphs, associated both with condition attributes and with decision attributes, change a look at approximations of sets. Now, we extend those investigations to the qualitative assessment of the accuracy of approximation in case of adding information on semantic relations between decision attribute values.

For the qualitative assessment, we propose to use a simple taxonomy distinguishing pseudo rough sets, marginally rough sets, restrainedly rough sets, and considerable rough sets. It is worth noting that the taxonomy can be extended to a more sophisticated one. In the classic approach, the accuracy of approximation is characterized quantitatively (numerically) [16]. In Sect. 3, we show an example in which the classic approach, omitting information about semantic relations between decision attribute values, leads to determining the same accuracy of approximations for all considered cases, whereas the qualitative assessment on the basis of semantic relations enables us additionally to differentiate those cases. Therefore, the qualitative assessment is a new look at approximations delivering some additional characteristic of rough sets. Moreover, one can see that the proposed taxonomy for the qualitative assessment may be considered in terms of a linguistic variable (a variable whose values are words or sentences in a natural or artificial language) what can lead to combining the presented approach with fuzzy logic (cf. [20]). Fuzzy logic plays a pivotal role in the "computing with words" methodology proposed by Zadeh [21]. There is a number of papers showing that approaches relying on the use of linguistic variables have a lot of applications, e.g. analysis of complex systems and decision processes [23], approximate reasoning [20], artificial intelligence [22], cognitive computing [6], etc.

The presented approach is based on the definitions of ontology given by Neches et al. [12] and Köhler et al. [8]. That is, ontology is constructed on the basis of a controlled vocabulary and the relationships of the concepts in the controlled vocabulary. Formally, the ontology can be represented by means of graph structures. The graph representing the ontology is called the ontological graph. In such a graph, each node represents one concept from the ontology, whereas each edge represents a semantic relation between two concepts. Relations are very important components in ontology modeling as they describe the relationships that can be established between concepts.

The rest of the paper is organized as follows. Section 2 recalls a series of definitions concerning rough set theory. Section 3 describes a new approach to the qualitative assessment of the accuracy of approximation used in rough set theory including a proper example. Finally, Sect. 4 provides conclusions and some directions for future work.

## 2    Basic Notions of Rough Set Theory

A series of definitions concerning rough set theory is recalled in this section (cf. [16,17]).

### 2.1    Binary Relations

Let $U$ be a non-empty set of objects. Any subset $R \subseteq U \times U$ is called a binary relation in $U$. By $R(u)$, where $u \in U$, we will denote the set of all $v \in U$ such that $(u, v) \in R$. The statement $(u, v) \in R$ is read "$u$ is $R$-related to $v$". In the paper, we consider relations which may be:

- reflexive, i.e., $(u, u) \in R$ for each $u \in U$,
- symmetric, i.e., whenever $(u, v) \in R$ then $(v, u) \in R$,
- transitive, i.e., whenever $(u, v) \in R$ and $(v, w) \in R$ then $(u, w) \in R$.

### 2.2    Information and Decision Systems

An information system $IS$ is a quadruple $IS = (U, A, V, f)$, where $U$ is a nonempty, finite set of objects, $A$ is a nonempty, finite set of attributes, $V = \bigcup_{a \in A} V_a$, where $V_a$ is a set of values of the attribute $a$, and $f : A \times U \to V$ is an information function such that $f(a, u) \in V_a$ for each $a \in A$ and $u \in U$.

A decision system $DS$ is a tuple $DS = (U, C, D, V_{con}, V_{dec}, f_{inf}, f_{dec})$, where $U$ is a nonempty, finite set of objects, $C$ is a nonempty, finite set of condition attributes, $D$ is a nonempty, finite set of decision attributes, $V_{con} = \bigcup_{c \in C} V_c$, where $V_c$ is a set of values of the condition attribute $c$, $V_{dec} = \bigcup_{d \in D} V_d$, where $V_d$ is a set of values of the decision attribute $d$, $f_{inf} : C \times U \to V_{con}$ is an information function such that $f_{inf}(c, u) \in V_c$ for each $c \in C$ and $u \in U$, $f_{dec} : D \times U \to V_{dec}$ is a decision function such that $f_{dec}(d, u) \in V_d$ for each $d \in D$ and $u \in U$.

### 2.3    Approximation of Sets

Let $IS = (U, A, V, f)$ be an information system. Each subset $B \subseteq A$ of attributes determines an equivalence relation on $U$, called an indiscernibility relation $IR_B$, defined as $IR_B = \{(u, v) \in U \times U : \forall_{a \in B} f(a, u) = f(a, v)\}$. The equivalence class containing $u \in U$ will be denoted by $IR_B(u)$. Any set of all indiscernible objects is called an elementary set, and forms a basic granule of knowledge about the universe. Equivalence classes of an indiscernibility relation $IR_B$ are referred to as $B$-elementary granules in $IS$. Analogously, we can define an indiscernibility relation $IR_B$ and $B$-elementary granules for a decision system $DS = (U, C, D, V_{con}, V_{dec}, f_{inf}, f_{dec})$, where $B \subseteq C \cup D$.

Let $X \subseteq U$ and $B \subseteq A$. We may characterize $X$ with respect to $B$ using the basic notions of rough set theory given below.

– The $B$-lower approximation, $\underline{B}(X)$, of a set $X$ with respect to $B$:

$$\underline{B}(X) = \{u \in U : IR_B(u) \subseteq X\}.$$

The $B$-lower approximation is a set of objects from $U$ which can be certainly classified into $X$ with respect to $B$.
– The $B$-upper approximation, $\overline{B}(X)$, of a set $X$ with respect to $B$:

$$\overline{B}(X) = \{u \in U : IR_B(u) \cap X \neq \emptyset\}.$$

The $B$-upper approximation is a set of objects from $U$ which can be possibly classified into $X$ with respect to $B$.
– The $B$-boundary region, $BN_B(X)$, of a set $X$ with respect to $B$:

$$BN_B(X) = \overline{B}(X) - \underline{B}(X).$$

The $B$-boundary region is a set of objects from $U$ which can be exactly classified neither as $X$ nor as not $X$ with respect to $B$.

The accuracy of $B$-approximation of $X$ can be expressed as a coefficient:

$$\alpha_B(X) = \frac{card(\underline{B}(X))}{card(\overline{B}(X))},$$

where $card$ denotes the cardinality of a set. Obviously, $0 \leq \alpha_B(X) \leq 1$. If $\alpha_B(X) = 1$, then $X$ is exact with respect to $B$, otherwise, $X$ is rough with respect to $B$.

## 3     Qualitative Assessment of Rough Sets

In [14], we proposed to consider attribute values in the ontological (semantic) spaces. That approach is based on the definitions of ontology given by Neches et al. [12] and Köhler et al. [8]. Formally, the ontology can be represented by means of graph structures. In our approach, the graph representing the ontology $\mathcal{O}$ is called the ontological graph. In such a graph, each node represents one concept from $\mathcal{O}$, whereas each edge represents a semantic relation between two concepts from $\mathcal{O}$.

Let $\mathcal{O}$ be a given ontology. An ontological graph is a quadruple $OG = (\mathcal{C}, E, \mathcal{R}, \rho)$, where $\mathcal{C}$ is a nonempty, finite set of nodes representing concepts in the ontology $\mathcal{O}$, $E \subseteq \mathcal{C} \times \mathcal{C}$ is a finite set of edges representing semantic relations between concepts from $\mathcal{C}$, $\mathcal{R}$ is a family of semantic descriptions (in a natural language) of types of relations (represented by edges) between concepts, and $\rho : E \to \mathcal{R}$ is a function assigning a semantic description of the relation to each edge.

Relations are very important components in ontology modeling as they describe the relationships that can be established between concepts. In the literature, a variety of taxonomies of different types of semantic relations has been proposed, e.g. [2,3,11,18,19]. In our approach, we will be interested in the following taxonomy of types of semantic relations (which is modeled on the project called Wikisaurus [1] aiming at creating a thesaurus of semantically related terms):

- synonymy,
- antonymy,
- hyponymy/hyperonymy.

Synonymy concerns concepts with a meaning that is the same as, or very similar to, the other concepts. Antonymy concerns concepts which have the opposite meaning to the other ones. Hyponymy concerns more specific concepts than the other ones. Hyperonymy concerns more general concepts than the other ones.

For simplicity, we will use the following labels and reading of semantic relations:

- $R_\sim$ - synonymy, $(u, v) \in R_\sim$ is read "$u$ is a synonym of $v$",
- $R_\leftrightarrow$ - antonymy, $(u, v) \in R_\leftrightarrow$ is read "$u$ is an antonym of $v$",
- $R_\lhd$ - hyponymy, $(u, v) \in R_\lhd$ is read "$u$ is a hyponym of $v$",
- $R_\rhd$ - hyperonymy, $(u, v) \in R_\rhd$ is read "$u$ is a hyperonym of $v$".

The labels above will be used instead of semantic descriptions (in a natural language) of types of relations assigned to edges in ontological graphs.

In the graphical representation of the ontological graph, for readability, we will omit reflexivity of relations. However, some of the above relations are reflexive, i.e., a given concept is a synonym of itself, a given concept is a hyponym of itself, a given concept is a hyperonym of itself.

We can create decision systems over ontological graphs. In this paper, we will use a definition given in [13]. In this approach, both, condition and decision attribute values of a given decision system are concepts from ontological graphs assigned to attributes. A similar approach, with respect to decision attributes, was considered in [10]. Decision attribute values were placed in directed acyclic graph spaces determined by subclass/superclass relations.

**Definition 1.** *A simple decision system $SDS^{OG}$ over ontological graphs is a tuple $SDS^{OG} = (U, C, D, \{OG_a\}_{a \in C \cup D}, f_{inf}, f_{dec})$, where:*

- *$U$ is a nonempty, finite set of objects,*
- *$C$ is a nonempty, finite set of condition attributes,*
- *$D$ is a nonempty, finite set of decision attributes,*
- *$\{OG_a\}_{a \in C \cup D}$ is a family of ontological graphs associated with condition and decision attributes,*
- *$f_{inf} : C \times U \to \mathcal{C}_C$, $\mathcal{C}_C = \bigcup_{c \in C} \mathcal{C}_c$, is an information function such that $f_{inf}(c, u) \in \mathcal{C}_c$ for each $c \in C$ and $u \in U$, $\mathcal{C}_c$ is a set of concepts from the graph $OG_c$*
- *$f_{dec} : D \times U \to \mathcal{C}_D$, $\mathcal{C}_D = \bigcup_{d \in D} \mathcal{C}_d$, is a decision function such that $f_{dec}(d, u) \in \mathcal{C}_d$ for each $d \in D$ and $u \in U$, $\mathcal{C}_d$ is a set of concepts from the graph $OG_d$.*

It is not necessary for condition and decision functions to be onto functions, i.e., $f_{inf} : C \times U \to \mathcal{C}_C^* \subseteq \mathcal{C}_C$ and $f_{dec} : D \times U \to \mathcal{C}_D^* \subseteq \mathcal{C}_D$.

In our approach, we propose to consider some relations defined over sets of attribute values in simple decision systems over ontological graphs (cf. [14,15]).

In defined relations, we use some additional knowledge about semantic relations between attribute values which is included in ontological graphs.

Let $OG_a = (C_a, E_a, \mathcal{R}, \rho_a)$, where $\mathcal{R} = \{R_\sim, R_\leftrightarrow, R_\triangleleft, R_\triangleright\}$, be an ontological graph associated with the attribute $a$ in a simple decision system over ontological graphs. Later on, we will use the following notation: $[v_i, v_j]$ is a simple path in $OG_a$ between $v_i \in C$ and $v_j \in C$, $\mathcal{E}([v_i, v_j])$ is a set of all edges from $E$ belonging to the simple path $[v_i, v_j]$, and $\mathcal{P}(OG_a)$ - is a set of all simple paths in $OG_a$. In the literature, there are different definitions for a simple path in the graph. In this paper, we follow the definition in which a path is simple if no node or edge is repeated, with the possible exception that the first node is the same as the last. Therefore, the path $[v_i, v_j]$, where $v_i, v_j \in C$ and $v_i = v_j$, can also be a simple path in $OG_a$.

We are interested in the following relations defined over sets of attribute values in simple decision systems over ontological graphs:

- An exact meaning relation between $v_1, v_2 \in C_a$ is defined as $EMR_a = \{(v_1, v_2) \in C_a \times C_a : v_1 = v_2\}$.
- A synonymous meaning relation between $v_1, v_2 \in C_a$ is defined as $SMR_a = \{(v_1, v_2) \in C_a \times C_a : (v_1, v_2) \in E_a \wedge \rho_a((v_1, v_2)) = R_\sim\}$.
- An antonymous meaning relation between $v_1, v_2 \in C_a$ is defined as $AMR_a = \{(v_1, v_2) \in C_a \times C_a : (v_1, v_2) \in E_a \wedge \rho_a((v_1, v_2)) = R_\leftrightarrow\}$.
- A hyperonymous meaning relation $HprMR_a^k$ of at most $k$-th order is a set of all pairs $(v_1, v_2) \in C_a \times C_a$ satisfying the following condition. There exists $v_3 \in C_a$ such that the following holds:

$$\exists_{[v_1, v_3] \in \mathcal{P}(OG_a)} \left[ \left( \forall_{e \in \mathcal{E}([v_1, v_3])} \rho_a(e) \in \{R_\sim, R_\triangleleft\} \right) \right.$$
$$\wedge$$
$$\left. card(\{e' \in \mathcal{E}([v_1, v_3]) : \rho_a(e') = R_\triangleleft\}) \leq k \right],$$

and

$$\exists_{[v_2, v_3] \in \mathcal{P}(OG_a)} \left[ \left( \forall_{e \in \mathcal{E}([v_2, v_3])} \rho_a(e) \in \{R_\sim, R_\triangleleft\} \right) \right.$$
$$\wedge$$
$$\left. card(\{e' \in \mathcal{E}([v_2, v_3]) : \rho_a(e') = R_\triangleleft\}) \leq k \right],$$

where $card$ denotes the cardinality of a set. If $v_1$ is $HprMR_a^k$-related to $v_2$, then there exists $v_3 \in C_a$ such that $v_3$ is a hyperonym (direct or indirect) of $v_1$ and $v_3$ is a hyperonym (direct or indirect) of $v_2$, both through at most $k$ concepts. In the hyperonymous path, we also take into consideration synonyms, but they do not affect the order of the relation.

In case of a hyperonymous meaning relation, we are interested in how far the hyperonym is. A kind of distance can be expressed by the order $k$ of the relation. In simple case used in this paper, we will distinguish far, middle-far, and close hyperonymous meaning relations.

**Definition 2.** *Let $OG_a = (C_a, E_a, \mathcal{R}, \rho_a)$ be an ontological graph associated with the attribute $a$, $HprMR_a^k$ be a hyperonymous meaning relation of at most $k$-th order defined over sets of values of the attribute $a$, and $\tau_1, \tau_2$ be two fixed positive integer values, where $\tau_1 < \tau_2$. $HprMR_a^k$ is called:*

- *a far hyperonymous meaning relation $FHprMR_a$ if $k > \tau_2$,*
- *a middle-far hyperonymous meaning relation $MHprMR_a$ if $\tau_1 < k \leq \tau_2$,*
- *a close hyperonymous meaning relation $CHprMR_a$ otherwise.*

We assume that the qualitative assessment of rough sets in simple decision systems over ontological graphs is made on the basis of semantic relations between values of a given decision attribute. It is worth noting that, in the standard approach in rough set theory, the approximations of a given set and the boundary region are calculated on the basis of the indiscernibility relation defined over sets of values of condition attributes (cf. Sect. 2.3). However, we can replace the indiscernibility relation with semantic relations delivered by ontological graphs associated with condition attributes. Such replacement influences lower and upper approximations of a given set and its boundary region (cf. [15]). Therefore, the qualitative assessment of the accuracy of approximation can also be made in terms of condition attribute values placed in semantic spaces of ontological graphs. That problem is worth considering in the future investigations.

To assess qualitatively rough sets in terms of decision attribute values, we propose to use a simple taxonomy distinguishing pseudo rough sets, marginally rough sets, restrainedly rough sets, and considerable rough sets.

Let $SDS^{OG} = (U, C, D, \{OG_a\}_{a \in C \cup D}, f_{inf}, f_{dec})$ be a simple decision system over ontological graphs, $X \subseteq U$, $B \subseteq C$, and $D = \{d\}$. Later on, we will use the following notation:

- $Dec_d(X) = \{f_{dec}(d, u) : u \in X\}$,
- $Dec_d(X^*) = \{f_{dec}(d, u) : u \in BN_B(X) - X\}$.

**Definition 3.** *Let $SDS^{OG} = (U, C, D, \{OG_a\}_{a \in C \cup D}, f_{inf}, f_{dec})$ be a simple decision system over ontological graphs, $B \subseteq C$, $D = \{d\}$, $X_v = \{u \in U : f_{dec}(d, u) = v\}$. $X_v$ is:*

- *a pseudo rough set if and only if*

$$\forall_{v' \in Dec_d(X_v^*)} \left[ (v, v') \in EMR_d \vee (v, v') \in SMR_d \vee (v, v') \in CHprMR_d \right],$$

- *a marginally rough set if and only if*

$$\forall_{v' \in Dec_d(X_v^*)} \left[ (v, v') \notin AMR_d \wedge (v, v') \notin FHprMR_d \right]$$

*and*

$$\exists_{v' \in Dec_d(X_v^*)} (v, v') \in MHprMR_d,$$

- *a restrainedly rough set if and only if*

$$\forall_{v' \in Dec_d(X_v^*)} (v, v') \notin AMR_d$$

*and*

$$\exists_{v' \in Dec_d(X_v^*)} (v, v') \in FHprMR_d,$$

– *a considerable rough set if and only if*

$$\exists_{v' \in Dec_d(X_v^*)} (v, v') \in AMR_d.$$

It is easy to see that the assessment depends on semantic meanings of decision attribute values (concepts) of the remaining objects (i.e., other than those belonging to the approximated set) from the boundary region.

*Example 1.* Let us consider a decision system, concerning the employment of persons, shown in Table 1. In this system, $U = \{u_1, u_2, ..., u_{12}\}$ is a set of twelve persons described with respect to their employment, $C = \{gender, education, abode\}$ is a set of condition attributes describing selected persons, $D = \{employment\}$ is a set of decision attributes. Both an information function $f_{inf}$ and a decision function $f_{dec}$ can be obtained from the table.

**Table 1.** A decision system concerning the employment of persons

| U/A | gender | education | abode | employment |
|-----|--------|-----------|-------|------------|
| $u_1$ | male | primary | village | unemployed |
| $u_2$ | male | primary | village | jobless |
| $u_3$ | male | higher | town | employed under a contract |
| $u_4$ | male | higher | town | freelance |
| $u_5$ | female | higher | city | freelance |
| $u_6$ | female | higher | city | full − time employed |
| $u_7$ | female | primary | village | jobless |
| $u_8$ | female | primary | village | working |
| $u_9$ | female | primary | town | unemployed |
| $u_{10}$ | male | higher | city | working |
| $u_{11}$ | male | secondary | town | employed under a contract |
| $u_{12}$ | male | secondary | city | full − time employed |

We are interested in the following sets:

1. $X_{unemployed} = \{u \in U : f_{dec}(employment, u) = unemployed\}$.
2. $X_{working} = \{u \in U : f_{dec}(employment, u) = working\}$.
3. $X_{contract} = \{u \in U : f_{dec}(employment, u) = employed\ under\ a\ contract\}$.
4. $X_{full-time} = \{u \in U : f_{dec}(employment, u) = full - time\ employed\}$.

Moreover, we assume that $B = A$.

If we consider a classic approach to rough sets, then:

1. $\underline{B}X_{unemployed} = \{u_9\}$, $\overline{B}X_{unemployed} = \{u_1, u_2, u_9\}$, therefore

$$\alpha_B(X_{unemployed}) = \frac{1}{3}.$$

2. $\underline{B}X_{working} = \{u_{10}\}$, $\overline{B}X_{working} = \{u_7, u_8, u_{10}\}$, therefore

$$\alpha_B(X_{working}) = \frac{1}{3}.$$

3. $\underline{B}X_{contract} = \{u_{11}\}$, $\overline{B}X_{contract} = \{u_3, u_4, u_{11}\}$, therefore

$$\alpha_B(X_{contract}) = \frac{1}{3}.$$

4. $\underline{B}X_{full-time} = \{u_{12}\}$, $\overline{B}X_{full-time} = \{u_5, u_6, u_{12}\}$, therefore

$$\alpha_B(X_{full-time}) = \frac{1}{3}.$$

It is easy to see that the accuracy of $B$-approximation of all the sets above is the same. From the point of view of the quantitative assessment of the accuracy of approximation, all situations are indiscernible.

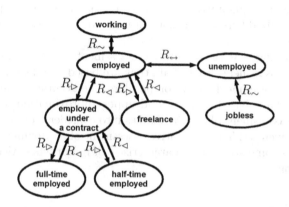

**Fig. 1.** An ontological graph $OG_{employment}$, representing semantic relations between concepts, associated with the attribute *employment*.

If we add information about semantics of values of the decision attribute (see the ontological graph $OG_{employment}$ in Fig. 1), then:

1. $X_{unemployed}$ is a pseudo rough set, because

$$Dec_{employment}(X^*_{unemployed}) = \{jobless\}$$

and

$$(unemployed, jobless) \in SMR_{employment}.$$

2. $X_{working}$ is a considerable rough set, because

$$Dec_{employment}(X^*_{working}) = \{jobless\}$$

and

$$(working, jobless) \in AMR_{employment}.$$

3. $X_{contract}$ is a marginally rough set, because

$$Dec_{employment}(X^*_{contract}) = \{freelance\}$$

and

$$(employed\ under\ a\ contract, freelance) \in MHprMR_{employment}.$$

4. $X_{full-time}$ is a restrainedly rough set, because

$$Dec_{employment}(X^*_{full-time}) = \{freelance\}$$

and

$$(full-time\ employed, freelance) \in FHprMR_{employment}.$$

It is necessary to note that the indiscernibility relation has been used for condition attributes to determine approximations of sets. Moreover $\tau_1 = 0$ and $\tau_2 = 1$. One can see that the qualitative assessment of the accuracy of approximation gives us significantly different situations in terms of roughness than the quantitative assessment.

It is worth noting that the qualitative assessment is not intended to replace the quantitative assessment. First of all, they are based on different assumptions and therefore they are not directly comparable. The accuracy of approximation recalled in Sect. 2.3 is based on lower and upper approximations of a given set. The qualitative assessment of accuracy takes into consideration only the boundary region of a given set. The qualitative assessment of approximation should be rather treated as some additional characteristic of rough sets enriching our look at approximations.

## 4    Conclusions

In the paper, we have shown that dealing with semantic relations enriches our look at approximations of sets defined in decision systems. The presented approach enables us to determine qualitatively the accuracy of approximation using a proper taxonomy. The proposed approach may be combined in fuzzy logic and soft computing methods based on linguistic variables. The main task in further work is to consider more sophisticated semantic relations (cf. [3,4]) between attribute values (concepts), to take into consideration ontological graphs associated with condition attributes, and to propose a wider taxonomy of assessment.

**Acknowledgments.** The author would like to thank the anonymous reviewers for their critical remarks and useful suggestions that greatly contributed to improving the final version of the paper.

# References

1. The Wikisaurus Homepage: http://en.wiktionary.org/wiki/Wiktionary: Wikisaurus
2. Brachman, R.: What IS-A is and isn't: an analysis of taxonomic links in semantic networks. Computer **16**(10), 30–36 (1983)
3. Chaffin, R., Herrmann, D.J.: The nature of semantic relations: a comparison of two approaches. In: Evens, M. (ed.) Relational Models of the Lexicon: Representing Knowledge in Semantic Networks, pp. 289–334. Cambridge University Press, New York (1988)
4. Cruse, D.: Lexical Semantics. Cambridge University Press, Cambridge (1986)
5. Greco, S., Matarazzo, B., Słowiński, R.: Rough sets theory for multicriteria decision analysis. Eur. J. Oper. Res. **129**(1), 1–47 (2001)
6. Gupta, M.M.: On fuzzy logic and cognitive computing: some perspectives. Scientia Iranica **18**(3), 590–592 (2011)
7. Ishizu, S., Gehrmann, A., Nagai, Y., Inukai, Y.: Rough ontology: extension of ontologies by rough sets. In: Smith, M.J., Salvendy, G. (eds.) HCII 2007. LNCS, vol. 4557, pp. 456–462. Springer, Heidelberg (2007)
8. Köhler, J., Philippi, S., Specht, M., Rüegg, A.: Ontology based text indexing and querying for the semantic web. Knowl.-Based Syst. **19**, 744–754 (2006)
9. Lukaszewski, T., Józefowska, J., Lawrynowicz, A.: Attribute value ontology - using semantics in data mining. In: Maciaszek, L.A., Cuzzocrea, A., Cordeiro, J. (eds.) Proceedings of the 14th International Conference on Enterprise Information Systems, pp. 329–334. Wroclaw, Poland (2012)
10. Midelfart, H., Komorowski, J.: A rough set framework for learning in a directed acyclic graph. In: Alpigini, J.J., Peters, J.F., Skowron, A., Zhong, N. (eds.) RSCTC 2002. LNCS (LNAI), vol. 2475, pp. 144–155. Springer, Heidelberg (2002)
11. Milstead, J.L.: Standards for relationships between subject indexing terms. In: Bean, C.A., Green, R. (eds.) Relationships in the organization of knowledge, pp. 53–66. Kluwer Academic Publishers, Dordrecht (2001)
12. Neches, R., Fikes, R., Finin, T., Gruber, T., Patil, R., Senator, T., Swartout, W.: Enabling technology for knowledge sharing. AI Mag. **12**(3), 36–56 (1991)
13. Pancerz, K.: Dominance-based rough set approach for decision systems over ontological graphs. In: Ganzha, M., Maciaszek, L., Paprzycki, M. (eds.) Proceedings of the FedCSIS 2012, pp. 323–330. Wroclaw, Poland (2012)
14. Pancerz, K.: Toward information systems over ontological graphs. In: Yao, J.T., Yang, Y., Słowiński, R., Greco, S., Li, H., Mitra, S., Polkowski, L. (eds.) RSCTC 2012. LNCS, vol. 7413, pp. 243–248. Springer, Heidelberg (2012)
15. Pancerz, K.: Semantic relationships and approximations of sets: an ontological graph based approach. In: Proceedings of the HSI 2013, pp. 62–69. Sopot, Poland (2013)
16. Pawlak, Z.: Rough Sets: Theoretical Aspects of Reasoning about Data. Kluwer Academic Publishers, Dordrecht (1991)
17. Pawlak, Z., Skowron, A.: Rudiments of rough sets. Inf. Sci. **177**, 3–27 (2007)
18. Storey, V.C.: Understanding semantic relationships. VLDB J. **2**, 455–488 (1993)
19. Winston, M.E., Chaffin, R., Herrmann, D.: A taxonomy of part-whole relations. Cogn. Sci. **11**(4), 417–444 (1987)
20. Zadeh, L.: The concept of a linguistic variable and its application to approximate reasoning - I. Inf. Sci. **8**(3), 199–249 (1975)

21. Zadeh, L.: Fuzzy logic = computing with words. IEEE Trans. Fuzzy Syst. **4**(2), 103–111 (1996)
22. Zadeh, L.: A new direction in AI: toward a computational theory of perceptions. AI Mag. **22**(1), 73–84 (2001)
23. Zadeh, L.A.: Outline of a new approach to the analysis of complex systems and decision processes. IEEE Trans. Syst. Man Cybern. SMC **3**(1), 28–44 (1973)

# Predicting the Presence of Serious Coronary Artery Disease Based on 24 Hour Holter ECG Monitoring

Jan G. Bazan[1], Sylwia Buregwa-Czuma[1(✉)], Przemysław Wiktor Pardel[1],
Stanisława Bazan-Socha[2], Barbara Sokołowska[2], and Sylwia Dziedzina[2]

[1] Institute of Computer Science, University of Rzeszów,
Pigonia 1, 35-959 Rzeszów, Poland
{bazan,sczuma,ppardel}@ur.edu.pl
[2] II Department of Internal Medicine, Jagiellonian University Medical College,
Skawińska 8 Street, 31-066 Kraków, Poland
mmsocha@cyf-kr.edu.pl, basiasok1@gmail.com, sylwiaz49@poczta.fm

**Abstract.** The purpose of this study was to evaluate the usefulness of classification methods in recognizing a cardiovascular pathology. Based on clinical and electrocardiographic (ECG) Holter data we propose a method for predicting a coronary stenosis demanding revascularization in patients with a diagnosis of a stable coronary heart disease. A possible solution of this problem has been set in a context of rough set theory and methods. The rough set theory introduced by Zdzisław Pawlak during the early 1980s provides a foundation for the construction of classifiers. From the rough set perspective, classifiers presented in the paper are based on a decision tree calculated on a basis of a local discretization method, related to the problem of reducts computation. We present a new modification of a tree building method which emphasizes the discernibility of objects belonging to decision classes indicated by human experts. The presented method may be used to assess the need for the coronary revascularization. The paper includes results of experiments that have been performed on medical data obtained from Second Department of Internal Medicine, Collegium Medicum, Jagiellonian University, Kraków, Poland.

**Keywords:** Rough sets · Discretization · Classifiers · Stable angina pectoris · Morbus ischaemicus cordis · ECG Holter

## 1 Introduction

Coronary heart disease (CHD) is a major health problem worldwide and is one of the leading causes of high mortality rates in industrialized countries (see [15]).

This work was partially supported by two grants of the Polish National Science Centre: DEC-2012/05/B/ST6/03215 and DEC-2013/09/B/ST6/01568, and also by the Centre for Innovation and Transfer of Natural Sciences and Engineering Knowledge of University of Rzeszów, Poland.

© Springer-Verlag Berlin Heidelberg 2015
J. Peters et al. (Eds.): TRS XIX, LNCS 8988, pp. 95–113, 2015.
DOI: 10.1007/978-3-662-47815-8_7

It is called angina, due to one of its main symptoms - chest pain, arising from ischemia of the heart muscle. The term angina pectoris derives from the Greek word *ankhoné* ("strangling") and the Latin *pectus* ("chest"), and can be translated as "a strangling feeling in the chest". Acute angina, called unstable refers to acute coronary syndrome (ACS), and when its course is chronic, it is called stable. The consequences depend largely on the number, degree and localization of artery stenosis. The current diagnostic standard of anatomic coronary vessel evaluation is an invasive angiography which permits the determination of the therapeutic plan and prognosis. In the case of unaltered coronary flow the pharmacological treatment is applied, otherwise revascularization is also needed. However the coronary angiography (coronarography) is a very sensitive method, it has its limitations. As an invasive investigation, it is relatively expensive, and it carries risks including a mortality rate of approximately 1 in 2000 (see [9]).

It would not be appropriate or practical to perform invasive investigations on all patients with a coronary heart disease diagnosis. Given the high incidence and prevalence of CHD, a non invasive test to reliably assess the coronary arteries would be clinically desirable.

We propose applying clinical data together with electrocardiographic (ECG) Holter recordings as prospective candidate data for coronary artery stenosis prediction. The proposed method helps to determine the management of patients with stable angina, including those needing coronary intervention, without performing invasive diagnostic procedure like angiography. It could also work as a screening tool for all patients with CHD.

The presented subject employs classifier building for temporal data sets, where a *classifier* is an algorithm which enables us to forecast repeatedly on the basis of accumulated knowledge in new situations (see [2] for more details). There are many suitable methods for classification: e.g. classical and modern statistical techniques, neural networks, decision trees, decision rules and inductive logic programming (see [2] for more details). Classifiers were constructed also for temporal data (see [2,11] for more details). In this paper, an approach for solving problems has been found in the context of rough set theory and its methods. Rough set theory introduced by Zdzisław Pawlak during the early 1980s provides the foundation for solving real life problems including medical decision making (see [23]), as well for the construction of temporal data set classifiers (see [2,4,5,19]).

We present a method of classifier construction that is based on features which aggregate time points (see [2]). The patients are characterized by parameters (sensors), measured in time points for a period, called *a time window*. In the ECG recording context, the exemplary parameters are the number of QRS complexes, ST interval elevations and depressions or the total power of the heart rate variability (HRV) spectrum. The aggregation of time points is performed by special functions called *temporal patterns* (see [2,20,25]), that are tools for numerical characterization of values from a selected sensor during the whole time window. Temporal patterns are constructed on the basis of the domain knowledge by human experts. Next, a collection of attributes is constructed on the basis of temporal patterns. Having such attributes, a classifier is constructed approximating a temporal concept. In studied subjects, the temporal concept means the presence of coronary stenosis. The classification is performed using a decision tree that is calculated on the basis of the local discretization (see [4, 18]).

To illustrate the method and to verify the effectiveness of presented classifiers, we have performed several experiments with the data sets obtained from Second Department of Internal Medicine, Collegium Medicum, Jagiellonian University, Kraków, Poland (see Sect. 5).

## 2   Medical Background

CHD refers to the narrowing (stenosis) or occlusion of the artery supplying blood to the heart, caused by plaque composed predominantly of cholesterol and fatty deposits within the vessel wall. The accumulation of a plaque is a process known as atherosclerosis developing slowly over many years. The presence of atherosclerotic plaque within the vessel wall, decreases the luminal cross-sectional area of the artery, reduces coronary flow and leads to imbalance between myocardial oxygen supply and consumption. Patients with stable angina are at risk of developing an acute coronary syndrome, including unstable angina and myocardial infarction (MI): non-ST-elevation MI or ST-elevation MI. Because of advancing age of populace and dissemination of risk factors, like: obesity, diabetes mellitus, hypertension and hypercholesterolemia, the incidence of angina pectoris is still increasing.

Patients suffering from stable coronary heart disease require the determination of appropriate treatment, soon after the diagnosis is made, in order to avoid irreversible heart damages. The therapy depends on several conditions, such as: coronary perfusion changes, heart valves failure and the heart's pumping function.

The diagnosis and assessment of angina involves a patient's medical history, physical examination, laboratory tests and specific cardiac investigations. Non-invasive investigations include a resting 12-lead ECG, ECG stress testing, resting two-dimensional and doppler echocardiography, and ECG Holter monitoring. Invasive techniques used in coronary anatomy assessment are: coronary arteriography and intravascular ultrasound. Using coronary arteriography the disease can be classified into one vessel, two vessel, three vessel, or left main coronary artery disease (LM CAD).

An electrocardiogram is a measurement of the electrical activity of the heart on the body's surface. Holter ECG monitoring is a continuous recording of the ECG, done over a period of 24 h or more. Several electrodes are placed on the patient's chest and connected by wires to the recorder. The patient goes about his or her usual daily activities.

Modern Holter devices record data onto digital flash memory devices. When the recording of the ECG signal is completed, the data is uploaded into a computer system which then automatically analyzes the input. Commercial Holter software carries out an integrated automatic analysis process which counts i.a.

ECG complexes, calculates summary statistics such as average, minimum and maximum heart rate and determines different kinds of heart beats and rhythms. It provides information about heart beat morphology, interval measurements, HRV and rhythm overview.

For many years, ECG analysis was adjusted and succeeded in methods for determining plentiful signal features in the background for the establishment of a diagnosis. The physicians had to gain experience in using these properties in diagnosing. Difficulties arise when there is a need to predict the necessity of revascularization. No distinct expert knowledge is available in this field, and the predictions based on non-invasive tests are conjectural. Due to the invasive character of revascularization which can exposure the patient to danger and is a costly procedure, it would be beneficial to know whether it's required in advance.

There are numerous systems analyzing ECG recordings. In our study the data was acquired using Aspel's HolCARD 24W application. Computer systems enable the processing and aggregation of data by means of existing signal analyzing methods. Based on the raw signal (voltage levels) the parameters are automatically calculated in a given period of time, i.e. maximum heart rate, or the number of ventricular tachycardia (low-level aggregation). Any such period of time corresponds to one timepoint in our analysis. Assuming a 24-h Holter recording and hourly period of time, there are 24 time points for which the parameter values are calculated (e.g. 24 maximum heart rate values). The parameter value of all windows is aggregated to one value with temporal patterns (e.g. the average of the maximum heart rate values). We took advantage of aggregated data to predict coronary angiography outcomes in regard to the necessity of revascularization. The data still remain temporal regardless data aggregation to points representing, e.g. an hour of ECG recording. The study answers the question whether the application of such inputs will be utilizable.

## 2.1   Coronary Arteriography

Coronary arteriography is a diagnostic invasive procedure that requires the percutaneous insertion of a catheter into the vessels and heart. The catheter is introduced into the body through a vein or an artery, and its progress is monitored by x-ray. Injected dye (contrast medium, CM or IV dye) enables the evaluation of heart valves functioning, coronary blood flow and allows for the identification of the presence, localization and degree of stenosis in the coronary arteries.

Coronary arteriography is considered a relatively safe procedure, but in some patients complications arise. Most of them are minor, resulting in no long-term consequences and include nausea, vomiting, allergic skin rashes and mild arrhythmias. In patients with kidney dysfunction, the supply of an excessive quantity of CM may worsen kidney functioning. There may be bleeding at the catheter insertion site, which might develop into swelling. Less frequently, angiography may be associated with more serious complications. These include blood vessel damages, formation of blood clots, infections, arrhythmias, MI (myocardial infarction) or a stroke. The risk of major complications associated with coronary angiography is determined to be up to 2 % (see [1]). Fatalities are extremely rare and may be caused by a perforation of the heart or surrounding vessels, arrhythmias, a heart attack or a severe allergic reaction to CM (see [14]).

Application of contrast medium during angiography additionally exposes patients to the adverse effects of supervention. Reactions to IV dye are relatively common, occurring in 1 to 12 % of patients (see [8]). Most of these reactions are mild, and include a feeling of warmth, nausea and vomiting. Generally, these symptoms last for only a short period of time and do not require treatment. Moderate reactions, including severe vomiting, urticaria and swelling, occur in 1 % of people receiving CM and frequently require treatment. Severe, life-threatening reactions, including anaphylaxis, are reported in 0,03 to 0,16 % of patients, with an expected death rate of one to three per 100 000 contrast administrations (see [10]).

Diagnostic cardiac angiography plays an important role in the evaluation of patients with coronary heart disease. It is used to assess the presence and degree of coronary artery stenosis, heart valve and muscle dysfunction. In some cases the catheterization cannot be carried out. These cases involve health centers with limited access to diagnostic procedures or tight budget and patients with allergy to CM and other contraindications to angiography.

There are no routine noninvasive diagnostic procedures to assess coronary flow disturbances and when there is no opportunity to perform coronary angiography, alternative solutions to the problem are needed. The application of the proposed methods could select potential candidates for myocardial revascularization. We suggest using clinical data, derived from a patient's history, and laboratory test outcomes together with ECG Holter recordings as prospective candidate data for coronary artery stenosis prediction.

## 2.2   Management of Angina

Once the diagnosis of CHD is made, it is important to define the treatment and the need for revascularization. The aim of CHD treatment is to prevent myocardial infarction and death by reducing the incidence of acute thrombotic events and the development of ventricular dysfunction. Main therapeutic methods are: lifestyle changes, pharmacotherapy and revascularization. Particular attention should be paid to the lifestyle (physical activity, smoking, dietary habits) which may influence prognosis. Pharmacological treatment should reduce plaque progression, stabilize plaque by reducing inflammation and by preventing thrombosis when endothelial failure or plaque rupture occurs. These all reduce the severity and frequency of symptoms and improve the prognosis leading to quality of life improvement. With appropriate management, the symptoms usually can be controlled and the prognosis improved.

Coronary arteriography in conjunction with a cardiovascular examination can appropriately select patients for coronary revascularization which means the restoration of the blood supply to ischemic myocardium. Modes of revascularization include thrombolysis with drugs, percutaneous coronary intervention (PCI) mainly by way of angioplasty, and coronary artery bypass grafting (CABG). PCI restores blood flow, usually with a balloon, inserted by a catheter through the peripheral artery, with or without stent placement. CABG refers to an "open heart" surgery where a peripheral vein is used to bypass the occlusion in the coronary artery.

Myocardial revascularization procedures require diagnosis which should indicate the localization, extent and severity of the disease, the presence and significance of the collateral circulation and the status of the left ventricular myocardium. For many years, the evaluation of the extent and severity of coronary artery disease has been mainly anatomical, carried out by a coronary angiography. However, this technique has methodological limitations and the interobserver variability is considerable. Intravascular ultrasounds (IVUS) have an indisputable advantage in determining lesion characteristics. But the only noninvasive technique that allows for quantitative assessment is a positron-emission tomography (PET), but it is highly complex and expensive, so its use is strictly limited.

## 3   Prediction of Coronary Atherosclerosis Presence

Forecasting coronary stenosis in patients without performing an angiography requires classifier construction, which on the basis of available knowledge assigns objects (patients) to defined decision classes. Considered decision classes are: *patients with unaltered arteries who do not need invasive treatment* (decision class: *NO*) and *patients with coronary stenosis who may need revascularization* (decision class: *YES*). Classification thus permits decision making about coronary stenosis and therapy management.

### 3.1   Temporal Concepts

The problem of forecasting coronary stenosis presence can be treated as an example of a concept approximation problem, where the term *concept* means *mental picture of a group of objects*. Such problems can often be modeled by systems of complex objects and their parts changing and interacting over time. The objects are usually linked by some dependencies, sometimes they can cooperate between themselves and are able to perform flexible complex autonomous actions (operations, changes). Such systems are identified as *complex dynamical systems* or *autonomous multiagent systems* (see [2] for more details). For example, in the problem of coronary stenosis prediction, a given patient can be treated as an investigated complex dynamical system, whilst diseases of this patient are treated as complex objects changing and interacting over time.

The concepts and methods of their approximation are usually useful tools for an efficient monitoring of a complex dynamic system (see [2]). Any concept can be understood as a way to represent some features of complex objects. An approximation of such concepts can be made using parameters (sensor values) registered for a given set of complex objects. However, a perception of composite features of complex objects requires observation of objects over a period called a *time window*. For construction of the features *temporal patterns* are used. In this paper, we consider temporal patterns as a numerical characterization of values of selected sensors from a time window (e.g., the minimal, maximal or mean value of a selected sensor, the initial and final values of a selected sensor, the deviation of selected sensor values).

One can see that any temporal pattern is determined directly by the values of some sensors. For example, in the case of coronary disease one can consider temporal patters such as minimal heart rate and estimated QT dispersion within a time window. We assume that any temporal pattern should to be defined by a human expert using domain knowledge accumulated for the given complex dynamical system.

The temporal patterns can be apply for defining new features that can be used to approximate more complex concepts, that we call *temporal concepts*. We assume that temporal concepts are specified by a human expert. Temporal concepts are usually used in queries about the status of some objects in a particular temporal window. Answers to such queries can be of the form of *Yes, No* or *Does not concern*. For example, in the case of the main problem in this paper we define a complex concept by using the following query: "Was the stenosis of a coronary artery detected for a given patient?".

## 3.2   Temporal Pattern Table

The approximation of temporal concepts can be defined by classifiers, which are usually constructed on the basis of decision tables. Hence, if we want to apply classifiers for the approximation of temporal concepts, we have to construct a suitable decision table called a *temporal pattern table* (PT) (see Fig. 1).

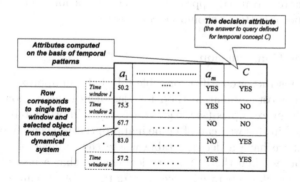

**Fig. 1.** The scheme of the temporal pattern table (PT)

A temporal pattern table is constructed from a table T consisting of registered information about objects (patients) occurring in a complex dynamical system. Any row of table T represents information about the parameters of a single object registered in a time window (see Fig. 2).

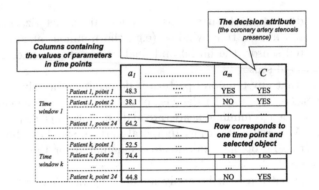

**Fig. 2.** The scheme of the table T

Such a table can be treated as a data set accumulated from the observations of the behavior of a complex dynamical system. Assume, for example, that we want to approximate temporal concept C using table (data set) T. Initially, we construct a temporal pattern table PT as follows.

- Construct table PT with the same objects that are contained in table T.
- Any condition attribute of table PT is computed using temporal patterns defined by a human expert for the approximation of concept C.
- Values of the decision attribute (the characteristic function of concept C) are proposed by the human expert.

We assume that any temporal pattern is given by a formula defined by an expert. In a more advanced approach, the classifiers for condition attributes related to temporal patterns should be constructed.

### 3.3   Classifier Construction

Next, we can construct a classifier for table PT that can approximate temporal concept C. The most popular method for classifiers construction is based on learning rules from examples (see, e.g., [2,4,5,19]). Unfortunately, the decision rules constructed in this way can often be inappropriate to classify unseen cases. For instance, if we have a decision table where the number of values is high for some attributes, then there is a very low chance that a new object will be recognized by the rules generated directly from this table, because the attribute value vector of a new object will not match any of these rules. Therefore, some discretization strategies are built for decision tables with such numeric attributes to obtain a higher quality classifiers. This problem is intensively studied and we consider discretization methods developed by Hung S. Nguyen (see [4,18] for more details). These methods are based on rough set techniques and boolean reasoning.

In this paper we use the local strategy of discretization (see [4]). One of the most important notion of this strategy is the notion of *a cut*. Formally, the cut is a pair $(a, c)$ defined for a given *decision table* $\mathbf{A} = (U, A \cup \{d\})$ in Pawlak's sense (see [19]), where $a \in A$ ($A$ is a set of attributes or columns in the data set) and $c$, defines a partition of $V_a$ into *left-hand-side* and *right-hand-side interval* ($V_a$ is a set of values of the attribute $a$) (see Fig. 3).

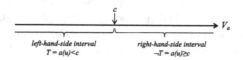

**Fig. 3.** The discretization of attribute $a \in A$ defined by the cut $(a, c)$

In other words, any cut $(a, c)$ is associated with a new binary attribute (feature) $f_{(a,c)} : U \to \{0, 1\}$ such that for any $u \in U$:

$$f_{(a,c)}(u) = \begin{cases} 0 & \text{if } a(u) < c \\ 1 & \text{otherwise} \end{cases} \tag{1}$$

Moreover, any cut $(a, c)$ defines two templates, where a template is understood as a description of some set of objects. The first template defined by a cut $(a, c)$ is a formula $T = (a(u) < c)$, while the second pattern defined by a cut $(a, c)$ is a formula $\neg T = (a(u) \geq c)$.

By means of the chosen attribute and its value the data set is divided into two groups of objects (e.g. patients), matching to both templates mentioned above for a given cut. For instance, for a numerical attribute $a$ (with plentiful organized values), the partition of the patients may be performed using the value $v$ of a given attribute, in a manner that patients with values of the attribute $a$ greater or equal to $v$ belong to one group, and the another group consists of patients whose values of the attribute are less then $v$. Let's notice that the partition of the object set may also take place using a symbolic attribute (non numerical, with a modest amount of values). For example, for the attribute $b$ with symbolic values the split may be performed using some value $v$ in that way, that patients whose value of the attribute $b$ is equal to $v$ belong to one group, and the patients with the value of the attribute $b$ different from $v$ to the another one. The way of the selection of an attribute and its value applying for the partition, is a key element of the discussed method for local discretization tree building and should be related to the analysis of the decision attribute values for training objects. The best cut in the sense of some measure is searched for.

The quality of a given cut can be computed as a number of object pairs discerned by this cut and belonging to different decision classes. For instance, when a given cut $c$ divides the objects into two groups of $M$ and $N$ size and the number of objects with $C_0$ and $C_1$ class equals $M_0$ and $M_1$ in one group, and the other group contains $N_0$ and $N_1$ objects of $C_0$ and $C_1$ class respectively, then the number of object pairs discriminated by the cut $c$ amounts to:

$$M_0 N_1 + M_1 N_0 \tag{2}$$

If we compute the value of this measure for all potential pairs (attribute, value), then we can greedily choose one of pairs and divide the whole data set into two parts based on it. The partition of $\mathbf{A}$ can be done by $n$ operations, if the values are sorted. Due to the sorting operation, the algorithm runs in time $O(n \cdot log\ n)$, where $n$ is the number of objects.

Such an approach was used in [3] and the classifier constructed through use of this method, will be called here as the *RSH-classic* classifier. However, in this paper a new method of a cut quality computation is introduced. It is based on special weights obtained for pairs of patient on the basic of domain knowledge (see Sect. 4). This method allowed to significantly improve the results of our experiments relatively to the results from [3] (see Sect. 5).

The quality of cuts may be computed for any subset of a given set of objects. Accordingly, a root of a tree contains the entire set of objects. In the local strategy of discretization, after finding the best cut and dividing the object set into two subsets of objects, this procedure is repeated for each object set separately until some stop condition holds. Using repeated divisions of a given data set the binary tree is constructed. The stopping criterion for a division is constructed in a manner that a given part is not split (becomes a leaf) when it contains the objects with single decision class (alternatively the objects of a given class constitute specified percent which is treated as a parameter of the method) or further partitions do not yield any results (all potential cuts do not discern the pairs of the objects with distinct classes any more).

In this paper, we assume that the division stops when all objects from the current set of objects belong to the same decision class. Hence, the local strategy can be realized by using a *decision tree* (see Fig. 4).

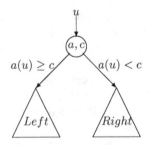

**Fig. 4.** The decision tree used in local discretization

The decision tree computed during local discretization can be treated as a classifier for the concept $C$ represented by a decision attribute from a given decision table $\mathbf{A}$. Let $u$ be a new object and $\mathbf{A}(T)$ be a subtable containing all objects matching the template $T$ defined by the cut from the current node of a given decision tree (at the beginning of algorithm work $T$ is the template defined by the cut from the root). We classify object $u$ starting from the root of the tree as follows:

---

**Algorithm 1.** Classification by decision tree (see [4])

---

**Input**: Table **A**, template **T**, object $u$
**Output**: A decision value for $u$

*Step 1:*
**if** $u$ *matches template* $T$ *found for* **A**
**then**
| go to subtree related to $\mathbf{A}(T)$
**else**
| go to subtree related to $\mathbf{A}(\neg T)$
**end**
*Step 2:*
**if** $u$ *is at the leaf of the tree*
**then**
| go to 3
**else**
| repeat 1-2 substituting $\mathbf{A}(T)$
**end**
*Step 3:*
Classify $u$ using decision value attached to the leaf.

---

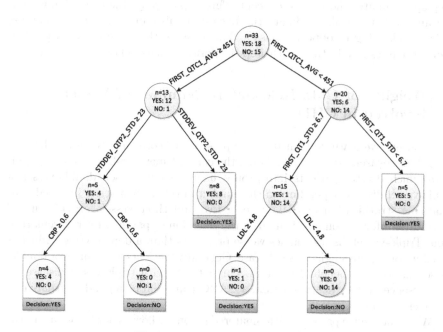

**Fig. 5.** The decision tree in CHD

Figure 5 presents a decision tree computed for the problem of forecasting the presence of coronary stenosis on the basis of a medical data set (see Sect. 5). The tree was generated based on all of the objects for illustrative purpose.

A sample application of the tree would be the classification of real life objects. For example, for a patient with the average of QTc interval in first time window ($FIRST\_QTC1\_AVG$) equal to 380 ms and QT standard deviation in first time window ($FIRST_QT1_STD$) equal 5.8, we flow from the root of the tree, down to the right subtree, as the patient suits a template $FIRST\_QTC1\_AVG < 451$. Then, in the next step we tread again along the right tree, which consists of one node, called a leaf, where we stop. The fitting path indicates that the coronary arteries of that patient are narrowed by atherosclerosis.

Our findings have clinical relevance. Generated cuts concern the QT interval which represents the duration of ventricular depolarization and subsequent repolarization. Prolongation of the QT interval reflects a delay in cardiac repolarization and is associated with the increased risk of potentially fatal ventricular arrhythmias. Because of the QT inverse relationship to heart rate, it is routinely corrected by means of various formulae to a less heart rate dependent value known as the QTc interval. The duration of QTc interval should normally be up to 450 ms for women, and 430 ms for men (see [22]).

The proposed method can predict coronary stenosis using the multi-step approach. In the first step, the ECG Holter data are preprocessed and consolidated with clinical data. The second step generates temporal patterns chosen by an expert. Using a decision tree with the local strategy of discretization, the third step approximates the studied concept. This step employs the transition weighting approach to locally estimate the biggest number of differentiated pairs of objects belonging to opposite classes. This ensures that received cuts produce more stable results. In the fourth step the performance of classification is tested.

## 4    Weights of Cuts Relevant to Number of Vessels Involved in CHD

The concept we learn is defined as the presence of coronary stenosis indicating a need for revascularization. We based the concept membership on angiographic data which divides patients into groups having single-vessel, double-vessel or triple-vessel disease depending on the number of coronary arteries involved. The anatomic stratification of CHD to one, two and three vessels provides useful prognostic information and is used in the selection of patients for revascularization. Triple-vessel disease carries worse prospects than a double-vessel, which is usually worse than single-vessel disease. Generally, patients with single or double-vessel disease can benefit from PCI. For patients with triple-vessel disease, or the presence of poor heart function, a CABG can often be a good alternative or a better treatment option.

We acknowledge patients with unaltered coronary flow as not belonging to the concept, while patients with one, two or three vessel disease as concept positive examples. However, this criterion is too simple and not limitless. Dissociation often exists between the coronarography and clinical manifestations. The group with one, two and three vessel disease is inhomogeneous. It distinguishes patients based only on the number of vessels, and treats a multidimensional problem as a one-dimensional. Some physiological facts are fundamental for correct interpretation of the study results. On the one hand in patients with angiographically confirmed stenosis, the collateral circulation and coronary self-regulation may compensate for the blood pressure decrease caused by stenosis in order to maintain constant coronary flow. In coronarography, the role of collateral circulation can be underestimated.

Some previous studies revealed that the ECG's may appear inconsistent with stenosis disclosed by arteriography. Sometimes the angiographic picture appears much more alarming than what is anticipated from the ECG. In patients with complete occlusion the resting ECG may be completely normal [12]. It was especially evident in the case of right coronary artery when severe luminal reduction, or even complete occlusions, was often accompanied by normal or only slightly abnormal ECG. In certain rare cases it may be difficult to demonstrate and evaluate the presence of coronary stenosis, despite a technically satisfactory arteriogam (e.g. [7]). On the other hand, some cases of myocardial infarction with normal coronary arteries are reported (e.g. [6]). Many theories have been proposed explaining these inconsistencies, such as syndrome X, microvascular angina, and non-atherosclerotic myocardial ischemia. The presence of muscular bridges may also bring some uncertainty although in these cases during the diastolic phases, the artery appears normal.

Several studies reported that ECG changes were not good indicators of coronary artery involvement with 51.5 % sensitivity in correctly detecting significant stenosis [16]. All things considered, the ECG data is still attractive as a prospective candidate for predicting CAD because it is noninvasive, easy to use, a relatively inexpensive tool offering safety and patient convenience.

Given the limited ability of ECG to distinguish patients correctly, we made an attempt to emphasize the differences between groups using weights for cuts discerning pairs of objects. It is a way to make the border between positive and negative examples more exposed. We are interested in distinguishing patients with normal and narrowed (one, two or three) vessels, that's why we put the biggest weights between negative examples of the concept and the remainder. Class differences are the most subtle between patients with normal arteries and with only one vessel changed, compared with the divergence between normal and tree-vessel disease. The cut discriminating patients with normal and one vessel disease is assigned the biggest weight. Thereupon we propose weights as presented in Table 1, where the first row and the first column signify the number of affected arteries. Increasing the values of weights, while maintaining assigned relations between weights, no longer improves the quality of classification.

A classifier constructed by using the method based on the weights from Table 1 will be refered to from here as the *RSH-weights* classifier. It is easy to see, that the method of cuts weights computation used in the *RSH-classic* classifier (see Sect. 3.3) can be treated as a special case of the *RSH-weights* method, that is, a case when there are only two values in the Table 1 (values: 0 or 1).

In the *RSH-classic* approach, the quality of a cut is a sum of the pairs of objects belonging to opposite decision classes, whereas in the *RSH-weights*

**Table 1.** Diversifying weights relevant to number of vessels involved in CHD

|   | 0 | 1 | 2 | 3 |
|---|---|---|---|---|
| 0 | 0 | 3 | 2 | 1 |
| 1 | 3 | 0 | 1 | 1 |
| 2 | 2 | 1 | 0 | 1 |
| 3 | 1 | 1 | 1 | 0 |

method it is a summation of the weights of each pair of objects belonging to different decision classes, with 1, 2 or 3 vessel disease distinction. For example, when a given cut $c$ divides the set of objects into two subsets of $M$ and $N$ size and the number of objects with normal arteries (class $NO$) and with one, two and three vessel disease (class $YES$) equals respectively $M_0$, $M_1$, $M_2$ and $M_3$ in one group, and the other group contains $N_0$, $N_1$, $N_2$ and $N_3$ objects of class $NO$ and class $YES$ with one, two and three vessel disease respectively, then the quality of the cut is equal to:

$$\sum_{i=0}^{3} \sum_{j=0}^{3} w_{ij} M_i N_j \tag{3}$$

where $w_{ij}$ for $i = 0, 1, 2, 3, j = 0, 1, 2, 3$ means the weight of a pair of objects belonging to different decision classes with 0, 1, 2 or 3 changed vessels.

## 5    Experiments and Results

To verify the effectiveness of classifiers based on temporal patterns, we have implemented the algorithms from the library RSH-lib, which is an extension of the RSES-lib library forming the kernel of the RSES system [5].

The experiments have been performed using the medical data set obtained from Second Department of Internal Medicine, Collegium Medicum, Jagiellonian University, Kraków, Poland. The data was collected between 2006 and 2009. Two part 48-h Holter ECG recordings were performed using Aspel's HolCARD 24W system. There was a coronary angiography after the first part of the Holter ECG (after first 24-h recording). In the paper, we reported the results of the experiments performed for the first part of Holter ECG recordings. The data set includes a detailed description of clinical status (age, sex, diagnosis), coexistent diseases, pharmacological management, the laboratory tests outcomes (level of cholesterol, troponin I, LDL - low density lipoproteins) and various Holter-based indices such as: ST interval deviations, HRV, arrythmias or QT dispersion. Moreover, for Holter-based indices a data aggregation was performed resulting in points describing one hour of recording. Our group of 33 patients with normal rhythm, underwent coronary angiography and 24.2 % of them required additional angioplasty, whereas 24.2 % were qualified for CAGB.

The acquired data of patients was recorded as binary files. In the first step we verified the completeness of patient data (dates of tests, results, etc.) and a database of patients and medical data was created in the Infobright Community Edition (ICE) environment (see [13]). ICE is an open source software solution designed to deliver a scalable data warehouse, optimized for analytic queries (data volumes up to 50 TB, market-leading data compression, from 10:1 to over 40:1).

Data files for each test were imported into the database using an importer created in Java environment. In this process the text data was converted to the corresponding data formats (reviewing the various parameters we defined formats allowing its efficient storage in the database) to allow for storage of patient data without loss of information (such as float, integer, long). After internal preprocessing in the ICE environment (e.g., a data aggregation of Holter-based indices as was mentioned above) for further processing data have been imported into Java environment.

In the data analyzed, there was no inconsistency in the knowledge (a large number of values for continuous attributes), but in the case of its occurrence, the table should be brought to the consistency, using for example, generalized decision attribute.

The aim of the conducted experiments was to check the effectiveness of the algorithm described in this paper in order to predict the stenosis in coronary arteries. Here we present the experimental results of presented method. For testing quality of classifiers we applied the leave-one-out (LOO) technique, that is usually employed when the size of a given data set is small. The LOO technique involves a single object from the original data set as the validation data, and the remaining observations as the training data. This is repeated such that each observation in the sample is used once as the validation data. The LOO provides a repeatable objects selection and uses the knowledge of all remaining objects in a learning process, and thus is comparable with other classification methods. Knowing however it is not free from disadvantages, we plan to use a bootstrap method, which allows to assess the statistical significance of the results.

As a measure of classification success (or failure) we used the following parameters which are well known from the literature: the accuracy, the coverage, the accuracy for positive examples (Sensitivity, SN), the coverage for positive examples, the precision for positive examples (Positive Predictive Value, PPV), the accuracy for negative examples (Specificity, SP), the coverage for negative examples and the precision for negative examples, also called Negative Predictive Value, NPV (see, e.g., [2]).

The number of objects correctly and incorrectly classified is contained in Table 2. Table 3 shows the results of applying the considered algorithm (*RSH-weights* classifier) for the concept related to the presence of coronary artery stenosis in patients with stable angina.

The method correctly identifies 94.4 % of all patients with stenosis (SN) and 80 % of those who did not have stenosis (SP). With PPV value equal 85 %, a positive screen test is good at confirming coronary stenosis, however a negative

**Table 2.** The confusion matrix

|  |  | Predicted | |
|---|---|---|---|
|  |  | No | Yes |
|  | No | 12 | 3 |
| Actual | Yes | 1 | 17 |

**Table 3.** Results of experiments for coronary stenosis in CHD

| Decision class | Accuracy | Precision |
|---|---|---|
| Yes | 0.944 | 0.85 |
| No | 0.8 | 0.923 |
| All classes (Yes + No) | 0.879 | - |

result is also good as a screening tool at affirming that a patient does not have stenosis (NPV equal 92.3 %).

During LOO procedure, the most of generated trees revealed the same topology as final decision tree preserving siblings and ancestors order. The topology of the rest of trees was similar, that is, there were some differences in case of attribute values in the tree nodes and sometimes in attributes at the lower levels of generated trees.

In Table 4 we gave the results of the experiments in applying other classification methods to our data (the same set of input data was used in all cases). Those methods were developed in the following systems well known from the literature: WEKA [24], RSES [5], ROSE2 [21] (we used an early implementation of ModLEM algorithm [17] that is available in ROSE2), and our previous approach called *RSH-classic* classifier (see [3]). The coverage of all tested methods was equal 1.0 (every object was classified).

It should be mentioned that the results for WEKA and ROSE2 were generated using a set of standard parameters.

Experimental results showed that the presented method of stenosis prediction in coronary arteries gives good results and the results are comparable with results of other systems. Interestingly, the tests classified 94.4 % of patients with narrowed vessels as stenosis presence, which is an expected property of the method, meaning that the number of false negatives should be the lowest.

## 5.1   Limitations of the Study

The main limitation of the study was the size of the study population. However, our results can be applicable for a similar patient population. Another limitation of our study was using only the number of affected vessels as a determinant of CHD severity, as used in most of the studies in the literature. It is known that resting ECG has a limited value in determining coronary artery lesion characteristics. Despite the low computational complexity, the current implementation

**Table 4.** Comparison results of alternative classification systems

| Method | Accuracy | | | Precision | |
|---|---|---|---|---|---|
| | All classes | Yes | No | Yes | No |
| C4.5 (WEKA) | 0.545 | 0.611 | 0.467 | 0.579 | 0.500 |
| NaiveBayes (WEKA) | 0.394 | 0.611 | 0.133 | 0.458 | 0.222 |
| SVM (WEKA) | 0.545 | 0.611 | 0.467 | 0.579 | 0.500 |
| k-NN (WEKA) | 0.667 | 0.833 | 0.467 | 0.652 | 0.700 |
| RandomForest (WEKA) | 0.515 | 0.722 | 0.267 | 0.542 | 0.444 |
| Multilayer Perceptron (WEKA) | 0.548 | 0.611 | 0.467 | 0.579 | 0.500 |
| Global discretization + all rules (RSES) | 0.667 | 0.611 | 0.733 | 0.733 | 0.611 |
| Local discretization + all rules (RSES) | 0.758 | 0.778 | 0.733 | 0.778 | 0.733 |
| ModLEM (ROSE2) | 0.576 | 0.556 | 0.600 | 0.625 | 0.529 |
| RSH-classic | 0.758 | 0.778 | 0.733 | 0.778 | 0.733 |
| RSH-weights | 0.879 | 0.944 | 0.8 | 0.85 | 0.923 |

of presented methods may need to improve scalability. For very large input data we propose a distributed computing using the apportionment of objects or attributes.

# 6 Conclusion

In the present work, clinical and ECG data were used to build a predictive model for the diagnosis of CHD. We believe that the method can be very useful to clinicians in managing patients with CHD. Patients with positive tests may be strongly considered for revascularization, even if other tests results indicate a moderate or weak risk range. For negative tests, the clinician may observe the patient continuing pharmacotherapy.

The most attractive aspect of the method is that it can be employed with easy to obtain clinical, laboratory, and electrocardiographic parameters. Estimating the coronary anatomy before angiography could be useful when deciding on diagnostic and therapeutic interventions.

The number of studies dealing with the relationship between electrocardiogram and the severity of CHD are limited. To the best of our knowledge, no study investigating the relationship between the presence of coronary artery stenosis and ECG Holter monitoring has been demonstrated until now. The proposed work is very important for medical practitioners who treat patients with CHD in every day practice. The prediction of coronary arteries stenosis may help in better and tailored CHD management and treatment.

Further investigation is needed to assess whether proposed method leads to a meaningful change in clinical outcomes and may be used as a more routine, screening test for stenosis prediction. To that end we submitted an application to

The National Science Centre in Poland for the research grant, which could enable the investigations on more numerous study population. Moreover, it would be of great benefit for patients, to develop a method predicting the precise number of narrowed arteries.

# References

1. ACC/AHA Guidelines for Coronary Angiography: Executive Summary and Recommendations. A Report of the American College of Cardiology/American Heart Association Task Force on Practice Guidelines. Circulation **99**, 2345–2357 (1999)
2. Bazan, J.G.: Hierarchical classifiers for complex spatio-temporal concepts. In: Peters, J.F., Skowron, A., Rybiński, H. (eds.) Transactions on Rough Sets IX. LNCS, vol. 5390, pp. 474–750. Springer, Heidelberg (2008)
3. Bazan, J.G., Bazan-Socha, S., Buregwa-Czuma, S., Pardel, P.W., Sokolowska, B.: Prediction of coronary arteriosclerosis in stable coronary heart disease. In: Greco, S., Bouchon-Meunier, B., Coletti, G., Fedrizzi, M., Matarazzo, B., Yager, R.R. (eds.) IPMU 2012, Part II. CCIS, vol. 298, pp. 550–559. Springer, Heidelberg (2012)
4. Bazan, J.G., Nguyen, H.S., Nguyen, S.H., Synak, P., Wróblewski, J.: Rough set algorithms in classification problems. In: Polkowski, L., Lin, T.Y., Tsumoto, S. (eds.) Rough Set Methods and Applications. STUDFUZZ, vol. 56, pp. 49–88. Springer, Heidelberg (2000)
5. Bazan, J.G., Szczuka, M.: The rough set exploration system. In: Peters, J.F., Skowron, A. (eds.) Transactions on Rough Sets III. LNCS, vol. 3400, pp. 37–56. Springer, Heidelberg (2005)
6. Bossarte, R.M., Brown, M.J., Jones, R.L.: Myocardial infarction with normal coronary arteries: a role for MRI? Clin. Chem. **5**, 995–996 (2007)
7. Bugiardini, R., Badimon, L., Collins, P., Erbel, R., Fox, K., Hamm, C., Pinto, F., Rosengren, A., Stefanadis, C., Wallentin, L., Van de Werf, F.: Angina, "normal" coronary angiography, and vascular dysfunction: risk assessment strategies. PLoS Med. **4**, 252–255 (2007)
8. Canter, L.M.: Anaphylactoid reactions to radiocontrast media. Allergy Asthma Proc. **26**, 199–203 (2005)
9. Chaubey, S., Davies, S.W., Moat, N.: Invasive investigations and revascularisation. Br. Med. Bull. **59**, 45–53 (2001)
10. Cochran, S.T.: Anaphylactoid reactions to radiocontrast media. Curr. Allergy Asthma Rep. **5**, 28–31 (2005)
11. Douzal-Chouakria, A., Amblard, C.: Classification trees for time series. Pattern Recogn. **45**(3), 1076–1091 (2011)
12. Gensini, G.G., Buonanno, C.: Coronary arteriography: a study of 100 cases with angiographically proved coronary artery disease. Dis. Chest. **54**, 90–99 (1968)
13. The Infobright Community Edition (ICE). http://www.infobright.org/
14. Lange, R.A., Hillis, L.D.: Diagnostic cardiac catheterization. Circulation **107**, e111–e113 (2003)
15. Mackay, J., Mensah, G.A.: The Atlas of Heart Disease and Stroke. World Health Organization, Geneva (2004)
16. Mahmoodzadeh, S., Moazenzadeh, M., Rashidinejad, H., Sheikhvatan, M.: Diagnostic performance of electrocardiography in the assessment of significant coronary artery disease and its anatomical size in comparison with coronary angiography. J. Res. Med. Sci. **16**(6), 750–755 (2011)

17. Napierała, K., Stefanowski, J.: Argument based generalization of MODLEM rule induction algorithm. In: Szczuka, M., Kryszkiewicz, M., Ramanna, S., Jensen, R., Hu, Q. (eds.) RSCTC 2010. LNCS, vol. 6086, pp. 138–147. Springer, Heidelberg (2010)

18. Nguyen, H.S.: Approximate boolean reasoning: foundations and applications in data mining. In: Peters, J.F., Skowron, A. (eds.) Transactions on Rough Sets V. LNCS, vol. 4100, pp. 334–506. Springer, Heidelberg (2006)

19. Pawlak, Z., Skowron, A.: Rudiments of rough sets. Inf. Sci. **177**, 3–27 (2007)

20. Peters, J.F.: Classification of objects by means of features. In: Fogel, D., Mendel, J., Yao, X., Omori, T. (eds.) Proceedings of the 1st IEEE Symposium on Foundations of Computational Intelligence (FOCI 2007), Honolulu, Hawaii, pp. 1–8 (2007)

21. The Rough Sets Data Explorer (ROSE2). http://idss.cs.put.poznan.pl/site/rose.html

22. Straus, S.M.J.M., Kors, J.A., De Bruin, M.L., van der Hooft, C.S., Hofman, A., Heeringa, J., Deckers, J.W., Kingma, J.H., Sturkenboom, M.C.J.M., Stricker, B.H.Ch., Witteman, J.C.M.: Prolonged QTc interval and risk of sudden cardiac death in a population of older adults. J. Am. Coll. Cardiol. **47**(2), 362–367 (2006)

23. Paszek, P., Wakulicz-Deja, A.: Applying rough set theory to medical diagnosing. In: Kryszkiewicz, M., Peters, J.F., Rybiński, H., Skowron, A. (eds.) RSEISP 2007. LNCS (LNAI), vol. 4585, pp. 427–435. Springer, Heidelberg (2007)

24. The Weka 3 - Data Mining Software in Java (WEKA). http://www.cs.waikato.ac.nz/ml/weka/

25. Wolski, M.: Perception and classification. A note on near sets and rough sets. Fundamenta Informaticae **101**(1), 143–155 (2010)

# Interface of Rough Set Systems and Modal Logics: A Survey

Pulak Samanta[1](✉) and Mihir Kumar Chakraborty[2]

[1] Department of Mathematics, Katwa College,
Katwa, Burdwan, West Bengal, India
pulak_samanta06@yahoo.co.in
[2] Visiting Professor, School of Cognitive Science,
Jadavpur University, Kolkata, India
mihirc4@gmail.com

**Abstract.** In this paper the relationship between rough set theory and modal logic has been discussed. Pawlakian rough set theory has obvious connection with modal logic system $S_5$. With the introduction of various other lower and upper approximation operators, other modal systems come into picture. Besides, the possibility of new modal systems also crop up. Some of these issues are focused here.

**Keywords:** Rough sets · Covering · Modal logic · Rough logic

## 1  Introduction

The relationship between the theory of Rough Sets and Modal Logic had been apparent from the very inception of the theory by Pawlak in 1982 [19]. Literature studying the interconnection has been abundant [3,4,16–18,20,27–29]. Pawlak's Rough Sets are developed on approximation space, which is a set $X$ endowed with an equivalence relation $R$. The relation $R$ emerges usually from an information system consisting of a list of attributes and values for objects belonging to the set $X$. But to develop the mathematics of Rough Sets, the source of the relation is not essential. This pair $< X, R >$ is precisely the Kripke frame for the modal system $S_5$ [12]. The standard normal modal systems are $K, D, T, S_4, B, S_5$. These are classical propositional logics enhanced by modal operators $L$ (necessity) and $M$ (possibility). Along with the axioms of propositional logic, modal axioms are added to define the systems in a hierarchical manner as given below:

**System $K$:** Propositional logic axioms $+ L(\alpha \to \beta) \to (L\alpha \to L\beta)$ (axiom $K$)
**System $D$:** System $K + (L\alpha \to M\alpha)$ (axiom $D$) where $M \equiv \sim L \sim$
**System $T$:** System $K + (L\alpha \to \alpha)$ (axiom $T$)
**System $S_4$:** System $T + (L\alpha \to LL\alpha)$ (axiom $S_4$)
**System $B$:** System $T + (\alpha \to LM\alpha)$ (axiom $B$)
**System $S_5$:** System $T + (ML\alpha \to L\alpha)$ (axiom $S_5$)

J. Peters et al. (Eds.): TRS XIX, LNCS 8988, pp. 114–137, 2015.
DOI: 10.1007/978-3-662-47815-8_8

In all the above mentioned modal systems $\alpha$, $\beta$ are wffs and the rules of inference are:

$$\frac{\alpha,\ \alpha \to \beta}{\beta} \quad \text{(ModusPonens)} \qquad and \qquad \frac{\alpha}{L\alpha} \quad \text{(Necessitation)}$$

Semantics for modal systems is usually given in terms of a Kripke frame $< X, R >$ where $X$ is a non-empty set and $R$ is a binary relation in $X$ called the accessibility relation. For the system $S_5$, to be sound and complete, $R$ has to be an equivalence relation, that is $< X, R >$ becomes an approximation space of Pawlak. Any subset $A$ of $X$, in other words, a triple $< X, R, A >$ may be considered as an interpretation of a wff of $S_5$. In this paper, symbols $\sim$, $\wedge$, $\vee$, $\to$ are taken as the Boolean propositional connectives and $L$, $M$ are the modal operators as mentioned before. For well formed formulae $\alpha$ and $\beta$ interpreted as $< X, R, A >$ and $< X, R, B >$, wffs $\sim \alpha$, $L\alpha$, $M\alpha$, $\alpha \wedge \beta$, $\alpha \vee \beta$, are interpreted as $< X, R, A^c >$, $< X, R, \underline{A} >$, $< X, R, \overline{A} >$, $< X, R, A \cap B >$ and $< X, R, A \cup B >$ respectively. Here $A^c$, $\underline{A}$, $\overline{A}$ denote the complement of $A$ (in $X$), lower approximation of $A$ in $< X, R >$ and upper approximation of $A$ in $< X, R >$ respectively.

The motivation of this paper is to explore into the interplay between the modal logic systems and rough set systems taking into consideration the latter's recent developments. Especially, some features of modal logic systems hitherto unknown or at least unheeded, that have surfaced from the studies in rough set theory would be focused on.

In Sect. 2 some new theorems, non-theorems and derived rules in $S_5$ which are related with rough sets are demonstrated. Section 3 deals with the outcome of relation based rough sets. In Sect. 4 covering based rough sets and their relationship with modal systems are discussed. This section contains a whole category of open issues. Section 5 focused on non-dual rough set systems and in Sect. 6 rough consequence logics as extensions of modal systems are revisited.

## 2  Theorems and Rules in $S_5$

With respect to the semantics given in the introduction, $S_5$ is sound and complete [12]. This means that all the queries about theoremhood or otherwise of the wffs of $S_5$ can be answered by looking at the properties of the triples $< X, R, A >$. It is interesting to observe that in the text books of modal logics the undermentioned wffs seldom appear within the listed theorems. Only after the advent of rough set theory some theorems and non-theorems have come to fore. A few such examples are given below.

(i)  $\vdash_{S_5} L\alpha \wedge L\beta \leftrightarrow L(\alpha \sqcap \beta)$ and $\vdash_{S_5} M\alpha \wedge M\beta \leftrightarrow M(\alpha \sqcap \beta)$ where $\alpha \sqcap \beta$ stands for $(\alpha \wedge \beta) \vee (\alpha \wedge M\beta \wedge \sim \overset{.}{M}(\alpha \wedge \beta))$.

(ii)  $\vdash_{S_5} L\alpha \vee L\beta \leftrightarrow L(\alpha \sqcup \beta)$ and $\vdash_{S_5} M\alpha \vee M\beta \leftrightarrow M(\alpha \sqcup \beta)$ where $\alpha \sqcup \beta$ stands for $(\alpha \vee \beta) \wedge (\alpha \vee L\beta \vee \sim L(\alpha \wedge \beta))$.

**(iii)** $\vdash_{S_5} L(\alpha \sqcup \beta) \leftrightarrow L(\beta \sqcup \alpha)$
  $\vdash_{S_5} L(\alpha \sqcap \beta) \leftrightarrow L(\beta \sqcap \alpha)$
  $\vdash_{S_5} M(\alpha \sqcup \beta) \leftrightarrow M(\beta \sqcup \alpha)$
  $\vdash_{S_5} M(\alpha \sqcap \beta) \leftrightarrow M(\beta \sqcap \alpha)$

**(iv)** $\nvdash_{S_5} (\alpha \sqcup \beta) \leftrightarrow (\beta \sqcup \alpha)$
  $\nvdash_{S_5} (\alpha \sqcap \beta) \leftrightarrow (\beta \sqcap \alpha)$

**(v)** $\vdash_{S_5} L(\alpha \sqcup \beta) \leftrightarrow L\alpha \sqcup L\beta$
  $\vdash_{S_5} L(\alpha \sqcap \beta) \leftrightarrow L\alpha \sqcap L\beta$
  $\vdash_{S_5} M(\alpha \sqcup \beta) \leftrightarrow M\alpha \sqcup M\beta$
  $\vdash_{S_5} M(\alpha \sqcap \beta) \leftrightarrow M\alpha \sqcap M\beta$

**(vi)** $\nvdash_{S_5} \alpha \sqcup \neg\alpha$
  $\vdash_{S_5} L\alpha \sqcup \neg L\alpha$
  $\vdash_{S_5} M\alpha \sqcup \neg M\alpha$

**(vii)** $\nvdash_{S_5} \alpha \sqcup (\beta \sqcup \gamma) \leftrightarrow (\alpha \sqcup \beta) \sqcup \gamma$
  $\nvdash_{S_5} \alpha \sqcap (\beta \sqcap \gamma) \leftrightarrow (\alpha \sqcap \beta) \sqcap \gamma$

**(viii)** $\vdash_{S_5} (\alpha \wedge \beta) \rightarrow (\alpha \sqcap \beta)$
  $\nvdash_{S_5} (\alpha \sqcap \beta) \rightarrow (\alpha \wedge \beta)$
  $\nvdash_{S_5} (\alpha \vee \beta) \rightarrow (\alpha \sqcup \beta)$
  $\vdash_{S_5} (\alpha \sqcup \beta) \rightarrow (\alpha \vee \beta)$

The interpretation of the wff $\alpha \sqcap \beta$ gives a subset $P$ of $X$ of the approximation space $< X, R >$ such that $\underline{P} = \underline{A} \cap \underline{B}$ and $\overline{P} = \overline{A} \cap \overline{B}$ where $A, B$ are subsets of $X$ assigned to wffs $\alpha, \beta$ under the same interpretation.

Similarly, $\alpha \sqcup \beta$ corresponds to a subset $Q$ of $X$ such that $\underline{Q} = \underline{A} \cap \underline{B}$ and $\overline{Q} = \overline{A} \cap \overline{B}$.

In modal logic literature, there is an implication called the 'strict implication' which is denoted by the symbol $\prec$, $\alpha \prec \beta$ being equivalent to $L(\alpha \rightarrow \beta)$ and read as 'necessarily $\alpha$ implies $\beta$'. The significance of strict implication may be obtained in various books e.g. [12]. From the studies on rough sets another kind of implication has emerged which may be called the 'rough implication'. This is denoted by the symbol $\Rightarrow$ and defined as

$$\alpha \Rightarrow \beta \equiv (L\alpha \rightarrow L\beta) \wedge (M\alpha \rightarrow M\beta).$$

This may be read as 'necessarily $\alpha$ implies necessarily $\beta$ and possibly $\alpha$ implies possibly $\beta$'. In rough set theoretic terms this formula stands for 'rough inclusion' [19]. If $\alpha \Leftrightarrow \beta$ denotes $(\alpha \Rightarrow \beta) \wedge (\beta \Rightarrow \alpha)$ (which stands for rough equality [19]) then some syntactic properties of the operators $\Rightarrow$ and $\Leftrightarrow$ are given by the following list. It should be noted that these properties constitute some fundamental results of rough set theory and rough logics [3, 19].

**(ix)** $\vdash_{S_5} (\alpha \sqcap \beta) \Rightarrow (\alpha \wedge \beta)$
  $\vdash_{S_5} (\alpha \vee \beta) \Rightarrow (\alpha \sqcup \beta)$

**(x)** $\vdash_{S_5} L(\alpha \sqcup \beta) \Leftrightarrow L\alpha \sqcup L\beta$
  $\vdash_{S_5} L(\alpha \sqcap \beta) \Leftrightarrow L\alpha \sqcap L\beta$
  $\vdash_{S_5} M(\alpha \sqcup \beta) \Leftrightarrow M\alpha \sqcup M\beta$
  $\vdash_{S_5} M(\alpha \sqcap \beta) \Leftrightarrow M\alpha \sqcap M\beta$

**(xi)** $\vdash_{S_5} \alpha \sqcup (\beta \sqcup \gamma) \Leftrightarrow (\alpha \sqcup \beta) \sqcup \gamma$

$\vdash_{S_5} \alpha \sqcap (\beta \sqcap \gamma) \Leftrightarrow (\alpha \sqcap \beta) \sqcap \gamma$

**(xii)** $\vdash_{S_5} ((\alpha \Rightarrow \beta) \sqcap (\beta \Rightarrow \alpha)) \leftrightarrow ((\alpha \Rightarrow \beta) \wedge (\beta \Rightarrow \alpha))$

## Rules of inference:

With respect to this new implication the following valid rules are obtained.

1. $\dfrac{\vdash_{S_5} \alpha, \vdash_{S_5} \alpha \Rightarrow \beta}{\vdash_{S_5} \beta}$

2. $\dfrac{\vdash_{S_5} \alpha \Rightarrow \beta, \vdash_{S_5} \beta \Rightarrow \gamma}{\vdash_{S_5} \alpha \Rightarrow \gamma}$

3. $\dfrac{\vdash_{S_5} \alpha}{\vdash_{S_5} \beta \Rightarrow \alpha}$

4. $\dfrac{\vdash_{S_5} \alpha \Rightarrow \beta}{\vdash_{S_5} \neg\beta \Rightarrow \neg\alpha}$

5. $\dfrac{\vdash_{S_5} \alpha \Rightarrow \beta, \vdash_{S_5} \alpha \Rightarrow \gamma}{\vdash_{S_5} \alpha \Rightarrow \beta \sqcap \gamma}$

6. $\dfrac{\vdash_{S_5} \alpha \Rightarrow \beta, \vdash_{S_5} \beta \Rightarrow \alpha, \vdash_{S_5} \gamma \Rightarrow \delta, \vdash_{S_5} \delta \Rightarrow \gamma}{\vdash_{S_5} (\alpha \Rightarrow \gamma) \Rightarrow (\beta \Rightarrow \delta)}$

7. $\dfrac{\vdash_{S_5} \alpha \Rightarrow \beta}{\vdash_{S_5} L\alpha \Rightarrow L\beta}$

8. $\dfrac{\vdash_{S_5} L\alpha \Rightarrow L\beta, \vdash_{S_5} M\alpha \Rightarrow M\beta}{\vdash_{S_5} \alpha \Rightarrow \beta}$

9. $\dfrac{\vdash_{S_5} \alpha \rightarrow \beta}{\vdash_{S_5} \alpha \Rightarrow \beta}$

10. $\dfrac{\vdash_{S_5} \alpha \wedge \beta}{\vdash_{S_5} \alpha \sqcap \beta}$

11. $\dfrac{\vdash_{S_5} \alpha \sqcap \beta}{\vdash_{S_5} \alpha \wedge \beta}$

12. $\dfrac{\vdash_{S_5} \alpha \sqcup \beta}{\vdash_{S_5} \alpha \vee \beta}$

We will see in Sect. 6 that various other implications are developed quite 'naturally' from the study of rough sets by using other wffs e.g. $L\alpha \rightarrow L\beta$, $L\alpha \rightarrow \beta$, $L\alpha \rightarrow M\beta$ etc.

Rough meet ($\sqcap$), rough join ($\sqcup$) and rough implication ($\Rightarrow$) operators emerged during the study of rough sets in [2,3]. Most of the results mentioned above are available in the same papers but not necessarily in the present form.

In this sense Rough Set Theory has been contributing to the development of Modal Logic Systems.

## 3   Relation Based Rough Sets and Modality

With the emergence of research on generalized Rough Sets, particularly, because of the development of lower and upper approximations of a set based not essentially on equivalence relation (or equivalently a partition of the universe), the modal logical counterpart has extended into a new, wider and colourful dimension. One kind of generalization is achieved by taking in the universe $X$ any binary relation $R$ which is not necessarily an equivalence relation (as in Pawlakian Rough Set). The scheme is as follows :

Let for $x \in X$, $R_x = \{y \mid xRy\}$. Let $A \subseteq X$.

Then the lower approximation $\underline{A} = \{x \mid R_x \subseteq A\}$ and the upper approximation $\overline{A} = \{x \mid R_x \cap A \neq \phi\}$.

$x$ belongs to $R_x$ only if $R$ is reflexive.

$\underline{A}$ and $\overline{A}$ generated thus shall also be denoted by $\mathbf{L_R}(A)$ and $\mathbf{M_R}(A)$ respectively.

With these definitions one can proceed towards their properties and depending on various properties (e.g. reflexivity, symmetry, transitivity, seriality and their various combinations) of the relation $R$ various properties of the lower and upper approximations are obtained. The following table may be noted where the suffixes of $R$ namely $r$, $s$ and $t$ or their combinations indicate that the relation is reflexive, symmetric and transitive respectively or their combinations. $R_{ser}$ denotes a serial relation.

**Table 1.** Properties of relation based approximations

|  | $R$ | $R_r$ | $R_s$ | $R_t$ | $R_{rs}$ | $R_{rt}$ | $R_{st}$ | $R_{rst}$ | $R_{ser}$ |
|---|---|---|---|---|---|---|---|---|---|
| Duality of $\underline{A}$ , $\overline{A}$ | Y | Y | Y | Y | Y | Y | Y | Y | Y |
| $\underline{\phi} = \phi$ | N | N | N | N | Y | Y | N | Y | Y |
| $\overline{\phi} = \phi$ | Y | Y | Y | Y | Y | Y | Y | Y | Y |
| $\underline{X} = X$ | Y | Y | Y | Y | Y | Y | Y | Y | Y |
| $\overline{X} = X$ | N | N | N | N | Y | Y | N | Y | Y |
| $\underline{A \cap B} \subseteq \underline{A} \cap \underline{B}$ | Y | Y | Y | Y | Y | Y | Y | Y | Y |
| $\underline{A} \cap \underline{B} \subseteq \underline{A \cap B}$ | Y | Y | Y | Y | Y | Y | Y | Y | Y |
| $\overline{A \cup B} \subseteq \overline{A} \cup \overline{B}$ | Y | Y | Y | Y | Y | Y | Y | Y | Y |
| $\overline{A} \cup \overline{B} \subseteq \overline{A \cup B}$ | Y | Y | Y | Y | Y | Y | Y | Y | Y |
| $A \subseteq B$ implies $\underline{A} \subseteq \underline{B}$ | Y | Y | Y | Y | Y | Y | Y | Y | Y |
| $A \subseteq B$ implies $\overline{A} \subseteq \overline{B}$ | Y | Y | Y | Y | Y | Y | Y | Y | Y |
| $\underline{A} \subseteq A$ | N | Y | N | N | Y | Y | N | Y | N |
| $A \subseteq \overline{A}$ | N | Y | N | N | Y | Y | N | Y | N |
| $\underline{A} \subseteq \overline{A}$ | N | Y | N | N | Y | Y | N | Y | Y |
| $A \subseteq \underline{(\overline{A})}$ | N | N | Y | N | Y | N | Y | Y | N |
| $\overline{(\underline{A})} \subseteq A$ | N | N | Y | N | Y | N | Y | Y | N |
| $\overline{A} \subseteq \overline{(\overline{A})}$ | N | N | N | Y | N | Y | Y | Y | N |
| $\underline{(\underline{A})} \subseteq \underline{A}$ | N | N | N | Y | N | Y | Y | Y | N |
| $\overline{A} \subseteq \underline{(\overline{A})}$ | N | N | N | N | N | N | N | Y | N |
| $\overline{(\underline{A})} \subseteq \underline{A}$ | N | N | N | N | N | N | N | Y | N |

The depiction in the above table is nothing novel. These correspond to the results on modalities of modal systems based on various conditions on the accessibility relation in the Kripke frames of modal systems. What is interesting is their re-appearance in the context of Rough Set theory [29]. Besides, what is revealing is a bunch of the uniqueness results of accessibility relation on the Kripke frame derived from the theorems by Zhu and Wang [38].

It is known that besides the interpretation of modal operators in terms of accessibility relation, there are other kinds of semantics too. The operator based semantics for a modal system $S$ is given in the algebraic structure $(P(X), ^c, \mathbf{L_S}, \mathbf{M_S}, \cap, \cup, \phi, X)$ where $\mathbf{L_S}$, $\mathbf{M_S}$ are mappings from $P(X)$ to $P(X)$ which are interpretations of the modalities $L_S$ and $M_S$ satisfying the axioms and rules of the system $S$. Operators for $^c, \cap, \cup$ are fixed viz. the complementation, intersection and union of sets. But operators of $\mathbf{L_S}$ and $\mathbf{M_S}$ may vary. There can be more than one functions $\mathbf{L_S}$ and $\mathbf{M_S}$ that satisfy the stipulated conditions. Recent results in rough set theory throw some light on the sufficient conditions for the appearance of accessibility relation in semantics. We paraphrase below the results by Zhu and Wang [38] in the modal logic framework. It is important to note that the duality between $L_S$ and $M_S$ is not generally assumed.

**Proposition 1.** *Let $S$ be a modal logic system with modality $L_S$ such that*

(i) $\dfrac{\vdash_S \alpha}{\vdash_S L_S \alpha}$ *and*

(ii) $\vdash_S L_s(\alpha \wedge \beta) \leftrightarrow (L_s \alpha \wedge L_s \beta)$.

*Then for each model $(P(X), ^c, \mathbf{L_S}, \cap, \cup, \phi, X)$, there exists a unique accessibility relation $R \subseteq X \times X$ such that $\mathbf{L_R} = \mathbf{L_S}$.*

*Proof outline*: In any model because of (i) and (ii), $\mathbf{L_S}$ satisfies the conditions $\mathbf{L_S}(X) = X$ and $\mathbf{L_S}(A \cap B) = \mathbf{L_S}(A) \cap \mathbf{L_S}(B)$, $A, B \subseteq X$.

Then there exists an accessibility relation $R$ in $X$ such that $\mathbf{L_R}(A) = \mathbf{L_S}(A)$ for all $A \subseteq X$ [30].

In [38], the uniqueness of $R$ is proved thus.

Let there be two relations $R_1$ and $R_2$ such that $\mathbf{L_{R_1}}(A) = \mathbf{L_S}(A) = \mathbf{L_{R_2}}(A)$ for all $A \subseteq X$.

Let $x R_2 y$. Then $x \notin \mathbf{L_{R_2}}(\{y\}^c)$. So, $x \notin \mathbf{L_{R_1}}(\{y\}^c)$, so, $x R_1 y$ and hence $R_2 \subseteq R_1$.

Similarly $R_1 \subseteq R_2$.

**Proposition 2.** *Let $S$ be a modal logic system with modality $M_S$ such that*

(i) $\dfrac{\vdash_S \sim \alpha}{\vdash_S \sim M_S \alpha}$ *and*

(ii) $\vdash_S M_s(\alpha \vee \beta) \leftrightarrow (M_s \alpha \vee M_s \beta)$.

*Then for each model $(P(X), ^c, \mathbf{M_S}, \cap, \cup, \phi, X)$, there exists a unique accessibility relation $R' \subseteq X \times X$ such that $\mathbf{M_{R'}} = \mathbf{M_S}$.*

*Note*: When $L_S$ and $M_S$ are dual operators, conditions in Propositions 1 and 2 become dual and one gets $R = R'$.

**Proposition 3.** *Let $S$ be a modal logic system with modality $L_S$ such that it satisfies (i) and (ii) of Proposition 1 and (iii) $\vdash_S L_s \alpha \rightarrow L_s(\sim L_s(\sim \alpha))$.*

*Then for each model $(P(X), ^c, \mathbf{L_S}, \cap, \cup, \phi, X)$, there is a unique symmetric accessibility relation $R \subseteq X \times X$ such that $\mathbf{L_R} = \mathbf{L_S}$.*

**Proposition 4.** *Let $S$ be a modal logic system with modality $M_S$ such that it satisfies (i) and (ii) of Proposition 2 and (iii) $\vdash_S M_s(\sim M_s\alpha) \to M_s(\sim \alpha)$.*

*Then for each model $(P(X),^c, \mathbf{M_S}, \cap, \cup, \phi, X)$, there is a unique symmetric accessibility relation $R' \subseteq X \times X$ such that $\mathbf{M_{R'}} = \mathbf{M_S}$.*

**Proposition 5.** *Let $S$ be a modal logic system with modality $L_S$ such that it satisfies (i) and (ii) of Proposition 1 and (iii) $\vdash_S L_s\alpha \to L_s L_s\alpha$.*

*Then for each model $(P(X),^c, \mathbf{L_S}, \cap, \cup, \phi, X)$, there is a unique transitive accessibility relation $R \subseteq X \times X$ such that $\mathbf{L_R} = \mathbf{L_S}$.*

**Proposition 6.** *Let $S$ be a modal logic system with modality $M_S$ such that it satisfies (i) and (ii) of Proposition 2 and (iii) $\vdash_S M_s M_s\alpha \to M_s\alpha$.*

*Then for each model $(P(X),^C, \mathbf{M_S}, \cap, \cup, \phi, X)$, there is a unique transitive accessibility relation $R' \subseteq X \times X$ such that $\mathbf{M_{R'}} = \mathbf{M_S}$.*

The note above applies in these cases also.

## 4    Covering Based Rough Sets and Modality

The scenario turns out to be more interesting in the context of covering based Rough Sets. A covering $\mathcal{C}$ of the universe $X$ is a collection $\mathcal{C}$ of subsets such that $\cup \mathcal{C} = X$. The partition of $X$ generated by the equivalence relation $R$ in the case of Pawlakian Rough Sets is a special covering. The collection $\{R_x | x \in X\}$ may or may not form a covering but is so if $R$ is reflexive. The first work on covering based RST is perhaps due to Pomykala [21]. Later years saw a great spurt of activity in this area. Emergence of research on covering based Rough Sets has taken place not only from the theoretical stand point of generalization but also because of its practical import. Basic granules of objects which form the building blocks of concepts overlap in most cases of practical life. Coverings rather than well defined partitions reflect the real life situations in a better way. An excellent work is done by Yao [32] in which he has suggested some methodology of categorization of various approaches that have emerged in the area. Prior to this, Samanta and Chakraborty [24,25] carried out a survey. In the present work we shall focus on modal properties, that is the properties of lower and upper approximation operators and categorize the rough set models accordingly. The properties we shall concentrate upon are already present in Table 1. It is necessary to restate the methods of construction [24,25] of various approximation operators by taking a covering instead of a partition only.

Let $\mathcal{C} = \{C_i\}$ be a covering of $X$ and let us consider the following sets.

$N_x^{\mathcal{C}} = \cup \{C_i : x \in C_i\} = Friends(x)$ [15,22]

$P_x^{\mathcal{C}} = \{y \in X : \forall C_i(x \in C_i \Leftrightarrow y \in C_i)\}$ (Partition generated by a covering) [19,22,36]

$N(x) = \cap \{C_i : x \in C_i\} = Neighbour(x)$ [14,33,36]

$e.f(x) = X - Friends(x)$ [15]

$Md(x) = \{C_i : x \in C_i \wedge (\forall S \in \mathcal{C} \wedge x \in S \subseteq C_i \Rightarrow C_i = S)\}$ [8,23].

Various pairs of lower and upper approximations of a set $A$ are available in rough set literature.

$$\underline{P_1}(A) = \{x : N_x^{\mathcal{C}} \subseteq A\}$$
$$\overline{P}^1(A) = \cup\{C_i : C_i \cap A \neq \phi\} \quad [22,31]$$

$$\underline{P_2}(A) = \cup\{N_x^{\mathcal{C}} : N_x^{\mathcal{C}} \subseteq A\}$$
$$\overline{P}^2(A) = \{z : \forall y(z \in N_y^{\mathcal{C}} \Rightarrow N_y^{\mathcal{C}} \cap A \neq \phi)\} \quad [22]$$

$$\underline{P_3}(A) = \cup\{C_i : C_i \subseteq A\}$$
$$\overline{P}^3(A) = \{y : \forall C_i(y \in Ci \Rightarrow C_i \cap A \neq \phi)\} \quad [14,22,26,31,36]$$

$$\underline{P_4}(A) = \cup\{P_x^{\mathcal{C}} : P_x^{\mathcal{C}} \subseteq A\}$$
$$\overline{P}^4(A) = \cup\{P_x^{\mathcal{C}} : P_x^{\mathcal{C}} \cap A \neq \phi\} \quad [7,14,19,22,23,26,30,33,36]$$

$$\underline{C_1}(A) = \cup\{C_i : C_i \subseteq A\}$$
$$\overline{C}^1(A) = \sim \underline{C_1}(\sim A) = \cap\{\sim C_i : C_i \cap A = \phi\} \quad [23]$$

$$\underline{C_2}(A) = \{x \in X : N(x) \subseteq A\}$$
$$\overline{C}^2(A) = \{x \in X : N(x) \cap A \neq \phi\} \quad [14,23]$$

$$\underline{C_3}(A) = \{x \in X : \exists u(u \in N(x) \wedge N(u) \subseteq A)\}$$
$$\overline{C}^3(A) = \{x \in X : \forall u(u \in N(x) \to N(u) \cap A \neq \phi)\} \quad [23]$$

$$\underline{C_4}(A) = \{x \in X : \forall u(x \in N(u) \to N(u) \subseteq A)\}$$
$$\overline{C}^4(A) = \cup\{N(x) : N(x) \cap A \neq \phi\} \quad [23]$$

$$\underline{C_5}(A) = \{x \in X : \forall u(x \in N(u) \to u \in A)\}$$
$$\overline{C}^5(A) = \cup\{N(x) : x \in A\} \quad [23]$$

With the same lower approximation there are a few different upper approximations.

$$\underline{C_*}(A) = \underline{C_-}(A) = \underline{C_\#}(A) = \underline{C_@}(A) = \underline{C_+}(A) = \underline{C_\%}(A)$$
$$= \cup\{C_i : C_i \subseteq A\} \equiv \underline{P_3}(A) \quad [15]$$

$$\overline{C}^*(A) = \underline{C_*}(A) \cup \{Md(x) : x \in A \setminus A_*\} \quad [8,15,36,37]$$

$$\overline{C}^-(A) = \cup\{C_i : C_i \cap A \neq \phi\} \quad [15,35]$$

$$\overline{C}^\#(A) = \cup\{Md(x) : x \in A\} \quad [15,36]$$

$$\overline{C}^@(A) = \underline{C_@}(A) \cup \{C_i : C_i \cap (A \setminus \underline{C_@}(A)) \neq \phi\} \quad [15]$$

$$\overline{C}^+(A) = \underline{C_+}(A) \cup \{Neighbour(x) : x \in A \setminus \underline{C_+}(A)\} \quad [15,33]$$

$$\overline{C}^\%(A) = \underline{C_\%}(A) \cup \{\sim \cup\{Friends(y) : x \in A \setminus \underline{C_\%}(A), y \in e.f(x)\}\} \quad [15]$$

Two other types of lower and upper approximations are defined with the help of covering.

(1) Let, $Gr_*(A) = \cup\{C_i : C_i \subseteq A\} \equiv \underline{P}_3(A)$.
This is taken as lower approximation of $A$ and is denoted by $\underline{C}_{Gr}(A)$.
$Gr^*(A) = \cup\{C_i : C_i \cap A \neq \phi\} \equiv \overline{P}^1(A)$.
The upper approximation is defined by $\overline{C}^{Gr}(A) = Gr^*(A) \setminus NEG_{Gr}(A)$, where, $NEG_{Gr}(A) = \underline{C}_{Gr}(\sim A)$, $\sim A$ being the complement of $A$ [26].

(2) A set $D$ is said to be definable iff there exists a set $A$ ($\subseteq X$) such that $D = \bigcup_{x \in A} N(x)$. Let $\mathcal{D} = \{D \subseteq X : D \text{ is definable}\}$. $\underline{C}_t, \overline{C}_t : P(X) \to P(X)$ are such that $\underline{C}_t(A) = \bigcup\{D \in \mathcal{D} : D \subseteq A\}$ and $\overline{C}_t(A) = \bigcap\{D \in \mathcal{D} : A \subseteq D\}$. [13]. It may be observed that $\bigcup\{D \in \mathcal{D} : D \subseteq A\} = \bigcup\{N(x) : N(x) \subseteq A\} = \{x \in X : N(x) \subseteq A\} = \underline{C}_t(A)$ and $\bigcap\{D \in \mathcal{D} : A \subseteq D\} = \bigcup\{N(x) : x \in A\} = \overline{C}_t(A)$. The following chart summarizes the properties.

**Table 2.** Properties of covering based approximations

| | $P_1$ | $P_2$ | $P_3$ | $P_4$ | $C_1$ | $C_2$ | $C_3$ | $C_4$ | $C_5$ | $C_{Gr}$ | $C_*$ | $C_-$ | $C_\#$ | $C_@$ | $C_+$ | $C_\%$ | $C_t$ |
|---|---|---|---|---|---|---|---|---|---|---|---|---|---|---|---|---|---|
| Duality of $\underline{A}, \overline{A}$ | Y | Y | Y | Y | Y | Y | Y | Y | Y | Y | N | N | N | N | N | N | N |
| $\underline{\phi} = \phi = \overline{\phi}$ | Y | Y | Y | Y | Y | Y | Y | Y | Y | Y | Y | Y | Y | Y | Y | Y | Y |
| $\underline{X} = X = \overline{X}$ | Y | Y | Y | Y | Y | Y | Y | Y | Y | Y | Y | Y | Y | Y | Y | Y | Y |
| $\underline{A \cap B} \subseteq \underline{A} \cap \underline{B}$ | Y | Y | Y | Y | Y | Y | Y | Y | Y | Y | Y | Y | Y | Y | Y | Y | Y |
| $\overline{A \cap B} \subseteq \overline{A} \cap \overline{B}$ | Y | N | N | Y | N | Y | N | Y | Y | N | N | N | N | N | N | N | Y |
| $\underline{A \cup B} \subseteq \underline{A} \cup \overline{B}$ | Y | N | N | Y | N | Y | N | Y | Y | N | N | Y | N | Y | Y | N | Y |
| $\overline{A \cup B} \subseteq \overline{A} \cup \overline{B}$ | Y | Y | Y | Y | Y | Y | N | Y | Y | Y | Y | Y | Y | Y | N | N | Y |
| $A \subseteq B$ implies $\underline{A} \subseteq \underline{B}$ | Y | Y | Y | Y | Y | Y | Y | Y | Y | Y | Y | Y | Y | Y | Y | Y | Y |
| $A \subseteq B$ implies $\overline{A} \subseteq \overline{B}$ | Y | Y | Y | Y | Y | Y | Y | Y | Y | Y | N | Y | Y | N | Y | Y | Y |
| $\underline{A} \subseteq A$ | Y | Y | Y | Y | Y | Y | N | Y | Y | Y | Y | Y | Y | Y | Y | N | Y |
| $A \subseteq \overline{A}$ | Y | Y | Y | Y | Y | Y | N | Y | Y | Y | Y | Y | Y | Y | Y | N | Y |
| $\underline{A} \subseteq \overline{A}$ | Y | Y | Y | Y | Y | Y | N | Y | Y | Y | Y | Y | Y | Y | Y | Y | Y |
| $\underline{A} \subseteq \overline{(\underline{A})}$ | Y | N | N | Y | N | N | N | Y | N | N | Y | Y | Y | Y | N | N | Y |
| $\overline{(\underline{A})} \subseteq A$ | Y | N | N | Y | N | N | N | Y | N | N | N | N | Y | Y | Y | N | Y |
| $A \subseteq \underline{(\overline{A})}$ | N | Y | Y | Y | Y | Y | N | N | Y | Y | Y | Y | Y | Y | Y | Y | Y |
| $\underline{(\overline{A})} \subseteq \overline{A}$ | N | Y | Y | Y | Y | Y | N | N | Y | Y | Y | N | N | Y | Y | Y | Y |
| $\overline{A} \subseteq \overline{(\overline{A})}$ | N | N | N | Y | N | N | N | N | N | N | Y | Y | Y | Y | N | N | Y |
| $\underline{(\underline{A})} \subseteq \underline{A}$ | N | N | N | Y | N | N | N | N | N | N | N | N | Y | Y | Y | N | Y |
| $(A^C \cup B) \subseteq (\underline{A})^C \cup \overline{B}$ | Y | N | N | Y | N | Y | N | Y | Y | N | N | N | N | N | N | N | Y |

Two points should be noted. First that the properties taken are not independent. But we have preferred to retain all in order to place them before our eyes. Second, and which is not properly noticed, is that some of the approximations, though defined differently, are in fact the same. Such a case has been presented in Proposition 7.

**Proposition 7.** $\overline{P_3}(A) = \overline{C_1}(A) = \overline{C_{Gr}}(A)$

All these being duals to $\underline{P_3}(A) = \underline{C_1}(A) = \underline{C_{Gr}}(A) = \bigcup\{C_i : C_i \subseteq A\}$ they should be equal. Yet we give below a direct proof of the equality of the three upper approximations.

*Proof.* $\overline{P}^3(A) = \{y : \forall C_i(y \in C_i \Rightarrow C_i \cap A \neq \phi)\}$

$\overline{C}^1(A) = \cap\{C_i^C : C_i \cap A = \phi\}$

$\overline{C}^{Gr}(A) = Gr^*(A) \setminus NEG_{Gr}(A)$

$= \bigcup\{C_i : C_i \cap A \neq \phi\} \setminus \bigcup\{C_j : C_j \subseteq A^C\}$

Let, $x \in \overline{C_1}(A) = \cap\{C_i^C : C_i \cap A = \phi\}$

Let, $x \in C_i$ for any $C_i$.

Two cases : (i) $C_i \cap A = \phi$, (ii) $C_i \cap A \neq \phi$.

In case (i), by assumption $x \in C_i^C$ which contradicts $x \in C_i$ for any $i$.

So, $x \in C_i \Rightarrow C_i \cap A \neq \phi$, for any $C_i$. So, $x \in \overline{P_3}(A)$.

As a result, $\overline{C_1}(A) \subseteq \overline{P_3}(A)$ ... (i)

Conversely, $\forall C_i(x \in C_i \Rightarrow C_i \cap A \neq \phi)$

iff $\forall C_i(C_i \cap A = \phi \Rightarrow x \notin C_i)$

iff $\forall C_i(C_i \cap A = \phi \Rightarrow x \in C_i^C)$.

This implies $x \in \cap\{C_i^C : C_i \cap A = \phi\}$.

So, $\overline{P_3}(A) \subseteq \overline{C_1}(A)$ ... (ii).

So, $\overline{P_3}(A) = \overline{C_1}(A)$ follows from (i) and (ii) ... (1).

Now, $\overline{C}^{Gr}(A) = Gr^*(A) \setminus NEG_{Gr}(A)$

$= \bigcup\{C_i : C_i \cap A \neq \phi\} \setminus \bigcup\{C_j : C_j \subseteq A^C\}$

$= \bigcup\{C_i : C_i \cap A \neq \phi\} \setminus \bigcup\{C_j : C_j \cap A = \phi\}$.

So, $x \in \overline{C}^{Gr}(A)$ iff $x \in C_i : C_i \cap A \neq \phi$, for some $C_i$   and $x \notin C_j$ for all $C_j$ s.t. $C_j \cap A = \phi(\equiv C_j \subseteq A^C)$.

Let $x \in \cap\{C_i^C : C_i \cap A = \phi\}(= \overline{C}^1(A))$

iff $x \in [\bigcup\{C_j : C_j \cap A = \phi\}]^C$

iff $x \notin \bigcup\{C_j : C_j \cap A = \phi\}$.

Now $\exists C_i$ s.t. $x \in C_i$. So, $C_i \cap A \neq \phi$.

So, $x \in \bigcup\{C_i : C_i \cap A \neq \phi\} \setminus \bigcup\{C_j : C_j \cap A = \phi\} = \overline{C}^{Gr}(A)$.

So, $\overline{C}^1(A) = \overline{C}^{Gr}(A)$ ... (2).

From (1) and (2) the proposition follows.    □

*Remark 1.* If the basic formation of approximations of two covering-systems are the same, columns below them should be obviously identical such as systems $P_3$, $C_1$ and $C_{Gr}$ above. But the converse may not be true as would be clear from the following observations:

**Observation 1:** In Table 2, $\underline{P_1}$ and $\underline{C_4}$ have identical columns. Nonetheless $\underline{P_1}$ and $\underline{C_4}$ are different operators, as is shown by the following example.

*Example 1.* sLet $X = \{1, 2, 3, 4, 5, 6\}$ and $\mathcal{C} = \{C_1, C_2, C_3, C_4\}$ where
$C_1 = \{1, 2\}$, $C_2 = \{2, 3, 4\}$, $C_3 = \{4, 5\}$, $C_4 = \{6\}$.
Then, $N_1^C = \{1, 2\}$, $N_2^C = \{2\}$, $N_3^C = \{2, 3, 4\}$, $N_4^C = \{1, 2, 3, 4, 5\}$,
$N_5^C = \{4, 5\}$, $N_6^C = \{6\}$, and
$N(1) = \{1, 2\}$, $N(2) = \{2\}$, $N(3) = \{2, 3, 4\}$, $N(4) = \{4\}$,
$N(5) = \{4, 5\}$, $N(6) = \{6\}$.
Let, $A = \{4, 5\}$. Then $\underline{P}_1(A) = \{5\}$ and $\underline{C}_4(A) = \phi$.
So, $\underline{P}_1$ and $\underline{C}_4$ are different.

**Observation 2:** In Table 2, $\underline{C}_2$ and $\underline{C}_5$ have identical columns. Nonetheless they are different operators. For example:

*Example 2.* Let $X = \{1, 2, 3\}$ and $\mathcal{C} = \{C_1, C_2\}$ where $C_1 = \{1, 2\}$, $C_2 = \{2, 3\}$.
Then $N(1) = \{1, 2\}$, $N(2) = \{2\}$ and $N(3) = \{2, 3\}$.
Let $A = \{1, 2\}$. Then $\underline{C}_2(A) = \{1, 2\}$ and $\underline{C}_5(A) = \{1\}$.
So, $\underline{C}_2$ and $\underline{C}_5$ are different.

**An Analysis of the Table**

It is immediately observed that there are two broad groups: systems in which the approximation operators are dual to each others and in which they are non-dual.

Yao and Yao in [32] present some twenty pairs of approximation operators some of which are not present in our study here. This is because we are interested mostly in modal logical aspects of these operators.

In case of the dual operators, the dual property will automatically be available if either one is present. So for systems $P_1, ... C_{Gr}$, we would like to take a reduced list of properties.

| | Rough set theoretic properties | Corresponding modal properties |
|---|---|---|
| | $X = \overline{X}$ | Rule $N$ |
| (i) .. | $\underline{A \cap B} \subseteq \underline{A} \cap \underline{B}$ | $L(\alpha \wedge \beta) \to (L\alpha \wedge L\beta)$ |
| (ii) .. | $\underline{A} \cap \underline{B} \subseteq \underline{A \cap B}$ | $(L\alpha \wedge L\beta) \to L(\alpha \wedge \beta)$ |
| (iii) .. | $A \subseteq B$ implies $\underline{A} \subseteq \underline{B}$ | $\dfrac{\vdash \alpha \to \beta}{\vdash L\alpha \to L\beta}$ |
| | $\underline{A} \subseteq A$ | $L\alpha \to \alpha \quad (T)$ |
| | $\underline{A} \subseteq \overline{A}$ | $L\alpha \to M\alpha \quad (D)$ |
| | $A \subseteq \underline{(\overline{A})}$ | $\alpha \to LM\alpha \quad (B)$ |
| | $\underline{A} \subseteq \underline{(\underline{A})}$ | $L\alpha \to LL\alpha \quad (S_4)$ |
| | $\overline{(\underline{A})} \subseteq A$ | $ML\alpha \to L\alpha \quad (S_5)$ |
| | $\underline{A^C \cup B} \subseteq (\underline{A})^C \cup \underline{B}$ | $L(\alpha \to \beta) \to (L\alpha \to L\beta) \quad (K)$ |

One can recognize that the above list consists of set theoretic counterparts of modal axioms, theorems and rules. For example, the first is the necessitation rule.

(i) and (ii) are theorems $\vdash_S (L\alpha \wedge L\beta) \to L(\alpha \wedge \beta)$ and $\vdash_S L(\alpha \wedge \beta) \to (L\alpha \wedge L\beta)$ respectively. (iii) is the derived rule $\frac{\vdash \alpha \to \beta}{\vdash L\alpha \to L\beta}$. The last one is axiom $K$.

First thing to observe is that all the systems (dual or non dual) satisfy the necessitation rule $(N)$, (i) and (iii). It is also to be noted that the rule $MP$ i.e. $A \cap (A^C \cup B) \subseteq B$ naturally holds in the power set algebra. Now, rule $N$, $MP$ and axiom $K$ imply (i), (ii) and (iii). And $MP$, (ii) and (iii) imply $K$. So, in the present context, i.e. for the systems in the table, (ii) and $K$ are equivalent. This is visible also in the table : the respective rows are identical.

Below we give another sufficient condition for $K$ to hold. This condition depends exactly on the nature of construction of the lower approximation of a set.

**Proposition 8.** *If the lower approximation of a covering system is defined in terms of a set $S(x)$, $x \in X$ by $\underline{A} = \{x : S(x) \subseteq A\}$ then $\underline{A^c \cup B} \subseteq (\underline{A})^c \cup \underline{B}$, which is axiom $K$.*

*Proof.* $x \in \underline{(A^c \cup B)}$ iff $S(x) \subseteq A^c \cup B$. .... (1)
Also $x \in (\underline{A})^c \cup \underline{B}$ iff $x \in (\underline{A})^c$ or $x \in \underline{B}$
    iff $x \notin \underline{A}$ or $x \in \underline{B}$
    iff $S(x) \nsubseteq A$ or $S(x) \subseteq B$.
So, (K) does not hold iff there exists some $x$ such that $x \in \underline{(A^c \cup B)}$ and not $(S(x) \nsubseteq A$ or $S(x) \subseteq B)$ i.e. $S(x) \subseteq A$ and $S(x) \nsubseteq B$.
    But this is impossible since,
    $S(x) \subseteq A$ implies $S(x) \cap A^c = \phi$. So $S(x) \subseteq B$ by (1).
    But we also require $S(x) \nsubseteq B$.                                    □

*Remark 2.* One can verify from the construction of lower approximations that this is in fact the case for all the rough-set systems discussed here.

We give below a table of dual systems, covering based as well as relation based, to make the picture clear with respect to the standard modal axioms.

**Table 3.** Table of dual systems

|       | $P_1$ | $P_2$ | $P_3$ | $P_4$ | $C_1$ | $C_2$ | $C_3$ | $C_4$ | $C_5$ | $C_{Gr}$ | $R$ | $R_r$ | $R_s$ | $R_t$ | $R_{rs}$ | $R_{rt}$ | $R_{st}$ | $R_{rst}$ |
|-------|-------|-------|-------|-------|-------|-------|-------|-------|-------|----------|-----|-------|-------|-------|----------|----------|----------|-----------|
| $K$   | Y | N | N | Y | N | Y | N | Y | Y | N | Y | Y | Y | Y | Y | Y | Y | Y |
| $T$   | Y | Y | Y | Y | Y | Y | N | Y | Y | Y | N | Y | N | N | Y | Y | N | Y |
| $D$   | Y | Y | Y | Y | Y | Y | N | Y | Y | Y | N | Y | N | N | Y | Y | N | Y |
| $S_4$ | N | Y | Y | Y | Y | Y | N | N | Y | Y | N | N | N | Y | N | Y | Y | Y |
| $B$   | Y | N | N | Y | N | N | N | Y | N | N | N | N | Y | N | Y | N | Y | Y |
| $S_5$ | N | N | N | Y | N | N | N | N | N | N | N | N | N | N | N | N | N | Y |

Let us now focus only on the systems which possess $K$. Then depending on the identity of the columns below, the systems are clustered in the following groups: $\{P_1, C_4, R_{rs}\}$, $\{P_4, R_{rst}\}$ and $\{C_2, C_5, R_{rt}\}$.

Identity of $P_4$ and $R_{rst}$ was evident right from the beginning of Rough Set theory. From the other two groups we can say that $P_1$ and $C_4$ are at least modal system $B$ (because of the presence of $R_{rs}$ in the group) and not $S_5$. Similarly, systems $C_2$ and $C_5$ are at least $S_4$ and not $S_5$. The following natural question may be asked now (Table 3).

**(a)** Are the lower approximations formed by the above systems the same?
**(b)** Can the systems be differentiated by some modal axioms?

From the Remark 1, we see that the answer to the first question is negative. As for the second question, it will be shown below that with respect to the following axioms intermediate of $S_4$ and $S_5$ as present in Hughes and Cresswell [12], systems $C_2$ and $C_5$ cannot be differentiated. But the question still remains open.

The axioms that are considered here are $N_1, J_1, H_1, G_1, M$ and $R_1$. It will be shown below that none of the axioms hold either in $C_2$ or in $C_5$.

The proof is through the following examples.

$N_1 = L(L(\alpha \to L\alpha) \to \alpha) \to (ML\alpha \to \alpha)$ i.e. $(((A^c \cup \underline{A}))^c \cup A) \subseteq (\overline{(\underline{A})})^c \cup A$

$J_1 = L(L(\alpha \to L\alpha) \to \alpha) \to \alpha$ i.e. $(((A^c \cup \underline{A}))^c \cup A) \subseteq A$

*Example 3.* Let, $X = \{1,2,3,4,5,6,7,8\}$ and $C_1 = \{1,2,3\}, C_2 = \{3,4\},$
   $C_3 = \{5,6\}, C_4 = \{5,7\}, C_5 = \{4,5\}, C_6 = \{8\}$
   Then, $N(1) = \{1,2,3\} = N(2),\ N(3) = \{3\},\ N(4) = \{4\},\ N(5) = \{5\},$
   $N(6) = \{5,6\},\ N(7) = \{5,7\},\ N(8) = \{8\}$

Case $C_2$.
   Let, $A = \{2,3,6,7,8\} \ldots$ (i). Then $\underline{A} = \{3,8\},\ A^c = \{1,4,5\}$.
   $(A^c \cup \underline{A}) = \{1,3,4,5,8\},\ \overline{(A^c \cup \underline{A})} = \{3,4,5,8\},\ (\overline{(A^c \cup \underline{A})})^c = \{1,2,6,7\},$
   $(\overline{(A^c \cup \underline{A})})^c \cup A = \{1,2,3,6,7,8\},\ (((A^c \cup \underline{A}))^c \cup A) = \{1,2,3,8\} \ldots$ (ii)
   $\overline{(\underline{A})} = \{1,2,3,8\},\ (\overline{(\underline{A})})^c = \{4,5,6,7\}$
   $(\overline{(\underline{A})})^c \cup A = \{2,3,4,5,6,7,8\} \ldots$ (iii)

From (ii) and (iii), $N_1$ and from (i) and (iii), $J_1$ does not hold for $C_2$.

Case $C_5$
   $A = \{2,3,5,7,8\} \ldots$ (iv). Then $\underline{A} = \{7,8\},\ A^c = \{1,4,6\}$.
   $(A^c \cup \underline{A}) = \{1,4,6,7,8\},\ \overline{(A^c \cup \underline{A})} = \{4,8\},\ (\overline{(A^c \cup \underline{A})})^c = \{1,2,3,5,6,7\},$
   $(\overline{(A^c \cup \underline{A})})^c \cup A = \{1,2,3,5,6,7,8\},\ (((A^c \cup \underline{A}))^c \cup A) = \{1,2,3,8\} \ldots$ (v)
   From (iv) and (v), $J_1$ does not hold for $C_5$.

$H_1 = \alpha \to L(M\alpha \to \alpha)$ i.e. $A \subseteq (\overline{(\overline{A})})^c \cup A)$

*Example 4.* Let $X = \{1,2,3,4,5\}$ and $C_1 = \{1,2\}, C_2 = \{2,3\}, C_3 = \{4,5\}$.
   Then, $N(1) = \{1,2\}, N(2) = \{2\}, N(3) = \{2,3\}, N(4) = \{4,5\} = N(5)$.
Case $C_2$

$A = \{1,4\}$ ... (i). Then, $\overline{A} = \{1,4,5\}$, $(\overline{A})^c = \{2,3\}$,
$(\overline{A})^c \cup A = \{1,2,3,4\}$, $\underline{(\overline{A})^c \cup A} = \{1,2,3\}$ ... (ii)
From (i) and (ii), $H_1$ does not hold for $C_2$.

*Example 5.* Let, $X = \{1,2,3\}$ and $C_1 = \{1,2\}, C_2 = \{2,3\}$.
   Then, $N(1) = \{1,2\}, N(2) = \{2\}, N(3) = \{2,3\}$.

Case $C_5$

   Let, $A = \{1,2\}$ ... (i). Then, $\underline{A} = \{1\}$ and $\overline{A} = \{1,2,3\}$.
   $(\overline{A})^c = \phi$, $(\overline{A})^c \cup A = \phi \cup A = A$, $\underline{((\overline{A})^c \cup A)} = \underline{A} = \{1\}$ ... (ii).
   (i) and (ii) implies $H_1$ does not hold for $C_5$.

$G_1 = ML\alpha \rightarrow LM\alpha$ i.e. $\overline{(\underline{A})} \subseteq \underline{(\overline{A})}$.

*Example 6.* $X = \{1,2,3,4,5,6\}$ and $C_1 = \{1,2,3\}, C_2 = \{2,3,4\}, C_3 = \{5,6\}, C_4 = \{3,5\}, C_5 = \{2,5\}$.
   Then $N(1) = \{1,2,3\}, N(2) = \{2\}, N(3) = \{3\}, N(4) = \{2,3,4\}, N(5) = \{5\}, N(6) = \{5,6\}$.

Case $C_2$

   Let $A = \{2,5\}$. Then, $\underline{A} = \{2,5\}, \overline{(\underline{A})} = \{1,2,4,5,6\}$ ... (i).
   $\overline{A} = \{1,2,4,5,6\}$, $\underline{(\overline{A})} = \{2,5,6\}$ ... (ii)
   From (i) and (ii) $G_1$ does not hold for $C_2$.

*Example 7.* Let $X = \{1,2,3,4,5\}$ and $C_1 = \{1,2\}, C_2 = \{2,3\}, C_3 = \{4,5\}$.
   Then, $N(1) = \{1,2\}, N(2) = \{2\}, N(3) = \{2,3\}, N(4) = \{4,5\} = N(5)$.

Case $C_5$

   Let, $A = \{5\}$. Then, $\underline{A} = \phi, \overline{A} = \{4,5\}$
   $\overline{(\underline{A})} = \overline{(\phi)} = \phi$ ... (i) and $\underline{(\overline{A})} = \{4,5\}$ ... (ii).
   (i) and (ii) implies $G_1$ does not hold for $C_5$.

$M = LM\alpha \rightarrow ML\alpha$ i.e. $\underline{(\overline{A})} \subseteq \overline{(\underline{A})}$.

*Example 8.* Let $X = \{1,2,3,4,5\}$ and $C_1 = \{1,2\}, C_2 = \{2,3\}, C_3 = \{4,5\}$.
   Then, $N(1) = \{1,2\}, N(2) = \{2\}, N(3) = \{2,3\}, N(4) = \{4,5\} = N(5)$.

Case $C_2$

   Let, $A = \{1,4\}$. Then, $\overline{A} = \{1,4,5\}, \underline{A} = \phi$.
   $\overline{(\underline{A})} = \overline{(\phi)} = \phi$ ... (i) and $\underline{(\overline{A})} = \{4,5\}$ ... (ii).
   (i) and (ii) implies $M$ does not hold for $C_2$.

*Example 9.* Let, $A = \{1,2,3\}$ and $C_1 = \{1,2\}, C_2 = \{2,3\}$.
   Then, $N(1) = \{1,2\}, N(2) = \{2\}, N(3) = \{2,3\}$.

Case $C_5$

   Let, $A = \{1,2\}$. Then, $\underline{A} = \{1\}$ and $\overline{(\underline{A})} = \{1,2\}$ ... (i).
   $\overline{A} = \{1,2,3\}$ and $\underline{(\overline{A})} = \{1,2,3\}$ ... (ii).
   (i) and (ii) implies $M$ does not hold for $C_5$.

$R_1 = ML\alpha \rightarrow (\alpha \rightarrow L\alpha)$ i.e. $\overline{(\underline{A})} \subseteq (A^c \cup \underline{A})$.

*Example 10.* Let, $X = \{1, 2, 3, 4, 5, 6, 7, 8\}$ and $C_1 = \{1, 2, 3\}, C_2 = \{3, 4\}, C_3 = \{5, 6\}, C_4 = \{5, 7\}, C_5 = \{4, 5\}, C_6 = \{8\}$

Then, $N(1) = \{1, 2, 3\} = N(2), N(3) = \{3\}, N(4) = \{4\}, N(5) = \{5\}, N(6) = \{5, 6\}, N(7) = \{5, 7\}, N(8) = \{8\}$.

Case $C_2$.

Let, $A = \{2, 3, 6, 7, 8\}$. Then $\underline{A} = \{3, 8\}, A^c = \{1, 4, 5\}$.

$(A^c \cup \underline{A}) = \{1, 3, 4, 5, 8\} \ldots$ (i).

$(\overline{A}) = \{1, 2, 3, 8\} \ldots$ (ii).

(i) and (ii) implies $R_1$ does not hold for $C_2$.

*Example 11.* Let, $X = \{1, 2, 3\}$ and $C_1 = \{1, 2\}, C_2 = \{2, 3\}$.

Then, $N(1) = \{1, 2\}, N(2) = \{2\}, N(3) = \{2, 3\}$.

Case $C_5$

Let, $A = \{1, 2\}$. Then, $A^c = \{3\}, \underline{A} = \{1\}$ and $(\overline{A}) = \{1, 2\} \ldots$ (i).

$(A^c \cup \underline{A}) = \{1, 3\} \ldots$ (ii).

(i) and (ii) implies $R_1$ does not hold for $C_5$.

*Remark 3.* Similar questions in the case of $P_1$ and $C_4$ may be asked, but not investigated.

However, we make the following conjecture.

*Conjecture 1.*

(a) $P_1$ and $C_4$ give rise to the same modal systems.

(b) $C_2$ and $C_5$ give rise to the same modal systems.

From a covering $\{C_i\}$ a relation $R$ can be defined in the following three natural ways (cf. Propositions 9, 10 and 11). The question is: what kind of covering system is generated back by $R$ by defining the approximations in the standard way in terms of granules $R_x$ as mentioned in Sect. 3? The following propositions give the answer.

**Proposition 9.** *Let the relation $R$ be defined by $xRy$ iff $\forall C_i, x \in C_i$ implies $y \in C_i$. Then the system generated by $R$ is $P_1$.*

*Proof.* $y \in \underline{A}$

iff $R_y \subseteq A$

iff $\{x : yRx\} \subseteq A$

iff $\{x : y \in C_i \Rightarrow x \in C_i\} \subseteq A$

iff $\bigcup\{C_i : y \in C_i\} \subseteq A$.

So, $y \in \underline{P_1}(A) \Leftrightarrow N_y \subseteq A \Leftrightarrow \bigcup\{C_i : y \in C_i\} \subseteq A$.

**Proposition 10.** *Let the relation $R$ be defined by $xRy$ iff $\exists C_i$ s.t. $x \in C_i$ and $y \in C_i$. Then the system generated by $R$ is $C_2$.*

*Proof.* $y \in \underline{A}$

iff $R_y \subseteq A$

iff $\{x : yRx\} \subseteq A$

iff $\{x : \exists C_i$ s.t. $y \in C_i$ and $x \in C_i\} \subseteq A$

iff $\bigcap\{C_i : y \in C_i\} \subseteq A$.

So, $y \in \underline{C_2}(A) \Leftrightarrow N(y) \subseteq A \Leftrightarrow \bigcap\{C_i : y \in C_i\} \subseteq A$.

**Proposition 11.** *Let us consider a relation $R$ s.t. $xRy$ iff $\forall C_i, x \in C_i$ iff $y \in C_i$. Then the system generated by $R$ is $P_4$.*

The proof is immediate.

Interesting R-S Systems (i.e. rough set systems with lower and upper approximations defined as in Section 4) are $P_2(\equiv P_3)$ and $C_3$ in which the approximation operators are dual but Axiom $K$ fails. More precisely, in $P_2$, Axioms $T, D, S_4$ hold, $K, B, S_5$ do not hold. In $C_3$, the rule RN, $\frac{\vdash \alpha \rightarrow \beta}{\vdash L\alpha \rightarrow L\beta}$, $\frac{\vdash \alpha \rightarrow \beta}{\vdash M\alpha \rightarrow M\beta}$ and Axiom $L(\alpha \wedge \beta) \rightarrow (L\alpha \wedge L\beta)$ hold. Considering all aspects $P_2$ turns out to be the most interesting case.

Now, the question is, does there exist a non-normal modal system (non-normal, since $K$ fails) whose model could be $P_2$? Since $T$ implies $D$, we in fact need a modal system which is a subsystem of the modal system $S_4$, $P_2$ being one of its models.

## 5 Non-Dual RS-Systems

As shown in Table 2, there are quite a number of non-dual approximation operators within the extant rough set literature. From the angle of modal logic these may be considered as models of bi-modal systems. The common features of all the RS-systems in terms of modal logic formulae are the following

- $\frac{\vdash \alpha}{\vdash L\alpha}$, $N$-rule
- $L(\alpha \wedge \beta) \rightarrow (L\alpha \wedge L\beta)$
- $\frac{\vdash \alpha \rightarrow \beta}{\vdash L\alpha \rightarrow L\beta}$
- $L\alpha \rightarrow \alpha$
- $L\alpha \rightarrow M\alpha$
- $L\alpha \rightarrow LL\alpha$

where, $L, M$ are not duals.

Of all these RS-Systems $C_t$ is the closest to Pawlakian system $P_4(\equiv S_5)$: while in case of $P_4$ all the entries below are 'Y', in case of $C_t$ there is only one 'N' and that is regarding duality. This is an important feature, additionally the approximation operators seem to be quite natural.

Interestingly $C_t$ satisfies the modal logic axiom $K$ which is not satisfied by any other non-dual systems. This places the RS-System $C_t$ at a special position.

The next covering RS-System with minimal deviations from $C_t$ is $C_@$ having 3 deviations. So, it seems that modal systems corresponding to $C_t$ and $C_@$ deserve some investigation.

Another direction of growth may be located in the (proposal for) generation of so called Rough Consequence Logics. These constitute a cluster of logics extending Modal Logic System $S_5$. They arose naturally out of the issues connected with Rough Sets as well as Approximate Reasoning.

# 6   Rough Consequence Logics Revisited

In this section an overview of what has been termed as 'rough consequence logics' [10] based on rough modus ponens rule (RMP) will be presented. First appearance of the idea may be traced back in [11]. Further steps were taken in [9]. A full presentation, appeared in [10]. We shall, however, present the idea in its full generality and thus, the earlier concepts and results will be generalized. The idea is to graft a logic on top of a modal system S with the help of a new rule of inference.

### Rough Logics

Let $S$ be a modal system with consequence relation $\vdash_s$. Based on $S$ two other systems $L_r$ and $L_r^+$ are defined axiomatically by using Rough consequence relation $|\sim$ as follows:

### $L_r$:

(i) $\vdash_S \alpha$ implies $\Gamma \mid\sim \alpha$
(ii) $\{\alpha\} \mid\sim \alpha$
(iii) $\Gamma \mid\sim \alpha$ implies $\Gamma \cup \Delta \mid\sim \alpha$
(iv) RMP may be applied.

### $L_r^+$:

(i), (ii), (iii) as in $L_r$ and
$(iv)^+$ : rule RMP$^+$ may be applied.

It is therefore necessary to present rules RMP and RMP$^+$.
There is a bunch of rules within the category RMP viz.

$$\frac{\Gamma \mid\sim \beta \to \gamma, \quad \vdash_S \aleph(\alpha, \beta)}{\Gamma \mid\sim \gamma}$$

where $\aleph(\alpha, \beta)$ is any one of the following list of wffs.

### List of wffs

|  |  |  |  |
|---|---|---|---|
| (i) $L\alpha \to L\beta$ | (iv) $\alpha \to L\beta$ | (vii) $M\alpha \to L\beta$ | (x) $M(\alpha \to \beta)$ |
| (ii) $L\alpha \to \beta$ | (v) $\alpha \to \beta$ | (viii) $M\alpha \to \beta$ | (xi) $L(\alpha \to \beta)$ |
| (iii) $L\alpha \to M\beta$ | (vi) $\alpha \to M\beta$ | (ix) $M\alpha \to M\beta$ | (xii) $\alpha \Rightarrow \beta$ |
|  |  |  | (xiii) $\alpha \Leftrightarrow \beta$ |

The rules under the bunch RMP$^+$ have only a change in the third component viz. $\Gamma \vdash_S \aleph(\alpha, \beta)$ instead of $\vdash_S \aleph(\alpha, \beta)$.

For the definitions of $\Rightarrow$ and $\Leftrightarrow$ we refer to Sect. 2.

The following observations $3 - 7$, are true for all the logics $L_r$ and $L_r^+$.

**Observation 3:** Conditions (ii) and (iii) to define $|\sim$ are together equivalent to the condition overlap viz.

$$\alpha \in \Gamma \text{ implies } \Gamma \mid\sim \alpha (overlap).$$

**Observation 4:** $\Gamma \mid\sim \alpha$ implies $\delta_1, \delta_2, ... \delta_n \mid\sim \alpha$ for some $\delta_1, \delta_2, ... \delta_n \in \Gamma$. (Compactness).

**Observation 5:** $\Gamma, \alpha \mid\sim \beta$ and $\Delta \mid\sim \alpha$ imply $\Gamma \cup \Delta \mid\sim \beta$ (Cut).

Observations 3 and 4 follow by induction on the length of derivation of $\Gamma \mid\sim \alpha$ and observation 5 follows by induction on the derivation of $\Gamma, \alpha \mid\sim \beta$.

**Observation 6:**

(i) $\dfrac{\Gamma|\sim\alpha, \ \vdash_S \aleph(\alpha,\beta)}{\Gamma|\sim\beta}$

(ii) $\dfrac{\Gamma,\beta|\sim\gamma, \ \vdash_S \aleph(\alpha,\beta)}{\Gamma,\alpha|\sim\gamma}$

**Observation 7:** Ordinary MP rule for $|\sim$ viz.

$$\frac{\Gamma \mid\sim \alpha, \ \Gamma \mid\sim \alpha \to \gamma}{\Gamma \mid\sim \gamma}$$

may be derived as a special case of all the RMP rules for which $\vdash_S \aleph(\alpha, \alpha)$ holds.

Now, depending on the base modal system the above set of thirteen formulae will be clustered into equivalence classes by the equivalence relation defined by: $\vdash_S \varphi$ iff $\vdash_S \psi$.

For example if $S$ is the modal system $S_5$ we get the following classes: $\{(i), (ii)\}, \{(iii), (x)\}, \{(iv), (vii), (viii)\}, \{(v), (xi)\} \ \{(vi), (ix)\}, \{(xii)\}, \{(xiii)\}.$ So, in this case seven RMP rules would be available viz.

RMP$_1$ $\dfrac{\Gamma|\sim\alpha, \ \Gamma|\sim\beta\to\gamma, \ \vdash L\alpha\to L\beta}{\Gamma|\sim\gamma}$

RMP$_2$ $\dfrac{\Gamma|\sim\alpha, \ \Gamma|\sim\beta\to\gamma, \ \vdash L\alpha\to M\beta}{\Gamma|\sim\gamma}$

RMP$_3$ $\dfrac{\Gamma|\sim\alpha, \ \Gamma|\sim\beta\to\gamma, \ \vdash M\alpha\to\beta}{\Gamma|\sim\gamma}$

RMP$_4$ $\dfrac{\Gamma|\sim\alpha, \ \Gamma|\sim\beta\to\gamma, \ \vdash\alpha\to\beta}{\Gamma|\sim\gamma}$

RMP$_5$ $\dfrac{\Gamma|\sim\alpha, \ \Gamma|\sim\beta\to\gamma, \ \vdash M\alpha\to M\beta}{\Gamma|\sim\gamma}$

RMP$_6$ $\dfrac{\Gamma|\sim\alpha, \ \Gamma|\sim\beta\to\gamma, \ \vdash\alpha\Rightarrow\beta}{\Gamma|\sim\gamma}$

RMP$_7$ $\dfrac{\Gamma|\sim\alpha, \ \Gamma|\sim\beta\to\gamma, \ \vdash\alpha\Leftrightarrow\beta}{\Gamma|\sim\gamma}$

Besides these, two more rules, also called rough MP, had been proposed in [1] which are defined as

$$R_1 \quad \frac{\Gamma|\sim\alpha, \ \Gamma\vdash_{S_5} M\alpha\to M\gamma}{\Gamma|\sim\gamma}$$

and

$$R_2 \quad \frac{\Gamma|\sim M\alpha, \ \Gamma|\sim M\gamma}{\Gamma|\sim\gamma}$$

The RMP rules are related by the following hierarchical relations:

$$RMP_1 \Rightarrow RMP_4 \Rightarrow RMP_3,$$
$$RMP_2 \Rightarrow RMP_5 \quad \text{and}$$
$$RMP_1 \Rightarrow RMP_6 \Rightarrow RMP_7$$

where $RMP_i \Rightarrow RMP_j$ means that the ith rule implies the jth one. The relation between the corresponding logics $L_{r_i}$ and $L_{r_j}$ will be reverse inclusion, $L_{r_j} \preceq L_{r_i}$.

A detailed study of the systems $L_{r_i}$ and $L_{r_i}^+$ when $S$ is $S_5$ with rules $RMP_i$ and $RMP_i^+$ is done in [10]. We present below a diagram (Fig. 1) depicting the relevant portion of that study after making few modifications.

In the following diagram (Fig. 1) $\sim$ means equivalence, connection by a line means the lower logical system is proper subsystem of the upper one and connection by dotted line means the corresponding systems are mutually independent.

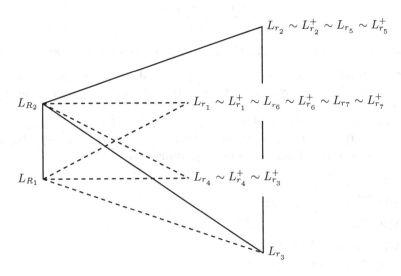

**Fig. 1.** Hierarchy of rough logics

In this Section we present some parallel results with base logics $S_4$ and $B$.

**Base logic $S_4$:**

In this case the clusters of implications are:

$\{(i), (ii)\}, \{(iii), (x)\}, \{(iv)\}, \{(v), (xi)\}, \{(vi), (ix)\}, \{(vii)\}, \{(viii)\}, \{(xii)\}, \{(xiii)\}$.

The nine RMP rules are:

$$\mathrm{RMP}_1 \quad \frac{\Gamma|\sim\alpha,\ \Gamma|\sim\beta\rightarrow\gamma,\ \vdash L\alpha\rightarrow L\beta}{\Gamma|\sim\gamma}$$

$$\mathrm{RMP}_2 \quad \frac{\Gamma|\sim\alpha,\ \Gamma|\sim\beta\rightarrow\gamma,\ \vdash L\alpha\rightarrow M\beta}{\Gamma|\sim\gamma}$$

$$\mathrm{RMP}_3 \quad \frac{\Gamma|\sim\alpha,\ \Gamma|\sim\beta\rightarrow\gamma,\ \vdash\alpha\rightarrow L\beta}{\Gamma|\sim\gamma}$$

$$\mathrm{RMP}_4 \quad \frac{\Gamma|\sim\alpha,\ \Gamma|\sim\beta\rightarrow\gamma,\ \vdash\alpha\rightarrow\beta}{\Gamma|\sim\gamma}$$

$$\mathrm{RMP}_5 \quad \frac{\Gamma|\sim\alpha,\ \Gamma|\sim\beta\rightarrow\gamma,\ \vdash M\alpha\rightarrow M\beta}{\Gamma|\sim\gamma}$$

$$\mathrm{RMP}_6 \quad \frac{\Gamma|\sim\alpha,\ \Gamma|\sim\beta\rightarrow\gamma,\ \vdash M\alpha\rightarrow L\beta}{\Gamma|\sim\gamma}$$

$$\mathrm{RMP}_7 \quad \frac{\Gamma|\sim\alpha,\ \Gamma|\sim\beta\rightarrow\gamma,\ \vdash M\alpha\rightarrow\beta}{\Gamma|\sim\gamma}$$

$$\mathrm{RMP}_8 \quad \frac{\Gamma|\sim\alpha,\ \Gamma|\sim\beta\rightarrow\gamma,\ \vdash\alpha\Rightarrow\beta}{\Gamma|\sim\gamma}$$

$$\mathrm{RMP}_9 \quad \frac{\Gamma|\sim\alpha,\ \Gamma|\sim\beta\rightarrow\gamma,\ \vdash\alpha\Leftrightarrow\beta}{\Gamma|\sim\gamma}$$

The hierarchy of RMP rules are given by:

$$\mathrm{RMP}_7 \Rightarrow \mathrm{RMP}_4 \Rightarrow \mathrm{RMP}_3 \Rightarrow \mathrm{RMP}_1 \Rightarrow \mathrm{RMP}_5,$$
$$\mathrm{RMP}_7 \Rightarrow \mathrm{RMP}_6 \Rightarrow \mathrm{RMP}_3 \Rightarrow \mathrm{RMP}_2 \Rightarrow \mathrm{RMP}_5,$$
$$\mathrm{RMP}_4 \Rightarrow \mathrm{RMP}_8 \quad \text{and}$$
$$\mathrm{RMP}_9 \Rightarrow \mathrm{RMP}_8.$$

So, the hierarchy of corresponding rough logics are:

$$\mathrm{Lr}_5 \preceq \mathrm{Lr}_1 \preceq \mathrm{Lr}_3 \preceq \mathrm{Lr}_4 \preceq \mathrm{Lr}_7,$$
$$\mathrm{Lr}_5 \preceq \mathrm{Lr}_2 \preceq \mathrm{Lr}_3 \preceq \mathrm{Lr}_6 \preceq \mathrm{Lr}_7,$$
$$\mathrm{Lr}_8 \preceq \mathrm{Lr}_4 \quad \text{and}$$
$$\mathrm{Lr}_8 \preceq \mathrm{Lr}_9.$$

**Base logic $B$:**

In this case the clusters of implications are:

$\{(i)\}, \{(ii)\}, \{(iii), (x)\}, \{(iv), (viii)\}, \{(v),(xi)\}, \{(vi)\}, \{(vii)\}, \{(ix)\}, \{(xii)\}, \{(xiii)\}$.

Ten RMP rules are:

$$\mathrm{RMP}_1 \quad \frac{\Gamma|\sim\alpha,\ \Gamma|\sim\beta\rightarrow\gamma,\ \vdash L\alpha\rightarrow L\beta}{\Gamma|\sim\gamma}$$

$$\mathrm{RMP}_2 \quad \frac{\Gamma|\sim\alpha,\ \Gamma|\sim\beta\rightarrow\gamma,\ \vdash L\alpha\rightarrow\beta}{\Gamma|\sim\gamma}$$

$$\mathrm{RMP}_3 \quad \frac{\Gamma|\sim\alpha,\ \Gamma|\sim\beta\rightarrow\gamma,\ \vdash L\alpha\rightarrow M\beta}{\Gamma|\sim\gamma}$$

$$\mathrm{RMP}_4 \quad \frac{\Gamma|\sim\alpha,\ \Gamma|\sim\beta\rightarrow\gamma,\ \vdash\alpha\rightarrow L\beta}{\Gamma|\sim\gamma}$$

$$\mathrm{RMP}_5 \quad \frac{\Gamma|\sim\alpha,\ \Gamma|\sim\beta\rightarrow\gamma,\ \vdash\alpha\rightarrow\beta}{\Gamma|\sim\gamma}$$

$$\text{RMP}_6 \ \frac{\Gamma|\sim\alpha,\ \Gamma|\sim\beta\to\gamma,\ \vdash\alpha\to M\beta}{\Gamma|\sim\gamma}$$

$$\text{RMP}_7 \ \frac{\Gamma|\sim\alpha,\ \Gamma|\sim\beta\to\gamma,\ \vdash M\alpha\to L\beta}{\Gamma|\sim\gamma}$$

$$\text{RMP}_8 \ \frac{\Gamma|\sim\alpha,\ \Gamma|\sim\beta\to\gamma,\ \vdash M\alpha\to M\beta}{\Gamma|\sim\gamma}$$

$$\text{RMP}_9 \ \frac{\Gamma|\sim\alpha,\ \Gamma|\sim\beta\to\gamma,\ \vdash\alpha\Rightarrow\beta}{\Gamma|\sim\gamma}$$

$$\text{RMP}_{10} \ \frac{\Gamma|\sim\alpha,\ \Gamma|\sim\beta\to\gamma,\ \vdash\alpha\Leftrightarrow\beta}{\Gamma|\sim\gamma}$$

The hierarchy of RMP rules are:

$$\text{RMP}_7 \Rightarrow \text{RMP}_4 \Rightarrow \text{RMP}_3 \Rightarrow \text{RMP}_1 \Rightarrow \text{RMP}_6 \Rightarrow \text{RMP}_5 \Rightarrow \text{RMP}_9 \ ,$$
$$\text{RMP}_3 \Rightarrow \text{RMP}_2 \Rightarrow \text{RMP}_8 \Rightarrow \text{RMP}_5 \quad \text{and}$$
$$\text{RMP}_{10} \Rightarrow \text{RMP}_9.$$

The hierarchy of Logics are:

$$\text{Lr}_9 \preceq \text{Lr}_5 \preceq \text{Lr}_6 \preceq \text{Lr}_1 \preceq \text{Lr}_3 \preceq \text{Lr}_4 \preceq \text{Lr}_7 \ ,$$
$$\text{Lr}_5 \preceq \text{Lr}_8 \preceq \text{Lr}_2 \preceq \text{Lr}_3 \quad \text{and}$$
$$\text{Lr}_9 \preceq \text{Lr}_{10}.$$

Here the study of logics with RMP rules with modal systems $S_4$ and $B$ as basis is too incomplete : firstly, unlike the case with $S_5$ as base it has not been investigated whether one logic is strictly stronger than the other. Neither the corresponding $L_{r_i}^+$ logics have been investigated. The semantic investigation is totally untouched. However, we wanted to focus on the point that RMP and RMP$^+$ rules open up a new direction of study in the area of modal logic. The section of the base modal systems $B$ and $S_4$ is only because they are standard modal systems other than $S_5$ and more complex than $T$. This gives a general method applicable to any modal system. Besides, these rules may have wide application in approximate reasoning which is left for future work.

# 7    Concluding Remarks

In this paper we have endeavoured to highlight some interrelations between modal logics on one hand and rough set theory on the other. The interrelation of the two areas was evident from the beginning of the theory. In this review work the focus is primary on some new relations and on identification of some consequences of rough set studies in modal logic. We have presented some wffs which are theorems of $S_5$ and certain rules; these are not available in standard text books, even an exercise. A few propositions showing the uniqueness of accessibility relation in a certain respect have been established. An analysis of various covering based rough sets has been carried out focusing on the nature of modalities they give rise to. Possibility of modal systems with covering based semantics has been indicated and a few conjectures have been made. Finally a general method for extending modal systems with rough modus ponens (RMP) rules is presented. Several unanswered questions have been raised that may seem interesting for future research. Very recently, we have received from Martin Bunder a copy of his paper *Rough consequence and other modal logics, Australasian Journal of Logic (12:1) 2015*, that contains further addition to the research on rough consequence. This paper might have some overlap with the current work.

**Acknowledgement.** We would like to thank Prof. Mohua Banerjee, IIT Kanpur for her valuable comments and suggestions.

# References

1. Banerjee, M.: Logic for rough truth. Fundamenta Informaticae **71**(2,3), 139–151 (2006)
2. Banerjee, M., Chakraborty, M.K.: Rough algebra. Bull. Polish Acad. Sc. (Math.) **41**(4), 293–297 (1993)
3. Banerjee, M., Chakraborty, M.K.: Rough sets through algebraic logic. Fundamenta Informatica **28**(3–4), 211–221 (1996)
4. Banerjee, M., Chakraborty, M.K.: Rough logics : a survey with further directions. In: Orloska, E. (ed.) Incomplete Information: Rough Set Analysis Studies in Fuzziness and Soft Computing, vol. 13, pp. 578–600. Physica-Verlag, Heidelberg (1998)
5. Banerjee, M., Chakraborty, M.K.: Algebras from rough sets. In: Pal, S., Polkowski, L., Skowron, A. (eds.) Rough-Neural Computing. Cognitive Technologies, pp. 157–184. Springer, Heidelberg (2004)
6. Banerjee, M., Yao, Y.: A categorial basis for granular computing. In: An, A., Stefanowski, J., Ramanna, S., Butz, C.J., Pedrycz, W., Wang, G. (eds.) RSFDGrC 2007. LNCS (LNAI), vol. 4482, pp. 427–434. Springer, Heidelberg (2007)
7. Bonikowski, Z.: A certain conception of the calculus of rough sets. Notre Dame J. Formal logic **33**, 412–421 (1992)
8. Bonikowski, Z., Bryniarski, E., Urszula, W.S.: Extensions and intentions in the rough set theory. J. Inf. Sci. **107**, 149–167 (1998)
9. Bunder, M.W.: Rough consequence and Jaskowskis D2 logics. Technical report 2/04, School of Mathematics and Applied Statistics, University of Wollongong, Australia (2004)
10. Bunder, M.W., Banerjee, M., Chakraborty, M.K.: Some rough consequence logics and their interrelations. In: Peters, J.F., Skowron, A. (eds.) Transactions on Rough Sets VIII. LNCS, vol. 5084, pp. 1–20. Springer, Heidelberg (2008)
11. Chakraborty, M.K., Banerjee, M.: Rough consequence. Bull. Polish Acad. Sc(Math.) **41**(4), 299–304 (1993)
12. Hughes, G.E., Cresswell, M.J.: A New Introduction to Modal Logic. Routledge, London (1996)
13. Kumar, A., Banerjee, M.: Definable and rough sets in covering-based approximation spaces. In: Li, T., Nguyen, H.S., Wang, G., Grzymala-Busse, J., Janicki, R., Hassanien, A.E., Yu, H. (eds.) RSKT 2012. LNCS, vol. 7414, pp. 488–495. Springer, Heidelberg (2012)
14. Li, T.-J.: Rough approximation operators in covering approximation spaces. In: Greco, S., Hata, Y., Hirano, S., Inuiguchi, M., Miyamoto, S., Nguyen, H.S., Słowiński, R. (eds.) RSCTC 2006. LNCS (LNAI), vol. 4259, pp. 174–182. Springer, Heidelberg (2006)

15. Liu, J., Liao, Z.: The sixth type of covering-based rough sets. In: Granular Computing - GrC, Hangzhou 26–28 August 2008. Published in IEEE International Conference on Granular Computing 2008, pp. 438–441 (2008)
16. Orlowska, E.: A Logic of indiscernibility relations. In: Skowron, A. (ed.) Computation Theory. LNCS, vol. 208, pp. 177–186. Springer, Heidelberg (1985)
17. Orlowska, E.: Logic for nondeterministic information. Stud. Logica **XLIV**, 93–102 (1985)
18. Orlowska, E.: Logic for reasoning about knowledge. Zeitschr. f. math. Logik und Grundlagen d.Math. **35**, 559–572 (1989)
19. Pawlak, Z.: Rough Sets - Theoretical Aspects of Reasoning About Data. Kluwer Academic Publishers, Dordrecht (1991)
20. Pagliani, P., Chakraborty, M.K.: A geometry of Approximation - Rough Set theory: Logic Algebra and Topology of Conceptual Patterns. Springer, Netherlands (2008)
21. Pomykala, J.A.: Approximation operations in approximation space. Bull. Pol. Acad. Sci. Math. **35**, 653–662 (1987)
22. Pomykala, J.A.: Approximation, Similarity and Rough Constructions. ILLC Prepublication Series for Computation and Complexity Theory CT-93-07, University of Amsterdam (1993)
23. Qin, K., Gao, Y., Pei, Z.: On covering rough sets. In: Yao, J.T., Lingras, P., Wu, W.-Z., Szczuka, M.S., Cercone, N.J., Ślęzak, D. (eds.) RSKT 2007. LNCS (LNAI), vol. 4481, pp. 34–41. Springer, Heidelberg (2007)
24. Samanta, P., Chakraborty, M.K.: Covering based approaches to rough sets and implication lattices. In: Sakai, H., Chakraborty, M.K., Hassanien, A.E., Slezak, D., Zhu, W. (eds.) RSFDGrC 2009. LNCS, vol. 5908, pp. 127–134. Springer, Heidelberg (2009)
25. Samanta, P., Chakraborty, M.K.: Generalized rough sets and implication lattices. In: Peters, J.F., Skowron, A., Sakai, H., Chakraborty, M.K., Slezak, D., Hassanien, A.E., Zhu, W. (eds.) Transactions on Rough Sets XIV. LNCS, vol. 6600, pp. 183–201. Springer, Heidelberg (2011)
26. Ślęzak, D., Wasilewski, P.: Granular sets – foundations and case study of tolerance spaces. In: An, A., Stefanowski, J., Ramanna, S., Butz, C.J., Pedrycz, W., Wang, G. (eds.) RSFDGrC 2007. LNCS (LNAI), vol. 4482, pp. 435–442. Springer, Heidelberg (2007)
27. Vakarelov, D.: A modal logic for similarity relations in Pawlak knowledge representation systems. Fundamenta Informaticae **XV**, 61–79 (1991)
28. Vakarelov, D.: Similarity relations and modal logics. In: Orlowska, E. (ed.) Incomplete Information : Rough Set Analysis Studies in Fuzziness and Soft Computing, vol. 13, pp. 492–550. Physica-Verlag, Heidelberg (1998)
29. Yao, Y.Y., Lin, T.Y.: Generalization of rough sets using model logic. Intell. Automaton Soft Comput. **2**(2), 103–120 (1996)
30. Yao, Y.Y.: Constructive and algebraic methods of the theory of rough sets. J. Inf. Sci. **109**, 21–47 (1998)
31. Yao, Y.Y.: On generalizing rough set theory. In: Wang, G., Liu, Q., Yao, Y., Skowron, A. (eds.) Rough set, Fuzzy sets, Data Mining and Granular Computing. Lecture Notes in Computer Science, vol. 2639, pp. 44–51. Springer, Heidelberg (2003)
32. Yao, Y., Yao, B.: Covering based rough set approximations. Inf. Sci. **200**, 91–107 (2012)
33. Zhu, W.: Topological approaches to covering rough sets. Sci. Direct Inf. Sci. **177**, 1499–1508 (2007)

34. Zhu, W.: Relationship between generalized rough sets based on binary relation and covering. Inf. Sci. **179**, 210–225 (2009)
35. Zhu, W.: Properties of the second type of covering-based rough sets. In: Butz, C.J., Nguyen, N.T., Takama, Y., Cheung, W., Cheung, Y.M. (eds.) IEEE/WIC/ACM International Conference on Web Intelligence and Intelligent Agent Technology Hong Kong, China 18–22 December 2006 Proceedings of the - WI-IAT 2006 Workshops, pp. 494–497. IEEE (2006)
36. Zhu, W., Wang, F.Y.: Relationship among three types of covering rough sets. In: Proceedings of the IEEE International Conference on Grannular Computing-GrC, pp. 43–48 (2006)
37. Zhu, W., Wang, F.Y.: Properties of the first type of covering-based rough sets. In: Tsumoto, S., Clifton, C.W., Zhong, N., Xindong Wu, Liu, J., Wah, B.W., Cheung, Y.M. (eds.) Proceedings of the Sixth IEEE International Conference on Data Mining - Workshops, ICDMW Hong Kong, China December 2006, pp. 407–411. IEEE Computer Society (2006)
38. Zhu, W., Wang, F.-Y.: Binary relation based rough sets. In: Wang, L., Jiao, L., Shi, G., Li, X., Liu, J. (eds.) FSKD 2006. LNCS (LNAI), vol. 4223, pp. 276–285. Springer, Heidelberg (2006)

# A Semantic Text Retrieval for Indonesian Using Tolerance Rough Sets Models

Gloria Virginia[1](✉) and Hung Son Nguyen[2]

[1] Informatics Engineering Department, Duta Wacana Christian University,
Dr. Wahidin Sudirohusodo 5-25, 55224 Yogyakarta, Indonesia
virginia@staff.ukdw.ac.id
[2] Institute of Mathematics, University of Warsaw, Banacha 2, 02-097 Warsaw, Poland

**Abstract.** The research of Tolerance Rough Sets Model (TRSM) ever conducted acted in accordance with the rational approach of AI perspective. This article presented studies who complied with the contrary path, i.e. a cognitive approach, for an objective of a modular framework of semantic text retrieval system based on TRSM specifically for Indonesian. In addition to the proposed framework, this article proposes three methods based on TRSM, which are the automatic tolerance value generator, thesaurus optimization, and lexicon-based document representation. All methods were developed by the use of our own corpus, namely ICL-corpus, and evaluated by employing an available Indonesian corpus, called Kompas-corpus. The endeavor of a semantic information retrieval system is the effort to retrieve information and not merely terms with similar meaning. This article is a baby step toward the objective.

**Keywords:** Information retrieval · Tolerance rough sets model · Text mining

## 1 Introduction

### 1.1 Information Retrieval

The percentage of individuals using the Internet continues to grow worldwide and in developing countries the numbers doubled between 2007 and 2011[1]. Accessing information by utilizing search systems becomes one habitual activity of million of people in facilitating their business, education, and entertainment in their daily life. The applications, such as web search engines which providing access to information over the Internet, are the most usual applications heavily use information retrieval (IR) service.

Information retrieval is concerned with representing, searching, and manipulating large collections of electronic text and other human-language data [1, p. 2].

---

This work was partially supported by the Polish National Science Centre grant DEC-2012/05/B/ST6/03215.
[1] Key statistical highlights: ITU data release June 2012. URL: http://www.itu.int. Accessed on 25 October 2012.

© Springer-Verlag Berlin Heidelberg 2015
J. Peters et al. (Eds.): TRS XIX, LNCS 8988, pp. 138–224, 2015.
DOI: 10.1007/978-3-662-47815-8_9

Clustering systems, categorization systems, summarization systems, information extraction systems, topic detection systems, question answering systems, and multimedia information retrieval systems are other applications utilize IR service.

The main task of information retrieval is to retrieve relevant documents in response to a query [2, p. 85]. In a common search application, an *ad hoc* retrieval mode is applied in which a query is submitted (by a user) and then evaluated against a relatively static document collection. A set of query identifying the possible interest to the user may also be supplied in advanced and then evaluated against newly created or discovered documents. This operational mode where the queries remain relatively static is called *filtering*.

Documents (i.e. electronic texts and other human-language data) are normally modeled based on the positive occurrence of words while the query is modeled based on the positive words of interest clearly specified. Both models then are examined in similarity basis using a devoted ranking algorithm and the output of information retrieval system (IRS) will be an ordered list of documents considered pertinent to the query at hand.

In the keyword search technique commonly used, the similarity between documents and query is measured based on the occurrence of query words in the documents. Thus, if the query is given by a user, then the relevant documents are those who contain literally one or more words expressed by him/her. The fact is, text documents (and query) highly probable come up in the form of natural language. While human seems effortless to understand and construct sentences, which may consist of ambiguous or colloquial words, it becomes a big challenge for an IRS. The keyword search technique is lack of capability to capture the meaning of words, wherefore the meaning of sentences, semantically on documents and query because it represents the information content as a syntactical structure which is lack of semantical relationship. For example, a document contains words *choir*, *performance*, and *ticket* may talk about a *choir concert*, in spite of the fact that the word *concert* is never mentioned on that particular document. When a user inputs the word *concert* to define his/her information need, the IRS which approximate the documents and query in a set of occurrence words may deliver lots of irrelevant results instead of corresponding documents.

We may expect better effectiveness to IRS by mimicking the human capability of language understanding. We should move from *keyword* to *semantic* search technique, hence the semantic IRS.

## 1.2 Philosophical Background

Semantic is the study of linguistic meaning [3, p. 1]. Sentence and word meaning can be analyzed in terms of what speakers (or utterers) mean of his/her utterances[2] [4]. With regard to the intended IRS, we devoted our study to written

---

[2] Utterances may include sound, marks, gesture, grunts, and groans (anything that can signal an intention).

document, which might be seen as an extension of speech. Hence, a text semantic retrieval system should know to some extent the meaning of words of texts
being processed, so to speak.

## Intentionality

Among others, Searl [5] and Grice [6] have been on a debate, namely the role
of intentionality in the theory of meaning. Intentionality (in Latin: *intendere*;
meaning aiming in a certain direction, directing thoughts to something, on the
analogy to drawing a bow at a target) has been used to name the property
of minds of having content, *aboutness*, being *about* something [7, p. 89]. Thus,
mental states such as beliefs, fears, hopes, and desires are intentional because
they are directed at an object. For example, if I have a belief, it must be a
belief of something, or if I have a fear, it must be a fear of something. However, mental states such as undirected anxiety, depression and elation are not
intentional because they are undirected at an object (e.g. I may anxious without
being anxious about anything), but the directed cases (e.g. I am anxious about
something) are intentional.

In addition that intentional is directed, another important characteristic of
intentional was proposed by Searl [5] that every intentional state consists of
an *intentional content* in a *psychological mode*. The intentional content is a
whole proposition which determines a set of condition of satisfaction and the
psychological mode (e.g. belief, desire, promise) determines a direction of fit
(i.e. mind-to-world or world-to-mind) of its propositional content. An example
should make this clear: If I make an assertive utterance that 'it is raining', then
the content of my belief is 'it is raining'. So, the conditions of satisfaction are 'it
is raining', and not, for example, that the ground is wet or the water is falling
out of the sky[3]. And, in my assertive utterance, the psychological mode is a
'belief' of the state in question, so the direction of fit is 'mind-to-world'[4].

Further, Searl claimed [5, pp. 19–21] that intentional contents do not determine their condition of satisfaction in isolation, rather they are internally related
in a holistic way to: (*a*) other intentional contents in the Network of intentional
states; and (*b*) a Background of nonrepresentational mental capacities. The following is Searl's example to describe the role of Network: Suppose there is a
man who forms the intention to run for the Presidency of the United States. In
order that his desire be a desire to run for the Presidency he must have a whole
lot of beliefs such as: the belief that the United States is a republic, that it has

---

[3] The reason is, in the context of speech act, we do not concern about whether the
belief of a speaker is true or not, rather we concern about the intention of speaker
what he/she wants to represent by his/her utterance. Thus, it might be the case
that a speaker represents his/her false belief as a true belief to the audience, e.g. a
speaker utters 'it is raining', while in fact 'it is a sunny day'.

[4] In other words, 'the mind to fit the world'. It is because a belief is like a statement,
can be true or false; if the statement is false then it is the fault of the statement, not
the world. The *world-to-mind* direction of fit is applied for the psychological mode
such as desire or promise; if the promise is broken, it is the fault of the promiser.

a presidential system of government, that it has periodic elections, and so on. And he would normally desire that he receives the nomination of his party, that people work for his candidacy, that voters cast votes for him, and so on. So, in short, we can see that his intention 'refers' to these other intentional states.

The Background is the set of practice, skills, habits, and stance that enable intentional contents to work in various ways. Consider these sentences: 'Berto opened his book to page 37' and 'The chairman opened the meeting'. The semantic content contributed by the word 'open' is the same in each sentence, but we understand the sentences quite differently. It is because the differences in the Background of practice (and in the Network) produce different understanding of the same verb.

## Meaning

Language is one of the vehicles of mental states, hence linguistic meaning is a form of derived intentionality.

According to Searle, *meaning* is a notion that literally applies to sentences and speech acts. He mentioned that the problem of meaning in its most general form is the problem of how do we get from the physics to the semantics. For this purpose, there are two aspects to meaning intentions: (*a*) the intention to represent; and (*b*) the intention to communicate. Here, representing intention is prior to communication intention and the converse is not the case. Hence, we can intend to represent something without intending to communicate it, but we cannot intend to communicate something without intending to represent it before. So to speak, in order to inform anyone that 'it is raining' we need to represent it in our mind that 'it is raining' then utter it. Conversely, we cannot inform anyone anything, i.e. that 'it is raining', when we do not make any representation of the state of affairs of the weather in our mind.

For Grice, when a speaker *mean* something by an utterance, he/she intends to produce certain effects on his/her audience and intends the audience to recognize the intention behind the utterance. By this definition, it seems that Grice has overlooked the intention to represent and overemphasized the intention to communicate. However, a careful analyses showed that Grice's account goes along with Searl's account [8], i.e. representing intention is prior to communication intention. Moreover, Grice definition makes a point that a successful speech act is both meaningful and communicative, i.e. the audience understands nothing when the audience does not recognize the intention behind the utterance, which can be happen when the speaker makes an utterance without intending to mean anything or fails to communicate it.

## The Importance of Knowledge

Based on Searl's and Grice's accounts, it should be clear that there is distinction between intentional content and the form of its externalization. To ask for the meaning is to ask for an intentional content that goes with the form of externalization [5]. It is maintained that for a successful speech act, a speaker normally chooses an expression which is conventionally fixed, i.e. by the community at large, to convey a certain meaning. Thus, before the selection process of appropriate expressions, it is fundamental for a speaker to know about the expression in order to produce an utterance, and consequently the audience is required to be familiar with those conventional expressions in order to understand the utterance.

We may infer now that Searl's and Grice's accounts pertaining the *meaning* suggest knowledge for language production and understanding. This knowledge should consists of concepts who are interrelated and commonly agreed by the community. The communication is satisfied when both sides are active participants and the audience experiences effects at some degree.

## 1.3    Challenges in Indonesian

### Indonesian Studies

Knowledge specifically for Indonesian is fundamental for a semantic retrieval system which processing Indonesian texts. The implication of this claim is far reaching, in particular because each language is unique. There are numerous aspects of monolingual text retrieval should be investigated for Indonesian, those including indexing and relevance assessment process, i.e. tasks such as tokenization, stopping, stemming, parsing, and similarity functions, are few to mention.

Considerable effort with regard to information retrieval for Indonesian is showed by a research community in University of Indonesia (UI) since mid of 1990s. They reported [9] that their studies range in area of computational lexicography (i.e. creating dictionary and spell-checking), morphological analysis (i.e. creating stemming algorithms and parser), semantic and discourse analysis (i.e. based on lexical semantics and text semantic analysis), document summarization, question-answering, information extraction, cross language retrieval, and geographic information retrieval. Other significant studies conducted by Asian which proposed an effective techniques for Indonesian text retrieval [10] and published the first Indonesian testbed [11]. It is worth to mention that despite the long list of works ever mentioned, only limited number of the results is available publicly and among those Indonesian studies, it is hardly to find a work pertaining to automatic ontology constructor specifically.

### Indonesian Speakers

The latest data released by Statistics Board of Indonesia (BPS-Statistics Indonesia)[5] pertaining the population of Indonesia, showed that the number reached 237.6 million for the 2010 census. With the population growth rate 1.49 % per year, the estimation of Indonesia population in 2012 is 245 million. This number ranked Indonesia on the forth most populous country in the world after China, India, and United States[6].

---

[5] BPS-Statistics Indonesia. URL: http://www.bps.go.id/. Accessed on 25 October 2012.

[6] July 2012 estimation of The World Factbook. URL: https://www.cia.gov. Accessed on 25 October 2012.

The incredible number is not only related to the population. Indonesia, which is an archipelago country, has around 6,000 inhibited island over 17,508[7]. Administratively, Indonesia consists of 33 provinces in which there are number of ethnics groups comes from each province which has its own regional language; according to Sneddon [12, p. 196], Indonesia has about 550 languages which is roughly one-tenth of all the languages in the world today. However, chosen as the national language, *Bahasa Indonesia* or Indonesian language is taught at all level of education and officially used in domains of formal activity, e.g. mass media, all government business, education, and law. Nowadays, most Indonesians are proficient in using the language; the number of speaker of Indonesian is approaching 100 % [12, p. 201]. Therefore, it is not overstated to consider Indonesian language as one of the large number of speakers in the world.

### Indonesian Internet Users

Another significant challenge pertains to the growth of Internet users. As the global trend, the percentage of individuals using the Internet continues to grow worldwide and in developing countries the numbers doubled between 2007 and 2011[8]. For Indonesia, the Internet World Stats[9] recorded that there are about 55 million internet users (with 22.4 % penetration rate) and 43 million Facebook users (with 17.7 % penetration rate) as of Dec. 31, 2011. Figure 1 shows the rapid growth of internet users in Indonesia during some previous years[10]. These facts are some indicators of the digital media usage proliferation in Indonesia which is considered to keep on growing.

### 1.4   Tolerance Rough Sets Model at Glance

Basically, an information retrieval system consists of three main tasks: *(1)* modeling the document; *(2)* modeling the query; and *(3)* measure the degree of correlation between document and query models. Thus, the endeavor of improving an IRS revolves around those three tasks. One of the effort is a method called tolerance rough set model (TRSM) which has performed positive results on some studies pertaining to information retrieval. In spite of the fact that TRSM does not require complex linguistic process, it has not been investigated at large extent.

Since it was formulated, tolerance rough sets model (TRSM) is accepted as a tool to model a document in a *richer* way than the base representation which is represented by a vector of TF*IDF-weight terms[11] (let us call it TFIDF-

---

[7] Portal Nasional Indonesia (National Portal of Indonesia). URL: http://www.indonesia.go.id. Accessed on 25 October 2012.

[8] Key statistical highlights: International Telecommunication Union (ITU) data release June 2012. URL: http://www.itu.int. Accessed on 25 October 2012.

[9] URL: http://www.internetworldstats.com. Accessed on 25 October 2012.

[10] The graph was taken from the International Telecommunication Union (ITU). URL: http://www.itu.int/ITU-D/ict/statistics/explorer/index.html. Accessed on 25 October 2012.

[11] Appendix A provides an explanation about the TF*IDF weighting scheme.

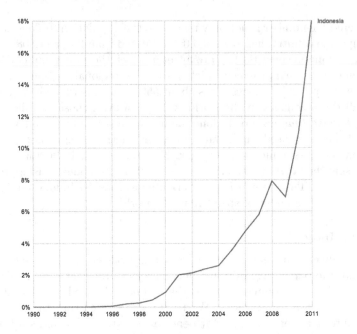

**Fig. 1.** The growth of internet users in Indonesia. The figure shows the growth of internet users in Indonesia since 1990 to 2011. On 2011, the penetration rate was close to 18 %.

representation). The richness of the document representation produced by applying the TRSM (let us call it TRSM-representation) is indicated by the number of index terms put into the model. That is to say, there are more terms belong to TRSM-representation than its base representation.

The power of TRSM is grounded on the knowledge, i.e. thesaurus, which is comprised by index terms and the relationships between them. In TRSM, each set of terms considered as semantically related with a single term $t_j$ is called the tolerance class of a term $I_\theta(t_j)$, hence the thesaurus contains tolerance classes of all index terms. The semantic relatedness is signified by the terms co-occurrence in a corpus in which a tolerance value $\theta$ is set to define the threshold of co-occurrence frequency.

### 1.5    Research Objective and Approach

The research aims to investigate the tolerance rough sets model in order to propose a framework for a semantic text retrieval system. The proposed framework is intended for Indonesian language specifically hence we are working with Indonesian corpora and applying tools for Indonesian, e.g. Indonesian stemmer, in all of the studies.

The researches of TRSM ever conducted pertaining to information retrieval have focused on the system performance and involved a combination of mathematics and engineering in their studies [13–17]. In this article, we are trying to look at TRSM from a quite different viewpoint. We are going to do empirical studies involving observations and hypotheses of human behavior as well as experimental confirmation. According to the Artificial Intelligence (AI) view, our studies follow a human-centered approach, particularly the cognitive modeling[12], instead of the rationalist approach [19, pp. 1–2]. Analogous to two faces in a coin, both approaches would result in a comprehensive perspective of TRSM.

In implementing the cognitive approach, we start our analysis from the performance of an ad hoc retrieval system. It is not our intention to compare TRSM with other methods and determine the best solution. Rather, we will take the benefit of the experimental data to learn and understand more about the process and characteristic of TRSM. The results of this process function as the guidance for computational modeling of some TRSM's tasks and finally the framework of a semantic IRS with TRSM as its heart.

## 1.6    Structure of the Article

Our research falls under the information retrieval umbrella. The following chapter provides an explanation about the main tasks of information retrieval and the semantic indexing in order to establish a general understanding of semantic IRS.

Several questions are generated in order to assist us to scrutinize the TRSM. The issues behind the questions should be apparent when we proceed into the nature of TRSM that would be exposed on theoretical basis in Sect. 3. We have selected four subjects of question and will discuss them in the following order:

1. **Is TRSM a viable alternative for a semantic IRS?** The simplicity of characteristic and positive result of studies makes TRSM an intriguing method. However, before moving any further, we need to ensure that TRSM is reasonable to be the ground floor of the intended system. This issue will be the content of Sect. 4.
2. **How to generate the system knowledge automatically?** The richer representation of document yielded by TRSM is achieved fundamentally by means of a knowledge, which is a thesaurus. The thesaurus is manually created, in the sense that a parameter, namely *tolerance value* $\theta$, is required to be determined by hand. In Sect. 5 we would propose an algorithm to resolve the matter in question, i.e. to select a value for $\theta$ automatically.
3. **How to improve the quality of the thesaurus?** The thesaurus of TRSM is generated based on a collection of text documents functions as a data source. In other words, the quality of document representation should depend on the quality of data source at some degree. Speaking of which, the TRSM basically works based on the co-occurrence data, i.e. the raw frequency of terms co-occurrence, and it arises an assumption that other co-occurrence

---

[12] The cognitive modeling is an approach employed in the Cognitive Science (CS). Cognitive science is an interdisciplinary study of mental representations and computations and of the physical systems that support those processes [18, p. xv].

data might bring a benefit for the effort to optimize the thesaurus. These presumptions would be reviewed and discussed in Sect. 6.

4. **How to improve the efficiency of the intended system?** The TRSM-representation is claimed to be richer in the sense that it consists of more terms than the base representation. Despite the fact that the terms of TRSM-representation are semantically related, more terms on document vector results in more cost of computation. In other words, system efficiency becomes the trade-off. We came into an idea of a compact document representation that would be explained in Sect. 7.

This article proposes three methods based on TRSM for the mentioned problems. All methods, which are discussed in Sects. 5 to 7, were developed by the use of our own corpus, namely ICL-corpus, and evaluated by employing an available Indonesian corpus, called Kompas-corpus[13]; Sect. 8 describes the evaluation process. The evaluation on the methods achieved satisfactory results, except for the compact document representation method; this last method seems to work only in limited domain.

The final chapter provides our conclusion of the research as well as discussion of some challenges that lead to advance studies in the future.

### 1.7 Contribution

The main contribution of this article is the modular framework of text retrieval system based on TRSM for Indonesian. Pertaining to the framework, we introduced novel strategies, which are the automatic tolerance value generator, thesaurus optimization, and lexicon-based document representation. An other contribution is a new Indonesian corpus (ICL-corpus), accompanied by a corpus consists of keywords defined by human experts (WORDS-corpus), in which both follow the format of Text REtrieval Conference (TREC)[14] [20] and ready to be used for an ad hoc evaluation of IRS. These contributions should open wider research directions pertinent to information retrieval.

## 2    Semantic Information Retrieval

### 2.1    Information Retrieval Models

The main problem of information retrieval system is the issue of determining the relevancy of a document with regard to the information need. The decision whether documents are relevant or not relies on the ranking algorithm being used which plays the role of calculating the degree of association between documents and the query as well as defining the order of documents by its degree of association, in which the top documents are considered as the most relevant ones.

---

[13] Explanation about all corpora used in this article is available in Appendix C.

[14] TREC is a forum for IR community which provides an infrastructure necessary to evaluate an IR system on a broad range of problems. URL: http://trec.nist.gov/.

In order to work, a ranking algorithm considers fundamental premises which are a set of representations of documents in given collection $D$, a set of representations for user information needs (user queries) $Q$, and a framework for modeling document/query representation $\mathcal{F}$. These basic premises, together with the ranking function $R$, determines the IR model as a quadruple $[D, Q, \mathcal{F}, R]$ [21, p. 23].

Baeza-Yates and Ribeiro-Neto [21] structured 15 IR models covered in their book into a taxonomy as well as discussed them theoretically and bibliographically. Figure 2 presents the summary of the taxonomy. A clear distinction is made on the way a user pursues information: by searching or by browsing. While browsing, a user might explore a document space which is constructed in a flat, hierarchical, or navigational organization. Another user might prefer to submit a query to the system and put the burden of searching process to the system. In order to accomplish the task, the system could analyze each document by reference to the document's content only or combination between the content and the structure of document. The *structured model* considers the latter while the *classic model* focusses on the former. The classic model is differentiated into three models with regard to the document representation: boolean, vector, and probabilistic. Respectively, in Boolean and probabilistic models, a document is represented based on set theory and probability theory, while vector model will represent a document as a vector in a high-dimensional space.

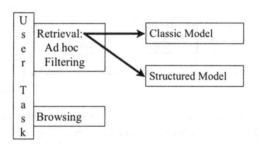

**Fig. 2.** A taxonomy of IR models. A summary of the IR models taxonomy structured by Baeza-Yates and Ribeiro-Neto.

In this article, we apply the classic vector model where document and query are represented as vectors in a high-dimensional space and each vector corresponds to a term in the vocabulary of the collection. The framework then is composed of a high-dimensional vectorial space and the standard linear algebra operations on vectors. The association degree of documents with regard to the query is quantified by the cosine of the angle between these two vectors[15].

---

[15] Appendix B provides explanation about Cosine similarity measure as a document ranking algorithm.

## 2.2   The Main Tasks of Information Retrieval

Suppose each text document conveys meaning expressed in the form of written words chosen specifically and subjectively by the writer. When text documents are fed into an IRS who employs vector space model, the text documents would be transformed into vectors of a space whose dimension is consistent with the number of index terms in the corpus. A query which conveys information need of a user could be considered as a pseudo-document, thereby analogous scenario and activities occur at user side. In the searching process, a ranking algorithm works over these two representations by measuring the degree of correlation between them.

By reference to its process, IR consists of three main tasks which are figured in Fig. 3 as filled rectangles: document modeling, query modeling, and matching process. The figure reflects that a successful matching process has two requirements: *(1)* models common to both query and document; and *(2)* system capability to construct a model which represent the information need of the user as well as the content of text document.

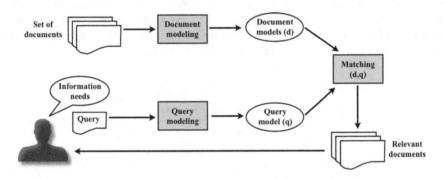

**Fig. 3.** The main tasks of information retrieval. Information retrieval consists of three tasks: (1)-document modeling; (2) query modeling; and (3) matching process.

Explained in the previous chapter, Searl's and Grice's accounts on meaning suggest knowledge shared by the speaker and its audience (i.e. the user and the system) for a successful communication. Suppose the IRS has knowledge corresponding at some degree to human, still the distinction between intentional content and the form of its externalization rises some complexity for IRS in order to construct representations of user's information need (in the query) and of author's idea (in the document). Language production and understanding are capabilities of most human, achieved through learning activities during his/her life and supported by the biologically mechanism genetically endowed [22], while none of those capabilities and support possessed by a computer system naturally. Reduced meaning retained by the representations of user's information need and of document's content become the consequence. It is highly probable that user's satisfaction of proper information with regard to his/her need then is sacrificed.

## 2.3  Semantic Indexing

Indexing is a process to construct a data structure over the text to speed up searching [21, p. 191]. The major data structure in IRS is inverted index (or *inverted file*) which provides a mapping between terms and their locations of occurrence in a text collection [1, p. 33].

The first paragraph of this chapter explained that in order to construct a model of IR, the representation of document (i.e. document indexing) as well as query should be first resolved before specifying the framework; and with these basis, an appropriate ranking function is determined. For a semantic IRS, shifting from traditional indexing into the semantic indexing hence becomes the first consideration. In case that the conventional retrieval strategies employ the bag-of-words representation of document and match directly on keywords, then the semantic indexing requires an enrichment of representation such that the IRS works with bag-of-concepts representation of documents and computes the concept similarity.

Several techniques function for enrichment of document representation are latent semantic indexing (LSI), explicit semantic analysis (ESA), and extended tolerance rough sets model (extended TRSM). These three techniques apply the classic vector space model (VSM) and thus it is possible to use conventional metrics (e.g. Cosine) in matching task. Further, they do not rely on any human-organized knowledge.

LSI, ESA, and extended TRSM naturally use statistical co-occurrence data in order to enrich the document representation, however LSI works by applying singular value decomposition (SVD), ESA relies on knowledge repository (e.g. Wikipedia), and the extended TRSM is based on rough sets theory. As a technique to dimensionality reduction, LSI identifies a number of most prominent dimensions in the data, perceived as the *latent concepts* since these concepts cannot be mapped into the natural concepts manipulated by humans or the concepts generated by system. An opposite condition happens for ESA and extended TRSM, thus the entries of their vectors are *explicit concepts*.

The following sections will describe LSI, ESA, and extended TRSM in the order given. For convenience of the explanation, a matrix is used as data structure where each entry defines the association strength between document and term. The most common measure used to calculate the strength value is the TF*IDF weighting scheme defined in Eq. (A.1).

## 2.4  Latent Semantic Indexing

Latent semantic indexing introduced by Furnas et al. [23] employs singular value decomposition (SVD) in 1988. By running SVD, it approximates the term-document matrix into a lower dimensional space hence removes some of the noise found in the document and locates two documents with similar semantic (whether or not they have matching terms) close to one another in a multi-dimensional space [24].

Running the SVD means that a term-document matrix $A$ is decomposed into the product of three other matrices such that

$$A_{m \times n} = U_{m \times s} D_{s \times s} V_{s \times n}^T. \tag{1}$$

Matrix $U$ is the left singular vectors matrix whose columns are eigenvectors of the $AA^T$ and holds the coordinates of term vectors. Matrix $V$ is the right singular vectors matrix whose columns are eigenvectors of the $A^T A$ and holds the coordinates of document vectors. Matrix $D$ refers to a diagonal matrix whose elements are the singular values of $A$, sorted by magnitude. $m$ is the total number of terms, $n$ is the total number of documents, and $s = min(m, n)$.

The *latent semantic* representation of $A$ is developed by keeping the top $k$ singular values of $D$ along with their corresponding columns in $U$ and $V^T$ matrices. The result is a $k$-rank matrix $A'$ which is closest in the least squares sense to matrix $A$; it contains less noisy dimensions and captures the major associational structure of the data [23]. Figure 4 presents the schematic of SVD for matrix $A$ and its reduced model.

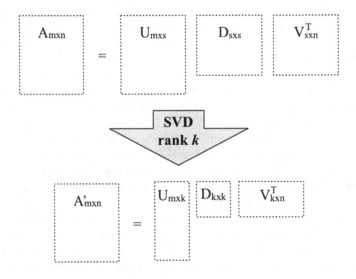

**Fig. 4.** The illustration of SVD. SVD illustration of a terms-by-documents matrix $A$ of rank $k$.

With regard to a query, its vector $q$ is treated similar to matrix $A$ by following this rule

$$q_{1 \times k} = q_{1 \times m}^T U_{m \times k} S_{k \times k}^{-1}. \tag{2}$$

After all, the matching process between query and documents is conducted by computing the similarity coefficient between $k$-rank query vector $q_k$ and corresponding columns of $k$-rank matrix $V_k$.

## 2.5   Explicit Semantic Analysis

In 2007, Gabrilovich and Markovitch introduced the notion of explicit semantic analysis (ESA) [25]. Later, Gottron et al. [26] showed that ESA is a variation of the generalized vector space model (GVSM)[16] [27] who considers term correlation.

ESA represents documents and query as vectors in a high dimensional of *concept space*, instead of *term space*, thus each dimension corresponds to a concept. Each coordinate of concept vector expresses the degree of association between the document and the corresponding concept. Suppose $D = \{d_1, \ldots, d_i, \ldots, d_N\}$ is a set of documents and $T = \{t_1, \ldots, t_j, \ldots, t_M\}$ is the vocabulary of terms, then the association value $u_{ik}$ between document $d_i$ and concept $c_k, k \in \{1, \ldots, K\}$ is defined as

$$u_{ik} = \sum_{t_j \in T} w_{ij} \times c_{jk} \qquad (3)$$

where $w_{ij}$ denotes the weight of term $t_j$ in document $d_i$ and $c_{jk}$ signifies the correlation between term $t_j$ and concept $c_k$.

Equation (3) describes the association value as the product of weight of term in document ($w_{ij}$) and weight of concept in knowledge base concept ($c_{jk}$), hence there are two computations need to be done in advance. Basically, both computations could be done using the TF*IDF weighting scheme, however the former is calculated over a corpus functioned as the system data, while the latter is calculated over a corpus functioned as the knowledge base; thus there are two corpora functioned differently. The merge of system data's and knowledge base's weights yields a new representation for the system data, i.e. bag-of-concepts representation.

Gabrilovich and Markovitch [25] suggests Wikipedia articles for the corpus functioned as the knowledge base considering that it is a vast amount of highly organized human knowledge and undergoes constant development. However, the main reason is Wikipedia treats each description as a separate article, thus each description is perceived as a single concept. By this definition, any collection of documents is possible to be used as the external knowledge base.

Figure 5 shows the computation process of ESA in order to convert the bag-of-words representation of system data into the bag-of-concepts representation by utilizing natural language definition of concepts from the knowledge base.

## 2.6   Extended Tolerance Rough Sets Model

As its name, the extended TRSM is an extension of TRSM proposed by Nguyen et al. [28] in 2012. Detail explanation about TRSM is available in the following chapter, hence in this section we focus only on the extension part of TRSM.

The study of Nguyen et al. [28,29] aimed to enrich the document representation worked in clustering task by incorporating other information than the index

---

[16] Consistent with VSM, GVSM interprets index term vectors as linearly independent, however they are not orthogonal.

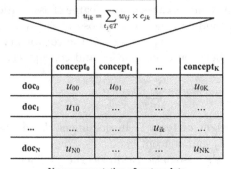

| | term$_0$ | term$_1$ | ... | term$_M$ |
|---|---|---|---|---|
| doc$_0$ | $w_{00}$ | $w_{01}$ | ... | $w_{0M}$ |
| doc$_1$ | $w_{10}$ | ... | ... | ... |
| ... | ... | ... | $w_{ij}$ | ... |
| doc$_N$ | $w_{N0}$ | ... | ... | $w_{NM}$ |

Representation of system data

| | concept$_0$ | concept$_1$ | ... | concept$_K$ |
|---|---|---|---|---|
| term$_0$ | $c_{00}$ | $c_{01}$ | ... | $c_{0K}$ |
| term$_1$ | $c_{10}$ | ... | ... | ... |
| ... | ... | ... | $c_{jk}$ | ... |
| term$_M$ | $c_{M0}$ | ... | ... | $c_{MK}$ |

Representation of knowledge base

$$u_{ik} = \sum_{t_j \in T} w_{ij} \times c_{jk}$$

| | concept$_0$ | concept$_1$ | ... | concept$_K$ |
|---|---|---|---|---|
| doc$_0$ | $u_{00}$ | $u_{01}$ | ... | $u_{0K}$ |
| doc$_1$ | $u_{10}$ | ... | ... | ... |
| ... | ... | ... | $u_{ik}$ | ... |
| doc$_N$ | $u_{N0}$ | ... | ... | $u_{NK}$ |

New representation of system data

**Fig. 5.** The ESA. Visualization of the semantic indexing process in ESA.

terms of document corpus, namely citation and semantic concept. The citation referred to the bibliography of a given scientific article while the semantic concept was constructed based on an additional knowledge source, i.e. DBpedia. Thereby, the extended TRSM was defined as a tuple

$$\mathcal{R}_{Final} = (\mathcal{R}_T, \mathcal{R}_B, \mathcal{R}_C, \alpha_n) \tag{4}$$

where $\mathcal{R}_T$, $\mathcal{R}_B$, and $\mathcal{R}_C$ denote the tolerance spaces which are determined respectively over the set of terms $T$, the set of bibliography items cited by a document $B$, and the set of concepts in the knowledge domain $C$. The function $\alpha_n : P(T) \longrightarrow P(C)$ is called the *semantic association* for terms, thus $\alpha_n(T_i)$ is the set of $n$ concepts most associated with $T_i$ for any $T_i \subset T$ [30].

In this model, each document $d_i \in D$ associated with a pair $(T_i, B_i)$ is represented by a triple

$$\mathbf{U}_\mathcal{R}(d_i) = \{\mathbf{U}_{\mathcal{R}_T}(d_i), \mathbf{U}_{\mathcal{R}_B}(d_i), \alpha_n(T_i)\} \tag{5}$$

where $T_i$ is the set of terms occurring in document $d_i$ and $B_i$ is the set of bibliography items cited by document $d_i$. The study of extended TRSM which presented with positive results indicated that the method would be effective to be realized in a real application.

It is obvious from Eqs. (4) and (5) that the extended TRSM accommodates different factors at once for a semantic indexing, instead of one factor such as in original TRSM as well as LSI and ESA. Further, the model is more nature considering the real life situation of information retrieval process.

# 3    Tolerance Rough Sets Model

## 3.1    Rough Sets Theory

In 1982, Pawlak introduced a method called rough sets theory [31] as a tool for data analysis and classification. During the years, this method has been studied and implemented successfully in numerous areas of real-life applications [32]. Basically, rough sets theory is a mathematical approach to vagueness which expresses the vagueness of a concept by means of the boundary region of a set; when the boundary region is empty, it is a crisp set. Otherwise, it is a rough set [33]. The central point of rough sets theory is an idea that any concept can be approximated by its *lower* and *upper approximations*, and the vagueness of concept is defined by the region between its upper and lower approximations. Consider Fig. 6 for illustration.

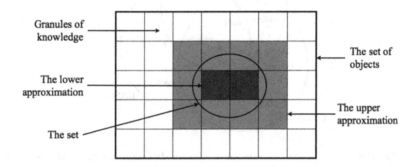

**Fig. 6.** Rough sets. Basic idea of rough sets theory as it is explained in [33]

Let us think of a concept as a subset $X$ of a universe $U$, $X \subseteq U$, then in a given *approximation space* $A = (U, R)$ we can denote the lower approximation of concept $X$ as $L_A(X)$ and the upper approximations of concept $X$ as $U_A(X)$. The boundary region, $BN_A(X)$, is the difference between the upper and lower approximations, hence

$$BN_A(X) = U_A(X) - L_A(X) \tag{6}$$

Let $R \subseteq U \times U$ be an *equivalence relation* that will partition the universe into *equivalence classes*, or *granules of knowledge*, thus formal definition of lower and upper approximations are

$$L_A(X) = \bigcup_{x \in U} \{R(x) : R(x) \subseteq X\} \tag{7}$$

$$U_A(X) = \bigcup_{x \in U} \{R(x) : R(x) \cap X \neq \emptyset\} \tag{8}$$

## 3.2   Tolerance Rough Sets Model

The equivalence relation $R \subseteq U \times U$ of classical rough sets theory required three properties [32]: reflexive ($xRx$), symmetric ($xRy \to yRx$), and transitive ($xRy \wedge yRz \to xRz$); for $\forall x, y, z \in U$, thus the universe of an object would be divided into disjoint classes. These requirements have been showed to be not suitable for some practical applications (viz. working on text data), because the association between terms was better viewed as overlapping classes (see Fig. 7), particularly when term co-occurrence was used to identify the semantic relatedness between terms [14].

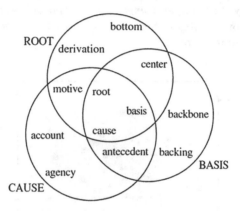

**Fig. 7.** Overlapping classes. Overlapping classes between terms *root*, *basis*, and *cause* [14]

The overlapping classes can be generated by a relation called *tolerance relation* which was introduced by Skowron and Stepaniuk [34] as a relation in *generalized approximation space*. The generalized approximation space is denoted as a quadruple $\mathcal{A} = (U, I, \nu, P)$, where $U$ is a non-empty universe of objects, $I$ is the uncertainty function, $\nu$ is the vague inclusion function, and $P$ is the structurality function.

Tolerance Rough Sets Model (TRSM) was introduced by Kawasaki, Nguyen, and Ho in 2000 [13] as a document representation model based on generalized approximation space. In the information retrieval context, we can assume a document as a concept. Thus, implementing TRSM means that we approximate concepts determined over the set of terms $T$ on a tolerance approximation space $\mathcal{R} = (T, I, \nu, P)$ by employing the tolerance relation.

In order to generate the document representation, which is claimed to be richer in terms of semantic relatedness, the TRSM needs to create tolerance classes of terms and approximations of subsets of documents. If $D = \{d_1, d_2, ..., d_N\}$ is a set of text documents and $T = \{t_1, t_2, ..., t_M\}$ is a set of index terms from $D$, then the tolerance classes of terms in $T$ is created based on the co-occurrence of index terms in all documents $D$. A document representation is represented as a vector of weight $d_i = \{w_{i,1}, w_{i,2}, ..., w_{i,M}\}$, where $w_{i,j}$ denotes the weight of term $t_j$ in document $d_i$ and calculated by considering the upper approximation of document $d_i$.

### Tolerance Approximation Space

The definitions of tolerance approximation space $\mathcal{R} = (T, I, \nu, P)$ are as follows

**Universe:** The universe $U$ is the set of index terms $T$

$$U = \{t_1, t_2, ..., t_M\} = T \tag{9}$$

**Tolerance class:** Skowron and Stepaniuk [34] maintain that an uncertainty function $I : U \to \mathbb{P}(U)$, where $\mathbb{P}(U)$ is a power set of $U$, is any function from $U$ into $\mathbb{P}(U)$ satisfying the conditions $x \in I(x)$ for $x \in U$ and $y \in I(x) \Leftrightarrow x \in I(y)$ for any $x, y \in U$. This means that we assume the relation $xIy \Leftrightarrow y \in I(x)$ is a tolerance relation and $I(x)$ is a tolerance class of $x$. The parameterized tolerance class $I_\theta$ is then defined as

$$I_\theta(t_i) = \{t_j \mid f_D(t_i, t_j) \geq \theta\} \cup \{t_i\} \tag{10}$$

where $\theta$ is a positive parameter and $f_D(t_i, t_j)$ denotes the number of documents in $D$ that contain both terms $t_i$ and $t_j$. From Eq. (10), it is clear that it satisfies the condition of being reflexive ($t_i \in I_\theta(t_i)$) and symmetric ($t_j \in I_\theta(t_i)$) required by a tolerance relation; the tolerance relation $R \subseteq T \times T$ can be defined by means of function $I_\theta$ as $t_i R t_j \Leftrightarrow t_j \in I_\theta(t_i)$.
Assuming that a term is a concept, then the tolerance class $I_\theta(t_i)$ consists of terms related to a concept $t_i$ and the precision of the concept determined might be tuned by varying the threshold $\theta$.

**Vague inclusion function:** the vague inclusion function $\nu : \mathbb{P}(U) \times \mathbb{P}(U) \to [0, 1]$ measures the degree of inclusion between two sets and is defined as

$$\nu(X, Y) = \frac{|X \cap Y|}{|X|} \tag{11}$$

where the function $\nu$ must be *monotone* w.r.t the second argument, i.e. if $Y \subseteq Z$ then $\nu(X, Y) \leq \nu(X, Z)$ for $X, Y, Z \subseteq U$. Hence, the vague inclusion function can determine the matter whether the tolerance class $I(x)$ of an object $x \in U$ is included in a set $X$.
Together with the uncertainty function $I$, the vague inclusion function $\nu$ defines the *rough membership function* for $x \in U, X \subseteq U$ as $\mu_{I,\nu}(x, X) = \nu(I(x), X)$. Therefore, the membership function $\mu$ for $t_i \in T, X \subseteq T$ is defined as

$$\mu(t_i, X) = \nu(I_\theta(t_i), X) = \frac{|I_\theta(t_i) \cap X|}{|I_\theta(t_i)|} \tag{12}$$

**Structurality function:** with structurality function $P : I(U) \rightarrow \{0,1\}$, where $I(U) = \{I(x) : x \in U\}$, one can construct two subsets based on value of $P(I(x))$, named *structural subset* and *nonstructural subset*, when $P(I(x)) = 1$ and $P(I(x)) = 0$ respectively. In TRSM, all tolerance classes of index terms are considered as structural subsets, hence for all $t_i \in T$

$$P(I_\theta(t_i)) = 1 \tag{13}$$

## Approximations

With the foregoing definitions, we can define the lower approximation $\mathbf{L}_\mathcal{R}(X)$, upper approximation $\mathbf{U}_\mathcal{R}(X)$, and boundary region $\mathbf{BN}_\mathcal{R}(X)$ of any subset $X \subseteq T$ in a tolerance space $\mathcal{R} = (T, I_\theta, \nu, P)$ as follows

$$L_\mathcal{R}(X) = \{t_i \in T \mid \nu(I_\theta(t_i), X) = 1\} \tag{14}$$
$$U_\mathcal{R}(X) = \{t_i \in T \mid \nu(I_\theta(t_i), X) > 0\} \tag{15}$$
$$BN_\mathcal{R}(X) = U_\mathcal{R}(X) - L_\mathcal{R}(X) \tag{16}$$

Refers to the basic idea of rough sets theory [33], for any set of $X$, intuitively we may assume the upper approximation as the set of concepts that share some semantic meanings with $X$, the lower approximation as the *core* concepts of $X$, while the boundary region consists of concepts that *cannot be classified uniquely* to the set or its complement, by employing available knowledge.

## TRSM Document Representation

After all, the richer representation of document $d_i \in D$ is achieved by simply representing the document with its upper approximation, i.e.

$$\mathbf{U}_\mathcal{R}(d_i) = \{t_i \in T \mid \nu(I_\theta(t_i), d_i) > 0\} \tag{17}$$

followed by calculating the weight vector using the extended weighting scheme, i.e.

$$w_{ij}^* = \frac{1}{S} \begin{cases} (1 + \log f_{d_i}(t_j)) \log \frac{N}{f_D(t_j)} & \text{if } t_j \in d_i \\ 0 & \text{if } t_j \notin \mathbf{U}_\mathcal{R}(d_i) \\ \min_{t_k \in d_i} w_{ik} \frac{\log \frac{N}{f_D(t_j)}}{1 + \log \frac{N}{f_D(t_j)}} & \text{otherwise} \end{cases} \tag{18}$$

where $S$ is a normalisation factor applied to all document vectors. The extended weighting scheme is defined from the standard TF*IDF weighting scheme and is necessary in order to ensures that each term occurring in the upper approximation of $d_i$ but not in the $d_i$ itself has a weight smaller that the weight of any terms in $d_i$.

By employing TRSM, the final document representation has less zero-valued similarities. This leads to a higher possibility of two documents having non-zero similarities although they do not share any terms. This is the main advantage the TRSM-based algorithm claims to have over traditional approaches.

### 3.3    The Challenges of TRSM

We identified that there are three fundamental components of TRSM to work which are dependent in sequence: *(1)* the tolerance classes of all index terms; *(2)* the upper document representation; and *(3)* the TRSM weighting scheme. Figure 8 displays the basic process of tolerance rough sets model which contains those three components.

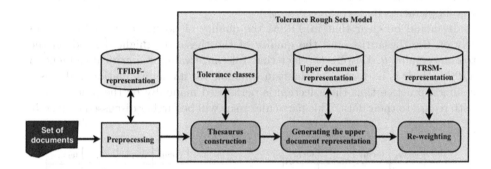

**Fig. 8.** Tolerance rough sets model. The process of tolerance rough sets model.

A document representation based on TRSM (TRSM-representation) can be seen as the revised version of a base representation which is recalculated using the TRSM weighting scheme. The base representation is modeled by calculating the term frequency (TF) and the inverse document frequency (IDF) of a term, i.e. commonly called TF*IDF weighting scheme; hence we dub this representation the TFIDF-representation. Suppose the representation of document produced by TRSM and TF*IDF are structured as matrices, thus Fig. 9 shows the relationship between them, where *tfidf* and *trsm* denote the weight of term computed by TF*IDF and TRSM weighting scheme respectively.

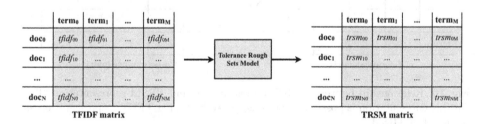

**Fig. 9.** Relationship between TFIDF-representation and TRSM-representation. The TRSM-representation is possible to be constructed by taking TFIDF-representation as the input of TRSM.

During term weight computation, TRSM consults the upper document representation, whereas the upper representation of a document is only possible to be generated when the tolerance classes of all index terms are available. Refer to Eq. (10), a tolerance class of a term $t_i$ consists of all index terms consider as semantically related with the term $t_i$ and the precision of relatedness between a pair of terms is defined by the tolerance value $\theta$. In other words, the importance of relationship between terms is determined by $\theta$ value.

Based on the nature of TRSM, tolerance classes can be categorized as a thesaurus; a lightweight ontology who reflects the relationship between terms [35]. As the heart of TRSM, thesaurus becomes the knowledge of the system who implements it.

It should be clear that in TRSM the quality of document modeling would rely on the thesaurus, and the quality of the thesaurus might depend on the tolerance value $\theta$. Despite the fact that tolerance value is a critical element in TRSM, there is no formal mechanism available for its determination and it is a common practice that the selection is performed manually by the practitioners with regard to their data. This particular issue will be further discussed in Sect. 5.

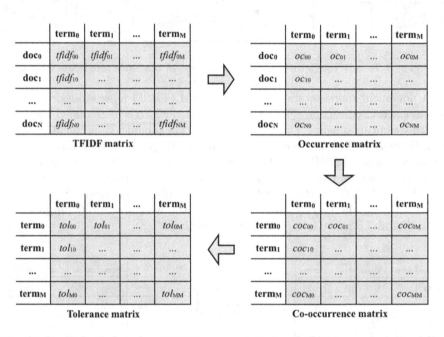

**Fig. 10.** Relationship between TFIDF-representation and TRSM-representation. The TRSM-representation is possible to be constructed by taking TFIDF-representation as the input of TRSM.

The thesaurus might be constructed based on an algorithm explained by Nguyen and Ho [15]. The algorithm takes a document-by-term matrix (i.e. the TFIDF matrix) as the input and yields the tolerance matrix, which is structured as a binary term-by-term matrix. Figure 10 shows the steps of the algorithm. Subsequently, the occurrence binary matrix *OC matrix*, the co-occurrence matrix *COC matrix*, and the tolerance matrix *TOL matrix* were generated in sequence manner by employing Eqs. (19), (20), and (21). Note that $tfidf_{i,j}$ denotes the weight of term $j$ computed by TF*IDF scheme in document $i$, card($OC^x$ AND $OC^y$) denotes the cardinality of two terms, $t_x$ and $t_y$, being occurred together in a collection, and $\theta$ is the co-occurrence threshold of terms.

$$oc_{i,j} = 1 \quad \Leftrightarrow \quad tfidf_{i,j} > 0 \ . \tag{19}$$

$$coc_{x,y} = \text{card}(OC^x \text{ AND } OC^y) \ . \tag{20}$$

$$tol_{x,y} = 1 \quad \Leftrightarrow \quad coc_{x,y} \geq \theta \ . \tag{21}$$

In addition to tolerance value, the algorithm demonstrated that a data source might have an impact on the thesaurus quality since it manipulates a set of documents functions as the only input. The effects might be as a consequence of the type of data source or the size of the collection.

Another important subjects relevant to the thesaurus of TRSM is pertaining to the fact that the thesaurus is created based on the quantity of two terms occur together, thereby employs tolerance value $\theta$ as the threshold of semantic relatedness. In fact, refer to the term weighting scheme, there are other alternatives of co-occurrence data who takes more factors into consideration, e.g. the TF*IDF weighting scheme. By that means, other similarity measures, i.e. cosine, might be applied. The presumptions pertaining to thesaurus optimization will be examined in Sect. 6.

Refer to the path of TRSM, its computation complexity is the aggregation of each task, i.e. thesaurus construction, upper representation generation, and re-weighting task. The first task requires $O(NM)$ [15], where $N$ defines the number of documents and $M$ defines the number of index terms, while the second and third tasks both require $O(M^2)$. Thus totally, the upper bound of TRSM implementation would be $O(NM^2)$. With regard to the system efficiency, minimizing the dimensionality of document vectors would be a practical alternative for the complexity. Using this as the starting point, in Sect. 7 we are going to introduce a novel model of document namely the lexicon-based representation.

## 4    The Potential of TRSM

### 4.1    Introduction

We may find studies showing the positive results of TRSM implementation for document clustering task [13–16], query expansion [17], and document retrieval task [36]. Those studies claimed that TRSM-representation was richer than the baseline representation (TFIDF-representation), however none has shown and explained empirically the richness.

It has been known that the richness of TRSM-representation is understood as having less zero-value similarities and having higher possibility that two documents holding non-zero similarities although they do not share any terms. The result of our study presented in this chapter confirmed those affirmations and add another fact. We found that the TRSM-representation consists of terms considered as important by human experts. Further, the study revealed that rough sets theory seems to work in accordance with the natural way of human thinking. Finally, the study showed that TRSM is a viable option for a semantic IRS.

## 4.2    Experiment Process

We used two corpora, ICL-corpus and WORDS-corpus[17], with 127 topics. We took an assumption that each topic given by human experts in annotation process was a concept, therefore we considered the keywords determined by the human experts[18] as the term variants that highly related with particular concept. These keywords are the content of text body in WORDS-corpus, hence each document of WORDS-corpus contains important terms of particular concept(s) selected by human experts. With regard to the automatic process of the system, we considered these keywords as the relevant terms for each document (which bear one or more topics) that should be selected by the system. Therefore, WORDS-corpus was treated as the ground truth of this study.

Figure 11 shows the general process of the study and the dashed rectangle identifies the focus of the experiment, which were performed twice, i.e. with stemming task and without stemming task.

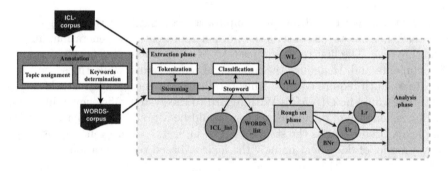

**Fig. 11.** Main phases of the study. There were 3 main phases: extraction, rough sets, and analysis. A rectangle represents a process while a circle represent a result.

**Extraction Phase.** The main objective of extraction phase was preprocessing both corpora. Confix-Stripping stemmer (CS stemmer), a version of Indonesian

---

[17] ICL-corpus consists of 1,000 documents taken from an Indonesian choral mailing list, while WORDS-corpus consists of 1,000 documents created from ICL-corpus in an annotation process conducted by human experts. Further explanation of these corpora is available in Appendix C.1.

[18] We collaborated with 3 choral experts during annotation process. Their backgrounds could be reviewed in Appendix C.3.

stemmer, was employed in the stemming task while Vega's stopword [37] was applied in the stopword task[19]. CS stemmer was introduced as a new confix-stripping approach for automatic Indonesian stemming and was showed as the most accurate stemmer among other automated Indonesian stemmers [38]. Vega's stopword has shown to produce the highest precision@10, R-precision, and recall values (although the differences without stopping words are not significant ($p < 0.05$), except for the recall value (0.038)), among other available Indonesian stopword lists [10].

Documents were tokenized based on character other than alphabetic. The resulted tokens were stemmed by the CS stemmer and then compared to the Vega's stopword. It yielded lists of unique terms and its frequency. There were 9,458 unique terms extracted from ICL-corpus and 3,390 unique terms extracted from WORDS-corpus; called *ICL_list* and *WORDS_list* respectively. When it was run without stemming process, we identified 12,363 unique terms in ICL_list and 4,281 unique terms in WORDS_list.

Both corpora were classified based on 127 topics yielded in preliminary process, i.e. annotation process[20]. Recall that we took an assumption that each topic was a concept and keywords determined by human experts were important variants of a concept hence aggregation of terms appeared in each class were taken as the terms for representative vector of each class. The set of classes resulted from ICL-corpus was called *IL* while the set of classes resulted from WORDS-corpus was called *WL*; each set of class, IL and WL, consists of 127 classes. So, technically speaking, instead of document-term matrix, we worked with topic-by-term matrix.

**Rough Set Phase.** This phase was conducted in order to generate the lower set, upper set, and boundary set of each class in IL. These sets were possible to be created using Eqs. (14), (15), and (16) when tolerance classes of all index terms were ready.

The tolerance classes was constructed by following the steps described in Fig. 10 of previous chapter, with an exception that in this experiment the algorithm took the topic-by-term frequency matrix as its input. Thereby, Fig. 12 displays the steps applied for the thesaurus generation of this particular study.

**Analysis Phase.** In analysis phase, we examined the mean recall and precision of upper set (US), lower set (LS), and boundary set (BS) of IL by taking the WL as the ground truth. The computations were run for co-occurrence threshold $\theta$ between 1 to 75.

Recall and precision are the most frequent and basic measures for information retrieval effectiveness [39]. Recall $R$ is the fraction of relevant documents that are retrieved while precision $P$ is the fraction of retrieved documents that are

---

[19] We used CS stemmer and Vega's stopword in all of our studies presented in this article.

[20] Please see Appendix C.1 for explanation of annotation process.

| | term$_0$ | term$_1$ | ... | term$_M$ |
|---|---|---|---|---|
| **topic$_0$** | tf$_{00}$ | tf$_{01}$ | ... | tf$_{0M}$ |
| **topic$_1$** | tf$_{10}$ | ... | ... | ... |
| ... | ... | ... | ... | ... |
| **topic$_N$** | tf$_{N0}$ | ... | ... | tf$_{NM}$ |

Topic-by-term matrix

| | term$_0$ | term$_1$ | ... | term$_M$ |
|---|---|---|---|---|
| **topic$_0$** | oc$_{00}$ | oc$_{01}$ | ... | oc$_{0M}$ |
| **topic$_1$** | oc$_{10}$ | ... | ... | ... |
| ... | ... | ... | ... | ... |
| **topic$_N$** | oc$_{N0}$ | ... | ... | oc$_{NM}$ |

Occurrence matrix

| | term$_0$ | term$_1$ | ... | term$_M$ |
|---|---|---|---|---|
| **term$_0$** | tol$_{00}$ | tol$_{01}$ | ... | tol$_{0M}$ |
| **term$_1$** | tol$_{10}$ | ... | ... | ... |
| ... | ... | ... | ... | ... |
| **term$_M$** | tol$_{M0}$ | ... | ... | tol$_{MM}$ |

Tolerance matrix

| | term$_0$ | term$_1$ | ... | term$_M$ |
|---|---|---|---|---|
| **term$_0$** | coc$_{00}$ | coc$_{01}$ | ... | coc$_{0M}$ |
| **term$_1$** | coc$_{10}$ | ... | ... | ... |
| ... | ... | ... | ... | ... |
| **term$_M$** | coc$_{M0}$ | ... | ... | coc$_{MM}$ |

Co-occurrence matrix

**Fig. 12.** Tolerance classes construction. The construction of tolerance classes in this study took topic-term matrix as the input and produced a term-by-term matrix as the output. Here, $M$ denotes the number of index term and $N$ denotes the number of topic. relevant. Suppose $Rel$ denotes relevant documents and $Ret$ denotes retrieved documents, then recall $R$ and precision $P$ are defined as follows

$$R = \frac{\sharp(relevant\ items\ retrieved)}{\sharp(relevant\ items)} = \frac{|Rel \cap Ret|}{|Rel|} \tag{22}$$

$$P = \frac{\sharp(relevant\ items\ retrieved)}{\sharp(retrieved\ items)} = \frac{|Rel \cap Ret|}{|Ret|} \tag{23}$$

In this study, both measures were used for the *terms* rather than *documents*. That is to say, by considering WORDS_list as the ground truth, then recall $R$ is the fraction of relevant terms that are retrieved while precision $P$ is the fraction of retrieved terms that are relevant. The formulas applied for recall and precision are displayed in the first and second rows of Table 1, where $US_{IL}$, $LS_{IL}$, and $BS_{IL}$ respectively denote the upper set, lower set, and boundary region of IL. The Recall$_{IL}$ of the third row is an additional calculation used for evaluating the recall of IL terms in each set. Based on the definition, better recall is preferred than better precision for the reason that better recall would ensure the availability of relevant terms in the set.

**Table 1.** Formulas for recall and precision calculations.

| | US | LS | BS |
|---|---|---|---|
| Recall$_{WL}$ | $\frac{|WL \cap US_{IL}|}{|WL|}$ | $\frac{|WL \cap LS_{IL}|}{|WL|}$ | $\frac{|WL \cap BS_{IL}|}{|WL|}$ |
| Precision$_{WL}$ | $\frac{|WL \cap US_{IL}|}{|US_{IL}|}$ | $\frac{|WL \cap LS_{IL}|}{|LS_{IL}|}$ | $\frac{|WL \cap BS_{IL}|}{|BS_{IL}|}$ |
| Recall$_{IL}$ | $\frac{|IL \cap US_{IL}|}{|IL|}$ | $\frac{|IL \cap LS_{IL}|}{|IL|}$ | $\frac{|IL \cap BS_{IL}|}{|IL|}$ |

## 4.3   Discussion

With regard to the process of developing WORDS-corpus, the fact that ICL_list could cover almost all WORDS_list terms was not surprising. It was interesting though that there were some terms of WORDS_list did not appear in ICL_list; 17 terms produced by the process without stemming task and 11 terms produced by the process with stemming task. By examining those terms, we found that the CS stemmer could only handle the formal terms (6 terms) and left the informal terms (5 terms) as well as the foreign term (1 term); the other terms caused by typographical error (5 terms) in ICL_corpus.

Despite the fact that CS stemmer succeeded in reducing the size of ICL_list for 23.50 % as well as of WORDS_list for 20.81 %, it reduced the mean recall of IL about 0.64 % for each class from 97.39 %. We noticed that the mean precision of IL was increased about 0.25 % for each class, however the values themselves were very small (14.56 % for process without stemming task and 14.81 % for process with stemming task). From these, we could say that the ICL_list was too noisy of containing numerous unimportant terms for particular topic.

Table 2 shows the mean values of recall and precision for the sets of IL (i.e. the upper set (US), lower set (LS), and boundary set (BS)) when they were run with and without stemming by considering WORDS_list (WL) as the ground truth. Exceptional is for the third row which is the recall of IL sets over the ICL_list (IL). All of these calculations performed by applying the formulas displayed in Table 1.

**Table 2.** Average recall and precision of ICL_list (IL) and WORDS_list (WL).

|  | With stemming | | | Without stemming | | |
|---|---|---|---|---|---|---|
|  | US (%) | LS (%) | BS (%) | US (%) | LS (%) | BS (%) |
| (1) | (2) | (3) | (4) | (5) | (6) | (7) |
| Recall$_{WL}$ | 97.64 | 5.55 | 92.08 | 97.55 | 4.64 | 92.91 |
| Precision$_{WL}$ | 13.77 | 27.49 | 13.50 | 14.13 | 26.30 | 13.75 |
| Recall$_{IL}$ | 100.00 | 5.00 | 95.00 | 100.00 | 4.43 | 95.57 |

## Recall

It was explained in Sect. 3.2, that for any set of X, the upper set might consist of terms that share some semantic meanings with X. Further, notice that in this study we used a specific domain of corpus, which is a choral corpus. Based on these, the values of Recall$_{IL}$-US in Table 2 for process with and without stemming task, which are 100 %, made us confident that the TRSM model has been employed correctly. The upper sets consist of all ICL_list terms due to the fact that generally all index terms are semantically related with choral domain.

One task of annotation process conducted by human experts was keyword determination[21]. It was a fact that during that task our human experts seemed to encounter difficulty in defining keywords for a topic many times. When they were in this position, they preferred to choose sentences on the text or even make their own sentences to describe the topic rather than listing the highly related keywords specifically. The consequence was they introduced numerous number of terms for particular topic. It explains why the value of Recall$_{WL}$-US for both with and without stemming in Table 2 are very high. It is important to be noted that we reckoned all the terms used by the human experts as relevant terms for the reason that those terms however selected to be used to describe a topic.

This human behavior is reflected by the rough sets theory. We may see on Table 2, that the mean recalls of WORDS_list (WL) in lower sets (LS) are very low while the mean recalls of WL in boundary regions (BS) are very high. Refer to Sect. 3.2, intuitively the lower set might consist of the *core* terms while the boundary region might consist of the *uncertain* terms. We can see similar results from the table for ICL-corpus, i.e. the mean recalls of ICL_list (IL) in lower sets (LS) are very low while the mean recalls of IL in boundary regions (BS) are very high. We might infer now that the rough sets theory mimics the natural way of human thinking.

With regard to stemming, we can see that all values in column 3 of Table 2 are higher than all values in column 6 while all values in column 4 are lower than all values in column 7. It seems that employing stemming task increases system's capability to retrieve the *core* terms of a concept and to avoid the *uncertain* terms at the same time. Further, the table also shows us that Recall$_{WL}$-US value of process with stemming is higher than the one without stemming, which leads us to an assumption that the stemming task is able to retrieve more relevant terms in general. It supports our confidence so far that stemming task with CS stemmer would bring more benefit in this framework of study.

**Precision**

Despite the fact that better recall is preferred than better precision, as we mentioned in Sect. 4.2, we notice that the values of Precision$_{WL}$-US are small (13.77 % and 14.13 %). With regard to Table 1, they were calculated using equation $P = \frac{|WL \cap US_{IL}|}{|US_{IL}|}$. Based on the formula, we may expect to improve the precision value by doing one, or both, of these:

1. increase the co-occurrence terms between $WL$ and $US_{IL}$; or
2. decrease the total number of $US_{IL}$.

Refer to Eq. (21), make the $\theta$ value higher will reduce the size of upper sets[22], and refer to Eq. (23) it will increase the mean precision of upper sets in WL_list. So, technically the total number of terms in an upper set is easily

---

[21] Please see Appendix C.1.

[22] If the size of tolerance classes are smaller then the size of upper sets will be smaller, and vice versa.

modified by altering the tolerance value $\theta$. However, it raises a typical question, i.e. what is the best $\theta$ value and how to set it up? As we have briefly explained the importance and the problem pertaining to $\theta$ value in Sect. 3.3 of previous chapter, this issue seems to support our argumentation that an algorithm to set the $\theta$ value automatically is significant.

The index term of WORDS-corpus is clearly constant for we took it as the ground truth, hence there is nothing we can do about WL. Suppose we have a constant number of US (after setting up the $\theta$ at a certain value), then the possibility to improve the precision lies on the cardinality of terms in $WL \cap US_{IL}$ set, or in other words on maximizing the availability of relevant terms in upper sets. Based on the nature of TRSM method, this could be happened when we have an optimized thesaurus which defines the relationship between terms appropriately. Knowing that a thesaurus is constructed by a set of documents functioned as data source then we might expect better thesaurus if we know the characteristic of data source we should have. Moreover, based on Eq. (10), another alternative could be related with the semantic relatedness measure applied in thesaurus construction process.

### Tolerance Value

Figure 13 shows the mean recall of WORDS_list in upper sets of ICL_list for a process with stemming task when $\theta$ value is altered from 1 to 75. It is clear from the figure that the number of relevant terms of WORDS_list drastically filtered out from the upper set of ICL_list at low $\theta$ values. However, at some points the changes starts to be stable; Taking one value, e.g. $\theta = 21$. The average number of terms in upper sets when $\theta = 21$ (733.79 terms) is interesting for it was reduced up to 92.24 % of the average number of terms in upper sets when $\theta = 0$ (9,458 terms). Whereas from Fig. 13, we can see that the mean recall at $\theta = 21$ is maintained to be high (97.58 %). By this manual inspection, we are confident to propose $\theta \geq 21$ to be used in similar framework of study.

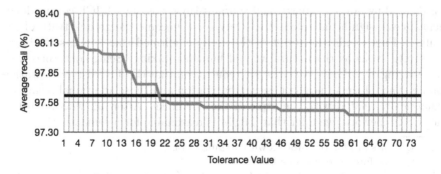

**Fig. 13.** The Recall$_{WL}$-US graph. This graph shows the average recall of the sets of WORDS_list in upper sets of ICL_list) for $\theta$ value 1 to 75.

We urge that the upper sets of ICL_list ($US_{IL}$) enrich the sets of ICL_list (IL). This assertion is based on two empirical data[23]:

1. the mean recall of $WL$ in $IL$ over 127 topics is 96.75 %; while
2. the mean recall of $WL$ in $US_{IL}$ over 127 topics when $\theta$ is altered between 1 to 75 is 97.64 %

Thus, we might infer now that the upper sets of ICL_list contain more relevant terms then the sets of ICL_list. In order to construct a document representation, TRSM considers the upper set of a document, hence we might expect that the resulted TRSM-representation consists of more terms and those terms are semantically related. This is a stronger assertion for the claim that tolerance rough sets model enriches the traditional representation of a document and this is a good indicator of TRSM as a feasible method for a semantic IRS.

### ICL_list vs. Lexicon

Lexicon is a vocabulary of terms [39]. The lexicon utilized by CS stemmer consists of 29,337 Indonesian base words. Comparison between ICL_list and Lexicon showed that there were 3,321 co-occurrence terms. In other words, 64.89 % of ICL_list (which is 6,137 terms in total) was different from Lexicon.

We analyzed all of the 6,137 terms with respect to the document frequency and identified that the biggest problem (36.47 %) was caused by foreign language[24]. Next two problems were the colloquial terms (27.03 %) and proper nouns (21.74 %). Combination of foreign and Indonesian terms, e.g. *workshop-nya*[25], was considered as colloquial terms. We also found that the CS stemmer should be improved as there were 48 formal terms left unstemmed in ICL_list.

### 4.4   Summary

We did a study in order to understand the meaning of *richness* claimed for the representation of document produced by TRSM. The WORDS-corpus who was created by human experts, and contains keywords of each ICL-corpus document, played significant role in the study, for it became the ground truth of the analysis. First of all, the result of the study confirmed that rough sets theory intuitively works as the natural way of human thinking. Being concerned with the meaning of *richness*, we came into conclusion that the TRSM-representation contains more terms than its base representation and those additional terms are semantically related with the topic of the document. After all, with regard to the IRS framework, we infer that TRSM is reasonable for a semantic IRS.

---

[23] These values are for the process with stemming task.

[24] Most of the foreign terms was English.

[25] It comes from an English term *workshop* and an Indonesian suffix *-nya*.

# 5  An Automatic Tolerance Value Generator

## 5.1  Introduction

Despite the fact, that the value of tolerance value $\theta$ is crucial for TRSM implementation, there is no consensus about how we can set a certain number as a $\theta$ value. It is usually chosen by the researcher or human expert based on manual inspection through the training data or his/her consideration about the data. It is not deniable that each datum is distinctive hence requires different treatment, however determining the $\theta$ value by hand is an exhaustive task before even starting the TRSM paths.

We did a study for an algorithm to generate a tolerance value $\theta$ automatically from a set of documents. The idea was based on the fundamental objective of tolerance rough sets model for having a richer representation than the base representation. We took an advantage from the singular value decomposition (SVD) method in order to project all document representations (i.e. TFIDF-representation and TRSM-representation) on a lower dimensional space and then computed the distance between them. The result, together with the analysis of system performance, helped us to understand the pattern of our data and to learn about the principle for a tolerance value determination. In the end, we came up with an intuitive algorithm.

## 5.2  Experiment Process

The experiment was conducted by following the four phases depicted in Fig. 14. Thus, basically we preprocessed the data, constructed the document representation based on TRSM, computed the SVD of TFIDF-representation and TRSM-representation, and finally analyzed them. In the figure, the dashed rectangle identifies the main parts of the experiment that would be run for $\theta = 1$ to 100. In implementation level, we applied the inverted index as the data structure of all document representations[26].

**Preprocessing Phase.** We used ICL-corpus and WORDS-corpus as the system data and came up with the TFIDF-representations for each corpus. We applied an information retrieval library freely available called Lucene[27] with some modifications in order to embed the Vega's stopword and the CS stemmer.

**TRSM Phase.** The tolerance rough sets model was implemented in this phase, which means we converted the TFIDF-representation into TRSM-representation by following these steps:

---

[26] Inverted index was applied for document representations in all experiments in this article.

[27] It is an open source project implemented in Java licensed under the liberal Apache Software License [40]. We used `Lucene 3.1.0` in our study. URL for download: http://lucene.apache.org/core/downloads.html.

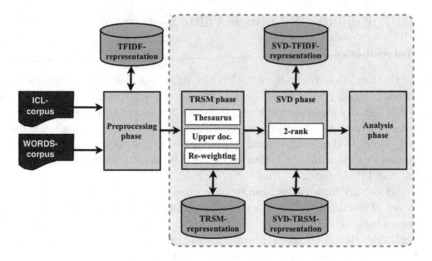

**Fig. 14.** Main phases of the study. This study consists of 4 main phases: preprocessing phase, TRSM phase, SVD phase, and analysis phase.

1. Construct the thesaurus based on Eq. (10).
2. Create the upper approximation of documents using Eq. (17).
3. Generate
   the TRSM-representation by recalculating the TFIDF-representation using Eq. (18) and considering the upper approximation of documents.

**SVD Phase.** The objective of this phase was to compress the high dimensional vector of document so it could be analyzed and plotted on a 2-dimensional graph. We implemented a Java package called JAMA[28] and calculated the SVD, where rank $= 2$, each for the base representation (TFIDF-representation) and the TRSM-representation; The resulted representation hence called *SVD-TFIDF-representation* and *SVD-TRSM-representation* respectively.

**Analysis Phase.** In analysis phase, we did two tasks. First of all, we calculated the mean distance and the largest distance between pairs of SVD-representations (i.e. SVD-TFIDF-representation and SVD-TRSM-representation). In order to calculate the distance, we applied the Euclidean function

$$d(V, U) = \sqrt{\sum_{i=0}^{M} (v_i - u_i)^2}$$

---

[28] JAMA has been developed by the MathWorks and NIST. It provides user-level classes for constructing and manipulating real, dense matrices. We used `JAMA 1.0.2` in this study. URL: http://math.nist.gov/javanumerics/jama/.

where $[v_i]_{i=0}^M$ and $[u_i]_{i=0}^M$ denote weight vectors of documents $V$ and $U$. Those distances then were plotted on graphs. We also drew a scatter graph of mean distance for several tolerance values.

Secondly, we generated the recall and precision of retrieval system by employing the 28 topics listed in Table C.3 as the information needs. The recall was computed based on Eq. (22), while for the precision we were interested in several measurements. First of all, it was the mean average precision (MAP) which is the arithmetic mean of average precision values for individual information needs, thus provides a single-figure measure of quality across recall levels [39]. The MAP is defined as follows

$$MAP(Q) = \frac{1}{|Q|} \sum_{j=1}^{|Q|} \frac{1}{|Rel|_j} \sum_{k=1}^{|Rel|_j} Precision(R_{jk}) \tag{24}$$

where $q_j \in Q$ is the $j^{th}$ information need, $|Rel|_j$ is the total number of relevant documents for $q_j$, and $R_{jk}$ is the set of ranked retrieval results from the top result until document $d_k$.

The other measurements of precision we used were Precision@10, Precision@20, and Precision@30, in which all of them are the measures at fixed low level of retrieved result and hence are referred as *precision at k* [39], where $k$ defines the amount of top documents would be examined that retrieved by the system. The last precision measure we concerned was R-Precision which is basically similar with the *precision at k* measures, except that the $k$ is the amount of relevant documents for each query.

In order to guarantee consistency with published results, we applied the *trec_eval*[29] program created and maintained by Chris Buckley to compute the recall and MAP of the retrieved documents.

### 5.3   Discussion

**Learning from WORDS-Corpus**

We kept the assumption that each document of WORDS-corpus consists of essential keywords, which should appear in corresponding document representation of ICL-corpus. The distance between document representations of both corpora measures how far an ICL-corpus document from a WORDS-corpus document is. Thus, the assumption brings us to a preference of smaller value of distance; When we had a smaller value of distance, we might expect more keywords appear in an ICL-corpus document.

Figure 15 depicts the distances between TRSM-representations of ICL-corpus and WORDS-corpus after they were reduced into 2 dimensions for tolerance value 0 to 50. Figure 15(a) shows the result of mean distance (let us call it `mean_distance`) which was calculated by taking the mean average of the distances of all TRSM documents at certain tolerance value. The largest distance

---

[29] We used the `trec_eval.9.0` which is publicly available on http://trec.nist.gov/ trec_eval/.

(let us call it largest_distance) displayed in Fig. 15(b) reveals the largest value of distance, hence it gives us a clue about the size of document cluster at each particular tolerance value. The mean_distance graph simply tells us that the higher tolerance value, the farther the distance, and thus the less relevant terms should appear in TRSM-representation of ICL-corpus. It seems that large largest_distance might lead to large mean_distance.

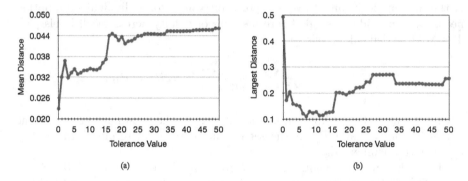

(a)                                           (b)

**Fig. 15.** Distances between document of ICL-corpus and WORDS-corpus. The distances between TRSM-representations of ICL-corpus and of WORDS-corpus where $1 \leq \theta \leq 50$. Graph (a) is the mean distance, while (b) is the largest distance.

Analyzing scatter graph of distance between each document of ICL-corpus and WORDS-corpus after the TRSM method should give us more understanding about the relationship between those corpora and the alteration of tolerance value. Figure 16 depicts the clusters of TRSM-documents of ICL-corpus which at certain distance with TRSM-documents of WORDS-corpus when tolerance values are set to 0, 10, 15, and 41.

Concerning that the graphs reflect the distances between ICL-corpus and WORDS-corpus, the ideal graph in Fig. 16 would be a single line on X-axis. In this situation, when the documents of ICL-corpus have zero distance with of WORDS-corpus, we might be certain that terms considered relevant in WORDS-corpus are successfully retrieved by TRSM method and put into the TRSM-representation of ICL documents while the other irrelevant ones are filtered out. Suppose we take the WORDS-corpus as the ground truth, then we might expect high recall in low tolerance value.

We know that the corpora we used in this study lie on a single domain specific[30], i.e. choral, hence all index terms from both corpora are generally semantically related, even though in a very remote relationship. Therefore, in Fig. 16, if the resulted cluster is a line-formed on X-axis, then we would have common documents which contain common terms. Similar circumstances should be happened at any line-formed clusters parallel to X-axis for the reason that

---

[30] WORDS-corpus is generated based on ICL-corpus hence they dwell in a single domain.

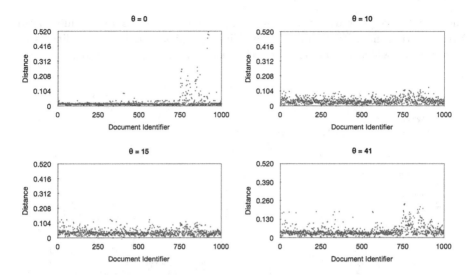

**Fig. 16.** Scatter graph of distance. The scatter graph of distance between TRSM documents of ICL-corpus and of WORDS-corpus when $\theta = 0$, $\theta = 10$, $\theta = 15$, and $\theta = 41$.

similar distance comes from similar document. In other words, we have the least discrimination power of document at this state. Based on these, a cluster with scattered documents inside should be preferred, and in order to have such cluster, its size should big enough.

The largest size of cluster in Fig. 16 occurs when $\theta = 0$, in which the documents are much less scattered and even tends to be a line-formed. Consider that $\theta$ value is a threshold to filter the index term out from document representation, $\theta = 0$ means that all index terms are determined to be semantically related to each other even though any pair of terms never occurs together. Consequently, the TRSM-representation yielded would have most of index terms within. As the result, we are standing in similar position of foregoing paragraph and it confirms that a parallel line with X-axis signify the commonality of documents in the cluster. Further, comparison between the cluster of $\theta = 0$ and the other clusters in Fig. 16 indicates that the tolerance value has a significant role in removing irrelevant terms as well as relevant terms, for the other clusters are smaller in size and the documents within are more disseminated.

Nevertheless, refer to the context of richness in TRSM method, merely having all index terms in the document representation is out of the intention. Therefore, $\theta = 0$ should be out of our consideration when determining a good tolerance value for any set of documents.

Pertaining to the relationship between mean_distance and largest_ distance, the four tolerance values, i.e. $\theta = 0$, 10, 15, and 41, were assumed to reflect four conditions. Those are the condition when we have, respectively: (*a*) small mean_distance and large largest_distance; (*b*) small mean_distance and small largest_distance; (*c*) large mean_distance and small `largest_distance`; and (*d*) large `mean_distance` and large `largest_distance`. To be more clear, Fig. 17 depicts these four conditions in extreme way which will be useful for the discussion in further sections.

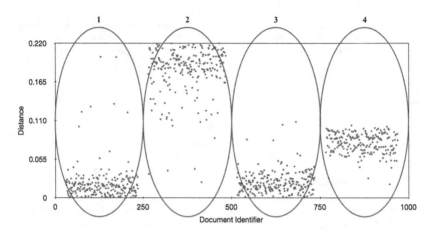

**Fig. 17.** Extreme conditions of mean distance and largest distance. The four extreme conditions of the mean distance and largest distance: (1) small `mean_distance` and large `largest_distance`; (2) large `mean_distance` and large `largest_distance`; (3) small `mean_distance` and small `largest_distance`; and (4) large `mean_distance` and small `largest_distance`.

## Learning from ICL-Corpus

In this section, we present and discuss results based on two grounds: *(1)* distance calculation; and *(2)* retrieval system performance

**Distance Calculation.** We computed the distances between TFIDF-representation and TRSM-representation of single corpus, i.e. ICL-corpus, after the dimensionality of those representations were reduced into 2 dimensions using the SVD method. Refer to the capability of TRSM which is to enrich a document representation, larger distance is preferred since it gives an indication that TRSM-representation is richer than the base representation. So, taking the characteristic of document representation into account, we should treat the distance value differently; When we are learning from WORD-corpus (as in previous section), we prefer smaller distance value, whereas when we are analysing ICL-corpus (as in this section) we prefer larger distance value.

In similar fashion with Figs. 15 and 18 displays the `mean_distance` and Fig. 19 shows the largest value of distances between TFIDF-representation and TRSM representation for each tolerance value ranging from 1 to 100. The green horizontal line in each figure reflect the average of `mean_distance` and of `largest_distance`, thus let us call this green lines as `average_distance`.

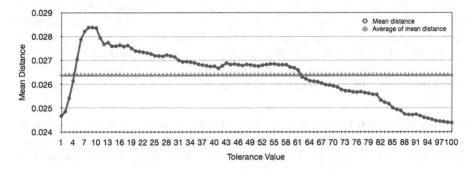

**Fig. 18.** Mean distance. The mean distance between TFIDF-representation and TRSM-representation of ICL-corpus for SVD 2-rank where $1 \leq \theta \leq 100$. The horizontal line is the average of the mean distance values.

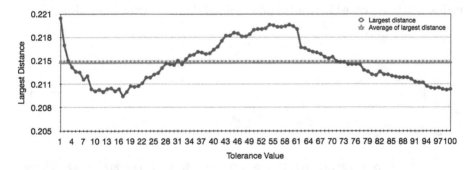

**Fig. 19.** Largest distance. The largest distance between TFIDF-representation and TRSM-representation of ICL-corpus for SVD 2-rank where $1 \leq \theta \leq 100$. The horizontal line is the average of the largest distance values.

Based on the nature of TRSM we prefer large value of mean_distance, and as learning from WORDS-corpus in Sect. 5.3, we were suggested to take large value of largest_distance. Further, our study in Sect. 4 proposed $\theta \geq 21$, owing to the fact that at $\theta = 21$ the average size of upper documents has been reduced sufficiently up to 92.24 % while the average recall (of WORDS-corpus' index terms in ICL-corpus documents) was kept high (97.58 %); The average sizes of upper documents were smaller afterward but the changes were observed not significant. Suppose we consider the *large* value for both mean_distance and largest_distance as having value more than or equal to its average_distance, thus Figs. 18 and 19 recommend us to focus on $31 \leq \theta \leq 61$.

**Retrieval System Performance.** We put the ICL-corpus into a framework of information retrieval system and generated several results based on the performance measures. Figures 20, 21, 22, 23 and 24 exhibit the results in the form of graphs which goes from the general level to the specific low level, all for tolerance value 1 to 100.

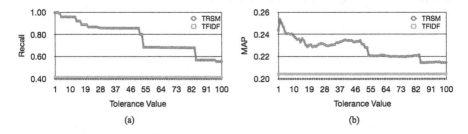

**Fig. 20.** Recall and MAP. The system performances while implementing TRSM method and base method in terms of (a) recall and (b) mean average precision (MAP).

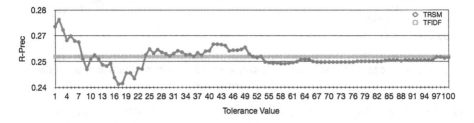

**Fig. 21.** R-Precision. The precision of ICL-corpus at top $|R|$ documents, where $|R|$ is the total of relevant documents for each topic.

The recall and MAP calculations shown by Fig. 20(a) and (b) clearly define that we can rely on TRSM method whose effectiveness is better then

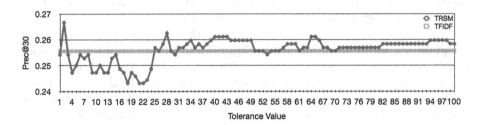

**Fig. 22.** Precision@30. The precision of ICL-corpus at top 30 documents.

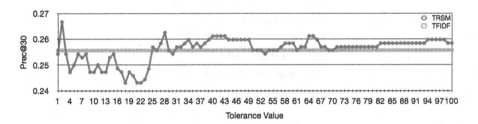

**Fig. 23.** Precision@20. The precision of ICL-corpus at top 20 documents.

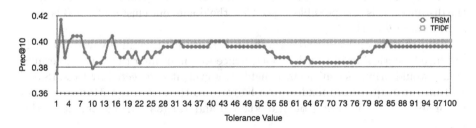

**Fig. 24.** Precision@10. The precision of ICL-corpus at top 10 documents.

the base method[31], nonetheless the performance of TRSM has a progressive decline at higher tolerance value. Correlated with the mean_distance and largest_distance Fig. 20 say no more, despite the recall graph confirms that we might have high value of document recall on lower $\theta$.

For analysis, we went further and came with the results of *precision at k* computations which are displayed in Figs. 21, 22, 23, and 24, for R-Precision, Precision@30, Precision@20, and Precision@10 sequentially. We applied our finding of distance calculation (i.e. that we should adjust our attention on tolerance values between 31 to 61) on those graphs by intersecting the tolerance values of each graph whose precision values (of TRSM method) are higher or equal to the base method (TFIDF method) with the tolerance values between 31 to 61. In conclusion, we have tolerance values between 40 to 43. Table 3 lists the tolerance values we manually observed whose values are high for several precision measurements.

**Table 3.** The tolerance values with high precision based on several measurements.

| Measures | Tolerance values |
|---|---|
| R-Precision | 31–53 |
| Precision@30 | 31–61, but 53 |
| Precision@20 | 34–61 |
| Precision@10 | 31–32, 40–43 |

With regard to the mean_distance and largest_distance graphs, at $40 \leq \theta \leq 43$ the distances are adjacent to their average_distance. Suppose we apply this into Fig. 17, instead of those extreme conditions, we would have considerable large of cluster in which the documents are scattered. In other words, at those tolerance values, the TRSM method might yield fairly richer documents representation and at the same time sill maintain the distinction between documents. Figure 25(a) and (b) are the scatter graph of distance when $\theta = 41$ and, for comparison purpose, $\theta = 0$. Despite the slight difference between the distances, it is still possible to see that the document cluster of $\theta = 0$ is more solid than of $\theta = 41$.

Examination on the scatter graph of distance for tolerance values 40 to 43 produced identical results with $\theta = 41$, hence we might infer that those tolerance values would bring us equivalent benefit. However, it is reflected by Figs. 18 and 19 that the graphs have tendency to be close to their average_distance, thus we prefer $\theta$ with the closest mean_distance and largest_distance, as for this case $\theta = 41$.

---

[31] Base method means that we employed the TF*IDF weighting scheme only without TRSM implementation.

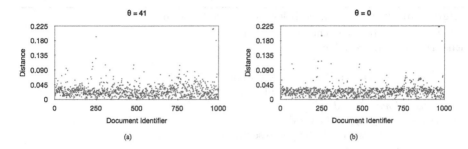

**Fig. 25.** Scatter graph of distance. The distance of TRSM-representation from TFIDF-representation when (a) $\theta = 41$ and (b) $\theta = 0$.

## 5.4 Tolerance Value Generator

We had already introduced the first version of the algorithm to generate a tolerance value $\theta$ automatically from a set of documents [41] in which we took the mean_distance as a single parameter for consideration. In this chapter, through a more careful analysis, we came into an understanding that both mean_distance and largest_distance have significant contribution in determining a single tolerance value from a set of documents.

Based on our analysis, a good tolerance value belongs to a fair size of document cluster in which its documents are scattered. Associated with the distances of TRSM-representation from TFIDF-representation, the preferred tolerance value is characterized by the mean_distance and the largest_distance whose distances are larger or equal to its average_distance, in which the closer the distances to its average_distance, the better.

Unfortunately, when this rules were applied on Figs. 18 and 19, we came with $\theta = 61$, whose mean_distance is the closest to its average_distance but the largest_distance is very large. On that account, the R-Precision and Precision@10 of TRSM are beneath the TFIDF. So we learned, when the size of document cluster is very large, it is an indication that the TRSM method a little bit out of line in discriminating the document.

For that reason, further restriction needs to be added for the acceptable limit of the largest_distance in order to ensure that the largest_distance will not have very large value. By observing Fig. 19, setting the maximum limit to half of the length between maximum distance and the average_distance seems to be appropriate.

Algorithm 1 presents the core idea of our algorithm. Line 1 up to 16 of Algorithm 1 are the initialization and the rest is the main process. The main process says that we choose the tolerance value (namely $finalTheta$) based on values of mean distance ($mean\_dist$) and of largest distance ($largest\_dist$) for certain range of $\theta$ whose distances to its average ($md\_toAverage$ for mean distance and $ld\_toAverage$ for largest distance) are the smallest. When searching the tolerance value, we only consider those whose value of $mean\_dist$ is larger than its average ($md\_avg$) and of $largest\_dist$ exists between its average ($ld\_avg$) and its limit

---

**Algorithm 1.** Main Idea of Tolerance Value Generator

---

**Require:** A set of documents as data source
**Ensure:** A tolerance value

1: *tfidf* ← construct TFIDF-representation
2: *svd_tfidf* ← construct SVD 2-rank of *tfidf*
3: *theta* ← *lowerBound*
4: **while** *theta* ≤ *upperBound* **do**
5:     *trsm* ← construct TRSM-representation
6:     *svd_trsm* ← construct SVD 2-rank of *trsm*
7:     *mean_dist* ← mean distance between *svd_tfidf* and *svd_trsm*
8:     *largest_dist* ← the largest distance between *svd_tfidf* and *svd_trsm*
9:     *theta* + +
10: **end while**
11: *md_avg* ← average of *mean_dist*
12: *md_toAverage_min* ← Integer.MAX_VALUE
13: *ld_max* ← maximum of *largest_dist*
14: *ld_avg* ← average of *largest_dist*
15: *ld_limit* ← *ld_avg* + (*ld_max* − *ld_avg*)/2
16: *ld_toAverage_min* ← Double.MAX_VALUE
17: **for** *i* ← 0, (|*mean_dist*| − 1) **do**
18:     *md_toAverage* ← *mean_dist*[*i*] − *md_avg*
19:     *ld_toAverage* ← *largest_dist*[*i*] − *ld_avg*
20:     **if** *md_avg* ≤ *mean_dist*[*i*] **and** *ld_avg* ≤ *largest_dist*[*i*] ≤ *ld_limit* **then**
21:         **if** *md_toAverage* == *md_toAverage_min* **and** *ld_toAverage* ≤ *ld_toAverage_min* **then**
22:             *finalTheta* ← theta of *mean_dist*[*i*]
23:             *ld_toAverage_min* ← *ld_toAverage*
24:         **else if** *md_toAverage* < *md_toAverage_min* **then**
25:             *finalTheta* ← theta of *mean_dist*[*i*]
26:             *md_toAverage_min* ← *md_toAverage*
27:             *ld_toAverage_min* ← *ld_toAverage*
28:         **end if**
29:     **end if**
30: **end for**
31: **return** *finalTheta*

---

(*ld_limit*). The limit is computed as the average plus half of the length between maximum value of distance and the average (*ld_avg* + (*ld_max* − *ld_avg*)/2).

Suppose we apply the Golub-Kahan SVD algorithm [42, p. 455] for dimensionality reduction of TFIDF-representation and TRSM-representation, then in order to compute singular values matrix and V matrix it needs $4MN^2 + 8N^3$ floating-point operations (flops) [42, p. 254], where $M$ is the number of index terms and $N$ is the number of documents. Whereas, for TRSM implementation, the complexity is $O(NM^2)$. Combining these together, the computation of Algorithm 1 does not grow faster than $O(N^3M^2K)$, where $K$ defines the number of tolerance values being examined.

## Training Documents

Naturally the number of input data for the algorithm should be all documents in the corpus, so the resulted thesaurus consists of all index terms occurs in the corpus and the chosen tolerance value might suggest the best relationship between those index terms. Our study in Sect. 6 showed that the number of documents used as the data source for thesaurus does not guarantee the thesaurus to be more qualified, but the total number of unique terms and the type of input documents should. Further study related to this issue is necessary in the future, particularly concerning the efficiency issue.

## Upper and Lower Bound

Recall that tolerance value is required in thesaurus construction in order to filter the index terms out based on the co-occurrence frequency between terms in the corpus. Consequently, the fair scenario is to evaluate all possibilities of tolerance value, i.e. by setting the lowerBound to 1 and the upperBound to the maximum number of co-occurrence between terms (namely maxCOC). The upperBound thus is subject to change with regard to the size of data source used.

We have no objection about setting the lowerBound to 1. Nevertheless, the upperBound needs to be determined prudently. Here are the details of three idea specifically for the upperBound.

We urge not to use the idea above (using the maximum number of co-occurrence frequency) alone for the upperBound, because it will give us an extensive search range. Take an example, for the ICL-corpus which consists of 1,000 documents and 9,742 index terms, the maximum number of co-occurrence frequency is 329. Thus, if we applied the idea, the upperBound is set to 329. For another reason, by manually observing the co-occurrence data, we identified that there were limited number of terms having the co-occurrence frequency bigger than 164 (about half of 329), and much less index terms to be preserved when we increased the tolerance value. This behavior might decrease the ability of TRSM to enrich the base document representation.

We took an advantage of the knowledge, that TRSM is able to enrich the base representation in terms of having more semantic terms, for the second alternative of upperBound. Technically speaking, enriching the base representation means that the TRSM-representation contains more terms than the TFIDF-representation. So, the comparison between the average length of TRSM-representation (in the algorithm it is the avgLengthTRSM) and TFIDF-representation (namely avgLengthTFIDF) would be good for the search termination process. This idea seems to be more affirmative than the use of co-occurrence frequency. However, it gives us an uncertainty state of the real search scope at some extent.

We came into the third idea for the upperBound based on our real experience when conducting the experiment. Initially, we set the upperBound in a low value and had the first result. Based on the analysis of the results, we decided whether to have another run with higher upperBound. This particular process might be happened for several times and we stopped the procedure when we were confident that there would be no other significant changes as if we had another run. In order to be confident, we tried to grasp the pattern of the mean_distance manually and decided to stop the procedure if we identified that the best mean_distance was located in about 2/3 of the range (namely certainty_range), which meant that the resulted tolerance value (namely finalTheta) was lower than a certain threshold (namely threshold_theta). If the resulted tolerance value was higher than the threshold, we set a new value to the upperBound as well as the threshold_theta and went for another run. Our experience is implemented mainly as described in Algorithm 2.

---

**Algorithm 2.** Set Up the upperBound

---

1: $finalTheta \leftarrow 0$
2: $lowerBound \leftarrow 1$
3: $range \leftarrow r$
4: $certainty\_range = (2/3) * range$
5: $upperBound = lowerBound + range$
6: $threshold\_theta = lowerBound + certainty\_range$
7: $theta \leftarrow lowerBound$
8: **while** $theta \leq upperBound$ **do**
9:     $finalTheta \leftarrow$ compute the final tolerance value
10:     **if** $finalTheta > threshold\_theta$ **then**
11:         $threshold\_theta = upperBound + certainty\_range$
12:         $upperBound+ = range$
13:     **end if**
14:     $theta + +$
15: **end while**

---

We put all those three alternatives of upperBound into our algorithm as it is shown in Algorithm 3 for reason. The experiment results suggest us to have a high tolerance value, however it is possible to have a low tolerance value in implementation, e.g. when we have a small number of index term in a set of documents for the data source. The third alternative of upperBound should be effective for this particular circumstance since it ensures us to have reasonable search range of tolerance value. The second alternative which make use of the comparison between the average length of TFIDF-representation and of TRSM-representation should guarantee that we would have a tolerance value whose TRSM-representation is richer than the base representation. At last, the maximum co-occurrence frequency might be useful as the final termination process.

Another advantage of knowing the maximum co-occurrence frequency is to set the range, for example, by setting it to about 1/3 of the maximum number. Suppose, for the ICL-corpus whose maximum co-occurrence frequency is 329, we set $range = 111$, then we will have maximum 3 times runs.

Eventually, putting all together, we came with Algorithm 3 which is the final version of our algorithm.

---

**Algorithm 3.** Tolerance Value Generator

---

**Require:** A set of documents as data source
**Ensure:** A tolerance value

1: $tfidf \leftarrow$ construct TFIDF-representation
2: $svd\_tfidf \leftarrow$ construct SVD 2-rank of $tfidf$
3: $finalTheta \leftarrow 0$
4: $lowerBound \leftarrow 1$
5: $range \leftarrow r$
6: $certainty\_range = (2/3) * range$
7: $upperBound = lowerBound + range$
8: $threshold\_theta = lowerBound + certainty\_range$
9: $avgLengthTRSM \leftarrow$ Integer.MAX_VALUE
10: $avgLengthTFIDF \leftarrow$ the average length of $tfidf$
11: $maxCOC \leftarrow$ the maximum co-occurrence frequency between terms
12: $theta \leftarrow lowerBound$
13: **while** $theta \leq upperBound$ **and** $avgLengthTFIDF < avgLengthTRSM$ **and** $theta \leq maxCOC$ **do**
14:     **while** $theta \leq upperBound$ **do**
15:         $trsm \leftarrow$ construct TRSM-representation
16:         $svd\_trsm \leftarrow$ construct SVD 2-rank of $trsm$
17:         $mean\_dist \leftarrow$ the mean distance between $svd\_tfidf$ and $svd\_trsm$
18:         $largest\_dist \leftarrow$ the largest distance between $svd\_tfidf$ and $svd\_trsm$
19:         $theta + +$
20:     **end while**
21:     $md\_avg \leftarrow$ average of $mean\_dist$
22:     $md\_toAverage\_min \leftarrow$ Integer.MAX_VALUE
23:     $ld\_max \leftarrow$ maximum of $largest\_dist$
24:     $ld\_avg \leftarrow$ average of $largest\_dist$
25:     $ld\_limit \leftarrow ld\_avg + (ld\_max - ld\_avg)/2$
26:     $ld\_toAverage\_min \leftarrow$ Double.MAX_VALUE

---

## 5.5  Summary

In this chapter we put forward a revised version of algorithm for defining a tolerance value automatically from a set of documents. The heart of the algorithm is measuring the distances between document representations of data source, i.e. one computed using base method while the other using TRSM method, in their 2-dimensional space which are constructed by utilizing the singular value decomposition (SVD) method over a range of $\theta$ values.

We learned from two corpora, ICL-corpus and WORDS-corpus, in order to generate some principles that served as the foundation for the algorithm. We found that we should consider both, the mean_distance as well as the largest_distance, for realizing a fairly big document cluster in which the documents are adequately scattered. Further, we discussed the parameters used in the algorithm.

---

**Algorithm 4.** Tolerance Value Generator (continued)

---

27:    **for** $i \leftarrow 0, (|mean\_dist| - 1)$ **do**

28:        $md\_toAverage \leftarrow mean\_dist[i] - md\_avg$

29:        $ld\_toAverage \leftarrow largest\_dist[i] - ld\_avg$

30:        **if** $md\_avg \leq mean\_dist[i]$ **and** $ld\_avg \leq largest\_dist[i] \leq ld\_limit$ **then**

31:            **if** $md\_toAverage \;==\; md\_toAverage\_min$ **and** $ld\_toAverage \leq ld\_toAverage\_min$ **then**

32:                $finalTheta \leftarrow$ theta of $mean\_dist[i]$

33:                $ld\_toAverage\_min \leftarrow ld\_toAverage$

34:            **else if** $md\_toAverage < md\_toAverage\_min$ **then**

35:                $finalTheta \leftarrow$ theta of $mean\_dist[i]$

36:                $md\_toAverage\_min \leftarrow md\_toAverage$

37:                $ld\_toAverage\_min \leftarrow ld\_toAverage$

38:            **end if**

39:        **end if**

40:    **end for**

41:    $avgLengthTRSM \leftarrow$ the average length of $trsm$ at $theta$

42:    **if** $finalTheta > threshold\_theta$ **then**

43:        $threshold\_theta = upperBound + certainty\_range$

44:        $upperBound+ = range$

45:    **end if**

46: **end while**

---

# 6    Optimizing the Thesaurus

## 6.1    Introduction

Based on the process of modeling a document in TRSM, thesaurus is the heart of TRSM, in which the relationship between terms in the thesaurus is determined by a tolerance value $\theta$. Thus, choosing the right $\theta$ value is essential in TRSM implementation. In the previous chapter, we have seen that it is possible to determine a value for $\theta$ by considering the mean and the largest distances between TRSM-representation and TFIDF-representation. We also proposed a new version of algorithm to generate the tolerance value automatically. In this chapter, we move the focus on the important issues relevant to the quality of the thesaurus.

We might find most of the graphs presented in the last chapter support the fact that the values of distances between TRSM-representation and TFIDF-representation vary with regard to the tolerance value, and so is the quality of the TRSM-representation. Therefore, it seems that the thesaurus, which stores information about terms relationship exploited to enrich a document representation, is influenced by the tolerance value. Moreover, the thesaurus of TRSM is created from a collection of text documents as a data source and relied on the co-occurrence of terms as the semantic relatedness measure. These facts imply that the data source and the semantic measure have capacity to produce effect on the quality of the thesaurus.

Tolerance rough sets model uses the frequency of co-occurrence in order to define the semantic relatedness between terms. Despite the raw frequency, there are several ways to calculate the degree of association between pairs of terms from co-occurrence data, i.e. Cosine, Dice, and Tanimoto measure. Further, with regard to the term-weighting scheme, the term frequency (TF) is the simplest approach to assign the weight for a term but it suffers from a critical problem that it considers all terms as equally important no matter how often it occurs in the set of documents. The inverse document frequency (IDF) is well known to be used in order to enhance the discriminating power of a term in determining relevance by considering the document frequency of the term in the corpus. Combination of term frequency and inverse document frequency produces a composite weight commonly assigned to a term, which is known as TF*IDF weighting scheme. After all, in spite of the raw frequency of co-occurrence, we might get different results from the same co-occurrence data with different formula.

We conducted a study to investigate the quality of the thesaurus of TRSM with regard to these three factors (i.e. tolerance value, data source of thesaurus, and semantic measure) in the framework of an information retrieval system (IRS). We used different corpus as data sources of the thesaurus, implemented different semantic relatedness measure, and altered the tolerance value. In order to analyze the results, we calculated the performance measure of an information retrieval system, i.e. recall and precision, and compared the results with the base representation (TFIDF-representation).

## 6.2   Experiment Process

We did two experiments. The first experiment focused on the data source of thesaurus, while the second experiment focused on the semantic measure of thesaurus.

For the first experiment, we maintained the frequency of co-occurrence as the measure of semantic relatedness in the thesaurus construction and we used our primary corpus, i.e. ICL-corpus, as the main data which was processed by the IRS. Specifically for the data source of thesaurus, we employed several corpora as listed in Table 4; *Total document* column defines the total number of documents in each set of documents which served as the data source, *Total unique term* column defines the total number of index terms, and *Total term* column is the total number of terms appear in the set.

ICL_1000 is actually the ICL-corpus which is a set of the first 1,000 emails of Indonesian Choral Lovers (ICL) Yahoo! Groups, while the ICL_2000 and ICL_3000 are the extension of ICL_1000, which contain the first 2,000 emails and the first 3,000 emails respectively. WORDS_1000 is the WORDS-corpus, hence it is a set of 1,000 documents which are the keywords defined by our choral experts with regard to each corresponding document in ICL-corpus. Finally, WIKI_1800 is a set of 1,800 short abstracts of Indonesian Wikipedia articles[32]

---

[32] Please see Appendix C.2.

**Table 4.** List of data source for thesaurus. This table presents the list of data source used specifically for thesaurus construction.

| No | Data source | Total document | Total unique term | Total term |
|---|---|---|---|---|
| 1. | ICL_1000 | 1,000 | 9,742 | 129,566 |
| 2. | ICL_2000 | 2,000 | 14,727 | 245,529 |
| 3. | ICL_3000 | 3,000 | 21,101 | 407,805 |
| 4. | ICL_1000 + WORDS_1000 | 2,000 | 9,754 | 146,980 |
| 5 | ICL_1000 + WIKI_1800 | 2,800 | 17,319 | 191,784 |

In the second experiment, we only used single corpus, which was the ICL-corpus that acted as the main data processed by the IRS as well as the data source. For the semantic measure in thesaurus construction, we considered the Cosine measure which is probably the similarity measure that has been most extensively used in information retrieval research. The Cosine was calculated over the TF*IDF weight of term.

For both experiments conducted in this study, we followed three phases displayed in Fig. 26 which are preprocessing phase, TRSM phase, and analysis phase; The main differences between experiments were on the TRSM phase. The dashed rectangle shows the central activities of the study, i.e. the TRSM phase and analysis phase, which were iterated for a range of tolerance value (i.e. for $\theta = 1$ to 100). Below are the description of each phase.

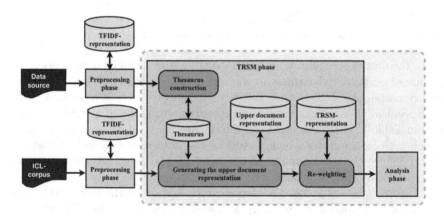

**Fig. 26.** Main phases of the study. This figure shows the three main phases of the study in the IRS framework, which are preprocessing phase, TRSM phase, and analysis phase.

**Preprocessing Phase.** There were no special treatment in this phase. What we did in this phase was similar with the preprocessing phase of the preceding study explained in Sect. 5, in the sense that the Lucene library was implemented and both the Vega's stopword and the CS stemmer were embedded in Lucene. In this study, we separated the data for IRS system from the data source for thesaurus construction, in which both operations yielded the TFIDF-representations.

**TRSM Phase.** Basically, we followed the three steps of tolerance rough sets model, which were constructing the thesaurus, creating the upper document representation, and re-weighting the TFIDF-representation using the TRSM-weighting scheme. However, we applied the first step only for the data source in order to generate the thesaurus, while the other two steps were applied to the system data, that is the ICL-corpus.

In the first experiment, the thesaurus were constructed from each data source listed in Table 4 based on the frequency of co-occurrence terms, while in the second experiment the thesaurus were constructed only using the ICL-corpus and calculated based on Cosine semantic measure. Then, for both experiments, the TRSM-representation was re-weighted by considering the TFIDF-representation and the upper document representation of ICL-corpus, in which the thesaurus became the bottom layer of the upper representation generation.

**Analysis Phase.** We applied the Cosine similarity[33] in order to retrieve documents from the corpus relevant to the 28 information needs. The queries, which were the 28 topics determined by our choral experts, were modeled into TRSM-query-representations based on the following rule

$$
w_j = \begin{cases} (1 + \log f_q(t_j)) \log \frac{N}{f_D(t_j)} & \text{if } t_j \in q \\ 1 & \text{if } t_j \notin \mathbf{L}_{\mathcal{R}}(q) \\ \frac{|I_\theta(t_j) \cap q|}{|I_\theta(t_j)|} & \text{if } t_j \notin \mathbf{U}_{\mathcal{R}}(q) \\ 0 & \text{otherwise} \end{cases} \tag{25}
$$

where $w_j$ defines the weight of term $t_j$ in a query, $f_q(t_j)$ is the occurrence frequency of term $t_j$ in the query, $f_D(t_j)$ is the document frequency of term $t_j$ in a corpus, $N$ is the total document in the corpus, and $\frac{|I_\theta(t_j) \cap q|}{|I_\theta(t_j)|}$ is the rough membership function between tolerance class of term $t_j$ and the query. We considered a query as a new document in a corpus, thus we add 1 to the total document $N$ and the document frequency $f_D(t_j)$, if term $t_j$ occurs in the query.

Our primary data to analyze the thesaurus were the calculation results of recall and precision of the TRSM-representations created. As the experiment in previous chapter, we calculated the recall and mean average precision (MAP) based on Eqs. (22) and (24) sequentially. We compared them to the computation result of the base representation, i.e. the TFIDF-representation.

---

[33] Explanation about Cosine as a document ranking is available in Appendix B.

## 6.3  Discussion

### Result of First Experiment: Data Source of Thesaurus

Figure 27 shows the recall values of ICL-corpus by implementing TRSM in which the thesaurus was generated based on the co-occurrence frequency between terms of data sources listed in Table 4 and the tolerance value was altered between 0 to 100. The TFIDF in the graph is the recall values of TFIDF-representation.

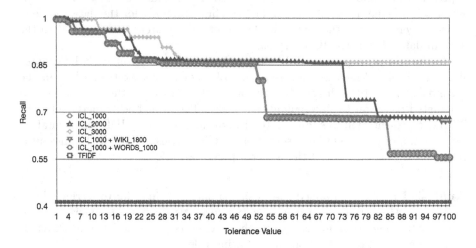

**Fig. 27.** Recall. Recall values based on several data sources of thesaurus.

Generally, all data sources perform similar pattern. When $\theta = 1$ they have very high recall values (0.9967 for ICL_1000 and ICL_1000 + WORDS_1000 data sources, and 0.9968 for ICL_2000, ICL_3000, and ICL_1000 + WIKI_1800 data sources) and the values are gradually decreased when the tolerance value is increased. It is also clear from the graph that all the recall values of TRSM-representations outperform the TFIDF-representation's recall value (0.4125).

Refer to the nature of tolerance rough sets model, the general result of recall values shown in Fig. 27 is predictable. When we set $\theta = 0$, we put all the index terms into all TRSM-representations that leads to the retrieval of all documents in the corpus, including the relevant ones, for all the queries. When $\theta$ is set to 1, a number of index terms which co-occur with no other index terms are removed. It reduces a number of index term appear in document vector at some degree and decreases the retrieval of relevant documents. If we set the $\theta$ even higher, more index terms are filtered out and lesser recall values are obtained.

A careful analysis to Fig. 27 gave us several interesting points. First, we cannot expect anything from adding the WORDS_1000 as the data source; It has the same result with the ICL_1000. The fact that it consists of keywords defined by human experts seems to be not significant for thesaurus optimization. Secondly, it is interesting that adding WIKI_1800 as data source unpredictably came with similar result to ICL_1000 up to $\theta = 85$. Thirdly, some improvement were achieved by adding the ICL_1000 with similar documents as the data source, as it is shown by the ICL_2000 and ICL_3000.

Considering Table 4, we learned that the number of unique terms and total terms in the set contribute more to the quality of thesaurus than the number of documents. Put our focus on adding the ICL_1000 with WORDS_1000 and WIKI_1800, it seems that the kind of unique terms in a set are also count. From Table 4, we can see that adding WORDS_1000 (which have 3,477 unique terms) for the data source gives us 12 new unique terms. It means that most of index terms contained in WORDS_1000 are also the index terms of ICL_1000 and we may infer by Fig. 27 that the condition brings no improvement to the thesaurus. On the contrary, the index terms of WIKI_1800 are different from the ICL_1000 to a considerable extent; From Table 4, we can see that the ICL_1000 has 9,742 unique terms and aggregation of ICL_1000 + WIKI_1800 has 17,319 unique terms, while there are 10,549 unique terms in WIKI_1800 solely. Refer to Fig. 27, this fact also gives no significant improvement.

The results are a little bit different by implementing the ICL_2000 and ICL_3000 as the data source for thesaurus. Compared to the ICL_1000, both of them have more unique terms as well as total terms in their sets, and we could be certain that most documents inside them are corresponding in topic, i.e. choral, with ICL_1000. As in Fig. 27, these conditions lead to some improvement in recall values. Thus, we may conclude that merely introducing new unique terms does not guarantee any improvement for thesaurus. It should be provided by terms in documents of related domain.

In similar fashion with Figs. 27 and 28 presents the mean average precision (MAP). One obvious note from Fig. 28 is all results of TRSM-representations outperform the result of TFIDF-representation. Specifically, ICL_1000 shows to have the highest MAP value in a very low tolerance value ($\theta = 2$) and its graph tends to decrease as the tolerance value is increased. With regard to the nature of TRSM, this fact is predictable with similar reason we explained in a paragraph above. However, we can see that there are some points where the graph looks to be stable for several tolerance values; After drastic changes in the beginning, the graph tends to be stable at $\theta = 18$, $\theta = 54$, and $\theta = 84$.

As the recall values, the MAP values of combining ICL_1000 with WORDS_1000 are the same with utilizing ICL_1000 separately[34]. It seems to confirm that a set of keywords defined by human experts does not serve as a contributor to the quality improvement of thesaurus.

The ICL_1000 + WIKI_1800 shows to be comparable with the others in high tolerance values ($\theta \geq 32$), even though in low tolerance values, its performance is the worst. The other data sources, ICL_2000 and ICL_3000, perform similar pattern where they both have tendency to decrease as the tolerance value is increased. However, their performances are more stable than the ICL_1000.

---

[34] In fact, we found the same result between ICL_1000 and ICL_1000 + WORDS_1000 in all calculations we made, such as in R-Precision, Precision@10, Precision@20, and Precision@30.

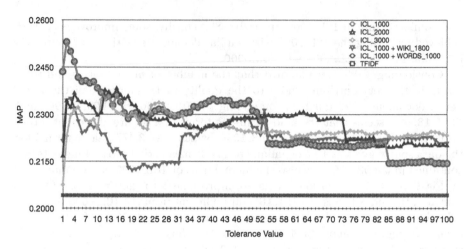

**Fig. 28.** Mean average precision. Mean average precision (MAP) values based on several data sources of thesaurus.

Based on these facts, Fig. 28 also indicates that documents in a corresponding domain with the system data (such as ICL_2000 and ICL_3000) may give some contribution to thesaurus improvement.

### Result of Second Experiment: Similarity Measure of Thesaurus

Instead of raw frequency of co-occurrence between terms, in the second experiment we considered the Cosine value based on TF*IDF weight of each term in order to define the semantic relatedness between terms of ICL-corpus. With regard to the nature of Cosine measure, the value of relatedness are between 0 to 1, hence in this experiment each $\theta$ value was divided by 100. Thus, for $\theta$ value 1 to 100, it was read by the thesaurus construction module as 0.01 to 1. Figures 29 and 30 display the recall and MAP values of ICL-corpus based on Cosine measure in thesaurus construction.

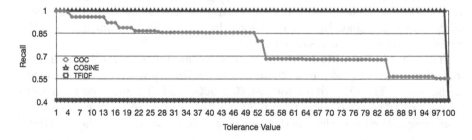

**Fig. 29.** Recall. Recall values where the co-occurrence (COC) and Cosine (COSINE) measures were applied to define the semantic relatedness between terms in thesaurus construction.

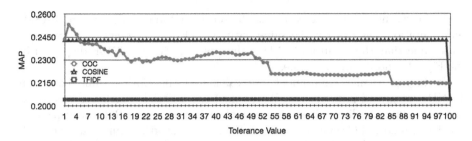

**Fig. 30.** Mean average precision. Mean average precision (MAP) values where the co-occurrence (COC) and Cosine (COSINE) measures were applied to define the semantic relatedness between terms in thesaurus construction.

**Table 5.** Total number of distinct vector length. This table presents the total number of distinct length of TRSM-representation based on Cosine measure for tolerance value 1 to 100.

| Tolerance value | Total distinction | Tolerance value | Total distinction | Tolerance value | Total distinction |
|---|---|---|---|---|---|
| 1–64 | 425 | 80 | 426 | 92 | 419 |
| 65–69 | 423 | 81–82 | 425 | 93 | 425 |
| 70–72 | 424 | 83 | 423 | 94 | 427 |
| 73 | 426 | 84 | 427 | 95 | 430 |
| 74–75 | 427 | 85 | 423 | 96 | 429 |
| 76–77 | 426 | 86 | 422 | 97 | 425 |
| 78 | 423 | 87–89 | 421 | 98 | 430 |
| 79 | 425 | 90–91 | 417 | 99 | 442 |
| | | | | 100 | 237 |

At first glance, both figures show perfect performances, where most of the results outperform the TFIDF-representation and TRSM-representation based on co-occurrence measure (COC) at tolerance values 1 to 99. Those performances obtained because most of the index terms occurred in almost all of TRSM-representation. In fact, there were more than 9,000 index terms out of 9,742 occurred in almost all of TRSM-representation, and the changes of amount of index terms occurred in TRSM-representation between tolerance value 1 to 99 were very small; Table 5 lists the total number of distinct length of document vector yielded by TRSM when the Cosine measure was implemented for tolerance value 1 to 100. It is not an ideal condition we are looking for. The condition signify that the TRSM has successfully enrich the base representation but it lessen the uniqueness of document at large extent.

Both figures also shows that the graphs are suddenly drop at $\theta = 100$ to the TFIDF level. In the thesaurus construction, when $\theta = 100$, the system filtered out index terms whose Cosine values less than 1. It made the tolerance class of each index terms consisted of the term itself and thus the final TRSM-representation was exactly the same with the TFIDF-representation. So, it is reasonable that the recall and MAP values of COSINE and TFIDF displayed in the Figs. 29 and 30 are the same.

In this particular experiment, we calculated the Cosine value based on TF*IDF weight of index terms. We applied the TF*IDF weighting scheme in order to refine the discriminating power of each index term. Refer to Eq. (B.2), the denominator of Cosine measure functions as the length-normalization of each vector being calculated in order to counterbalance the effect of various document length.

So, philosophically, the Cosine measure seems to be an ideal measure. Further, we found that implementing the Cosine measure in thesaurus construction has lessened the discrepancy of document in the corpus at large extent when Cosine value was less than 1. The fact that ICL-corpus is a set of documents in a specific domain (hence the index terms are generally related) might be the reason why most of the index terms occurred in the TRSM-representation. If this is the reason, it contradicts the result of our first experiment explained in Sect. 6.3 which indicated that we might expect having better contribution in order to improve the quality of thesaurus from a set of documents which was in the same domain with the system data.

Mathematically, the Cosine behavior might be explained by the nature of its equation, in which the association between pairs of terms is basically computed based on the co-occurrence data (even though in this particular experiment we have refined the raw frequency into the TF*IDF weight). Empirically, there were numerous pairs of terms occurred together in documents which leads to high values of Cosine and little changes in the values. Notice that conventionally a document is written using the common words of a subject. Thus, the fact that ICL-corpus came from a mailing lis of a specific domain confirms that its documents should contain general words of particular domain. Based on this, we urge that the characteristic of ICL-corpus is the primary cause of the Cosine behavior in this experiment.

After all, we might infer that Cosine measure is not appropriate to define the semantic relatedness between terms in thesaurus construction of tolerance rough sets model.

## 6.4  Summary

The result of the study confirmed that tolerance value, data source of thesaurus, and semantic measure influence the quality of the thesaurus. Even though we could not say affirmatively what kind of data source for an effective thesaurus, but empirically the result of study indicated that a set of documents in a corresponding domain with the system data might give better contribution to improve the quality of thesaurus. We also learned that the number of unique terms and total terms in the set contribute more to the quality of thesaurus than the total number of documents. Related to the semantic measure, we suggested to maintain the raw frequency of co-occurrence between terms rather than implementing the other measures, i.e. Cosine.

# 7 Lexicon-Based Document Representation

## 7.1 Introduction

TRSM employs a vector space model hence it represents the document as a vector of term weight in a high dimensional space. The richer representation claimed as the benefit of TRSM means that there is less zero-valued in document vector. Despite the fact that it can increase the possibility of two documents having non-zero similarity although they do not share any terms in original document, this fact leads us to a presumption that higher computational cost may become a significant trade-off in TRSM.

In Sect. 4, we showed that TRSM was able to fetch the important terms which should be retrieved by the automated process of the system. Nevertheless, based on comparison between the lexicon[35] and of the indexed terms, we identified 64.89 % did not occur in lexicon; the contributors were foreign terms (mostly in English), colloquial terms, and proper nouns. The following are the example of colloquial terms: *yoi* (it has the same meaning with word *iya* (in Indonesian) and *yes* (in English)), *terus* (it has the same meaning with word *lalu* (in Indonesian) and *and then* (in English)), *rekans* (it has the same meaning with word *teman-teman* (in Indonesian) and *friends* (in English)).

In this chapter, we propose a novel method, called a lexicon-based document representation, for a compact document representation. The heart of our method is the mapping process of terms occurring in TRSM-representation to terms occurring in lexicon, which gives us a new document representation consisting only of terms occurring in lexicon (we refer to this representation as LEX-representation) as an output. Consider Fig. 31 for depiction of the idea.

Hence this method represents a document as a vector in a lower dimensional space and eliminates all informal terms previously occurring in TRSM-representation. By this fact, we can expect less computational cost. For analysis, we take advantage of recall and precision commonly used in information retrieval research to measure the effectiveness of LEX-representation. We also did manual investigation into the list of terms considered as highly related with a particular concept in order to assess the quality of the representations.

## 7.2 Experiment Process

Experiment in this chapter used two corpora, i.e. ICL-corpus and WORDS-corpus, and employed two types of topic, i.e. 127 topics and 28 topics. The

---

[35] It is an Indonesian lexicon created by the University of Indonesia described in a study of Nazief and Adriani in 1996 [43] which consists of 29,337 Indonesian root words. The lexicon has been used in other studies [10,38].

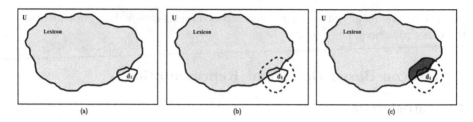

**Fig. 31.** The idea of mapping process. Picture (a) shows relation between lexicon and a TFIDF-representation ($d_1$), picture (b) shows relation between lexicon and a TRSM-representation (depicted by area inside dashed line), while picture (c) shows relation between lexicon and a LEX-representation (depicted by the darkest area).

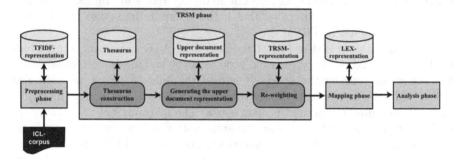

**Fig. 32.** Main phases of the study. The process consisted of 4 phases: preprocessing phase, TRSM phase, mapping phase, and analysis phase.

experiment was conducted by following four main phases which were preprocessing phase, TRSM phase, mapping phase and analysis phase as depicted in Fig. 32. Generally we did the first three phases over both corpora individually and analysed them in the analysis process.

**Preprocessing Phase.** The goal of this phase is to generate document representation based on the TF*IDF weighting scheme. So, the activities in this study basically the same with the preprocessing phase of experiments in Sects. 4 and 6. This is a phase when we did tokenisation, stopword elimination, stemming, and finally generated TFIDF-representation. This phase was powered by Lucene in which Vega's stopword list and CS stemmer were embedded in it.

In order to work, the CS stemmer requires a dictionary called DICT-UI which showed to produce more accurate results than the use of the other dictionary, i.e. *Kamus Besar Bahasa Indonesia* (KBBI)[36] [38]. The DICT-UI is actually the lexicon of this study.

---

[36] KBBI is a dictionary copyrighted by *Pusat Bahasa* (in English: Language Center), Indonesian Ministry of Education, which consists of 27,828 root words.

**TRSM Phase.** In this phase we acted in accordance with the consecutive steps of tolerance rough sets model and came up with TRSM-representation for both corpora. Let us call the TRSM-representation for ICL-corpus and WORDS-corpus as *ICL-TRSM-representation* and *WORDS-TRSM-representation* respectively. In thesaurus construction, we maintained the use of raw frequency of co-occurrence between terms and altered the tolerance value from 1 to 100.

**Mapping Phase.** Our intention in this phase is to map the index terms of TRSM-representation into the terms of the lexicon.

We noticed that the total number of terms in the lexicon (29,337 terms) was much bigger than the total number of index terms in ICL-corpus (9,742 terms) and WORDS-corpus (3,477 terms). We also noted that relationship between terms of tolerance classes were constructed based on term co-occurrence in a set of documents, hence there would be no relationship to other terms outside the corresponding set. However, there must be an intersection between lexicon and each document in a corpus because all documents must have some *formal* terms in order to be understood. Consequently, there would be no benefit in considering all terms in the lexicon during the mapping process.

In order to make the process faster, we intersected the lexicon with each corpus and called the result as *known-terms* $K$. Let $D = \{d_1, d_2, ..., d_N\}$ is a set of text documents, $T = \{t_1, t_2, ..., t_M\}$ is a set of index terms from $D$, and $B = \{b_1, b_2, ..., b_P\}$ is a set of terms in the lexicon, then $K = \{t_i \in T \mid t_i \cap b_j\} = \{k_1, k_2, ..., k_C\}$, for all $b_j \in B$. The terms appeared in known-terms then became the index terms of LEX-representation. The total number of known-terms for ICL-corpus and WORDS-corpus were 3,444 and 1,566 respectively. The mapping process was conducted as follows

**Input:**
  Matrix of TRSM-representation $TRSM_{matrix} = [trsm_{i,j}]_{NxM}$ for all $t_j \in T$ and $d_i \in D$, where $trsm_{i,j}$ denotes weight of term $t_j$ in document $d_i$.
**Output:**
  Matrix of LEX-representation $LEX_{matrix} = [lex_{i,l}]_{NxC}$ for all $k_l \in K$ and $d_i \in D$, where $lex_{i,l}$ denotes weight of term $k_l$ in document $d_i$.
**Process:**
  Generate $LEX_{matrix}$ based on Eq. (26) for all $t_j \in T$, $k_l \in K$, and $d_i \in D$

$$lex_{i,l} = \begin{cases} trsm_{i,j} & \text{if } k_l = t_j \\ 0 & \text{otherwise} \end{cases} \tag{26}$$

Even though we describe the document representations in terms of matrix, in implementation level of experiment we applied the inverted index as the data structure.

We should mention that, during the annotation process, we found that our human experts seemed to encounter difficulty in determining keywords. Thus, rather than listing the keywords, our experts chose sentence(s) from the text or made their own sentence(s). This action made the WORDS-corpus contain both formal and informal terms. Based on this fact, we decided to run the mapping process not only on ICL-corpus but also on WORDS-corpus, in order to remove the informal terms occurring in both corpora. Let us call the resulting representation *ICL-LEX-representation* for ICL-corpus and *WORDS-LEX-representation* for WORDS-corpus.

**Analysis Phase.** There were two tasks committed in the analysis phase. We named them categorisation and calculation. Categorisation was the task when we clustered documents of the same topics together. The motivation behind this task was based on the annotation process conducted by our human experts, i.e. keywords determination for each document. Thus, we perceived each topic as a concept and considered the keywords in WORDS-corpus as variants of terms semantically related with a particular concept. For this task, we used the 127-topics defined by our human experts, therefore we got 127 classes. Let us call the output of this process *ICL-topic-representation* and *WORDS-topic-representation* for each corpus. Technically, those representations were topic-term matrices.

In the calculation task, we used the notions of recall and precision, which are defined as Eqs. (22) and (23) respectively, in terms of calculating the *documents* as well as the *terms*. The first calculation computed the *terms* while the second calculation computed the *documents*. Thus, for the *terms-calculation*, the recall $R$ is the fraction of relevant *terms* that are retrieved while precision $P$ is the fraction of retrieved *terms* that are relevant. Notice that our WORDS-corpus consists of keywords defined by human experts, hence we considered WORDS-corpus as the *ground truth*, i.e. WORDS-corpus consists of *relevant terms* which should be retrieved by automated system.

Briefly, in the *terms-calculation*, we categorised LEX-representation of each corpus and then computed the recall and precision of topic-representations generated with and without a mapping process. Whereas, in the *documents-calculation*, we computed the standard recall and mean average precision (MAP) of all representations.

## 7.3   Discussion

### Calculating the Terms

Based on the *terms-calculation*, our findings are summarised by Fig. 33. Those graphs show that the mean of recall and precision values across 127 topics vary by the alternation of tolerance values $\theta$ and tend to be smaller as the tolerance value becomes higher.

We have mentioned in Sect. 7.2 that in the *terms-calculation* we focused on *terms* rather than *documents* when calculating recall and precision. Instead of document representation, the recall and precision values were computed over the terms of topic-representation. We measured the quality of topic-representation

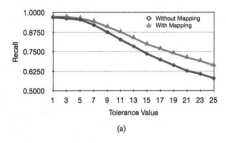 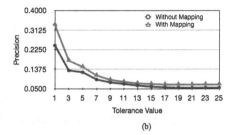

(a)                                              (b)

**Fig. 33.** Mean of Recall and Precision. Graph (a) shows the mean of recall values, while graph (b) shows the mean of precision values. The mean values were calculated over 127 topics.

of ICL-corpus based on the occurrence of relevant terms in it; the relevant terms were the index terms of topic-representation of WORDS-corpus.

Pertaining to the mapping process, we perceive the recall as a value which expresses the ability of the mapping process to keep the relevant terms out of the irrelevant ones. Thus, from Fig. 33(a) we can say that the mapping process outperforms the original TRSM method in terms of preserving the relevant terms. A gradual reduction of the ability is shown as the tolerance value $\theta$ gets higher, yet the mapping process seems to work better.

From another point of view, by the nature of TRSM method, a greater tolerance value should increase the number of index terms discarded from being introduced into the document representation. Considering Fig. 33(b), the behavior seems to shield not only the irrelevant index terms but also the relevant ones to be chosen to extend the base representation, even though at some point the change is not significant anymore, which happens at $\theta > 17$. However, the mapping process performs better once again in this figure.

## Calculating the Documents

In this task, the standard recall and precision were computed using the *trec_eval* program based on TFIDF-representation, TRSM-representation and LEX-representation of ICL-corpus over 28 topics for $\theta = 1$ to 100. Figure 34 is the graph of recall while Fig. 35 is the graph of mean average precision (MAP). In the figures, LEX is the LEX-representation, TRSM is the TRSM-representation, and TFIDF is the TFIDF-representation.

Figure 34 displays that LEX-representation works better than TFIDF-representation, even has slightly higher recall values than TRSM-representation at almost all level of $\theta = 1$ to 100. The trade-off to the recall values can be seen in Fig. 35. Here, the performance of LEX-representation is shown to be similar with TRSM-representation on low tolerance values ($\theta < 22$) and has slightly better precision at $\theta = 5$ to 15. Compared with TFIDF-representation, it performs better at $\theta < 85$.

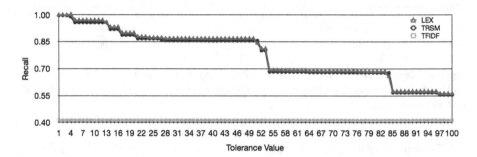

**Fig. 34.** Recall. This graph shows the recall values based on TFID-representation, TRSM-representation, and LEX-representation.

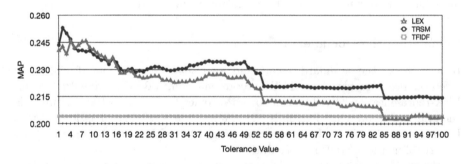

**Fig. 35.** Mean average precision. This graph shows the mean average precision (MAP) values based on TFID-representation, TRSM-representation, and LEX-representation.

The result depicted by Figs. 34 and 35 looks consistent with the result presented by Fig. 33. Those figures say that the increment of tolerance value leads to the less relevant terms in topic-representation and the more incapable the system to retrieve relevant documents. Even though the mapping process proved to be more capable to maintain the relevant terms and the recall value of LEX-representation have proportional result with of TRSM-representation, the mean average precision (MAP) value shows that TRSM-representation performs better in general. Figure 36, which presents the mean length of TRSM-representation and LEX-representation for tolerance values between 1 and 100, seems to explain that the vector length has contribution at some degree.

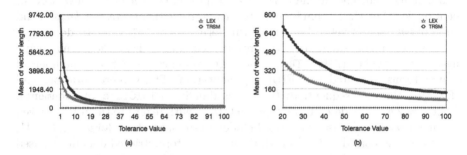

**Fig. 36.** Length of vector. Graph (a) shows the mean length of TRSM-representation and LEX-representation for $\theta = 1$ to 100 while graph (b) is the inset of graph (a) for $\theta = 20$ to 100.

Figure 36 tells us that document representation of TRSM tends to be longer than document representation yielded by mapping process. In fact, our observation through all document vectors for $\theta = 1$ to 100 yielded $\vec{x} \geq \vec{y}$, where $\vec{x}$ is document vector of TRSM and $\vec{y}$ is document vector of mapping process. It is not a surprising result due to the fact that mapping process conducted based on TRSM, in which the index terms of LEX-representation are those of TRSM-representation which appear in the lexicon.

The document ranking method we used in this study is the Cosine similarity method (B.2), which implies that the largest value of $similarity(Q, D)$ are obtained when the query $Q$ and the document $D$ are the same. Refer to this method, longer vector should have more benefit than the shorter one. Therefore, it is predictable that TRSM-representation outperforms the others when document vector of TRSM is the longest.

It is interesting though that at some levels between tolerance values 1 to 100 the LEX-representation has better performance than of TRSM. So, instead of the vector length, there must be another factor which give significant contribution to similarity computation based on Cosine method. The investigation went further to the tolerance classes which constructed the thesaurus.

## Tolerance Class

We picked 3 topics out of 28 which were the most frequent topics in ICL-corpus as it appears in Table C.3. These were *kompetisi* (in English: competition), *konser* (in English: concert), and *karya musik* (in English: musical work), and made an assumption that those topics were concepts which could be represented by a single term for each, namely *kompetisi, konser*, and *partitur* (in English: musical score)[37].

We generated the tolerance classes of those terms at several particular tolerance values, i.e. $\theta = 2$, $\theta = 8$, $\theta = 41$, and $\theta = 88$[38]. Specifically, we generated all terms considered semantically related with terms *kompetisi, konser*, and *partitur* (based on its occurrence in thesaurus) which appeared on the most relevant document retrieved by the system for each particular topic (i.e. *kompetisi, konser*, and *karya musik* respectively). Let us call this term sets as `TolClass_in_document`.

Tables 6 and 7 summarise the results; column 1 lists the terms being investigated, while `TFIDF`, `TRSM`, and `LEX` columns present the number of related terms appeared in TFIDF-representation, TRSM-representation, and LEX-representation sequentially (i.e. the cardinality of `TolClass_in_document`). When $\theta = 2$ we considered those representations with regard to the top-retrieved document calculated based on TRSM model. In similar fashion, for $\theta = 8$, $\theta = 41$, and $\theta = 88$, we considered ones with regard to the top retrieved document based on mapping process, TRSM method, and base model[39]. The `Total` column is the cardinality of particular tolerance class in thesaurus. In other words, it specifies the total terms defined semantically related with term *kompetisi, konser*, and *partitur* at $\theta = 8$, $\theta = 41$, and $\theta = 88$.

In a glance we should notice that document vector of TRSM consists of most related terms defined in thesaurus, even at high tolerance value ($\theta = 88$) it includes all of them. It is also clear in both tables that the cardinality of `TolClass_in_document` in TFIDF-representation (showed by the `TFIDF` columns) are mostly the least.

In order to assess the quality of document vector in terms of the relevant terms, we manually made a short list of terms we considered as semantically related with terms *kompetisi, konser*, and *partitur*. Table 8 displays the lists. By cross referencing our manual list with the `TolClass_in_document`, we found that `TolClass_in_document` consists of at least one term of our manual list. And as predicted, the `TolClass_in_document` of TRSM includes our terms the most.

---

[37] The index terms of thesaurus are in the form of single term, hence we choose term *partitur* as the representative of the *karya musik* concept.

[38] Figure 35 serves as a basis for the choice of $\theta$ values in which the TRSM-representation, LEX-representation, TRSM-representation, and TFIDF-representation outperform the other representations at $\theta = 2$, $\theta = 8$, $\theta = 41$, and $\theta = 88$ in respective order. However, particularly at $\theta = 88$, the TFIDF-representation only performs better than the LEX-representation.

[39] The base model means that we employed the TF*IDF weighting scheme without TRSM implementation nor the mapping process.

**Table 6.** Total number of terms considered as highly related with terms *kompetisi*, *konser*, and *partitur* at tolerance values 2 and 8 in a top-retrieved document representation generated based on TF*IDF weighting scheme (**TFIDF**), TRSM model (**TRSM**) and mapping process (**LEX**). The **Total** column is the total terms of respective tolerance class in thesaurus.

| Term | $\theta = 2$ | | | | $\theta = 8$ | | | |
|---|---|---|---|---|---|---|---|---|
| | TFIDF | TRSM | LEX | Total | TFIDF | TRSM | LEX | Total |
| *Kompetisi* | 54 | 1,587 | 883 | 1,589 | 31 | 315 | 203 | 320 |
| *Konser* | 37 | 3,508 | 1,664 | 3,513 | 23 | 902 | 513 | 909 |
| *Partitur* | 141 | 2,023 | 1,037 | 2,030 | 30 | 590 | 325 | 597 |

**Table 7.** The number of terms considered as highly related with terms *kompetisi*, *konser*, and *partitur* at tolerance values 41 and 88 in a top-retrieved document representation generated based on TF*IDF weighting scheme (**TFIDF**), TRSM model (**TRSM**) and mapping process (**LEX**). The **Total** column is the total terms of respective tolerance class in thesaurus.

| Term | $\theta = 41$ | | | | $\theta = 88$ | | | |
|---|---|---|---|---|---|---|---|---|
| | TFIDF | TRSM | LEX | Total | TFIDF | TRSM | LEX | Total |
| *Kompetisi* | 4 | 7 | 4 | 7 | 1 | 1 | 1 | 1 |
| *Konser* | 4 | 92 | 46 | 96 | 3 | 21 | 7 | 21 |
| *Partitur* | 18 | 54 | 23 | 56 | 1 | 4 | 2 | 4 |

Let us focus on Table 6 when tolerance value is 8. At $\theta = 8$, refers to Fig. 35, the LEX-representation performs better than the others, whereas refers to Fig. 36 the mean length of its vectors is shorter than of TRSM. Note that the cardinality of TolClass_in_document of mapping process (showed by the LEX column in Tables 6 and 7) for those three terms are smaller than of TRSM. A close observation to the vectors as well as the TolClass_in_document turned out that the length difference of both vectors were not too big and most of our manual terms (listed in Table 8) were found to sit on top ranks.

Indeed, based on the nature of mapping process, all of relevant terms we confronted occurred in TolClass_in_document of mapping process were always at higher rank than of TRSM. It happened because the index terms of LEX-representation were actually those of TRSM-representation which were not dropped out by the lexicon's.

Further, manual inspection yielded that numerous terms in TolClass_in_document of TRSM were remotely related to the terms *kompetisi*, *konser*, and *partitur*. With regard to the problem we mentioned in the beginning of this chapter (i.e. the existence of informal terms such as foreign terms, colloquial terms, and proper nouns), the LEX-representation had the most satisfactory result, i.e. it contained only the formal terms, which were index terms of lexicon.

**Table 8.** The list of index terms considered manually as highly related with terms *kompetisi, konser,* and *partitur.* The last column is the comparable English translation for each related index term mentioned in the middle column.

| Term | Related index terms | Comparable english translation (in respective order) |
|------|---------------------|------------------------------------------------------|
| *Kompetisi* | *kompetisi, festival, lomba, kategori, seleksi, juri, menang, juara, hasil, atur, nilai, jadwal, serta* | competition, festival, contest, category, selection, jury, win, champion, result, regulate, grade, schedule, participate |
| *Konser* | *konser, tiket, tonton, tampil, informasi, kontak, tempat, publikasi, poster, kritik, acara, panitia* | concert, ticket, watch, perform, information, contact, place, publication, poster, criticism, event, committee |
| *Partitur* | *partitur, lagu, karya, musik, koleksi, aransemen, interpretasi, komposisi, komposer* | musical score, song, creation, music, collection, arrangement, interpretation, composition, composer |

We may infer now, when the total terms in LEX-representation is not in big difference with the total terms in TRSM-representation, we might expect better performance from LEX-representation, which has shorter length but the same relevant terms whose ranks are higher, or in other words, which is more compact. It is practically feasible to improve the quality of LEX-representation by processing the terms more carefully in the preprocessing phase which have never been done by any of our experiments in this article.

## Time and Space Complexity

The computation cost of constructing the tolerance classes is $O(NM^2)$ [15]. In order to generate the LEX-representation, we need to construct the upper document representation and the TRSM-representation which are both $O(NM)$. Going from TRSM-representation to LEX-representation the computation cost is also $O(NM)$. After all, the total cost of mapping process is $O(NM^2)$.

We have mentioned before that the total number of index terms in ICL-corpus was 9,742 and WORDS-corpus was 3,477. As a result, the total numbers of index terms of TRSM-representations for ICL-corpus and WORDS-corpus were the same, 9,742 and 3,477 respectively. After the mapping process, we found that the total number of index terms in both corpora were reduced significantly, 64.65 % for ICL-corpus and 54.93 % for WORDS-corpus. The mapping process reduces the dimensionality of document vector quantitatively, thus we might expect more efficient computation when we further process the LEX-representation, e.g. for retrieval, categorization, or clustering process. The use of LEX-representation should give much benefit in applications when efficiency is put on the high priority.

## 7.4  Summary

We have presented a novel approach for an alternative to a document representation by employing the TRSM method and then run the mapping process, and finally come up with a compact representation of document. The mapping process is the process of mapping the index terms in TRSM-representation to terms in the lexicon.

We analyzed the LEX-representation based on the terms of topic-representation as well as of document representation. By a comparison between topic-representation with and without mapping we have seen that the mapping process should yield a better representation of document, concerning its nature ability to preserve the relevant terms. We have explained that the use of LEX-representation should lead to an effective process of retrieval due to the fact that the mean of recall and precision calculation gave comparable results with TRSM-representation. We might also expect a more efficient process of retrieval based on the finding that LEX-representation has much lower dimensional space than TRSM-representation. We conclude that the result of this study is promising.

# 8  Evaluation

## 8.1  Introduction

With regard to the intended retrieval system, we proposed some strategies pertaining to the implementation of tolerance rough sets model as we described in Sects. 5 to 7. All of the strategies were formulated by exploiting our domain specific testbed, namely ICL-corpus.

In this chapter, we are going to present our evaluation on those strategies when they were applied on a retrieval system with different corpus. The aim of evaluation is to validate all of our proposed strategies. Consecutively in following sections, we will discuss the effectiveness of tolerance value generator algorithm, the contributive factors of thesaurus optimization, and the lexicon-based document representation by means of employing another Indonesian corpus, called Kompas-corpus[40] [11], into the retrieval system.

Due to the fact that Kompas-corpus is the only Indonesian testbed available, we generated several corpora from Kompas-corpus as listed in Table 9. We named the variations using term Kompas_X, where X is a number specifies the amount of documents inside it, hence Kompas_3000 is the original Kompas-corpus who consists of 3,000 documents. In Kompas-corpus, not all documents are relevant with any topic defined in the topic file (i.e. information needs file) of Kompas-corpus. In fact, there are only 433 documents who have relevancy with at least one topic, and those 433 documents were assembled together into the Kompas_433. Respectively, Kompas_1000 and Kompas_2000 are composed of 1,000 and 2,000 documents in which all documents of Kompas_433 becomes part of them.

---

[40] Kompas-corpus is a TREC-like Indonesian testbed which is composed of 3,000 newswire articles and is accompanied by 20 topics. Please see Appendix C.4 for more explanation.

**Table 9.** The variation of Kompas-corpus.

| No | Variation | Total document | Total unique term | Total term |
|---|---|---|---|---|
| 1. | Kompas_433 | 433 | 8,245 | 85,063 |
| 2. | Kompas_1000 | 1,000 | 13,288 | 183,812 |
| 3. | Kompas_2000 | 2,000 | 19,766 | 370,472 |
| 4 | Kompas_3000 | 3,000 | 24,689 | 554,689 |

The evaluation data were acquired from experiments following a process depicted in Fig. 37 which is a schema for a retrieval system based on TRSM followed by calculating the LEX-representation. We employed all variations of Kompas-corpus listed in Table 9 for data source of the thesaurus and used only single corpus, Kompas_433, as the main data of the retrieval system for all runs. In the retrieval phase, the information needs and relevance judgments files were loaded in order to produce sets of ranked documents based on TFIDF-representation, TRSM-representation, and LEX-representation.

## 8.2   Evaluation on Tolerance Value Generator

In addition to retrieval system displayed in Fig. 37, we ran our tolerance value generator (let us call it *TolValGen* for short) for all variants of Kompas-corpus that served as the data source of thesaurus. Table 10 records the tolerance values provided by the TolValGen for each run of different variant.

Figure 38 shows the compilation of recall and MAP values of retrieval system for all data sources. From these graphs, we can see that the tolerance values yielded by TolValGen are appropriate since at each resulted $\theta$ value the associated corpus performs better then the TFIDF.

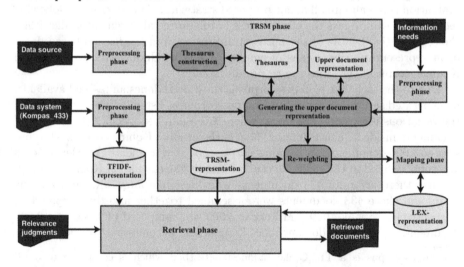

**Fig. 37.** IRS based on TRSM. The evaluation was conducted as a retrieval system in which tolerance rough sets model and the mapping process were implemented.

**Table 10.** Tolerance values generated by the TolValGen for each variant of Kompas-corpus functioned as the data source of thesaurus.

| No | Data system | Data source | Tolerance value |
|---|---|---|---|
| 1. | Kompas_433 | Kompas_433 | 37 |
| 2. | Kompas_433 | Kompas_1000 | 43 |
| 3. | Kompas_433 | Kompas_2000 | 46 |
| 4 | Kompas_433 | Kompas_3000 | 47 |

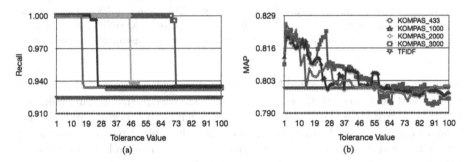

**Fig. 38.** Compilation values for TFIDF-representation and TRSM-representation. Graph (a) presents the recall values and graph (b) presents the MAP values of Kompas-corpus variants at $1 \leq \theta \leq 100$.

## 8.3 Evaluation on Thesaurus Optimization

In Sect. 6 we argued that tolerance value, data source, and semantic measure influence the quality of thesaurus in TRSM. Figure 38 that shows the compilation results of recall and MAP of the retrieval system for all variants of Kompas-corpus seems to agree with our argumentation. First of all, those graphs clearly confirm that we might have different quality of thesaurus that leads to different system performance by altering the tolerance value. This claim is supported by Obadi, et. al. [44] who did a study by implementing TRSM in a journal recommendation system based on topic search. They concluded that TRSM is very sensitive to parameter setting.

It is obvious from Table 9 that all corpora, Kompas_433, Kompas_1000, Kompas_2000, and Kompas_3000 have an increasing number of total term and distinct term respectively from one to another. Notice that all variants came from single corpus, hence those corpora are in the same domain with the data system. Considering the amount of terms in each corpus, Fig. 38 indicates that it agrees with our strategy in maximizing thesaurus quality by applying a set of corresponding documents whose total term and unique terms are larger in number.

With regard to the semantic measure in thesaurus construction, working with Kompas_433 as the main data as well as data source of the retrieval system while applying two different measures (i.e. raw frequency of co-occurrence and Cosine) produced recall and MAP graphs as they are depicted in Fig. 39. Put our concern

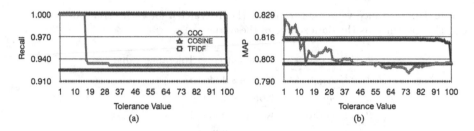

**Fig. 39.** Recall and MAP of different measures in thesaurus construction. Graph (a) presents the recall values and graph (b) presents the MAP values of `Kompas_433` corpus at $1 \leq \theta \leq 100$ in which the co-occurrence (COC) and Cosine (COSINE) measures were applied to define the semantic relatedness between terms in thesaurus construction.

on the Cosine measure (denoted by `COSINE` in the graphs), Fig. 39 is similar with Figs. 29 and 30 in Sect. 6. Based on this, we acknowledge that Cosine behavior occurs not only for a domain specific corpus such as ICL-corpus, but also for Kompas-corpus whose documents are more differ in topic. However, this finding affirms our assertion that the raw frequency of co-occurrence between terms is more suitable for thesaurus construction in TRSM.

### 8.4   Evaluation on Lexicon-Based Document Representation

The idea of document representation based on lexicon was confronted by the experimental results shown in Fig. 40 in which `Kompas_433` served as the data system and Kompas-corpus variants functioned as the data sources. It should come to our notice that the results are not as promising as ones of ICL-corpus.

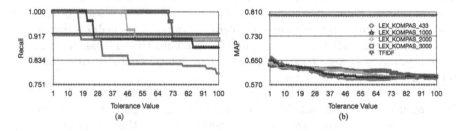

**Fig. 40.** Compilation of recall and MAP for LEX-representation. Graph (a) presents the recall values and graph (b) presents the MAP values of LEX-representation of Kompas-corpus variants at $1 \leq \theta \leq 100$.

We did the observation into tolerance classes of `Kompas_433` for terms *banjir* (in English: flood) and *sidang* (in English: trial) at several tolerance values, i.e. 1, 16, 17, and 37. The terms were chosen in order to represent topic *situasi banjir Jakarta* (in English: the flood situation in Jakarta) and topic *persidangan*

*Tommy Soeharto* (in English: the Tommy Soeharto's trial) which have document frequency 40 and 45 respectively. From the study, we found identical character-istic with of ICL-corpus in similar observation, that *(1)* TRSM-representation had most of related terms defined in thesaurus; and *(2)* the cardinality of TolClass_in_document in TFIDF-representation were mostly the least.

However, comparison between tolerance classes of TRSM-representation and of LEX-representation made us realize that the lexicon has removed some terms with high relevancy with the topic, which mostly were proper noun. For example, the topics *situasi banjir Jakarta* and *persidangan Tommy Soeharto* include signif-icant proper nouns *Jakarta* (which is the name of Indonesian's capital city) and *Tommy Soeharto* (which is the name of Indonesian second president's youngest son) respectively, and none of those proper nouns are part of the lexicon.

Table C.6 lists the 20 topics of Kompas-corpus and is comprised of 75 unique terms. First of all, it is obvious that almost all topics have proper nouns. Further, we identified that 26.6 % of the topic unique terms would be useless in retrieval phase because those terms have been removed from LEX-representation by the lexicon during mapping process, whereas most of the removed terms are proper nouns which are significant in defining the topics. The situation was quite dif-ferent with the ICL-corpus due to the fact that the topics of ICL-corpus which is comprised of 41 unique terms only have 1 proper noun, i.e. ICL, and thus yielded a compact LEX-representation.

For generalization, we acknowledge that this is a serious problem for LEX-representation for it might be corrupted and thus become much less reliable. Considering the fact that a lexicon consists of base words, we may infer that lexicon-based representation is not suitable for general use.

## 9  Conclusion

### 9.1  The TRSM-based Text Retrieval System

The research of extended TRSM, along with other researches of TRSM ever con-ducted, acted in accordance with the rational approach of AI perspective. This article presented studies who complied with the contrary path, i.e. a cognitive approach, for an objective of a modular framework of semantic text retrieval system based on TRSM specifically for Indonesian.

Figure 41 exhibits the schema of the intended framework which consists of three principal phases, namely preprocessing phase, TRSM phase, and retrieval phase. The framework supports a distinction between corpora functioned as data source and data system. In the framework, the query is converted into TRSM-representation by putting the thesaurus and the Eq. (25) to use while generating the upper approximation and re-weighting the query representation respectively. The mapping phase is included for an alternative and subject to change.

The proposed framework is in Java and takes a benefit of using Lucene 3.1 while indexing. Indonesian stemmer (i.e. CS stemmer), lexicon (i.e. created by University of Indonesia), and stopword (i.e. Vega's stopword) which are embed-ded make the framework works specifically for Indonesian language; altering

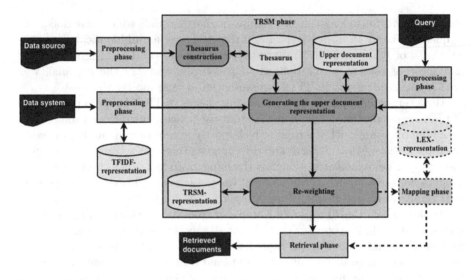

**Fig. 41.** The schema of the IRS. The schema of text retrieval system based on TRSM. The dashed line shapes are optional.

them specific to one language would make the framework dependent to that particular language.

It consists of 9 primary classes, in which one of it is the main class (i.e. IRS_TRSM), plus single class for the optional mapping phase. Three classes are included in preprocessing phase, two classes work for TRSM phase, and three classes are needed in retrieval phase. Figure 42 shows the classification of those classes based on the phases of the resulted IRS.

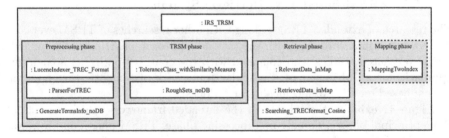

**Fig. 42.** Primary classes of the IRS. In total there are 10 primary classes in the proposed IRS, including the MappingTwoIndex for the optional mapping phase. The classes are classified based on the three phases of the IRS. The IRS_TRSM is the main class.

## 9.2   Novel Strategies for the TRSM-based Text Retrieval System

With regard to the framework of retrieval system, we delved into four issues based on the nature of TRSM. The very first issue questioned about the capacity of TRSM for the intended system, while the other three touched the system effectiveness.

In order to answer the first question, we did a feasibility study whose aim was to explain the meaning of *richness* of the TRSM-representation, rather than listing the strengths and weaknesses of TRSM. By working in close cooperation with human experts, we were able to reveal that the representation of document produced by TRSM does not merely contain more terms than the base representation, it rather contains more semantically related terms. Concerning our approach, we deem this as a stronger affirmation for the meaning of *richness* of the TRSM-representation as well as a satisfactory indicator in an endeavor to have a semantic retrieval system. Moreover, our analysis confirmed that rough sets theory intuitively works as the natural process of human thought.

Since the TRSM was introduced, no one has ever discussed or examined TRSM's parameter (i.e. tolerance value $\theta$) pertaining to its determination, whereas we consider it as fundamental for TRSM implementation. Obadi et. al. [44] seemed to realize this particular issue by stating that TRSM is very sensitive to parameter setting in their conclusion, however they did not explain or suggest anything about how to initiate it. In Sect. 5 we proposed a novel algorithm to define a tolerance value automatically by learning from a set of documents; and later we named it *TolValGen*. The algorithm was a result from careful observation and analysis performed through our corpora (i.e. ICL-corpus and WORDS-corpus) in which we learned some principles for a tolerance value resolution. The TolValGen was evaluated using another Indonesian corpus (i.e. Kompas-corpus) and yielded positive result. It was capable to produce an appropriate tolerance value for each variants of Kompas-corpus. Figure 43 displays the flowchart of TolValGen which works based on SVD.

We recognized that the thesaurus dominates TRSM in its work, hence optimizing the quality of thesaurus became another important issue we discussed. We admit that our idea to enhance the quality of thesaurus by adding more documents specifically for data source of thesaurus did not come up with a promising result as of Nguyen et al. [28, 29] which performed much more clever idea by extending the TRSM such that it accommodates more than one factors for a composite weight value of document vector. However, from the analysis carried out through several corpora (the variants of ICL-corpus and the Wiki_1800), we learned that tolerance value, data source of thesaurus, and semantic measure determine the quality of thesaurus. Specifically, a data source which is in a corresponding domain with the system data and is larger in number might bring more benefit. We also found that the total number of terms and index terms contribute more to the quality of thesaurus, despite the size of corpus. Finally, we suggested to keep the raw frequency of co-occurrence to define the semantic relatedness between terms for it gave better results in experiment rather than other measure, i.e. Cosine. All of these findings were validated by means of evaluation using Kompas-corpus.

The last issue discussed in this article associated with both the effectiveness and efficiency of system. It was motivated by a fact that the richer represen-

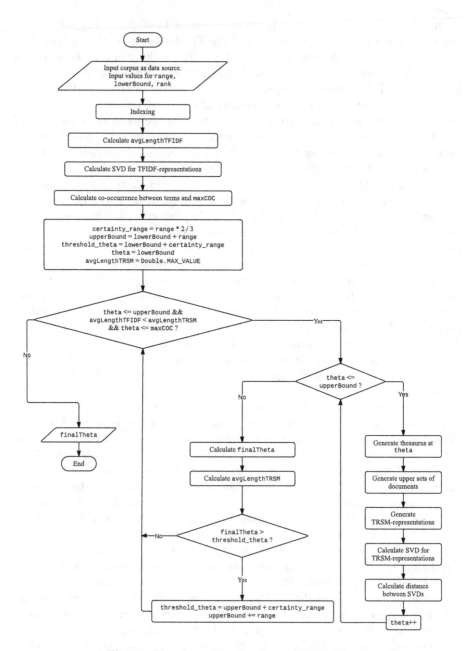

**Fig. 43.** Flowchart. The flowchart of TolValGen.

tation of TRSM is indicated by the larger number of index terms put into the model. Concerning the size of vector dimension, we came into an idea of a compact model of document based on the mapping process between index terms and lexicon after the document enriched by TRSM. The experimental data over ICL-corpus expressed a promising result, however the evaluation through Kompas-corpus remarked differently. Even though numerous irrelevant terms successfully removed from LEX-representation by the lexicon, we learned that our model cannot be applied for general use. The LEX-representation might be easily corrupted and thus become much less reliable when a query comprised of many terms which are not part of the lexicon and those terms are considered significant. Whereas, this particular situation is highly probable to occur in natural language.

### 9.3   Future Directions

The proposed framework is lack of comparison result. The studies presented in this article focused only on the use of TRSM which were compared to the result of TF*IDF. Comparison studies of methods, such those explained in Sect. 2 for semantic indexing, would put TRSM on certain position and bring some suggestion for further development.

The high complexity of our framework is the consequence of TRSM implementation. The application of Lucene module supports the indexing task in preprocessing phase of the framework, however we failed in the attempt to alter the index directly after TRSM phase which forced us to store the revised-index in different space. We found that it reduced the efficiency of IRS significantly, even though index file was applied. Studies focus on indexing in TRSM implementation is thus essential.

The proposed framework was developed for laboratory environment which is effective for restricted format and type of documents, i.e. follow the TREC-format and written in a *.txt* file. For a real application, our proposed framework should be extended to have the ability to deal with various format and type of documents. Much further, we should consider the recent phenomena of big data[41].

The TolValGen has showed to work on our corpora and their variations. However, it suffers from the expensive time and space to operate. In order to have cheaper complexity for tolerance value generator, further study on this theme with different methods is needed. We might expect some advantage by the use of machine learning method that accommodates the dynamic change of data source.

The lexicon-based document representation is an attempt on system efficiency. Despite the result of evaluation in Sect. 7 which signifies that it is lack of scalability, the fact that we did not implement any other linguistic methods arose our confident that those computations (such as tagging, feature selection,

---

[41] Big data is a term to describe the enormity of data, both structured and unstructured, in volume, velocity, and variety [45].

n-gram) might give us benefit in the effort of refining the thesaurus that serves as the basis of tolerance rough sets model, and thus the knowledge of our IRS.

In accordance with Searl's and Grice's accounts on meaning, Ingwersen [46, p. 33] defined that the concept of information, from a perspective of information science, has to satisfy dual requirements: *(1)* being the result of a transformation of generator's knowledge structures (by intentionality, model of recipients' states of knowledge, and in the form of signs); and *(2)* being something which when perceived, affects and transforms the recipients's state of knowledge. Thus, the endeavor of a semantic IRS is the effort to retrieve *information* and not merely terms with similar meaning. This article is a step toward the objective.

**Acknowledgments.** This work is partially supported by (1) Specific Grant Agreement Number-2008-4950/001-001-MUN-EWC from European Union Erasmus Mundus "External Cooperation Window" EMMA, (2) the National Centre for Research and Development (NCBiR) under Grant No. SP/I/1/77065/10 by the Strategic Scientific Research and Experimental Development Program: "Interdisciplinary System for Interactive Scientific and Scientific-Technical Information", (3) grant from Ministry of Science and Higher Education of the Republic of Poland N N516 077837, and (4) grant from Yayasan Arsari Djojohadikusumo (YAD) based on Addendum Agreement No. 029/C10/UKDW/2012. We thank Faculty of Computer Science, University of Indonesia, for the permission of using the CS stemmer.

# Appendix

## A    Weighting Scheme: The TF*IDF

Salton and Buckley summarised clearly in their paper [47] the insights gained in automatic term weighting and provided baseline single-term-indexing models with which other more elaborate content analysis procedures can be compared. The main function of a term-weighting system is the enhancement of retrieval effectiveness where this result depends crucially on the choice of effective term-weighting systems. *Recall* and *Precision* are two measures normally used to assess the ability of a system to retrieve the relevant and reject the non-relevant items of a collection. Considering the trade-off between recall and precision, in practice compromises are normally made by using terms that are broad enough to achieve a reasonable recall level without at the same time producing unreasonably low precision.

Salton and Buckley further explained that, with regard to the differing recall and precision requirements, three main considerations appear important:

1. *Term frequency* (tf). The frequent terms in individual documents appear to be useful as recall-enhancing devices.
2. *Inverse document frequency* (idf). The *idf* factor varies inversely with the number of documents $df_t$ to which a term $t$ is assigned in a collection of $N$ documents. It favors terms concentrated in a few documents of a collection and avoids the effect of high frequency terms which are widespread in the entirety of documents.

3. *Normalisation.* Normally, all relevant documents should be treated as equally important for retrieval purposes. The normalisation factor is suggested to equalise the length of the document vectors.

**Table A.1.** Term-weighting components with SMART notation [39]. Here, $tf_{t,d}$ is the term frequency of term $t$ in document $d$, N is the size of document collection, $df_t$ is document frequency of term $t$, $w_i$ is the weight of term $t$ in document $i$, $u$ is the number of unique terms in document $d$, and *CharLength* is the number of characters in the document.

| Term frequency component | |
| --- | --- |
| n (natural) | $tf_{t,d}$ |
| l (logarithm) | $1 + log(tf_{t,d})$ |
| a (augmented) | $0.5 + \frac{0.5 \times tf_{t,d}}{max_t(tf_{t,d})}$ |
| b (boolean) | $\begin{cases} 1 & \text{if } tf_{t,d} > 0 \\ 0 & \text{otherwise} \end{cases}$ |
| L (log ave) | $\frac{1+log(tf_{t,d})}{1+log(ave_{t \in d}(tf_{t,d}))}$ |
| Collection frequency component | |
| n (no) | $1$ |
| t (idf) | $log\frac{N}{df_t}$ |
| p (prob idf) | $max\{0, log\frac{N-df_t}{df_t}\}$ |
| Normalisation component | |
| n (none) | $1$ |
| c (cosine) | $\frac{1}{\sqrt{\sum_i (w_i)^2}}$ |
| u (pivoted unique) | $\frac{1}{u}$ |
| b (byte size) | $\frac{1}{CharLength^\alpha}, \alpha < 1$ |

Table A.1 summarises some of the term weighting schemes together with the mne-monic which is sometimes called SMART notation. One example of the mnemonic is *lnc.ltc*. The first triplet (i.e. *lnc*) represents the weighting combination for the document vector, while the second triplet (i.e. *ltc*) represents the weighting combination for the query vector. For each triplet, it describes the form of *tf* component, *idf* component, and *normalization* component being used. Thus, mnemonic *lnc.ltc* means that the document vector employs log-weighted term frequency, no idf for collection component, and cosine normalisation, while the query vector employs log-weighted term frequency, idf weighting for collection component, and cosine normalisation. Equation A.1 is the common weighting scheme used for a term in a document, i.e. mnemonic *ntn*, which is called TF*IDF weighting scheme.

$$w_{t,d} = tf \cdot idf = tf_{t,d} \cdot log\frac{N}{df_t} \tag{A.1}$$

# B    Document Ranking Method: The Cosine Measure

Manning et al. [39] stated that cosine similarity is fundamental to IR systems that use any form of vector space scoring. Given a query vector and a set of document vectors in a high dimensional space, we may rank the documents by comparing the angle between the query vector and each document vector; the smaller the angle, the more similar the vectors. In linear algebra, the angle $\theta$ between two vectors, $\vec{x}$ and $\vec{y}$, can be measured as follows:

$$\vec{x} \cdot \vec{y} = |\vec{x}| * |\vec{y}| * cos(\theta) \tag{B.1}$$

where $\vec{x} \cdot \vec{y}$ represents the *dot product* while $|\vec{x}|$ and $|\vec{y}|$ represent the lenght of the vectors. The dot product $\vec{x} \cdot \vec{y}$ of two vectors is defined as $\sum_{j=1}^{M} x_j * y_j$ and the Euclidean length of a vector $|\vec{x}|$ is defined as $\sqrt{\sum_{j=1}^{M}(x_j)^2}$. Thus, formula (B.2) can be used to measure the similarity between a query vector $Q$ and a document vector $D$:

$$similarity(Q, D) = \frac{\sum_{j=1}^{M} w_{qj} * w_{dj}}{\sqrt{\sum_{j=1}^{M}(w_{qj})^2 * \sum_{j=1}^{M}(w_{dj})^2}} \tag{B.2}$$

# C    The Corpora

## C.1    ICL-Corpus and WORDS-Corpus

Our original corpus, called ICL-corpus, consists of 1,000 first emails of Indonesian Choral Lovers (ICL) Yahoo! Groups and are formatted as of the Text REtrieval Conference (TREC) format [20]. Therefore our test collections consist of three parts (a set of documents, a set of information needs, and relevance judgments) and all documents are marked up in a TREC-like format, i.e. *each document* is marked up by <DOC> and </DOC> tags, the *document number* is marked up by <DOCNO> and </DOCNO> tags, the *subject of email* is marked up by <SUBJECT> and </SUBJECT> tags, the *date of email* is marked up by <DATE> and </DATE> tags, the *sender* is marked up by <FROM> and </FROM> tags, and the *text body* is marked up by <TEXT> and </TEXT> tags.

We worked with two choral experts intensively in the annotation process in order to construct the information needs and relevance judgments for our testbed. The annotation process consisted of two tasks which were a) topic assignment, where the human experts assigned topic(s) for each document within the original corpus; and b) keywords determination, where they determined terms considered as highly related with the topic(s) given. The annotation process aimed to grasp how the topic(s) could be assigned to a particular document which was mainly described by the keywords determined. We take benefit from these keywords as the list of terms closely related with the topic of document, as well as the document itself, and assume that the other terms not listed are

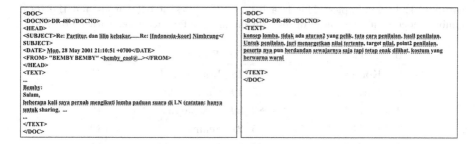

**Fig. C.1.** The content of corpora. Picture on the left is an example of ICL-corpus document which consists of original document, while picture on the right is an example of WORDS-corpus document which consists of keywords given by human expert manually for particular ICL-corpus document, i.e. in this case, the ICL-corpus document with number "DR-480" which is shown on the left.

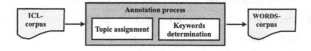

**Fig. C.2.** Corpus relationship. The WORDS-corpus was yielded by human expert in annotation process over ICL-corpus.

less important terms. The first step of topic assignment yielded 127 topics and the keywords determination yielded a new corpus, called WORDS-corpus.

Consult Fig. C.1 to see the content of both corpora. Notice that the main difference between documents in ICL-corpus and WORDS-corpus lies in the *text body*, i.e. the document of ICL-corpus consists of a body of emails while the document of WORDS-corpus consists of keywords defined by human experts. Figure C.2 shows the relationship between both corpora.

As we mentioned above, the topic assignment yielded 127 topics of which many have low document frequency; 81.10 % of them have document frequency < 10 and 32.28 % of them have document frequency 1. We further processed the 127-topics, as it is shown by Tables C.1 and C.2, and came up with 28 topics as listed in Table C.3. Thus, we have two version of relevance judgments *a*) relevance judgment which consists of 127 topics; and *b*) relevance judgment which consists of 28 topics.

For the 127-topics, distribution of topics is showed by Table C.4 while list of topics with document frequency ≥ 10 is showed by Table C.5. For all the tables here, *ID* column defines the topic identifier, *Topic* column is the topic in Indonesian, and *DF* column is the document frequency or total number of relevant documents with regard to the topic.

Refer to the TREC format, Fig. C.3 is an example of relevance judgment file while Fig. C.4 is an example of the information needs file. For the relevance judgment file, the first column defines the topic identifier, the third column defines the document identifier, and the fourth column defines the relevancy, i.e.

**Table C.1.** List of topics. This is a list of 127 topics of ICL-corpus and the total number (document frequency) of relevant documents for each topic with ID 0 to 35.

| ID | Topic | DF | ID | Topic | DF |
|----|-------|----|----|-------|----|
| 0 | Konser | 134 | 18 | Kompetisi PS | 15 |
| 1 | Partitur | 125 | 19 | Garpu tala | 14 |
| 2 | ICL | 80 | 20 | Lokakarya | 13 |
| 3 | ICL baru | 75 | 21 | Seminar | 12 |
| 4 | Lomba | 73 | 22 | Lagu sacred | 11 |
| 5 | Tanggapan konser | 46 | 23 | Publikasi | 10 |
| 6 | KPS Unpar | 39 | 24 | Hasil lomba | 9 |
| 7 | Pertemuan | 37 | 25 | Konser bersama | 9 |
| 8 | Dokumentasi | 35 | 26 | Koor gereja | 9 |
| 9 | Tanggapan lomba | 34 | 27 | Penilaian lomba | 9 |
| 10 | Media PS | 33 | 28 | PS sekolah | 9 |
| 11 | Manajemen dana | 32 | 29 | Spam | 9 |
| 12 | Aplikasi | 30 | 30 | Aturan spam | 8 |
| 13 | Buku vokal | 26 | 31 | Istilah musik | 8 |
| 14 | Teknikal milis | 25 | 32 | Manajemen PS | 8 |
| 15 | Festival | 17 | 33 | Peraturan lomba | 8 |
| 16 | Interpretasi | 17 | 34 | Manajemen penyanyi | 7 |
| 17 | Warna tangga nada | 17 | 35 | Organisasi PS | 7 |

1 if the document is relevant to the topic, and 0 otherwise. The second column is an arbitrary string and in this case brings no information. The information needs file consists of topics (string between <TITLE> and </TITLE> tags) with its description (string between <DESC> and </DESC> tags) and narrative (string between <NARR> and </NARR> tags). It follows the TREC format, thereby marked up by some tags in which each topic is enclosed by <TOP> and </TOP> tags.

## Annotation Process

We have mentioned above that the annotation process consisted of two tasks, namely topic assignment and keywords determination, and yielded WORDS-corpus and two lists of topics (127-topics and 28-topics). This was a collaborative work with three choral experts in which four phases were carried out as it is presented in Fig. C.5.

First of all, the first expert did the topic assignment and keyword determination for 1,000 documents of ICL-corpus. Considering his work (namely Result_1), we did the same thing and came with a different result (namely Result_2). Based on Result_1 and Result_2, the first expert did the revision

**Table C.2.** List of topics. This is a list of 127 topics of ICL-corpus and the total number (document frequency) of relevant documents for each topic with ID 36 to 126.

| ID | Topic | DF | ID | Topic | DF |
|---|---|---|---|---|---|
| 36 | Perkenalan | 7 | 83 | Usul | 2 |
| 37 | Analisa lagu | 6 | 84 | Website PS | 2 |
| 38 | Kualitas penyanyi | 6 | 85 | Workshop PS | 2 |
| 39 | Melatih PS | 6 | 86 | Agenda | 1 |
| 40 | UCV | 6 | 87 | Artikel konser | 1 |
| 41 | Bel Canto | 5 | 88 | Blocking | 1 |
| 42 | CKO | 5 | 89 | BMS | 1 |
| 43 | Impromptu | 5 | 90 | Children Choir Network | 1 |
| 44 | LPSAPTI | 5 | 91 | Choir building | 1 |
| 45 | Pakar PS anak | 5 | 92 | Database PS | 1 |
| 46 | Pembayaran | 5 | 93 | File | 1 |
| 47 | Penampilan | 5 | 94 | File konser | 1 |
| 48 | Piano | 5 | 95 | Forum | 1 |
| 49 | Ad Maiorem | 4 | 96 | FPS ITB | 1 |
| 50 | Choral sound | 4 | 97 | FPS Unpar | 1 |
| 51 | File uploaded | 4 | 98 | Harga lagu | 1 |
| 52 | ICL file | 4 | 99 | Hari haki | 1 |
| 53 | KCI | 4 | 100 | Himpunan seniman remaja bandung | 1 |
| 54 | Pembicara choir building | 4 | | | |
| 55 | Poling | 4 | 101 | ICL poling | 1 |
| 56 | PS anak | 4 | 102 | Informasi | 1 |
| 57 | Alamat | 3 | 103 | Interpretasi lagu | 1 |
| 58 | Arti konser | 3 | 104 | Jepang | 1 |
| 59 | Demam panggung | 3 | 105 | Kategori lomba | 1 |
| 60 | Folklore | 3 | 106 | Kategori PS | 1 |
| 61 | FX. Soetopo | 3 | 107 | Ketua PSM | 1 |
| 62 | Hak cipta | 3 | 108 | Lagu | 1 |
| 63 | ICL perkenalan | 3 | 109 | Lokakarya musik | 1 |
| 64 | Teknik pernafasan | 3 | 110 | Maria Luciana Dharmadi | 1 |
| 65 | Teknik vokal | 3 | 111 | Pemanasan | 1 |
| 66 | Tempat konser | 3 | 112 | Pesan foto | 1 |
| 67 | Tempat latihan | 3 | 113 | Poster | 1 |
| 68 | Tommy Prabowo | 3 | 114 | Poster konser | 1 |
| 69 | Acara | 2 | 115 | PS Perbanas | 1 |

(Continued)

**Table C.2.** (*Continued*)

| ID | Topic | DF | ID | Topic | DF |
|----|-------|----|----|-------|----|
| 70 | Aransemen | 2 | 116 | PS Petra | 1 |
| 71 | Berita | 2 | 117 | PSM Petra | 1 |
| 72 | Chamber choir | 2 | 118 | PSM UGM | 1 |
| 74 | Juri lomba | 2 | 119 | PSM Unpad | 1 |
| 75 | Memasyarakatkan PS | 2 | 120 | Respon ICL baru | 1 |
| 76 | Pembicara | 2 | 121 | Salam | 1 |
| 77 | Pertanyaan | 2 | 122 | Sponsor | 1 |
| 78 | Pitch | 2 | 123 | Tangga nada | 1 |
| 79 | PS GSS | 2 | 124 | Tiket | 1 |
| 80 | PS SD | 2 | 125 | VCD | 1 |
| 81 | Teknik pengucapan | 2 | 126 | VCD FPS ITB | 1 |
| 82 | Tiket konser | 2 | | | |

**Table C.3.** List of topics. This is a list of 28 topics of ICL-corpus and the total number (document frequency) of relevant documents for each topic.

| ID | Topic | DF | ID | Topic | DF |
|----|-------|----|----|-------|----|
| 0 | Komenter kegiatan | 80 | 14 | Orang | 16 |
| 1 | Internal milis ICL | 100 | 15 | Referensi | 27 |
| 2 | Kompetisi | 181 | 16 | Media paduan suara | 33 |
| 3 | Konser | 158 | 17 | Latihan | 12 |
| 4 | Karya musik | 125 | 18 | Pertemuan anggota milis ICL | 37 |
| 5 | Perkenalan anggota milis ICL | 87 | | | |
| | | | 19 | Spam | 14 |
| 6 | Manajemen | 46 | 20 | Instrumen | 19 |
| 7 | Kelompok musik | 52 | 21 | Genre | 14 |
| 8 | Aplikasi | 38 | 22 | Tangga nada | 18 |
| 9 | Hal teknis milis | 33 | 23 | Seminar atau pelatihan | 28 |
| 10 | Teknik vokal | 13 | 24 | Hak cipta | 11 |
| 11 | Performa | 14 | 25 | Terminologi | 11 |
| 12 | Dokumentasi | 38 | 26 | Forum | 15 |
| 13 | Interpretasi karya musik | 24 | 27 | Publikasi | 14 |

**Table C.4.** Topic distribution. This table shows the total number of topic which has document frequency < 10 out of 127 topics.

| Document frequency | 9 | 8 | 7 | 6 | 5 | 4 | 3 | 2 | 1 |
|--------------------|---|---|---|---|---|---|---|---|---|
| Number of topic | 6 | 4 | 3 | 4 | 8 | 8 | 12 | 17 | 41 |

**Table C.5.** List of topics. This table presents topics of ICL-corpus with document frequency $\geq$ 10 out of 127 topics.

| DF | Topic | DF | Topic | DF | Topic |
|---|---|---|---|---|---|
| 134 | Konser | 35 | Dokumentasi | 17 | Interpretasi |
| 125 | Partitur | 34 | Tanggapan lomba | 17 | Warna tangga nada |
| 80 | ICL | 33 | Media PS | 15 | Kompetisi PS |
| 75 | ICL baru | 32 | Manajemen dana | 14 | Garpu tala |
| 73 | Lomba | 30 | Aplikasi | 13 | Lokakarya |
| 46 | Tanggapan konser | 26 | Buku vokal | 12 | Seminar |
| 39 | KPS Unpar | 25 | Teknikal milis | 11 | Lagu sacred |
| 37 | Pertemuan | 17 | Festival | 10 | Publikasi |

```
0    0    DR-882    0
1    0    DR-882    1
2    0    DR-882    0
3    0    DR-882    0
4    0    DR-882    1
5    0    DR-882    0
6    0    DR-882    0
7    0    DR-882    0
```

**Fig. C.3.** The relevance judgment file. This picture is an inset of the relevance judgment file. Respectively, column 1 to 4 are the topic identifier, random string, document identifier, and document relevancy with topic.

```
<TOP>
<NUM>0</NUM>
<TITLE>komentar kegiatan</TITLE>
<DESC>laporan pandangan mata atau kesan atau tanggapan atau komentar atau kritik suatu
kegiatan misalnya konser atau kompetisi atau seminar atau pelatihan atau lokakarya atau choral
clinic</DESC>
<NARR>Dokumen yang relevan berisi tentang laporan pandangan mata, kesan, tanggapan,
kritik, dan saran tentang sebuah kegiatan (konser, kompetisi, seminar, pelatihan, lokakarya,
choral clinic) baik secara mendetil ataupun umum. Pembahasan bisa berkisar tentang lagu yang
dibawakan ketika konser atau kompetisi, penampilan penampil secara fisik maupun secara teknis
ketika bernyanyi, tempat penyelenggaraan, panggung, dekorasi, dan panitia penyelenggara,
materi yang dibawakan ketika seminar/pelatihan, pembicara atau pembawa materi dalam
seminar/pelatihan, cara pendaftaran suatu kegiatan, atau besarnya biaya.</NARR>
</TOP>


<TOP>
<NUM>1</NUM>
<TITLE>internal milis icl</TITLE>
<DESC>hal internal milis indonesian choral lovers atau icl</DESC>
<NARR>Dokumen yang relevan berisi informasi internal milis ICL secara khusus, misalnya
sejarah milis, penggagas milis, dan anggotanya. Dokumen yang berisi alamat surat elektronik
pribadi yang baru, rencana kegiatan anggota milis, atau surat pribadi antar anggota juga
termasuk relevan. Dokumen yang berisi ucapan selamat datang untuk anggota baru tidak
termasuk di sini.</NARR>
</TOP>
```

**Fig. C.4.** The information needs file. This picture is an inset of the information needs file.

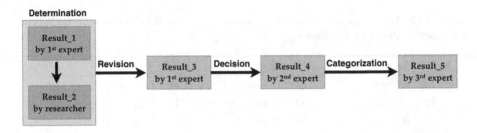

**Fig. C.5.** Annotation process. The annotation process had four phases: determination, revision, decision, and categorization.

of his previous result and produced the new result (namely `Result_3`). The second expert made a decision (`Result_4`) by analyzing `Result_1`, `Result_2`, and `Result_3`.

On this stage, we had 127 topics and decided to make the list smaller by categorizing it. Thus, we analyzed those topics and agreed on 28 topics. Refer to the 28-topics, the third expert reassigned each documents of ICL-corpus.

In addition to the construction process, this is another main difference of our corpus with an Indonesian corpus made by Jelita Asian from Kompas newswire articles (Kompas-corpus)[42]. In ICL-corpus, each document must be assigned by at least one topic while in Kompas-corpus it is not the case, i.e. there are documents that are not designated to any topics.

## C.2    WIKI_1800

WIKI-1800 is a corpus consists of 1,800 text documents in music domain which are the short abstract of Indonesian Wikipedia articles[43]. The full version of the corpus consists of 85,601 short abstracts in variety of topics and was downloaded from DBpedia[44]. The WIKI_1800 employed in this study was obtained by filtering out the 85,610 abstracts specifically based on music domain which was conducted by our third expert.

Figure C.6 shows a small chunk of WIKI-1800 document. Each document is represented as an RDF triple notation which contains three components (i.e. subject, predicate, and object), plus the URL of the Web page. In Fig. C.6, the <http://dbpedia.org/resource/Indonesia_Raya>, which acts as the subject, is an URI reference to the resource of *Indonesia Raya*. The

---

[42] Please see Appendix C.4 for more explanation about Kompas-corpus.

[43] Indonesian Wikipedia: http://id.wikipedia.org/wiki/Halaman_Utama.

[44] DBpedia is a community project which was started and is administered by research group from Universität Leipzig, Freie Universität Berlin, and OpenLink Software. The project is an effort to extract information from Wikipedia, make this information available on the Web under an open license, and interlink the DBpedia dataset with other open datasets on the Web. The Indonesian short abstracts of DBpedia was downloaded from http://downloads.dbpedia.org/3.7/id/.

<http://www.w3.org/2000/01/rdf-schema#comment> (or `rdfs:comment` for short), which acts as the predicate, is an URI reference that refers to the property used to provide a human-readable description of a resource; $R$ `rdfs:comment` $L$ states that $L$ is a human-readable description of $R$ [48]. Therefore, the string inside the quotes next to the `rdfs:comment` is the human-readable description of *Indonesia Raya*, which is actually the short abstract of the *Indonesia Raya* article. Finally, the <http://id.wikipedia.org/wiki/Indonesia_Raya#> is the URL that will go to the Web page of *Indonesia Raya*.

---

<http://dbpedia.org/resource/Indonesia_Raya> <http://www.w3.org/2000/01/rdf-schema#comment> "Indonesia Raya adalah lagu kebangsaan Republik Indonesia. Lagu ini pertama kali diperkenalkan oleh komponisnya, Wage Rudolf Soepratman, pada tanggal 28 Oktober 1928 pada saat Kongres Pemuda II di Batavia. Lagu ini menandakan kelahiran pergerakan nasionalisme seluruh nusantara di Indonesia yang mendukung ide satu \"Indonesia\" sebagai penerus Hindia Belanda, daripada dipecah menjadi beberapa koloni."@in <http://id.wikipedia.org/wiki/Indonesia_Raya#> .

<http://dbpedia.org/resource/Music> <http://www.w3.org/2000/01/rdf-schema#comment> "Musik adalah bunyi yang diterima oleh individu dan berbeda-beda berdasarkan sejarah, lokasi, budaya dan selera seseorang. Definisi sejati tentang musik juga bermacam-macam: Bunyi/kesan terhadap sesuatu yang ditangkap oleh indera pendengar Suatu karya seni dengan segenap unsur pokok dan pendukungnya. Segala bunyi yang dihasilkan secara sengaja oleh seseorang atau kumpulan dan disajikan sebagai musik Beberapa orang menganggap musik tidak berwujud sama sekali."@in <http://id.wikipedia.org/wiki/Musik#> .

---

**Fig. C.6.** WIKI-1800. An example of WIKI-1800 document.

## C.3  The Choral Experts

In data preparation of our study, we worked in collaboration with three people who have experiences in choral for years. They were Agastya Rama Listya, Kristoforus Kuntarahadi, and Inke Kusumastuti; in Appendix C.1, we called them the first expert, second expert, and third expert respectively. Figure C.7 displays the pictures of them.

Agastya Rama Listya was born in Yogyakarta on February 18, 1968, and now is living in Salatiga, Central Java, Indonesia. He obtained his Bachelor of Arts in Theory and Music Composition from the Indonesian Arts Institute, Yogyakarta, Indonesia, in 1992. In 2001, he received his Master of Sacred Music in Choral Conducting from Luther Seminary and St. Olaf College, Minnesota, USA. He was the Dean of the Faculty of Performing Arts, Satya Wacana Christian University at Salatiga for two periods (2009–2011) and was affiliated as the committee member of Badan Kerjasama Gereja-Gereja se-Salatiga (2007–2010), Lembaga Pengembangan Pesparawi Daerah Jawa Tengah (2007–2010), and Badan Pembina Seni Mahasiswa Indonesia Jawa Tengah (2008–2010). Agastya has published 7 books, 6 articles in journals, and 16 essays. He is a productive music composer

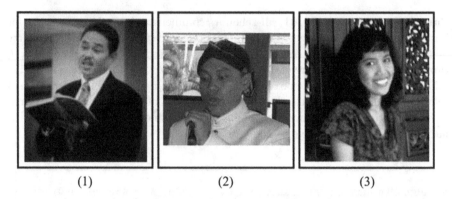

(1)                    (2)                    (3)

**Fig. C.7.** The choral experts. The choral experts involved in annotation process of our study: (1) Agastya Rama Listya, (2) Kristoforus Kuntarahadi, and (3) Inke Kusumastuti.

and arranger in which many of his choral works were performed by numerous choirs in Indonesia. He is also an active choral coach of a number of choirs where under his direction have made some prominent achievements regionally, nationally, and internationally. Individually, he was the winner of 4 different national choral composition contests during 1998–2009 and the winner of Yazeed Djamin Award for Piano Composition Contest in 2006. Agastya Rama Listya's name was included in the 30th Pearl Anniversary of Marquis Who's Who in The World (November 2012).

Kristoforus Kuntarahadi was born in Yogyakarta on January 14, 1979. He is now a staff in the office of Bishop's Conference of Indonesia, in Jakarta. He was the student of several well-known Indonesian vocalists and chorister, i.e. Avip Priatna, Lucia Kusumawardhani, Yoseph Chang, and Tommy Prabowo. He has been an active singers in some choirs since 1990, including the famous Indonesian choir, Batavia Madrigal Singers in Jakarta, and the tenor solo performer in some concerts. He obtained several achievements on regional singing festival during 1993–1997. Nationally, as a classical singer, he was the runner-up of Bintang Radio dan Televisi (a national radio and television singing competition) in 1995 and the third prize winner of PEKSIMINAS V (a national singing competition for student) in 1999. He received an award from Governor of Yogyakarta as an outstanding vocal artist in 1997.

Inke Kusumastuti is a medical doctor and currently continuing her education in Psychiatry in Udayana University, Denpasar, Bali. She was born in Blitar on April 17, 1986. She did not receive any formal education in music specifically, but she is practically a motivated self-learner when it comes to singing. She got numerous prizes in individual regional singing contests since she was in elementary school (1992–2001). In 2001–2004 she was involved in a band as the vocalist and the band won several regional competitions. In 2003, she experienced to be a cafe singer for a year. After that, while pursuing her medical education, she had been an active sopranos in some choirs, including the Eternal Choir, a well-known semi-professional small choir in Yogyakarta. As a chorister, she

**Table C.6.** List of topics. This is a list of 20 topics of Kompas-corpus and the document frequency, $DF$, of relevant documents for each topic.

| ID | Topic | DF |
|----|-------|----|
| 0 | Hubungan Indonesia Australia setelah Timor Timur | 11 |
| 1 | Dampak terorisme terhadap penurunan jumlah turis | 2 |
| 2 | Kecelakaan pesawat udara Indonesia | 22 |
| 3 | Pemberantasan narkoba | 18 |
| 4 | Pemilu presiden Prancis | 1 |
| 5 | Ulang tahun Megawati Sukarnoputri | 1 |
| 6 | Situasi banjir Jakarta | 40 |
| 7 | Duta besar Indonesia | 41 |
| 8 | Nama suami Megawati | 40 |
| 9 | Gejala dan penyebab asma | 1 |
| 10 | Pemenang pertandingan piala Thomas jenis apapun asal Indonesia | 8 |
| 11 | Nama bos Manchester United | 27 |
| 12 | Laporan piala dunia | 60 |
| 13 | Nilai tukar rupiah terhadap dolar AS | 74 |
| 14 | Aktor aktris calon atau pemenang Oscar | 3 |
| 15 | Akibat kenaikan harga BBM | 19 |
| 16 | Susunan kabinet Timor Leste | 1 |
| 17 | Persidangan Tommy Soeharto | 45 |
| 18 | Kunjungan luar negeri Megawati | 36 |
| 19 | Masa jabatan Gus Dur sebagai presiden | 3 |

was involved in numerous concerts and choral competitions and received some achievements. In 2007, she followed a *conducting* workshop given by Andrew deQuadros in the First Asian Choir Games and, in 2010, she joined a choral clinic given by Marc Anthony Carpio, a choirmaster of Phillippine Madrigal Singers. Recently, in 2012, she got the third prize winner in Bintang Radio RRI Jember (a singing contest conducted by national radio of Indonesia at Jember). Her favorite artist is The Real Group, a world-acclaimed Swedish-based a capella group, which has significantly shaped her current music interest, and her dream is to able to employ music as part of therapy for people with mental disorders.

## C.4    Kompas-Corpus

Kompas-corpus [11] is a set of newswire articles collected from a known Indonesian newspaper Kompas[45] published between January and June 2002. It consists of 3,000 documents constructed by following the TREC format, thereby accompanied by a file of information needs and a file of relevance judgments. There are 20 topics chosen by a native speaker after reading each documents in order

---

[45] Kompas. URL: http://www.kompas.com.

to represent the user information needs. Those topics are listed in Table C.6 as well as the total number of relevant documents for each topic. Out of 3,000, only 433 documents are assigned topic(s).

# References

1. Büttcher, S., Clarke, C.L.A., Cormack, G.V.: Information Retrieval: Implementing and Evaluating Search Engine. MIT Press, Cambridge (2010)
2. Weiss, S.M., Indurkhya, N., Zhang, T., Damerau, F.J.: Text Mining - Predictive Methods for Analyzing Unstructured Information. Springer, New York (2005)
3. Eifring, H., Theil, R.: Linguistics for Students of Asian and African Languages (2005)
4. Grandy, R.E., Warner, R.: Paul grice. http://plato.stanford.edu/entries/grice/, May 2006. Accessed 02 Oct 2012
5. Searle, J.R.: Intentionality: An Essay in the Philosophy of Mind. Cambridge University Press, Cambridge (1983)
6. Grice, H.P.: Studies in the Way of Words. Harvard University Press, Cambridge (1989)
7. Haugh, M., Jaszczolt, K.M.: Speaker intentions and intentionality. In: Allan, K., Jaszczolt, K.M. (eds.) The Cambridge Handbook of Pragmatics, pp. 87–112. Cambridge University Press, Cambridge (2012)
8. Akand, M.: Grice and searle on meaning. Copula - J. Philos. Dept **XXVIII**, 51–58 (2011)
9. Adriani, M., Manurung, R.: A survey of bahasa Indonesia NLP research conducted at the University of Indonesia. In: Proceedings of the 2nd International MALINDO Workshop (2008)
10. Asian, J.: Effective techniques for Indonesian text retrieval. Ph.D. thesis, School of Computer Science and Information Technology, RMIT University, Doctor of Philosophy Thesis (March 2007)
11. Asian, J., Williams, H.E., Tahaghoghi, S.M.M.: A testbed for Indonesian text retrieval. In: Bruza, P., Moffat, A., Turpin, A. (eds.) ADCS, pp. 55–58. University of Melbourne, Department of Computer Science (2004)
12. Sneddon, J.: The Indonesian Language: It's History and Role in Modern Society. UNSW Press, Sydney (2003)
13. Kawasaki, S., Nguyen, N.B., Ho, T.-B.: Hierarchical document clustering based on tolerance rough set model. In: Zighed, D.A., Komorowski, J., Żytkow, J.M. (eds.) PKDD 2000. LNCS (LNAI), vol. 1910, pp. 458–463. Springer, Heidelberg (2000)
14. Ho, T.B., Nguyen, N.B.: Nonhierarchical document clustering based on a tolerance rough set model. Int. J. Intell. Syst. **17**(2), 199–212 (2002)
15. Nguyen, H.S., Ho, T.B.: Rough document clustering and the internet. In: Handbook of Granular Computing, pp. 987–1003. Wiley, Hoboken (2008)
16. Wu, Y., Ding, Y., Wang, X., Xu, J.: On-line hot topic recommendation using tolerance rough set based topic clustering. J. Comput. **5**, 549–556 (2010)
17. Gaoxiang, Y., Heping, H., Zhengding, L., Ruixuan, L.: A novel web query automatic expansion based on rough set. Wuhan Univ. J. Nat. Sci. **11**(5), 1167–1171 (2006)
18. Bly, B.M., Rumelhart, D.E. (eds.): Cognitive Science: Handbook of Perception and Cognition, 2nd edn. Academic Press, Millbrae (1999)

19. Russell, S., Norvig, P.: Artificial Intelligence: A Modern Approach, 3rd edn. Pearson Education Inc., Upper Saddle River (2010)
20. Voorhees, E.M., Harman, D.: Overview of the ninth text retrieval conference (TREC-9). In: Proceedings of the Ninth Text Retrieval Conference (TREC-9), National Institute of Standards and Technology (NIST), pp. 1–14 (2000)
21. Baeza-Yates, R., Ribeiro-Neto, B.: Modern Information Retrieval. ACM Press, New York (1999)
22. Chomsky, N.: Language and Mind, 3rd edn. Cambridge University Press, New York (2006)
23. Furnas, G.W., Deerwester, S., Dumais, S.T., Landauer, T.K., Harshman, R.A., Streeter, L.A., Lochbaum, K.E.: Information retrieval using a singular value decomposition model of latent semantic structure. In: Proceedings of the 11th Annual International ACM SIGIR Conference on Research and Development in Information Retrieval. SIGIR 1988, New York, NY, USA, pp. 465–480. ACM (1988)
24. Grossman, D.A., Frieder, O.: Information Retrieval: Algorithms and Heuristics, 2nd edn. Springer, Netherlands (2004)
25. Gabrilovich, E., Markovitch, S.: Computing semantic relatedness using wikipedia-based explicit semantic analysis. In: Proceedings of the 20th International Joint Conference on Artificial intelligence. IJCAI 2007, San Francisco, CA, USA, pp. 1606–1611. Morgan Kaufmann Publishers Inc (2007)
26. Gottron, T., Anderka, M., Stein, B.: Insights into explicit semantic analysis. In: Proceedings of the 20th ACM International Conference on Information and Knowledge Management. CIKM 2011, New York, NY, USA, pp. 1961–1964. ACM (2011)
27. Wong, S.K.M., Ziarko, W., Wong, P.C.N.: Generalized vector spaces model in information retrieval. In: Proceedings of the 8th Annual International ACM SIGIR Conference on Research and Development in Information Retrieval. SIGIR 1985, New York, NY, USA, pp. 18–25. ACM (1985)
28. Nguyen, S.H., Świeboda, W., Jaśkiewicz, G.: Extended document representation for search result clustering. In: Bembenik, R., Skonieczny, L., Rybiński, H., Niezgodka, M. (eds.) Intelligent Tools for Building a Scient. Info. Plat. SCI, vol. 390, pp. 77–95. Springer, Heidelberg (2012)
29. Nguyen, S.H., Jaśkiewicz, G., Świeboda, W., Nguyen, H.S.: Enhancing search result clustering with semantic indexing. In: Proceedings of the Third Symposium on Information and Communication Technology. SoICT 2012, New York, NY, USA, pp. 71–80. ACM (2012)
30. Szczuka, M., Janusz, A., Herba, K.: Semantic clustering of scientific articles with use of DBpedia knowledge base. In: Bembenik, R., Skonieczny, L., Rybiński, H., Niezgodka, M. (eds.) Intelligent Tools for Building a Scient. Info. Plat. SCI, vol. 390, pp. 61–76. Springer, Heidelberg (2012)
31. Pawlak, Z.: Rough sets. Int. J. Comput. Inf. Sci. **11**(5), 341–356 (1982)
32. Komorowski, J., Pawlak, Z., Polkowski, L., Skowron, A.: Rough Sets: A Tutorial, pp. 3–98. Springer, Singapore (1998)
33. Pawlak, Z.: Some issues on rough sets. In: Peters, J.F., Skowron, A., Grzymała-Busse, J.W., Kostek, B., Swiniarski, R.W., Szczuka, M.S. (eds.) Transactions on Rough Sets I. LNCS, vol. 3100, pp. 1–58. Springer, Heidelberg (2004)
34. Skowron, A., Stepaniuk, J.: Tolerance approximation spaces. Fundam. Inf. **27**, 245–253 (1996)
35. Lassila, O., Mcguinness, D.: The role of frame-based representation on the semantic web. Technical report, Knowledge System Laboratory, Standford University (2001)
36. Virginia, G., Nguyen, H.S.: Lexicon-based document representation. Fundamenta Informaticae **124**, 27–45 (2013, to appear)

37. Vega, V.B.: Information retrieval for the Indonesian language. Master's thesis, National University of Singapore, Unpublished (2001)
38. Adriani, M., Asian, J., Nazief, B., Tahaghoghi, S.M.M., Williams, H.E.: Stemming indonesian: a confix-stripping approach. ACM Trans. Asian Lang. Inf. Process. **6**, 1–33 (2007)
39. Manning, C.D., Raghavan, P., Schütze, H.: Introduction to Information Retrieval. Cambridge University Press, New York (2008)
40. McCandless, M., Hatcher, E., Gospodnetić, O.: Lucene in Action. Manning Publications Co., Greenwich (2010)
41. Virginia, G., Nguyen, H.S.: An algorithm for tolerance value generator in tolerance rough sets model. In: Na, M.G., Toro, C., Posada, J., Howlett, R.J., Jain, L.C. (eds.) Advances in Knowledge-Based and Intelligent Information and Engineering Systems. KES 2012, Netherlands, pp. 595–604. IOS Press (2012)
42. Golub, G.H., Van Loan, C.F.: Matrix Computations, 3rd edn. Johns Hopkins University Press, Baltimore (1996)
43. Adriani, M., Nazief, B.: Confix-Stripping: Approach to Stemming Algorithm for Bahasa Indonesia. Internal Publication, Depok (1996)
44. Obadi, G., Drázdilová, P., Hlaváček, L., Martinovič, J., Snášel, V.: A tolerance rough set based overlapping clustering for the DBLP data. In: Proceedings of the 2010 IEEE/WIC/ACM International Conference on Web Intelligence and International Conference on Intelligent Agent Technology - Workshops. WI-IAT 2010, vol. 3, pp. 57–60. IEEE (2010)
45. Troester, M.: Big data meets big data analytics. http://www.sas.com/resources/whitepaper/wp_46345.pdf (2012). SAS Institute Inc. Accessed 22 Feb 2013
46. Ingwersen, P.: Information Retrieval Interaction, 1st edn. Taylor Graham, London (1992)
47. Salton, G., Buckley, C.: Term-weighting approaches in automatic text retrieval. Inf. Process. Manage. **24**(5), 513–523 (1988)
48. Manola, F., Miller, E.: Rdf primer. http://www.w3.org/TR/2004/REC-rdf-primer-20040210/ (2004). W3C. Accessed 12 Jan 2013

# Some Transportation Problems Under Uncertain Environments

Pradip Kundu[(✉)]

Department of Mathematics, National Institute of Technology Durgapur,
Durgapur 713209, India
kundu.maths@gmail.com

**Abstract.** Transportation problem (TP) is a very important area in
operations research and management science. TPs not only involve with
cost minimization, but also involve with many other goals such as profit
maximization, time minimization, minimization of total deterioration of
goods, etc. Also the available data of a transportation system such as
transportation costs, resources, demands, conveyance capacities are not
always crisp or precise but are uncertain. In this dissertation some transportation problems have been formulated and solved in different uncertain environments, e.g., fuzzy, type-2 fuzzy, rough and linguistic.

Section 1 is introductory. Some basic concepts and definitions of
fuzzy set, type-2 fuzzy set, rough set and variable are introduced in
Sect. 2. In Sect. 3, we have formulated and solved two solid transportation problems (STPs) with fuzzy parameters namely a multi-objective
STP with budget constraints and a multi-objective multi-item STP.
Section 4 presents some theoretical developments related to type-2 fuzzy
variables (T2 FVs) - a defuzzification method of T2 FVs and an interval approximation method of continuous T2 FVs. In this section, three
transportation models with type-2 fuzzy parameters have been formulated and solved. In Sect. 5, we have presented two transportation mode
selection problems with linguistic evaluations represented by fuzzy variables and interval type-2 fuzzy variables respectively. Here we have developed two fuzzy multi-criteria group decision making methods and these
methods are applied to solve the respective mode selection problems.
Section 6 presents a practical solid transportation model considering per
trip capacity for each type of conveyances. Also in this problem fluctuating cost parameters are represented by rough variables. Rough chance
constrained programming model, rough expected value model and rough
dependent-chance programming model are used to solve the problem
with rough cost parameters.

**Keywords:** Transportation problem · Solid transportation problem ·
Fuzzy set theory · Rough set theory · Multiple objective programming ·
Fuzzy programming

---

This article is an extract from the Author's Doctoral Dissertation.

© Springer-Verlag Berlin Heidelberg 2015
J. Peters et al. (Eds.): TRS XIX, LNCS 8988, pp. 225–365, 2015.
DOI: 10.1007/978-3-662-47815-8_10

# 1   Introduction

## 1.1   Transportation Problem (TP)

Transportation problem is one of the most important and practical application based area of operations research. The classical transportation problem (TP) is a distribution problem in which some goods/products are to be transported from some sources (factories, warehouses, etc.) to some destinations (demand points). The objective is to determine which routes to be considered for shipment and the amount of the shipment so that total transportation cost become minimum. Mathematically classical TP can be defined as a special type of linear programming problem.

**Basic Terminologies in TP.** The transportation systems depend on several parameters such as origin or source, destination, availability or resource, demand, unit transportation cost, conveyance, constraint, etc. Detailed descriptions on these parameters are available in the literature on transportation problems.

Origin or source: The places where the goods/products originate from, i.e. the goods are available (e.g., the plant, production center or warehouse etc.) are called the origins or the sources.

Destination: The places where the goods are to be transported are called destinations.

Availability or resource: The amount of goods available at some source that can be transported from the source is refereed as availability or resource of that source.

Demand: The amount of goods that is required at some destination is refereed as the demand of that destination.

Unit transportation cost: The cost of transportation of unit product from some source to some destination is called unit transportation cost of the product for that source-destination route.

Constraint: The availabilities as well as demands are limited to certain amount. Limitations on resource availability and fulfilment of demand of each destination form what are known as constraints.

Conveyance: Modes of transportation (e.g., trucks, goods trains, cargo flights, ships, etc.) are called conveyances.

**Different Types of Transportation Models:**

**Basic Transportation Problem (TP):** The classical transportation problem (TP) deals with transportation of goods from some sources (supply points) to some destinations (demand points) so that total transportation cost becomes minimum. Suppose there are $m$ origins (or sources) $O_i$, $(i = 1, 2, ..., m)$ and $n$ destinations (or demand points) $D_j$, $(j = 1, 2, ..., n)$ and $a_i$ be the amount of a homogeneous product available at $i$-th origin and $b_j$ be the demand at $j$-th destination. Let $c_{ij}$ is the cost for transportation of unit of product from source $i$ to destination $j$ and $x_{ij}$ be decision variable which represents the unknown quantity to be transported from $i$-th origin to $j$-th destination. Then the mathematical form of TP is

$$Min \ Z = \sum_{i=1}^{m} \sum_{j=1}^{n} c_{ij} \ x_{ij}, \tag{1}$$

$$s.t. \ \sum_{j=1}^{n} x_{ij} \ (=, \leq) \ a_i, \quad i = 1, 2, ..., m, \tag{2}$$

$$\sum_{i=1}^{m} x_{ij} \ (=, \geq) \ b_j, \quad j = 1, 2, ..., n, \tag{3}$$

$$x_{ij} \geq 0, \ \forall \ i, j, \ \sum_{i=1}^{m} a_i \ (=, \geq) \sum_{j=1}^{n} b_j. \tag{4}$$

The constraint (2) ensures that total transported amount to the destinations from some source must be equal or less than the availability of that source. The constraint (3) indicates that total transported amount from the sources should at least satisfy the demand of each destination. If the constraints (2) and (3) are of equality types and total available resources are equal to the total demands, then the problem is called balanced TP. However in some real systems, the balance condition does not always holds, i.e., it may happen that total available resources are greater or equal to the total demands. Then the constraints become inequality types and the problem is called unbalanced TP.

**Fixed Charge Transportation Problem (FCTP):** A transportation problem is often associated with additional costs (termed as fixed costs) besides transportation cost. This fixed costs may be due to permit fees, property taxes, toll charges etc. Suppose $d_{ij}$ be the fixed cost associated with route $(i, j)$. Mathematical formulation of FCTP is

$$MinZ = \sum_{i=1}^{m} \sum_{j=1}^{n} (c_{ij} \ x_{ij} + d_{ij} \ y_{ij}), \tag{5}$$

$$s.t. \ \sum_{j=1}^{n} x_{ij} \ (=, \leq) \ a_i, \quad i = 1, 2, ..., m \tag{6}$$

$$\sum_{i=1}^{m} x_{ij} \ (=, \geq) \ b_j, \quad j = 1, 2, ..., n, \tag{7}$$

$$\sum_{i=1}^{m} a_i \ (=, \geq) \sum_{j=1}^{n} b_j, \tag{8}$$

$$x_{ij} \geq 0, \ ; \ y_{ij} = \begin{cases} 1, \text{ if } x_{ij} > 0; \\ 0, \text{ otherwise.} \end{cases} \forall \ i, j \tag{9}$$

The notations $c_{ij}$, $a_i$, $b_j$ and $x_{ij}$ have the same meaning as in the above model. It is obvious that the fixed charge $d_{ij}$ will be costed for a route $(i, j)$ only if any transportation activity is assigned to that route. So $y_{ij}$ is defined such that if $x_{ij} > 0$ then $y_{ij} = 1$, otherwise it will be 0.

**Multi-objective Transportation Problem (MOTP):** If more than one objectives is to be optimized in an TP, then the problem is called multi-objective transportation problem (MOTP). The several objectives may be minimization of total transportation costs, maximization of profit, minimization of breakability, total delivery time, etc. If P objectives are to be optimized and $c_{ij}^p$ represents the unit transportation penalty (transportation cost, profit, breakability rate, distance, time etc.) for $p$-th objective ($p = 1, 2, ..., P$), then mathematical formulation is

$$Min/Max \ Z_p = \sum_{i=1}^{m} \sum_{j=1}^{n} c_{ij}^p \ x_{ij} \ p = 1, 2, ..., P, \tag{10}$$

$$s.t. \ \sum_{j=1}^{n} x_{ij} \ (=, \leq) \ a_i, \quad i = 1, 2, ..., m, \tag{11}$$

$$\sum_{i=1}^{m} x_{ij} \ (=, \geq) b_j, \quad j = 1, 2, ..., n, \tag{12}$$

$$x_{ij} \geq 0, \ \forall \ i, j, \ \sum_{i=1}^{m} a_i \ (=, \geq) \sum_{j=1}^{n} b_j. \tag{13}$$

**Multi-item Transportation Problem (MITP):** In multi-item TP, several types of items/goods are transported instead of one type of good. If $l$ items are to be transported and $c_{ij}^p$ be the unit transportation cost from $i$-th source to $j$-th destination for $p$-th ($p = 1, 2, ..., l$) item, then the mathematical formulation of MITP becomes

$$Min \ Z = \sum_{p=1}^{l} \sum_{i=1}^{m} \sum_{j=1}^{n} c_{ij}^p \ x_{ij}^p \tag{14}$$

$$s.t. \ \sum_{j=1}^{n} x_{ij}^p \ (=, \leq) a_i^p, \quad i = 1, 2, ..., m; \ p = 1, 2, .., l \tag{15}$$

$$\sum_{i=1}^{m} x_{ij}^p \ (=, \geq) b_j^p, \quad j = 1, 2, ..., n; \ p = 1, 2, .., l \tag{16}$$

$$x_{ij}^p \geq 0, \forall \ i, j, p; \ \sum_{i=1}^{m} a_i^p \ (=, \geq) \sum_{j=1}^{n} b_j^p, \ p = 1, 2, ..., l, \tag{17}$$

where, $a_i^p$ be the availability of $p$-th item at $i$-th origin, $b_j^p$ be the demand of $p$-th item at $j$-th destination and $x_{ij}^p$ be the decision variable represents the amount of $p$-th item to be transported from $i$-th origin to $j$-th destination.

**Solid Transportation Problem (STP):** Solid transportation problem (STP) is an extension of the basic TP. In a transportation system, there may be different types of mode of transport available, such as trucks, goods trains, cargo flights, ships, etc. In STP, modes of transportation are considered. STP deals with three type of constraints instead of two (source and destination) in a TP. This extra constraint is due to modes of transportation (conveyance). Mathematical formulation of STP is

$$Min\ Z = \sum_{i=1}^{m}\sum_{j=1}^{n}\sum_{k=1}^{K} c_{ijk}\ x_{ijk} \tag{18}$$

$$s.t.\ \sum_{j=1}^{n}\sum_{k=1}^{K} x_{ijk}\ (=,\leq)a_i,\quad i=1,2,...,m \tag{19}$$

$$\sum_{i=1}^{m}\sum_{k=1}^{K} x_{ijk}\ (=,\geq)b_j,\quad j=1,2,...,n \tag{20}$$

$$\sum_{i=1}^{m}\sum_{j=1}^{n} x_{ijk}\ (=,\leq)e_k,\quad k=1,2,...,K \tag{21}$$

$$x_{ijk}\geq 0, \forall\ i,j,k, \tag{22}$$

$$\sum_{i=1}^{m} a_i\ (=,\geq)\sum_{j=1}^{n} b_j\ and\ \sum_{k=1}^{K} e_k\ (=,\geq)\sum_{j=1}^{n} b_j \tag{23}$$

where, $c_{ijk}$ be the unit transportation cost from $i$-th origin to $j$-th destination through $k$-th conveyance, $x_{ijk}$ is the decision variable represents the amount of goods to be transported from $i$-th origin to $j$-th destination through $k$-th conveyance and $a_i$, $b_j$ have the same meaning as mentioned before. $e_k$ be the transportation capacity of conveyance $k$, so that the constraint (21) indicates that the total amount transported by conveyance $k$ is no more than its transportation capacity.

## 1.2   Uncertain Environment

In many real world problems the available data are not always exact or precise. Various types of uncertainties appear in those data due to various reason such as insufficient information, lack of evidence, fluctuating financial market, linguistic information, imperfect statistical analysis, etc. In order to describe and extract the useful information hidden in uncertain data and to use this data properly in practical problems, many researchers have proposed a number of improved theories such as fuzzy set, type-2 fuzzy set, random set, rough set etc. When some of or all the system parameters associated with a decision making problem are not exact or precisely defined, moreover those are represented by fuzzy, type-2 fuzzy, random or rough sets(/variables), etc., then it is called that the problem is defined in those uncertain environment respectively. Methodologies or Techniques to deal with such imprecision, uncertainty, partial truth, and approximation to achieve practicability, robustness and low solution cost is called Soft Computing.

## 1.3 Historical Review of Transportation Problems

**Historical Review of Transportation Problem in Crisp Environment:** The basic transportation problem (TP) was originally developed by Hitchcock [58] and later discussed in detail by Koopmans [69]. There are several methods introduced by many researchers for solving the basic transportation problems, such as the Vogel approximation method (VAM), the north-west corner method, the shortcut method, Russel's approximation method (Greig, [50]). Dantzig [34] formulated the transportation problem as a special case of linear programming problems and then developed a special case form of Simplex technique (Dantzig, [33]) taking advantage of the special nature of the coefficient matrix. Kirca and Satir [67] presented a heuristic algorithm for obtaining an initial solution for TP. Gass [47] described various aspects of TP methodologies and computational results. Ramakrishnan [128] improved Goyals modified VAM for finding an initial feasible solution for unbalanced transportation problem.

Balinski [11], Hirch and Dantzig [57] introduced fixed charge transportation problem (FCTP). Palekar et al. [119] introduced a branch-and-bound method for solving the FCTP. Adlakha and Kowalski [3] reviewed briefly the FCTP. Adlakha et al. [4] provided a more-for-less algorithm for solving FCTP. Kowalski and Lev [70] developed the fixed charge transportation problem as a nonlinear programming problem. Lee and Moore [79] studied the optimization of transportation problems with multiple objectives. To solve multi-objective transportation problem, Zimmerman [159,160] introduced and developed fuzzy linear programming. The solid transportation problem (STP) was first stated by Schell [130]. Haley [53] described a solution procedure of a solid transportation problem, which is an extension of the Modi method. Gen et al. [48] solved a bicriteria STP by genetic algorithm. Pandian and Anuradha [120] introduced a new method using the principle of zero point method for finding an optimal solution of STPs.

**Historical Review of Transportation Problem in Fuzzy Environment:** Several researchers studied various types of TPs with the parameters such as transportation costs, supplies, demands, conveyance capacities as fuzzy numbers(/variables). Chanas et al. [17] presented an FLP model for solving transportation problems with fuzzy supply and demand values. Chanas and Kuchta [18] studied transportation problem with fuzzy cost coefficients. Jiménez and Verdegay [60] considered two types of uncertain STP, one with interval numbers and other with fuzzy numbers. Jiménez and Verdegay [61] applied an evolutionary algorithm based parametric approach to solve fuzzy solid transportation problem. Bit et al. [13] applied fuzzy programming technique to multi-objective STP. Li and Lai [80], Waiel [135] applied fuzzy programming approach to multi-objective transportation problem. Saad and Abass [129] provided parametric study on the transportation problems in fuzzy environment. Liu and Kao [98] solved fuzzy transportation problems based on extension principle. Gao and Liu [46] developed the two-phase fuzzy algorithms for multi-objective transportation problem. Ammar and Youness [6] studied multi-objective transportation problem with unit transportation costs, supplies and demands as fuzzy numbers. Li et al. [81] presented a genetic algorithm for solving the multi-objective STP with coefficients of the objective function as fuzzy numbers. Pramanik and Roy [126] introduced a intuitionistic fuzzy goal programming approach for solving multi-objective transportation problems. Yang and Liu [153] presented expected value model, chance-constrained programming model and dependent chance programming for fixed charge STP with unit transportation costs, supplies, demands and conveyance capacities as fuzzy variables. Liu and Lin [93] solved a fuzzy fixed charge STP with chance constrained programming. Ojha et al. [118] studied entropy based STP with general fuzzy cost and time. Chakraborty and Chakraborty [15] considered a transportation problem having fuzzy parameters with minimization of transportation cost as well as time of transportation. Fegad et al. [42] found optimal solution of TP using interval and triangular membership functions. Kaur and Kumar [64] provided a new approach for solving TP with transportation costs as generalized trapezoidal fuzzy numbers. Kundu et al. [72] modeled a multi-objective multi-item STP with fuzzy parameters and solved it by using two different methods.

**Historical Review of Transportation Problem in Type-2 Fuzzy Environment:** Though type-2 fuzzy sets/varibles are used in various fields such as group decision making system (Chen et al. [21]; Chen et al. [26]), Portfolio selection problem (Hasuike and Ishi [54]), Pattern recognition (Mitchell, [112]), data envelopment analysis (Qin et al., [127]), neural network (Aliev et al. [5]), Ad hoc networks (Yuste et al. [155]) etc., Figueroa-Garca and Hernndez [43] first considered a transportation problem with interval type-2 fuzzy demands and supplies and we (Kundu et al. [75]) are the first to model and solve transportation problem with parameters as general type-2 fuzzy variables.

**Historical Review of Transportation Problem with Rough Sets/ Variables:** Tao and Xu [132] developed rough multi-objective programming for rough multi-objective solid transportation problem considering a appropriately large feasible region as a universe and equivalent relationship is induced to generate an approximate space. Kundu et al. [73] first developed some practical solid transportation models with transportation cost as rough variables.

**Historical Review of Transportation Mode Selection Problem:** Kiesmüller et al. [65] discussed transportation mode decision problem taken into account both distribution of goods and the manufacturing of products. Kumru and Kumru [71] considered a problem of selecting the most suitable way of transportation between two given locations for a logistic company and applied multi-criteria decision-making method to solve the problem. Tuzkaya and Önüt [134] applied fuzzy analytic network process to evaluate the most suitable transportation mode between Turkey and Germany. The evaluation ratings and the weights of the criteria in that problem are expressed in linguistic terms generated by triangular fuzzy numbers. There are also other several articles available related to transportation mode selection problem (Monahan and Berger [113]; Eskigun et al. [41]; Wang and Lee [138]).

## 1.4    Motivation and Objective of the Article

Motivation: Transportation problem (TP) is one of the most important and practical application based area of operations research. TP has vast economic importance because price of every commodity includes transportation cost. Transportation problems not only involve with economic optimization such as cost minimization, profit maximization but also involve with many other goals such as minimization of total deterioration of goods during transportation, time minimization, risk minimization etc.

The available data of a transportation system, such as unit transportation cost, supplies, demands, conveyance capacities are not always exact or precise but are uncertain or imprecise due to uncertainty in judgment, insufficient information, fluctuating financial market, linguistic information, uncertainty of availability of transportation vehicles etc. This motivated us to consider some innovative transportation problems (TPs) under uncertain environments like fuzzy, type-2 fuzzy, rough etc.

Many researchers developed TPs in stochastic and fuzzy (type-1) environments. However at the beginning of this research work, we observed that no TP with type-2 fuzzy, rough parameters was available the in literature though these improved uncertainty theories are applied in many other decision making fields. This motivated us to develop and solve some TPs with type-2 fuzzy, rough parameters.

Also appropriate transportation mode selection is a very important issue in a transportation system and human judgments are generally expressed in linguistic terms. These linguistic terms are generally of uncertain nature as a word does not have the same meaning to different people. This motivated us to consider some transportation mode selection problems with linguistic evaluations.

**Objective of the Article:** The main objectives of the presented thesis are:

– To formulate different types of transportation models: Some innovative and useful transportation models could have been formulated to deal with the rapidly growing financial competition, technological development, real-life situations, etc. Here we have formulated some different types of transportation models such as multi-objective multi-item solid transportation model, multi-item solid transportation model with restriction on conveyances and items, solid transportation models with limited vehicle capacity, etc.

- To consider transportation problems with type-1 fuzzy parameters: Though some research works have been done about transportation problem in fuzzy environment, however there are some scopes of research work in this field. This includes new improved methodologies/techniques to solve different types of TPs with fuzzy parameters. In this thesis, we have formulated and solved two different solid transportation models with type-1 fuzzy parameters using improved defuzzification and solution techniques.
- To consider transportation problems with type-2 fuzzy parameters: Decision making with type-2 fuzzy parameters is an emerging area. Type-2 fuzzy sets (/variables) give additional degrees of freedom to represent uncertainty. However computational complexity is very high to deal with type-2 fuzzy sets. Here we have contributed some theoretical development of type-2 fuzzy variables, formulated and solved two transportation models with parameters as type-2 fuzzy variables. To the best of our knowledge, very few TPs with type-2 fuzzy variables were developed.
- To consider transportation problems with rough parameters: Rough set theory is moderately new and growing field of uncertainty. For the first time we have formulated and solved a solid transportation model with unit transportation costs as rough variables.
- To consider transportation mode selection problem with linguistic evaluations: Linguistic judgments are always uncertain. Many researchers represented linguistic terms using type-1 fuzzy sets (/variables). Recently from literature it is known that modeling word by interval type-2 fuzzy set is more scientific and reasonable than by type-1 fuzzy set. Here we have developed two fuzzy multi-criteria group decision making methods and successfully applied to solve two transportation mode selection problems with linguistic evaluations represented by type-1 and interval type-2 fuzzy variables respectively.

## 1.5    Organization of the Article

This article is based on my Ph.D. thesis [77]. In this article, some transportation problems have been formulated and solved in different uncertain environments, e.g., fuzzy, type-2 fuzzy, rough and linguistic. We classified our thesis into the following sections:-

Section 1 is introductory. It contains brief discussion about different types of transportation problems, uncertain environments and historical review of transportation problems.

In Sect. 2, some basic concepts and definitions of fuzzy set and variable, type-2 fuzzy set and variable, rough set and variable and representation of linguistic terms are introduced. Some methodologies to solve single/multi-objective linear/nonlinear programming problems in crisp and various uncertain environments have been discussed.

Section 3 presents transportation problems with fuzzy (type-1) parameters. In this section, we have formulated and solved two solid transportation models with type-1 fuzzy parameters. The first model is a multi-objective solid transportation problem (MOSTP) with unit transportation penalties/costs, supplies, demands and conveyance capacities as fuzzy variables. Also, apart from source, demand and capacity constraints, an extra constraint on the total budget at each destination is imposed. The second model is a multi-objective multi-item solid transportation problem with fuzzy coefficients for the objectives and constraints. A defuzzifcation method based on fuzzy linear programming is applied for fuzzy supplies, demands and conveyance capacities, including the condition that both total supply and conveyance capacity must not fall below the total demand.

In Sect. 4, we have first provided some theoretical developments related to type-2 fuzzy variables. We have proposed a defuzzification method of type-2 fuzzy variables. An interval approximation method of continuous type-2 fuzzy variables is also introduced. We have formulated and solved three transportation problems with type-2 fuzzy parameters namely, fixed charge transportation problem with type-2 fuzzy cost parameters, fixed charge transportation problem with type-2 fuzzy costs, supplies and demands and multi-item solid transportation problem having restriction on conveyances with type-2 fuzzy parameters.

Section 5 contains problems related to transportation mode selection with respect to several criteria for a particular transportation system. Here we have developed two fuzzy multi-criteria (/attribute) group decision making (FMCGDM/FMAGDM) methods, the first one based on ranking fuzzy numbers and the second one based on ranking interval type-2 fuzzy variables. These proposed methods are applied to solve two transportation mode selection problems with the evaluation ratings of the alternative modes and weights of the selection criteria are presented in linguistic terms generated by fuzzy numbers and interval type-2 fuzzy variables respectively.

In Sect. 6 we have represented fluctuating cost parameters by rough variables and formulated solid transportation model with rough cost parameters. The formulated transportation model is applicable for the system in which full vehicles, e.g. trucks, rail coaches are to be booked for transportation of products so that transportation cost is determined on the basis of full conveyances. The presented model is extended including different constraints with respect to various situations like restriction on number of vehicles, utilization of vehicles, etc.

In Sect. 7, overall contribution of the article and possible future extensions of the models and methods are discussed.

# 2    Basic Concepts and Methods/Techniques

## 2.1    Classical Set Theory

Classical (crisp) set is defined as a well defined collection of elements or objects which can be finite, countable or infinite. Here 'well defined' means an element either definitely belongs to or not belongs to the set. In other words, for a given element, whether it belongs to the set or not should be clear. The word crisp means dichotomous, that is, yes-or-no type rather than more-or-less type. In set theory, an element can either belongs to a set or not; and in optimization, a solution is either feasible or not.

Subset: If every element of a set A is also an element of a set B, then A is called a subset of B and this is written as $A \subseteq B$. If $A \subseteq B$ and $B \subseteq A$, then we say that A and B are equal, written as A = B. A is called a proper subset of B, denoted by $A \subset B$ if A is a subset of B with $A \neq B$ and $A \neq \emptyset$, where $\emptyset$ denotes the empty set.

**Characteristic function:** Let A be a subset of X. The characteristic function of A is defined by

$$\chi(x) = \begin{cases} 1, \text{ if } x \in A; \\ 0, \text{ otherwise.} \end{cases}$$

Convex set: A subset $S \subset \Re^n$ is said to be convex, if for any two points $x_1$, $x_2$ in S, the line segment joining the points $x_1$ and $x_2$ is also contained in S. In other words, a subset $S \subset \Re^n$ is convex, if and only if $x_1, x_2 \in S \Rightarrow \lambda x_1 + (1 - \lambda)x_2 \in S; \ 0 \leq \lambda \leq 1$.

Interval arithmetic: Here we discussed for given two closed intervals in $\Re$, how to add, subtract, multiply and divide these intervals. Suppose $*$ be a binary operation such as +, -, $\cdot$, /etc. defined over $\Re$. If A and B are closed intervals, then $A * B = \{a * b : a \in A, b \in B\}$ defines a binary operation on the set of closed intervals (Moore [115]). Let $A = [a_1, a_2]$ and $B = [b_1, b_2]$ be two closed intervals in $\Re$. Then operations on the closed intervals A and B are defined as follows:

Addition: $A + B = [a_1, a_2] + [b_1, b_2] = [a_1 + b_1, a_2 + b_2]$

Subtraction: $A - B = [a_1, a_2] - [b_1, b_2] = [a_1 - b_1, a_2 - b_2]$

Multiplication:
$A \cdot B = [a_1, a_2] \cdot [b_1, b_2] = [\min(a_1b_1, a_1b_2, a_2b_1, a_2b_2), \max(a_1b_1, a_1b_2, a_2b_1, a_2b_2)]$
In particular if these intervals are in $\Re^+$, the set of positive real numbers, then the multiplication formula gets simplified to
$A \cdot B = [a_1, a_2] \cdot [b_1, b_2] = [a_1b_1, a_2b_2]$

Division:
$\frac{A}{B} = \frac{[a_1, a_2]}{[b_1, b_2]} = [a_1, a_2] \cdot [\frac{1}{b_2}, \frac{1}{b_1}] = [\min(\frac{a_1}{b_2}, \frac{a_1}{b_1}, \frac{a_2}{b_2}, \frac{a_2}{b_1}), \max(\frac{a_1}{b_2}, \frac{a_1}{b_1}, \frac{a_2}{b_2}, \frac{a_2}{b_1})]$, provided 0 not belongs to $[b_1, b_2]$.
In particular if these intervals are in $\Re^+$, the set of positive real numbers, then the division formula gets simplified to
$\frac{A}{B} = [\frac{a_1}{b_2}, \frac{a_2}{b_1}]$.

Scalar multiplication: For $k \in \Re^+$ the scalar multiplication $k \cdot A$ is defined as
$k \cdot A = k \cdot [a_1, a_2] = [ka_1, ka_2]$.

## 2.2 Fuzzy Set Theory

In the real world, various situations, concepts, value systems, human thinking, judgments are not always crisp and deterministic and cannot be described or represented precisely. Very often they are uncertain or vague. In real systems, there exist collection of objects so that those can not be certainly classified as a member of certain set. Zadeh [156] introduced the concept of fuzzy set in order to represent class of objects for which there is no sharp boundary between objects that belong to the class and those that do not. For example consider collection of real numbers close to 5. Then the number 4.5 can be taken as close to 5. The number 4.4 can also be taken as close to 5. Then how about the number 4.3 that smaller than 4.4 by only 0.1. Continuing in this way, it is difficult to determine an exact number beyond which a number is not close to 5. In fact there is no sharp boundary between close and not close to 5. Fuzzy sets describe such types of sets by assigning a number to every element in the universe, which indicates the degree (grade) to which the element belongs to the sets. This degree or grade is called membership degree or grade of the element in the fuzzy set. Mathematically a fuzzy set is defined as follows.

**Definition 2.1 (Fuzzy Set).** Let $X$ be a collection of objects and $x$ be an element of $X$, then a fuzzy set $\tilde{A}$ in $X$ is a set of ordered pairs

$$\tilde{A} = \{(x, \mu_{\tilde{A}}(x))| \ x \in X\},$$

where $\mu_{\tilde{A}}(x)$ is called the membership function or grade of membership of $x$ in $\tilde{A}$ which maps $X$ into the membership space $M$ which is a subset of nonnegative real numbers having finite supremum.

Generally the range of the membership function $\mu_{\tilde{A}}(x)$ is constructed as the close interval $[0, u]$, $0 < u \le 1$ and the representation of fuzzy set becomes (Mendel [101])

$$\tilde{A} = \{(x, \mu_{\tilde{A}}(x))| \mu_{\tilde{A}}(x) \in [0, 1], \forall x \in X\}.$$

A classical set $A$ can be described in this way by defining membership function $\mu_A(x)$ that takes only two values 0 and 1 such that $\mu_A(x) = 1$ or 0 indicates $x$ belongs to or does not belongs to $A$.

**Some Basic Definitions Related to Fuzzy Set:** The following definitions and properties are based on Zadeh [156], Klir and Yuan [68], Zimmermann [160], Kaufmann [62], Bector and Chandra [12] and Wang et al. [144].

Support: The support of a fuzzy set $\tilde{A}$ in $X$ is a crisp set $S(\tilde{A})$ defined as $S(\tilde{A}) = \{x \in X| \ \mu_{\tilde{A}}(x) > 0\}$.

Core: The core of a fuzzy set $\tilde{A}$ is a set of all points having unit membership degree in $\tilde{A}$ denoted by $Core(\tilde{A})$, and defined as $Core(\tilde{A}) = \{x \in X| \ \mu_{\tilde{A}}(x) = 1\}$

Centroid: The centroid $C(\tilde{A})$ of a fuzzy set $\tilde{A}$ is defined by $C(\tilde{A}) = \frac{\sum_x x\mu_{\tilde{A}}(x)}{\sum_x \mu_{\tilde{A}}(x)}$

for discrete case (discrete set of points) and $C(\tilde{A}) = \frac{\int_{-\infty}^{\infty} x\mu_{\tilde{A}}(x)dx}{\int_{-\infty}^{\infty} \mu_{\tilde{A}}(x)dx}$ for continuous case.

Height: The height of a fuzzy set $\tilde{A}$, denoted by $h(\tilde{A})$ is defined as $h(\tilde{A}) = \sup_{x \in X} \mu_{\tilde{A}}(x)$.

If $h(\tilde{A}) = 1$ for a fuzzy set $\tilde{A}$ then the fuzzy set $\tilde{A}$ is called a normal fuzzy set.

Complement: The complement of a fuzzy set $\tilde{A}$ is a fuzzy set denoted by $\tilde{A}^c$ is defined by the membership function $\mu_{\tilde{A}^c}(x)$, where $\mu_{\tilde{A}^c}(x) = h(\tilde{A}) - \mu_{\tilde{A}}(x)$, $\forall x \in X$. If $\tilde{A}$ is normal then obviously $\mu_{\tilde{A}^c}(x) = 1 - \mu_{\tilde{A}}(x)$, $\forall x \in X$.

**$\alpha$-cut:** $\alpha$-cut of a fuzzy set $\tilde{A}$ in $X$ where $\alpha \in (0, 1]$ is the crisp set $A_\alpha$ given by $A_\alpha = \{x \in X \mid \mu_{\tilde{A}}(x) \geq \alpha\}$.

**Some Properties of Fuzzy Set:** Union: The union of two fuzzy sets $\tilde{A}$ and $\tilde{B}$ is a fuzzy set $\tilde{C}$ whose membership function is given by $\mu_{\tilde{C}}(x) = \max(\mu_{\tilde{A}}(x), \mu_{\tilde{B}}(x))$, $\forall \, x \in X$. This is expressed as $\tilde{C} = \tilde{A} \cup \tilde{B}$.

Intersection: The intersection of two fuzzy sets $\tilde{A}$ and $\tilde{B}$ is a fuzzy set $\tilde{D}$ whose membership function is given by $\mu_{\tilde{D}}(x) = \min(\mu_{\tilde{A}}(x), \mu_{\tilde{B}}(x))$, $\forall \, x \in X$. This is expressed as $\tilde{D} = \tilde{A} \cap \tilde{B}$.

Convexity: A fuzzy set $\tilde{A}$ in $X$ is said to be convex if and only if for any $x_1$, $x_2 \in X$, $\mu_{\tilde{A}}(\lambda x_1 + (1 - \lambda)x_2) \geq \min(\mu_{\tilde{A}}(x_1), \mu_{\tilde{A}}(x_2))$ for $0 \leq \lambda \leq 1$. In terms of $\alpha$-cut, a fuzzy set is said to be convex if its $\alpha$-cuts $A_\alpha$ are convex for all $\alpha \in (0, 1]$.

Containment: A fuzzy set $\tilde{A}$ is contained in $\tilde{B}$ or a subset of $\tilde{B}$ if $\mu_{\tilde{A}}(x) \leq \mu_{\tilde{B}}(x)$, $\forall \, x \in X$. This is written as $\tilde{A} \subseteq \tilde{B}$.

Equality: Two fuzzy sets $\tilde{A}$ and $\tilde{B}$ in X is said to be equal if $\tilde{A} \subseteq \tilde{B}$ and $\tilde{B} \subseteq \tilde{A}$, i.e. if $\mu_{\tilde{A}}(x) = \mu_{\tilde{B}}(x)$, $\forall \, x \in X$.

**Fuzzy Number:** Fuzzy number can be taken as a generalization of interval of real numbers where rather than considering each point of an interval has the same importance or belongings, a membership grade in [0,1] is imposed to each element as in fuzzy set. This is done to handle a situation where one has to deal with approximate numbers or numbers that are close to a real number or around a interval of real numbers, etc. Let us consider set of numbers that are close to a real number $r$ and try to represent this set by a fuzzy set, say by $\tilde{A}$. That is $\tilde{A}$ would be defined as an interval around $r$ with each element having a membership grade that provided according to closeness of that point to $r$. Since the real number r is certainly close to r itself, so membership grade of $r$ in $\tilde{A}$ should be defined as $\mu_{\tilde{A}}(r) = 1$, i.e., $\tilde{A}$ should be a normal fuzzy set. Also the interval must be of finite length, i.e. support of $\tilde{A}$ need to be bounded. It is known that the only convex sets in $\Re$ are intervals. The fuzzy number is defined as follows:

**Definition 2.2 (Fuzzy Number).** A fuzzy subset $\tilde{A}$ of real number $\Re$ with membership function $\mu_{\tilde{A}} : \Re \to [0, 1]$ is said to be a fuzzy number (Grzegorzewski [52]) if

(i) $\tilde{A}$ is normal, i.e. $\exists$ an element $x_0$ s.t. $\mu_{\tilde{A}}(x_0) = 1$,

(ii) $\mu_{\tilde{A}}(x)$ is upper semi-continuous membership function,

(iii) $\tilde{A}$ is fuzzy convex, i.e. $\mu_{\tilde{A}}(\lambda x_1 + (1 - \lambda)x_2) \geq \mu_{\tilde{A}}(x_1) \wedge \mu_{\tilde{A}}(x_2)$ $\forall x_1, x_2 \in \Re$ and $\lambda \in [0, 1]$,

(iv) Support of $\tilde{A} = \{x \in \Re : \mu_{\tilde{A}}(x) > 0\}$ is bounded.

Klir and Yuan [68] proved the following theorem which gives characterization of a fuzzy number.

**Theorem 2.1** Let $\tilde{A}$ be a fuzzy set in $\Re$. Then $\tilde{A}$ is a fuzzy number if and only if there exists a closed interval (which may be singleton) $[a, b] \neq \phi$ such that

$$\mu_{\tilde{A}}(x) = \begin{cases} l(x), & \text{if } x \in (-\infty, a); \\ 1, & \text{if } x \in [a, b]; \\ r(x), & \text{if } x \in (b, \infty), \end{cases} \tag{24}$$

where (i) $l : (-\infty, a) \to [0, 1]$ is increasing, continuous from the right and $l(x) = 0$ for $x \in (-\infty, u)$, for some $u < a$ and (ii) $r : (b, \infty) \to [0, 1]$ is decreasing, continuous from the left and $r(x) = 0$ for $x \in (v, \infty)$, for some $v > b$.

In most of the practical applications the function $l(x)$ and $r(x)$ are continuous which give the continuity of the membership function.

$\alpha$-**cut of Fuzzy Number:** the $\alpha$-cut/$\alpha$ - level set of a fuzzy number $\tilde{A}$, i.e. $A_\alpha = \{x \in \Re \mid \mu_{\tilde{A}}(x) \geq \alpha\}$ is a nonempty bounded closed interval (Wu [145]) denoted by $[A_\alpha^L, A_\alpha^R]$ or $[A_\alpha^-, A_\alpha^+]$, where, $A_\alpha^L$ and $A_\alpha^R$ are the lower and upper bounds of the closed interval and

$$A_\alpha^L = \inf\{x \in \Re \mid \mu_{\tilde{A}}(x) \geq \alpha\}, \ A_\alpha^R = \sup\{x \in \Re \mid \mu_{\tilde{A}}(x) \geq \alpha\}$$

Now some particular type of fuzzy numbers with continuous $l(x)$ and $r(x)$ defined over the set of real numbers are given below.

**General Fuzzy Number (GFN):** A GFN $\tilde{A}$ is specified by four numbers $a_1$, $a_2$, $a_3$, $a_4 \in \Re$ and two functions $l(x)$ and $r(x)$ (as defined in Theorem 2.1) with the following membership function

$$\mu_{\tilde{A}}(x) = \begin{cases} 0, & \text{if } x < a_1; \\ l(x), & \text{if } a_1 \leq x < a_2; \\ 1, & \text{if } a_2 \leq x \leq a_3; \\ r(x), & \text{if } a_3 < x \leq a_4; \\ 0, & \text{if } x > a_4. \end{cases} \tag{25}$$

**Triangular Fuzzy Number** (TFN): A TFN $\tilde{A}$ is a fuzzy number fully determined by triplet $(a_1, a_2, a_3)$ of crisp numbers with $a_1 < a_2 < a_3$, whose membership function is given by

$$\mu_{\tilde{A}}(x) = \begin{cases} \frac{x-a_1}{a_2-a_1}, & \text{if } a_1 \leq x \leq a_2; \\ 1, & \text{if } x = a_2; \\ \frac{a_3-x}{a_3-a_2}, & \text{if } a_2 \leq x \leq a_3; \\ 0, & \text{otherwise.} \end{cases} \tag{26}$$

The TFN $\tilde{A}$ is depicted in Fig. 1.

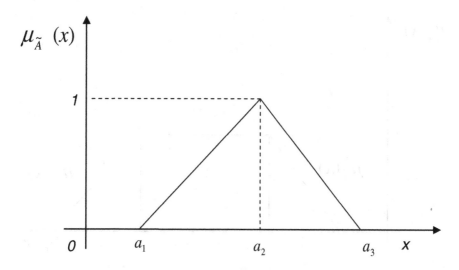

**Fig. 1.** TFN $\tilde{A}$

**Trapezoidal Fuzzy Number** (TrFN): A TrFN $\tilde{A}$ is a fuzzy number fully determined by quadruplet $(a_1, a_2, a_3, a_4)$ of crisp numbers with $a_1 < a_2 \le a_3 < a_4$, whose membership function is given by

$$\mu_{\tilde{A}}(x) = \begin{cases} \frac{x-a_1}{a_2-a_1}, & \text{if } a_1 \le x \le a_2; \\ 1, & \text{if } a_2 \le x \le a_3; \\ \frac{a_4-x}{a_4-a_3}, & \text{if } a_3 \le x \le a_4; \\ 0, & \text{otherwise.} \end{cases} \tag{27}$$

where $\frac{x-a_1}{a_2-a_1} = \mu_{\tilde{A}}^l(x)$ and $\frac{a_4-x}{a_4-a_3} = \mu_{\tilde{A}}^r(x)$ are called the left and right hand side of the membership function $\mu_{\tilde{A}}(x)$. The TrFN $\tilde{A}$ is depicted in Fig. 2. Obviously if $a_2 = a_3$ then TrFN becomes a TFN.

**Arithmetic of Fuzzy Numbers: Operation Based on the Zadeh's Extension Principle:** Arithmetical operations of fuzzy numbers can be performed by applying the Zadehs extension principle (Zadeh [158]). If $\tilde{A}$ and $\tilde{B}$ be two fuzzy numbers and $*$ be any operation then the fuzzy number $\tilde{A} * \tilde{B}$ is defined as

$$\mu_{\tilde{A}*\tilde{B}}(z) = \sup_{z=x*y} \min(\mu_{\tilde{A}}(x), \mu_{\tilde{B}}(y)), \ \forall z \in \Re.$$

So in particular we have

$$\mu_{\tilde{A}\oplus\tilde{B}}(z) = \sup_{z=x+y} \min(\mu_{\tilde{A}}(x), \mu_{\tilde{B}}(y)),$$

$$\mu_{\tilde{A}\ominus\tilde{B}}(z) = \sup_{z=x-y} \min(\mu_{\tilde{A}}(x), \mu_{\tilde{B}}(y)),$$

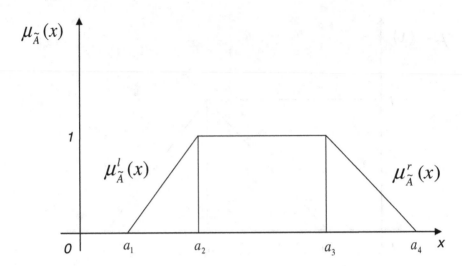

**Fig. 2.** TrFN $\tilde{A}$

$$\mu_{\tilde{A}\otimes\tilde{B}}(z) = \sup_{z=x\times y} \min(\mu_{\tilde{A}}(x), \mu_{\tilde{B}}(y)),$$

$$\mu_{\tilde{A}\oslash\tilde{B}}(z) = \sup_{z=x/y} \min(\mu_{\tilde{A}}(x), \mu_{\tilde{B}}(y)),$$

where $\oplus$, $\ominus$, $\otimes$ and $\oslash$ denote the addition, substraction, multiplication and division operations on fuzzy numbers.

**Operation Based on the $\alpha$-cuts:** Let $\tilde{A}$ and $\tilde{B}$ be two fuzzy numbers and $A_\alpha = [A_\alpha^L, A_\alpha^R]$, $B_\alpha = [B_\alpha^L, B_\alpha^R]$ be $\alpha$-cuts, $\alpha \in (0,1]$, of $\tilde{A}$ and $\tilde{B}$ respectively. Let $*$ denote any of the arithmetic operations $\oplus$, $\ominus$, $\otimes$, $\oslash$ of fuzzy numbers. Then the $*$ operation on fuzzy numbers $\tilde{A}$ and $\tilde{B}$, denoted by $\tilde{A} * \tilde{B}$, gives a fuzzy number in $\Re$ where

$$(\tilde{A} * \tilde{B})_\alpha = A_\alpha * B_\alpha, \ \alpha \in (0,1].$$

For particular operations we have

$$(\tilde{A} \oplus \tilde{B})_\alpha = A_\alpha \oplus B_\alpha = [A_\alpha^L + B_\alpha^L, A_\alpha^R + B_\alpha^R],$$

$$(\tilde{A} \ominus \tilde{B})_\alpha = A_\alpha \ominus B_\alpha = [A_\alpha^L - B_\alpha^R, A_\alpha^R - B_\alpha^L],$$

$$(\tilde{A} \otimes \tilde{B})_\alpha = A_\alpha \otimes B_\alpha =$$

$$= [\min\{A_\alpha^L B_\alpha^L, A_\alpha^L B_\alpha^R, A_\alpha^R B_\alpha^L, A_\alpha^R B_\alpha^R\}, \max\{A_\alpha^L B_\alpha^L, A_\alpha^L B_\alpha^R, A_\alpha^R B_\alpha^L, A_\alpha^R B_\alpha^R\}].$$

If the fuzzy numbers $\tilde{A}$ and $\tilde{B}$ in $\Re^+$, the set of positive real numbers, then the multiplication formula becomes

$$(\tilde{A} \otimes \tilde{B})_\alpha = A_\alpha \otimes B_\alpha = [A_\alpha^L B_\alpha^L, A_\alpha^R B_\alpha^R].$$

**Operations Under Function Principle:** Hsieh [59] presented Function Principle in fuzzy theory for computational model avoiding the complications which are caused by the operations using Extension Principle. The fuzzy arithmetical operations under Function Principle of two trapezoidal fuzzy numbers $\tilde{A} = (a_1, a_2, a_3, a_4)$ and $\tilde{B} = (b_1, b_2, b_3, b_4)$ are

(i) Addition: $\tilde{A} \oplus \tilde{B} = (a_1 + b_1, a_2 + b_2, a_3 + b_3, a_4 + b_4)$,
(ii) Substraction: $-\tilde{B} = (-b_4, -b_3, -b_2, -b_1)$ and $\tilde{A} \ominus \tilde{B} = (a_1 - b_4, a_2 - b_3, a_3 - b_2, a_4 - b_1)$,
(iii) Multiplication: $\tilde{A} \otimes \tilde{B} = (a_1 b_1, a_2 b_2, a_3 b_3, a_4 b_4)$, where $a_i$, $b_i$, for $i = 1, 2, 3, 4$ are positive real numbers.
(iv) Division: $\tilde{A} \oslash \tilde{B} = (a_1/b_4, a_2/b_3, a_3/b_2, a_4/b_1)$, where $a_i$, $b_i$, for $i = 1, 2, 3, 4$ are positive real numbers.
(v) $\lambda \otimes \tilde{A} = \begin{cases} (\lambda a_1, \lambda a_2, \lambda a_3, \lambda a_4), & \text{if } \lambda \geq 0; \\ (\lambda a_4, \lambda a_3, \lambda a_2, \lambda a_1), & \text{if } \lambda < 0. \end{cases}$

Here it should be mentioned that all the above operations can be defined using operations based on the $\alpha$-cuts of the fuzzy numbers that produce the same result.

**Defuzzification of Fuzzy Numbers:** Defuzzification methods/techniques of fuzzy numbers convert a fuzzy number or fuzzy quantity approximately to a crisp or deterministic value so that this can be used efficiently in practical applications. Some important defuzzification methods are presented below.

**Graded Mean and Modified Graded Mean:** Graded Mean (Chen and Hasieh [22]) Integration Representation method is based on the integral value of graded mean $\alpha$-level(cut) of generalized fuzzy number. For a fuzzy number $\tilde{A}$ the graded mean integration representation of $\tilde{A}$ is defined as

$$P(\tilde{A}) = \int_0^1 \alpha \Big[ \frac{A_\alpha^L + A_\alpha^R}{2} \Big] d\alpha \Big/ \int_0^1 \alpha \, d\alpha,$$

where $[A_\alpha^L, A_\alpha^R]$ is the $\alpha$-cut of $\tilde{A}$.
For example graded mean of a TrFN $\tilde{A} = (a_1, a_2, a_3, a_4)$ is $\frac{1}{6}[a_1 + 2a_2 + 2a_3 + a_4]$.

Here, equal weightage has been given to the lower and upper bounds of the $\alpha$-level of the fuzzy number. But the weightage may depends on the decision maker's preference or attitude. So, the modified graded mean $\alpha$-level value of the fuzzy number $\tilde{A}$ is $\alpha \big[ k A_\alpha^L + (1 - k) A_\alpha^R \big]$, where $k \in [0, 1]$ is called the decision makers attitude or optimism parameter. The value of $k$ closer to 0 implies that the decision maker is more pessimistic while the value of $k$ closer to 1 means that the decision maker is more optimistic. Therefore, the modified form of the above graded mean integration representation is

$$P(\tilde{A}) = \int_0^1 \alpha \Big[ \frac{k A_\alpha^L + (1 - k) A_\alpha^R}{2} \Big] d\alpha \Big/ \int_0^1 \alpha \, d\alpha.$$

For example modified graded mean of a TrFN $\tilde{A} = (a_1, a_2, a_3, a_4)$ is $\frac{1}{3}[k(a_1 + 2a_2) + (1 - k)(2a_3 + a_4)]$.

**Centroid Method:** The centroid $C(\tilde{A})$ of a fuzzy set $\tilde{A}$ is defined by

$$C(\tilde{A}) = \frac{\sum_x x\mu_{\tilde{A}}(x)}{\sum_x \mu_{\tilde{A}}(x)}$$

for discrete case and

$$C(\tilde{A}) = \frac{\int_{-\infty}^{\infty} x\mu_{\tilde{A}}(x)dx}{\int_{-\infty}^{\infty} \mu_{\tilde{A}}(x)dx}$$

for continuous case.

For example if $\tilde{A} = (a_1, a_2, a_3)$ is triangular fuzzy number then its centroid value is $C(\tilde{A}) = (a_1 + a_2 + a_3)/3$.

**Nearest Interval Approximation:** Grzegorzewski [52], presented a method to approximate a fuzzy number by a crisp interval. Suppose $\tilde{A}$ is a fuzzy number with $\alpha$-cut $[A_L(\alpha), A_R(\alpha)]$. Let $C_d(\tilde{A}) = [C_L, C_R]$ be the nearest interval approximation of the fuzzy number $\tilde{A}$ with distance metric $d$, where distance metric $d$ to measure distance of $\tilde{A}$ from $C_d(\tilde{A})$ is given by

$$d(\tilde{A}, C_d(\tilde{A})) = \sqrt{\int_0^1 \{A_L(\alpha) - C_L\}^2 d\alpha + \int_0^1 \{A_R(\alpha) - C_R\}^2 d\alpha}.$$

Now $C_d(\tilde{A})$ is optimum when $d(\tilde{A}, C_d(\tilde{A}))$ is minimum with respect to $C_L$ and $C_R$ and in this prospect the value of $C_L$ and $C_R$ are given by

$$C_L = \int_0^1 A_L(\alpha)d\alpha \ \text{and} \ C_R = \int_0^1 A_R(\alpha)d\alpha. \tag{28}$$

For example, $\alpha$−cut of a trapezoidal fuzzy number $(r_1, r_2, r_3, r_4)$ is $[r_1 + \alpha(r_2 - r_1), r_4 - \alpha(r_4 - r_3)]$ and its interval approximation is obtained as $[(r_1 + r_2)/2, (r_3 + r_4)/2]$.

**Fuzzy Variable:** Zadeh [158] introduced the possibility theory to interpret degree of uncertainty of members of a fuzzy set. The membership $\mu_{\tilde{A}}(x)$ of an element $x$ in a fuzzy set $\tilde{A}$ is then termed as degree of possibility that the element belongs to the set.

**Possibility Measure:** Suppose $\tilde{A}$ and $\tilde{B}$ be two fuzzy sets (/numbers) with memberships $\mu_{\tilde{A}}$ and $\mu_{\tilde{B}}$ respectively. Then possibility (Zadeh [158], Dubois and Prade [38], Liu and Iwamura [92]) of the fuzzy event $\tilde{A} \star \tilde{B}$ is defined as

$$Pos(\tilde{A} \star \tilde{B}) = \sup_{x \star y} \min(\mu_{\tilde{A}}(x), \mu_{\tilde{B}}(y)), \ x, y \in \Re, \tag{29}$$

where $\star$ is any operations like $<, >, =, \leq, \geq$, etc. Now for any real number $b$,

$$Pos(\tilde{A} \star b) = \sup_{x \star b}\{\mu_{\tilde{A}}(x), \ x \in \Re\}. \tag{30}$$

**Definition 2.3 (Possibility Space).** A triplet $(\Theta, p, Pos)$ is called a possibility space, where $\Theta$ is non-empty set of points, $p$ is power set of $\Theta$ and $Pos : \Theta \mapsto [0, 1]$ is a mapping, called possibility measure (Wang [136]) defined as

(i) $Pos(\emptyset) = 0$ and $Pos(\Theta) = 1$.
(ii) For any $\{A_i | i \in I\} \subset \Theta$, $Pos(\cup A_i) = \sup_i \ Pos(A_i)$.

**Definition 2.4 (Fuzzy Variable).** A fuzzy variable (Nahmias [117]) is defined as a function from the possibility space $(\Theta, p, Pos)$ to the set of real numbers $\Re$ to describe fuzzy phenomena, where possibility measure $(Pos)$ of a fuzzy event $\{\tilde{\xi} \in B\}$, $B \subset \Re$ is defined as $Pos\{\tilde{\xi} \in B\} = \sup_{x \in B} \ \mu_{\tilde{\xi}}(x)$, $\mu_{\tilde{\xi}}(x)$ is referred to as possibility distribution of $\tilde{\xi}$.

Necessity measure is dual of the possibility measure, the grade of necessity of an event is the grade of impossibility of the opposite event. Necessity measure $(Nes)$ of a fuzzy event $\{\tilde{\xi} \in B\}$, $B \subset \Re$ and $\sup_{x \in \Re} \ \mu_{\tilde{\xi}}(x) = 1$, is defined as $Nec\{\tilde{\xi} \in B\} = 1 - Pos\{\tilde{\xi} \in B^c\} = 1 - \sup_{x \in B^c} \ \mu_{\tilde{\xi}}(x)$.

**Credibility Theory:** Liu and Liu [94] introduced the concept of credibility measure. Liu [88] [90] presented credibility theory as a branch of mathematics for studying the behavior of fuzzy phenomena. Let $\Theta$ be a nonempty set, and $p$ the power set of $\Theta$. Each element in $p$ is called an event. For an event $A$, a number $Cr\{A\}$ which indicates the credibility that $A$ will occur has the following four axioms (Liu [90]):

1. Normality: $Cr\{\Theta\} = 1$.
2. Monotonicity: $Cr\{A\} \leq Cr\{B\}$ whenever $A \leq B$.
3. Self-Duality: $Cr\{A\} + Cr\{A^c\} = 1$ for any event $A$.
4. Maximality: $Cr\{\cup_i A_i\} = \sup_i Cr\{A_i\}$ for any events $\{A_i\}$ with $\sup_i Cr\{A_i\} < 0.5$.

**Definition 2.5 (Credibility Measure, Liu, [90]).** The set function $Cr$ is called a credibility measure if it satisfies the normality, monotonicity, self-duality, and maximality axioms.

For example let $\mu$ be a nonnegative function on $\Theta$ (for example, the set of real numbers) such that $\sup_{x \in \Theta} \ \mu(x) = 1$, then the set function defined by

$$Cr\{A\} = \frac{1}{2}(\sup_{x \in A} \ \mu(x) + 1 - \sup_{x \in A^c} \ \mu(x)) \qquad (31)$$

is a credibility measure on $\Theta$.
From it is clear that in case of a fuzzy variable $\tilde{\xi}$ with membership function (possibility distribution) $\mu_{\tilde{\xi}}$ and $B \subset \Re$, $\sup_{x \in \Re} \ \mu_{\tilde{\xi}}(x) = 1$, credibility measure is actually the average of possibility and necessity measure, i.e.

$$Cr\{\tilde{\xi} \in B\} = \frac{1}{2}(\sup_{x \in B} \ \mu_{\tilde{\xi}}(x) + 1 - \sup_{x \in B^c} \ \mu_{\tilde{\xi}}(x))$$

$$= \frac{1}{2}(Pos\{\tilde{\xi} \in B\} + Nec\{\tilde{\xi} \in B\}). \qquad (32)$$

**Example of Some Important Fuzzy Variables:**

**Equipossible Fuzzy Variable:** An equipossible fuzzy variable on [a, b] is a fuzzy variable whose membership function (possibility distribution) is given by

$$\mu_x = \begin{cases} 1, \text{ if } a \leq x \leq b; \\ 0, \text{ otherwise.} \end{cases}$$

**Trapezoidal and Triangular Fuzzy Variable:** Triangular fuzzy number and trapezoidal fuzzy number are two kinds of special fuzzy variables. As both trapezoidal and triangular fuzzy numbers are normal and defined over the set of real numbers $\Re$ so possibility, necessity as well as credibility measures are defined on them. So a trapezoidal fuzzy variable (TrFV) $\tilde{A}$ is a fuzzy variable fully determined by quadruplet $(a_1, a_2, a_3, a_4)$ of crisp numbers with $a_1 < a_2 \leq a_3 < a_4$, whose membership function is given by

$$\mu_{\tilde{A}}(x) = \begin{cases} \frac{x-a_1}{a_2-a_1}, & \text{if } a_1 \leq x \leq a_2; \\ 1, & \text{if } a_2 \leq x \leq a_3; \\ \frac{a_4-x}{a_4-a_3}, & \text{if } a_3 \leq x \leq a_4; \\ 0, & \text{otherwise.} \end{cases}$$

When $a_2 = a_3$, the trapezoidal fuzzy variable becomes a triangular fuzzy variable (TFV).

**Some Methodologies to Deal with Fuzzy Variables: Expected Value (Liu and Liu [94]):** Let $\tilde{\xi}$ be a fuzzy variable. Then the expected value of $\xi$ is defined as

$$E[\tilde{\xi}] = \int_0^{\infty} cr\{\tilde{\xi} \geq r\}dr - \int_{-\infty}^0 cr\{\tilde{\xi} \leq r\}dr \qquad (33)$$

provided that at least one of the two integrals is finite.

**Example 2.1.** Expected value of a triangular fuzzy variable $\tilde{\xi} = (r_1, r_2, r_3)$ is $E[\tilde{\xi}] = \frac{r_1 + 2r_2 + r_3}{4}$.

**Optimistic and Pessimistic Value (Liu [86,89]):** Let $\tilde{\xi}$ be a fuzzy variable and $\alpha \in [0, 1]$. Then

$$\xi_{sup}(\alpha) = \sup\{r : cr\{\tilde{\xi} \geq r\} \geq \alpha\} \qquad (34)$$

is called $\alpha$-optimistic value to $\tilde{\xi}$; and

$$\xi_{inf}(\alpha) = \inf\{r : cr\{\tilde{\xi} \leq r\} \geq \alpha\} \qquad (35)$$

is called $\alpha$-pessimistic value to $\tilde{\xi}$.

**Example 2.2.** Let $\tilde{\xi} = (r_1, r_2, r_3, r_4)$ be a trapezoidal fuzzy variable. Then its $\alpha$-optimistic and $\alpha$-pessimistic values are

$$\tilde{\xi}_{sup}(\alpha) = \begin{cases} 2\alpha r_3 + (1 - 2\alpha)r_4, & \text{if } \alpha \leq 0.5; \\ (2\alpha - 1)r_1 + 2(1 - \alpha)r_2, & \text{if } \alpha > 0.5. \end{cases} \tag{36}$$

$$\tilde{\xi}_{inf}(\alpha) = \begin{cases} (1 - 2\alpha)r_1 + 2\alpha r_2, & \text{if } \alpha \leq 0.5; \\ 2(1 - \alpha)r_3 + (2\alpha - 1)r_4, & \text{if } \alpha > 0.5. \end{cases} \tag{37}$$

## 2.3  Type-2 Fuzzy Set

So far in the Subsect. 2.2, we have discussed fuzzy sets with crisply defined membership functions, i.e., membership degree (/grade) of each of the points is an precise real number in [0,1]. However it is not always possible to represents uncertainty by a fuzzy set with crisp membership function, i.e., points having crisp membership grades. For instance, in rule-based fuzzy logic systems, the words that are used in the antecedents and consequents of rules can be uncertain as human judgements are not always precise and also a word does not have the same meaning or value to different people. Zadeh [157] introduced an extension of the concept of usual fuzzy set into a fuzzy set whose membership function itself is a fuzzy set. Then the usual fuzzy set with crisp membership function is termed as type-1 fuzzy set and the fuzzy set with fuzzy membership function is termed type-2 fuzzy set. So membership grade of each element of a type-2 fuzzy set is no longer a crisp value but a fuzzy set with a support bounded by the interval [0,1] which provides additional degrees of freedom for handling uncertainties. So because of fuzzy membership function a type-2 fuzzy set has three-dimensional nature. This membership function is called type-2 membership function.

**Definition 2.6 (Type-2 Fuzzy Set).** A type-2 fuzzy set (T2 FS) $\tilde{A}$ in $X$ is defined (Mendel and John [105,106]) as

$$\tilde{A} = \{((x, u), \mu_{\tilde{A}}(x, u)) : \forall x \in X, \forall u \in J_x \subseteq [0, 1]\},$$

where $0 \leq \mu_{\tilde{A}}(x, u) \leq 1$ is called the type-2 membership function, $J_x$ is the primary membership of $x \in X$ which is the domain of the secondary membership function $\tilde{\mu}_{\tilde{A}}(x)$ (defined below). The values $u \in J_x$ for $x \in X$ are called primary membership grades of $x$.

$\tilde{A}$ is also be expressed as

$$\tilde{A} = \int_{x \in X} \int_{u \in J_x} \mu_{\tilde{A}}(x, u)/(x, u) , \quad J_x \subseteq [0, 1], \tag{38}$$

where $\int \int$ denotes union over all admissible $x$ and $u$. For discrete universes of discourse $\int$ is replaced by $\sum$.

**Secondary Membership Function:** For each values of $x$, say $x = x'$, the secondary membership function (Mendel and John [105]), denoted by $\mu_{\tilde{A}}(x = x', u)$, $u \in J_{x'} \subseteq [0, 1]$ is defined as

$$\mu_{\tilde{A}}(x', u) \equiv \tilde{\mu}_{\tilde{A}}(x') = \int_{u \in J_{x'}} f_{x'}(u)/u, \tag{39}$$

where $0 \le f_{x'}(u) \le 1$. The amplitude of a secondary membership function is called a secondary grade. So for a particular $x = x'$ and $u = u' \in J_{x'}$, $f_{x'}(u') = \mu_{\tilde{A}}(x', u')$ is the secondary membership grade.

Now using (39), $\tilde{A}$ can be written an another way as $\tilde{A} = \{(x, \tilde{\mu}_{\tilde{A}}(x)) : x \in X\}$, i.e.

$$\tilde{A} = \int_{x \in X} \tilde{\mu}_{\tilde{A}}(x)/x = \int_{x \in X} \left[ \int_{u \in J_x} f_x(u)/u \right] / x. \tag{40}$$

**Example 2.3.** $X = \{4, 5, 6\}$ and the primary memberships of the points of X are $J_4 = \{0.3, 0.4, 0.6\}$, $J_5 = \{0.6, 0.8, 0.9\}$, $J_6 = \{0.5, 0.6, 0.7, 0.8\}$ respectively and the secondary membership functions of the points are
$\tilde{\mu}_{\tilde{A}}(4) = \mu_{\tilde{A}}(4, u) = (0.6/0.3) + (1/0.4) + (0.7/0.6)$
i.e., $\mu_{\tilde{A}}(4, 0.3) = 0.6$, $\mu_{\tilde{A}}(4, 0.4) = 1$ and $\mu_{\tilde{A}}(4, 0.6) = 0.7$. Here $\mu_{\tilde{A}}(4, 0.3) = 0.6$ means membership (secondary) grade that the point 4 has the membership (primary) 0.3 is 0.6.
$\tilde{\mu}_{\tilde{A}}(5) = \mu_{\tilde{A}}(5, u) = (0.7/0.6) + (1/0.8) + (0.8/0.9)$,
$\tilde{\mu}_{\tilde{A}}(6) = \mu_{\tilde{A}}(6, u) = (0.3/0.5) + (0.4/0.6) + (1/0.7) + (0.8/0.5)$.
So discrete type-2 fuzzy set $\tilde{A}$ is given by
$\tilde{A} = (0.6/0.3)/4 + (1/0.4)/4 + (0.7/0.6)/4 + (0.7/0.6)/5 + (1/0.8)/5 + (0.8/0.9)/5 + (0.3/0.5)/6 + (0.4/0.6)/6 + (1/0.7)/6 + (0.8/0.5)/6$.
$\tilde{A}$ is also written as

$$\tilde{A} \sim \begin{cases} 4, \text{ with membership } \tilde{\mu}_{\tilde{A}}(4); \\ 5, \text{ with membership } \tilde{\mu}_{\tilde{A}}(5); \\ 6, \text{ with membership } \tilde{\mu}_{\tilde{A}}(6). \end{cases}$$

The T2 FS $\tilde{A}$ is depicted in Fig. 3.

**Definition 2.7 (Interval Type-2 Fuzzy Set).** If all the secondary membership grades are 1 (i.e. $f_x(u) = \mu_{\tilde{A}}(x, u) = 1$, $\forall x, u$) then this T2 FS is called interval type-2 fuzzy set (IT2 FS) (Mendel et al. [107], Wu and Mendel [146]). The third dimension of the general T2 FS is not needed in this case and the IT2 FS can be expressed as a special case of the general T2 FS:

$$\tilde{A} = \int_{x \in X} \int_{u \in J_x} 1/(x, u) , \ J_x \subseteq [0, 1] \tag{41}$$

or, alternatively it can be represented as

$$\tilde{A} = \int_{x \in X} \tilde{\mu}_{\tilde{A}}(x)/x = \int_{x \in X} \left[ \int_{u \in J_x} 1/u \right] / x. \tag{42}$$

**Footprint of Uncertainty:** A IT2 FS is characterized by the footprint of uncertainty (FOU) which is the union of all of the primary memberships $J_x$, i.e. FOU of a IT2 FS $\tilde{A}$ is defined as

$$FOU(\tilde{A}) = \bigcup_{x \in X} J_x.$$

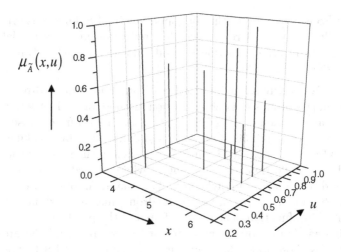

**Fig. 3.** Type-2 fuzzy set $\tilde{A}$

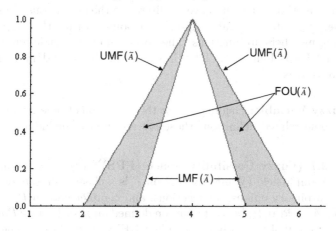

**Fig. 4.** Interval type-2 fuzzy set $\tilde{A}$

The FOU is bounded by an upper membership function $\bar{\mu}_{\tilde{A}}(x)$ (UMF) and a lower membership function $\underline{\mu}_{\tilde{A}}(x)$ (LMF), both are type-1 membership functions so that $J_x = [\underline{\mu}_{\tilde{A}}(x), \bar{\mu}_{\tilde{A}}(x)], \forall x \in X$. So the IT2 FS can be represented by $(\tilde{A}^U, \tilde{A}^L)$, where $\tilde{A}^U$ and $\tilde{A}^L$ are TIFSs.

For example consider a IT2 FS $\tilde{A}$ whose upper and lower membership functions are type-1 triangular membership functions and it is depicted in Fig. 4.

**Type Reduction:** We already knows that a type-2 fuzzy set (T2 FS) is a fuzzy set with fuzzy membership function. Due to fuzzyness in membership function of T2 FS, the computational complexity is very high to deal with T2 FS. For this reason to deal with T2 FS, generally a T2 FS is converted to a type-1 fuzzy set (T1 FS) by some type reduction methods. Type reduction is the procedure by which a T2 FS is converted to the corresponding T1 FS, known as type reduced set (TRS). But till now there are very few reduction methods available in the literature. Centroid type reduction (Kernik and Mendel [63]) is an example of such type reduction method.

**Geometric Defuzzification for T2 FS (Coupland and John [31]):** Coupland and John [31] proposed a defuzzification method for T2 FSs with the help of geometric representation of a T2 FS. A geometric T2 FS from a discrete T2 FS is constructed (Coupland [30], Coupland and John [31])) by breaking down the membership function of the T2 FS into five areas and then each of the five areas is modeled by a collection of 3-D triangles where the edges of these triangles are connected to form a 3-D polyhedron. The final defuzzified value is found by calculating the center of area of the polyhedron which approximates the type-2 fuzzy membership function. The center of area of the polyhedron is obtained by taking weighted average of $x$-component of the centroid and the area of each of the triangles those form the polyhedron. In case of T2 FS having continuous domains of primary or secondary membership function, to apply geometric defuzzification method, first one have to discretize the continuous domains into finite number of points (preferably equidistant points) within the support of the corresponding membership functions. The approach of geometric representation of discrete T2 FS is limited to T2 FSs where all the secondary membership functions are convex.

**Type-2 Fuzzy Variable:** Before going to the definition of type-2 fuzzy variable we present some related definitions those are required to define a type-2 fuzzy variable.

**Definition 2.7 (Fuzzy Possibility Space (FPS)).** Let $\Gamma$ be the universe of discourse. An ample field (Wang [136]) $\mathcal{A}$ on $\Gamma$ is a class of subsets of $\Gamma$ that is closed under arbitrary unions, intersections, and complements in $\Gamma$.

Let $\tilde{Pos} : \mathcal{A} \mapsto \Re([0,1])$ be a set function defined on $\mathcal{A}$ such that $\{\tilde{Pos}(A)|\mathcal{A} \ni A \text{ } atom\}$ is a family of mutually independent RFVs. Then $\tilde{Pos}$ is called a fuzzy possibility measure (Liu and Liu [99]) if it satisfies the following conditions:
(P1) $\tilde{Pos}(\emptyset) = \tilde{0}$.
(P2) For any subclass $\{A_i | i \in I\}$ of $\mathcal{A}$ (finite, countable or uncountable),

$$\tilde{Pos}\left(\bigcup_{i \in I} A_i\right) = sup_{i \in I}\tilde{Pos}(A_i).$$

The triplet $(\Gamma, \mathcal{A}, \tilde{Pos})$ is referred to as a fuzzy possibility space (FPS).

**Definition 2.8 (Regular Fuzzy Variable (RFV)).** For a possibility space $(\Theta, p, Pos)$, a regular fuzzy variable (Liu and Liu [99]) $\tilde{\xi}$ is defined as a measurable map from $\Theta$ to $[0,1]$ in the sense that for every $t \in [0,1]$, one has $\{\gamma \in \Theta \mid \tilde{\xi}(\gamma) \leq t\} \in p$.

A discrete RFV is represented as $\tilde{\xi} \sim \begin{pmatrix} r_1 & r_2 & \cdots & r_n \\ \mu_1 & \mu_2 & \cdots & \mu_n \end{pmatrix}$, where $r_i \in [0,1]$ and $\mu_i > 0$, $\forall i$ and $\max_i\{\mu_i\} = 1$.

If $\tilde{\xi} = (r_1, r_2, r_3, r_4)$ with $0 \leq r_1 < r_2 < r_3 < r_4 \leq 1$, then $\tilde{\xi}$ is called a trapezoidal RFV.

If $\tilde{\xi} = (r_1, r_2, r_3)$ with $0 \leq r_1 < r_2 < r_3 \leq 1$, then $\tilde{\xi}$ is called a triangular RFV.

**Definition 2.9 (Type-2 Fuzzy Variable).** As a fuzzy variable (type-1) is defined as a function from the possibility space to the set of real numbers, a type-2 fuzzy variable (T2 FV) is defined as a function from the fuzzy possibility space to the set of real numbers. If $(\Gamma, \mathcal{A}, \tilde{Pos})$ is a fuzzy possibility space (Liu and Liu [99]), then a type-2 fuzzy variable $\tilde{\xi}$ is defined as a map from $\Gamma$ to $\Re$ such that for any $t \in \Re$ the set $\{\gamma \in \Gamma \mid \tilde{\xi}(\gamma) \leq t\} \in \mathcal{A}$, i.e. a type-2 fuzzy variable (T2 FV) is a map from a fuzzy possibility space to the set of real numbers.

Then $\tilde{\mu}_{\tilde{\xi}}(x)$, called secondary possibility distribution function of $\tilde{\xi}$, is defined as a map $\Re \mapsto \Re[0, 1]$ such that $\tilde{\mu}_{\tilde{\xi}}(x) = \tilde{Pos}\{\gamma \in \Theta \mid \tilde{\xi}(\gamma) = x\}$, $x \in \Re$. $\mu_{\tilde{\xi}}(x, u)$, called type-2 possibility distribution function, is a map $\Re \times J_x \mapsto [0, 1]$, defined as $\mu_{\tilde{\xi}}(x, u) = Pos\{\tilde{\mu}_{\tilde{\xi}}(x) = u\}$, $(x, u) \in \Re \times J_x$, $J_x \subseteq [0, 1]$ is the domain or support of $\tilde{\mu}_{\tilde{\xi}}(x)$, i.e., $J_x = \{u \in [0, 1] \mid \mu_{\tilde{\xi}}(x, u) > 0\}$. Here $J_x$ may be called as primary possibility of the point $x$ and for a particular value of $x$, say $x = x'$, $\tilde{\mu}_{\tilde{\xi}}(x') \sim \mu_{\tilde{\xi}}(x', u)$, $u \in J_{x'}$ gives the secondary possibility of $x'$.

The secondary possibility distribution of a particular value $x = x'$, i.e. $\tilde{\mu}_{\tilde{\xi}}(x')$ actually represents a regular fuzzy variable (RFV).

**Definition 2.10 (Interval Type-2 Fuzzy Variable).** If for a type-2 fuzzy variable $\tilde{\xi}$ we call the $\mu_{\tilde{\xi}}(x', u')$ as secondary possibility degree for a point $x = x'$ and $u' \in J_{x'}$, then if secondary possibility degrees for all the points with respective primary possibilities are 1, $\tilde{\xi}$ is said to be interval type-2 fuzzy variable (IT2 FV).

**Example 2.4.** Let $\tilde{\xi}$ is a T2 FV defined as

$$\tilde{\xi} = \begin{cases} 5, \text{ with possibility } (0.2, 0.4, 0.6); \\ 6, \text{ with possibility } (0.4, 0.6, 0.8); \\ 7, \text{ with possibility } (0.1, 0.3, 0.5, 0.7). \end{cases} \tag{43}$$

i.e., the possibilities that $\tilde{\xi}$ has the values 5 and 6 are $\tilde{\mu}_{\tilde{\xi}}(5) = (0.2, 0.4, 0.6)$ and $\tilde{\mu}_{\tilde{\xi}}(6) = (0.4, 0.6, 0.8)$ respectively, each of which is triangular RFV and possibility that $\tilde{\xi}$ takes the value 7 is $\tilde{\mu}_{\tilde{\xi}}(7) = (0.1, 0.3, 0.5, 0.7)$ which is trapezoidal RFV. Obviously as $\tilde{\mu}_{\tilde{\xi}}(5) = (0.2, 0.4, 0.6)$ is triangular RFV, we have,

$$\mu_{\tilde{\xi}}(5, u) = \begin{cases} \frac{u - 0.2}{0.2}, & \text{if } 0.2 \leq u \leq 0.4; \\ 1, & \text{if } u = 0.4; \\ \frac{0.6 - u}{0.2}, & \text{if } 0.4 \leq u \leq 0.6. \\ 0, & \text{otherwise}; \end{cases}$$

from which we get the secondary possibilities for the point 5 and each values of $u$, $0.2 \leq u \leq 0.6$. $\tilde{\xi}$ is depicted in Fig. 5.

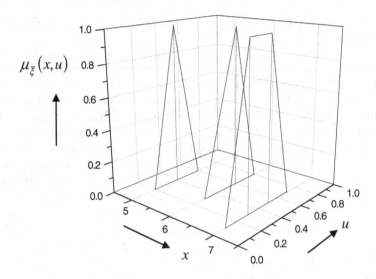

**Fig. 5.** Type-2 fuzzy variable $\tilde{\xi}$

**Example 2.5 (Type-2 Triangular Fuzzy Variable).** A type-2 triangular fuzzy variable (Qin et al. [127]) $\tilde{\xi}$ is represented by $(r_1, r_2, r_3; \theta_l, \theta_r)$, where $r_1$, $r_2$, $r_3$ are real numbers and $\theta_l, \theta_r \in [0, 1]$ are two parameters characterizing the degree of uncertainty that $\tilde{\xi}$ takes a value $x$ and the secondary possibility distribution function $\tilde{\mu}_{\tilde{\xi}}(x)$ of $\tilde{\xi}$ is defined by

$$\tilde{\mu}_{\tilde{\xi}}(x) = \Big(\frac{x - r_1}{r_2 - r_1} - \theta_l \min\Big\{\frac{x - r_1}{r_2 - r_1}, \frac{r_2 - x}{r_2 - r_1}\Big\}, \frac{x - r_1}{r_2 - r_1}, \frac{x - r_1}{r_2 - r_1} +$$
$$+ \theta_r \min\Big\{\frac{x - r_1}{r_2 - r_1}, \frac{r_2 - x}{r_2 - r_1}\Big\}\Big) \qquad (44)$$

for any $x \in [r_1, r_2]$, and

$$\tilde{\mu}_{\tilde{\xi}}(x) = \Big(\frac{r_3 - x}{r_3 - r_2} - \theta_l \min\Big\{\frac{r_3 - x}{r_3 - r_2}, \frac{x - r_2}{r_3 - r_2}\Big\}, \frac{r_3 - x}{r_3 - r_2}, \frac{r_3 - x}{r_3 - r_2} +$$
$$+ \theta_r \min\Big\{\frac{r_3 - x}{r_3 - r_2}, \frac{x - r_2}{r_3 - r_2}\Big\}\Big) \qquad (45)$$

for any $x \in (r_2, r_3]$.

A type-2 triangular fuzzy variable can be seen as an extension of a type-1 triangular fuzzy variable or simply a triangular fuzzy variable. In a triangular fuzzy variable (TFV) $(r_1, r_2, r_3)$, the membership grade (possibility degree) of each point is a fixed number in [0,1]. However in a type-2 triangular fuzzy variable $\tilde{\xi} = (r_1, r_2, r_3; \theta_l, \theta_r)$, the primary memberships (possibilities) of the points are no longer fixed values, instead they have a range between 0 and 1. Here $\theta_l$ and $\theta_r$ are used to represent the spreads of primary memberships of type-2 TFV. Obviously if $\theta_l = \theta_r = 0$, then type-2 TFV $\tilde{\xi}$ becomes a type-1 TFV and the

Eqs. (44) and (45) together become the membership function of a type-1 TFV. Now from Eqs. (44) and (45), $\tilde{\mu}_{\tilde{\xi}}(x)$ can be written as

$$
\tilde{\mu}_{\tilde{\xi}}(x) = \begin{cases} \left( \frac{x-r_1}{r_2-r_1} - \theta_l \frac{x-r_1}{r_2-r_1}, \frac{x-r_1}{r_2-r_1}, \frac{x-r_1}{r_2-r_1} + \theta_r \frac{x-r_1}{r_2-r_1} \right), & \text{if } x \in [r_1, \frac{r_1+r_2}{2}]; \\ \left( \frac{x-r_1}{r_2-r_1} - \theta_l \frac{r_2-x}{r_2-r_1}, \frac{x-r_1}{r_2-r_1}, \frac{x-r_1}{r_2-r_1} + \theta_r \frac{r_2-x}{r_2-r_1} \right), & \text{if } x \in (\frac{r_1+r_2}{2}, r_2]; \\ \left( \frac{r_3-x}{r_3-r_2} - \theta_l \frac{x-r_2}{r_3-r_2}, \frac{r_3-x}{r_3-r_2}, \frac{r_3-x}{r_3-r_2} + \theta_r \frac{x-r_2}{r_3-r_2} \right), & \text{if } x \in (r_2, \frac{r_2+r_3}{2}]; \\ \left( \frac{r_3-x}{r_3-r_2} - \theta_l \frac{r_3-x}{r_3-r_2}, \frac{r_3-x}{r_3-r_2}, \frac{r_3-x}{r_3-r_2} + \theta_r \frac{r_3-x}{r_3-r_2} \right), & \text{if } x \in (\frac{r_2+r_3}{2}, r_3]. \end{cases} \tag{46}
$$

Let us illustrate Example 2.5 numerically. Consider the type-2 triangular fuzzy variable $\tilde{\xi} = (2, 3, 4; 0.5, 0.8)$.

Then its secondary possibility distribution is given by

$$
\tilde{\mu}_{\tilde{\xi}}(x) = \begin{cases} (0.5(x-2), x-2, 1.8(x-2)), & \text{if } x \in [2, 2.5]; \\ ((x-2) - 0.5(3-x), x-2, (x-2) + 0.8(3-x)), & \text{if } x \in (2.5, 3]; \\ ((4-x) - 0.5(x-3), 4-x, (4-x) + 0.8(x-3)), & \text{if } x \in (3, 3.5]; \\ (0.5(4-x), 4-x, 1.8(4-x)), & \text{if } x \in (3.5, 4]. \end{cases}
$$

Here secondary possibility degree of each value of $x$ is a triangular fuzzy variable (more precisely a triangular RFV), e.g., $\tilde{\mu}_{\tilde{\xi}}(2.5) = (0.25, 0.5, 0.9)$, $\tilde{\mu}_{\tilde{\xi}}(3.2) = (0.7, 0.8, 0.96)$, etc. So the domain of secondary possibility $\tilde{\mu}_{\tilde{\xi}}(2.5)$ varies from 0.25 to 0.9 and that of $\tilde{\mu}_{\tilde{\xi}}(3.2)$ varies from 0.7 to 0.96.

The FOU of $\tilde{\xi}$ is depicted in Fig. 6.

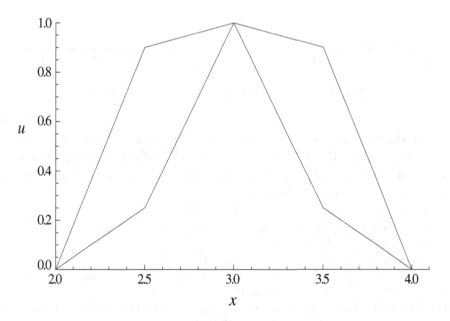

**Fig. 6.** FOU of $\tilde{\xi}$.

**Example 2.6 (Trapezoidal Interval Type-2 Fuzzy Variable).** A trapezoidal interval type-2 fuzzy variable $\tilde{A}$ in the universe of discourse X can be represented by $\tilde{A} = (\tilde{A}^U, \tilde{A}^L) = ((a_1^U, a_2^U, a_3^U, a_4^U; w^U), (a_1^L, a_2^L, a_3^L, a_4^L; w^L))$, where both $\tilde{A}^U$ and $\tilde{A}^L$ are trapezoidal fuzzy variables of height $w^U$ and $w^L$ respectively.

For example consider a trapezoidal IT2 FV $\tilde{A} = ((2, 4, 6, 8; 1), (3, 4.5, 5.5, 7; 0.8))$ which is depicted in Fig. 7.

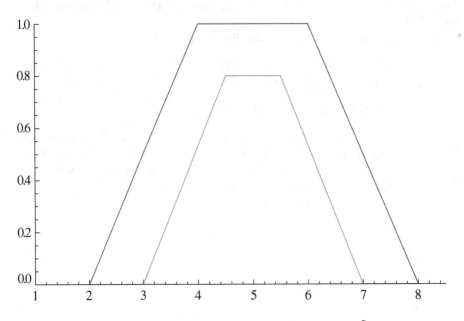

**Fig. 7.** Trapezoidal interval type-2 fuzzy variable $\tilde{A}$

The arithmetic operations between trapezoidal interval type-2 fuzzy variables $\tilde{A}_1 = (\tilde{A}_1^U, \tilde{A}_1^L) = ((a_{11}^U, a_{12}^U, a_{13}^U, a_{14}^U; w_1^U), (a_{11}^L, a_{12}^L, a_{13}^L, a_{14}^L; w_1^L))$ and $\tilde{A}_2 = (\tilde{A}_2^U, \tilde{A}_2^L) = ((a_{21}^U, a_{22}^U, a_{23}^U, a_{24}^U; w_2^U), (a_{21}^L, a_{22}^L, a_{23}^L, a_{24}^L; w_2^L))$ are defined based on Chen and Lee [23, 24] as follows:

Addition operation: $\tilde{A}_1 \oplus \tilde{A}_2 = (\tilde{A}_1^U, \tilde{A}_1^L) \oplus (\tilde{A}_2^U, \tilde{A}_2^L)$
$= ((a_{11}^U + a_{21}^U, a_{12}^U + a_{22}^U, a_{13}^U + a_{23}^U, a_{14}^U + a_{24}^U, min(w_1^U, w_2^U)), (a_{11}^L + a_{21}^L, a_{12}^L + a_{22}^L, a_{13}^L + a_{23}^L, a_{14}^L + a_{24}^L, min(w_1^L, w_2^L)))$,

Multiplication operation: $\tilde{A}_1 \otimes \tilde{A}_2 = (\tilde{A}_1^U, \tilde{A}_1^L) \otimes (\tilde{A}_2^U, \tilde{A}_2^L)$
$= ((a_{11}^U \times a_{21}^U, a_{12}^U \times a_{22}^U, a_{13}^U \times a_{23}^U, a_{14}^U \times a_{24}^U, min(w_1^U, w_2^U)), (a_{11}^L \times a_{21}^L, a_{12}^L \times a_{22}^L, a_{13}^L \times a_{23}^L, a_{14}^L \times a_{24}^L, min(w_1^L, w_2^L)))$.

The arithmetic operations between trapezoidal interval type-2 fuzzy variable $\tilde{A}_1$ and a crisp value $k(> 0)$ are defined as follows:
$k\tilde{A}_1 = ((k \times a_{11}^U, k \times a_{12}^U, k \times a_{13}^U, k \times a_{14}^U; w_1^U), (k \times a_{11}^L, k \times a_{12}^L, k \times a_{13}^L, k \times a_{14}^L; w_1^L))$,
$\frac{\tilde{A}_1}{k} = ((\frac{1}{k} \times a_{11}^U, \frac{1}{k} \times a_{12}^U, \frac{1}{k} \times a_{13}^U, \frac{1}{k} \times a_{14}^U; w_1^U), (\frac{1}{k} \times a_{11}^L, \frac{1}{k} \times a_{12}^L, \frac{1}{k} \times a_{13}^L, \frac{1}{k} \times a_{14}^L; w_1^L))$.

**Critical Value (CV)-Based Reduction Method for Type-2 Fuzzy Variables (Qin et al. [127]):** The CV-based reduction method is developed using the following definitions.

**Critical Values (CVs) for RFVs:** Qin et al. [127] introduced three kinds of critical values (CVs) of a RFV $\tilde{\xi}$. These are:

(i) the optimistic CV of $\tilde{\xi}$, denoted by $CV^*[\tilde{\xi}]$, is defined as

$$CV^*[\tilde{\xi}] = \sup_{\alpha \in [0,1]} [\alpha \wedge Pos\{\tilde{\xi} \geq \alpha\}] \tag{47}$$

(ii) the pessimistic CV of $\tilde{\xi}$, denoted by $CV_*[\tilde{\xi}]$, is defined as

$$CV_*[\tilde{\xi}] = \sup_{\alpha \in [0,1]} [\alpha \wedge Nec\{\tilde{\xi} \geq \alpha\}] \tag{48}$$

(iii) the CV of $\tilde{\xi}$, denoted by $CV[\tilde{\xi}]$, is defined as

$$CV[\tilde{\xi}] = \sup_{\alpha \in [0,1]} [\alpha \wedge Cr\{\tilde{\xi} \geq \alpha\}]. \tag{49}$$

**Example 2.7.** Let $\tilde{\xi}$ be a discrete RFV defined by

$$\tilde{\xi} = \begin{pmatrix} 0.2 \ 0.4 \ 0.5 \ 0.7 \\ 0.3 \ 0.7 \ 1.0 \ 0.6 \end{pmatrix}$$

Then for $\alpha \in [0,1]$,

$$Pos\{\tilde{\xi} \geq \alpha\} = \sup_{r \geq \alpha} \mu_{\tilde{\xi}}(r) = \begin{cases} 1, & \text{if } \alpha \leq 0.5; \\ 0.6, & \text{if } 0.5 < \alpha \leq 0.7; \\ 0, & \text{if } 0.7 < \alpha \leq 1. \end{cases}$$

$$Nec\{\tilde{\xi} \geq \alpha\} = 1 - \sup_{r < \alpha} \mu_{\tilde{\xi}}(r) = \begin{cases} 1, & \text{if } \alpha \leq 0.2; \\ 0.7, & \text{if } 0.2 < \alpha \leq 0.4; \\ 0.3, & \text{if } 0.4 < \alpha \leq 0.5; \\ 0, & \text{if } 0.5 < \alpha \leq 1. \end{cases}$$

and so,

$$Cr\{\tilde{\xi} \geq \alpha\} = \begin{cases} 1, & \text{if } \alpha \leq 0.2; \\ 0.85, & \text{if } 0.2 < \alpha \leq 0.4; \\ 0.65, & \text{if } 0.4 < \alpha \leq 0.5; \\ 0.3, & \text{if } 0.5 < \alpha \leq 0.7; \\ 0, & \text{if } 0.7 < \alpha \leq 1. \end{cases}$$

Now from (47), (48) and (49) we have

$$CV^*[\tilde{\xi}] = \sup_{\alpha \in [0,1]} [\alpha \wedge Pos\{\tilde{\xi} \geq \alpha\}]$$

$$= \sup_{\alpha \in [0,0.5]} [\alpha \wedge 1] \vee \sup_{\alpha \in (0.5,0.7]} [\alpha \wedge 0.6] \vee \sup_{\alpha \in (0.7,1]} [\alpha \wedge 0]$$

$$= 0.5 \vee 0.6 \vee 0 = 0.6$$

$$CV_*[\tilde{\xi}] = \sup_{\alpha \in [0,1]} [\alpha \wedge Nec\{\tilde{\xi} \geq \alpha\}]$$

$$= \sup_{\alpha \in [0,0.2]} [\alpha \wedge 1] \vee \sup_{\alpha \in (0.2,0.4]} [\alpha \wedge 0.7] \vee \sup_{\alpha \in (0.4,0.5]} [\alpha \wedge 0.3] \vee \sup_{\alpha \in (0.5,1]} [\alpha \wedge 0]$$

$$= 0.2 \vee 0.4 \vee 0.3 \vee 0 = 0.4$$

and

$$CV[\tilde{\xi}] = \sup_{\alpha \in [0,1]} [\alpha \wedge Pos\{\tilde{\xi} \geq \alpha\}]$$

$$= \sup_{\alpha \in [0,0.2]} [\alpha \wedge 1] \vee \sup_{\alpha \in (0.2,0.4]} [\alpha \wedge 0.85] \vee \sup_{\alpha \in (0.4,0.5]} [\alpha \wedge 0.65] \vee$$

$$\sup_{\alpha \in (0.5,0.7]} [\alpha \wedge 0.3] \vee \sup_{\alpha \in (0.7,1]} [\alpha \wedge 0]$$

$$= 0.2 \vee 0.4 \vee 0.5 \vee 0.3 \vee 0 = 0.5.$$

The following theorems introduce the critical values (CVs) of trapezoidal and triangular RFVs.

**Theorem 2.2** (Qin et al. [127]). Let $\tilde{\xi} = (r_1, r_2, r_3, r_4)$ be a trapezoidal RFV. Then we have

(i) the optimistic CV of $\tilde{\xi}$ is

$$CV^*[\tilde{\xi}] = r_4/(1 + r_4 - r_3), \tag{50}$$

(ii) the pessimistic CV of $\tilde{\xi}$ is

$$CV_*[\tilde{\xi}] = r_2/(1 + r_2 - r_1), \tag{51}$$

(iii) the CV of $\tilde{\xi}$ is

$$CV[\tilde{\xi}] = \begin{cases} \frac{2r_2 - r_1}{1 + 2(r_2 - r_1)}, & \text{if } r_2 > \frac{1}{2}; \\ \frac{1}{2}, & \text{if } r_2 \leq \frac{1}{2} < r_3; \\ \frac{r_4}{1 + 2(r_4 - r_3)}, & \text{if } r_3 \leq \frac{1}{2}. \end{cases} \tag{52}$$

**Theorem 2.3** (Qin et al. [127]). Let $\tilde{\xi} = (r_1, r_2, r_3)$ be a triangular RFV. Then we have

(i) the optimistic CV of $\tilde{\xi}$ is

$$CV^*[\tilde{\xi}] = r_3/(1 + r_3 - r_2), \tag{53}$$

(ii) the pessimistic CV of $\tilde{\xi}$ is

$$CV_*[\tilde{\xi}] = r_2/(1 + r_2 - r_1),\qquad(54)$$

(iii) the CV of $\tilde{\xi}$ is

$$CV[\tilde{\xi}] = \begin{cases} \frac{2r_2 - r_1}{1 + 2(r_2 - r_1)}, & \text{if } r_2 > \frac{1}{2}; \\ \frac{r_3}{1 + 2(r_3 - r_2)}, & r_2 \leq \frac{1}{2}. \end{cases}\qquad(55)$$

Now we discussed the CV-based reduction method.

**The CV-Based Reduction Method:** Because of fuzziness in membership function of T2 FS, computational complexity is very high to deal with T2 FS. A general idea to reduce its complexity is to convert a T2 FS into a T1 FS so that the methodologies to deal with T1 FSs can also be applied to T2 FSs. Qin et al. [127] proposed a CV-based reduction method which reduces a type-2 fuzzy variable to a type-1 fuzzy variable (may or may not be normal). Let $\tilde{\xi}$ be a T2 FV with secondary possibility distribution function $\tilde{\mu}_{\tilde{\xi}}(x)$ (which represents a RFV). The method is to introduce the critical values (CVs) as representing values for RFV $\tilde{\mu}_{\tilde{\xi}}(x)$, i.e., $CV^*[\tilde{\mu}_{\tilde{\xi}}(x)]$, $CV_*[\tilde{\mu}_{\tilde{\xi}}(x)]$ or $CV[\tilde{\mu}_{\tilde{\xi}}(x)]$ and so corresponding type-1 fuzzy variables (T1 FVs) are derived using these CVs of the secondary possibilities. Then these methods are respectively called optimistic CV reduction, pessimistic CV reduction and CV reduction method.

**Example 2.4 (Continued).** The possibilities of each point of the T2 FV $\tilde{\xi}$ in Example 2.4, are triangular or trapezoidal RFVs. So from Theorems 2.2 and 2.3 we obtain

$$CV^*[\tilde{\mu}_{\tilde{\xi}}(5)] = \tfrac{1}{2},\ CV^*[\tilde{\mu}_{\tilde{\xi}}(6)] = \tfrac{2}{3},\ CV^*[\tilde{\mu}_{\tilde{\xi}}(7)] = \tfrac{7}{12}.$$

$$CV_*[\tilde{\mu}_{\tilde{\xi}}(5)] = \tfrac{1}{3},\ CV_*[\tilde{\mu}_{\tilde{\xi}}(6)] = \tfrac{1}{2},\ CV_*[\tilde{\mu}_{\tilde{\xi}}(7)] = \tfrac{1}{4}.$$

$$CV[\tilde{\mu}_{\tilde{\xi}}(5)] = \tfrac{3}{7},\ CV[\tilde{\mu}_{\tilde{\xi}}(6)] = \tfrac{4}{7},\ CV[\tilde{\mu}_{\tilde{\xi}}(7)] = \tfrac{1}{2}.$$

Then by optimistic CV, pessimistic CV and CV reduction methods, the T2 FV $\tilde{\xi}$ is reduced respectively to the following T1 FVs

$$\begin{pmatrix} 5 & 6 & 7 \\ \tfrac{1}{2} & \tfrac{2}{3} & \tfrac{7}{12} \end{pmatrix},\ \begin{pmatrix} 5 & 6 & 7 \\ \tfrac{1}{3} & \tfrac{1}{2} & \tfrac{1}{4} \end{pmatrix}\ \text{and}\ \begin{pmatrix} 5 & 6 & 7 \\ \tfrac{3}{7} & \tfrac{4}{7} & \tfrac{1}{2} \end{pmatrix}.$$

In the following theorem the optimistic CV, pessimistic CV and CV reductions of a type-2 triangular fuzzy variable are obtained. Since the secondary possibility distribution of a type-2 triangular fuzzy variable is a triangular RFV, so applying Theorem 2.3, Qin et al. [127] established the following theorem in which a type-2 triangular fuzzy variable is reduced to a type-1 fuzzy variable.

**Theorem 2.4** (Qin et al. [127]). Let $\tilde{\xi}$ be a type-2 triangular fuzzy variable defined as $\tilde{\xi} = (r_1, r_2, r_3; \theta_l, \theta_r)$. Then we have:

(i) Using the optimistic CV reduction method, the reduction $\xi_1$ of $\tilde{\xi}$ has the following possibility distribution

$$\mu_{\xi_1}(x) = \begin{cases} \frac{(1+\theta_r)(x-r_1)}{r_2-r_1+\theta_r(x-r_1)}, & \text{if } x \in [r_1, \frac{r_1+r_2}{2}]; \\ \frac{(1-\theta_r)x+\theta_r r_2-r_1}{r_2-r_1+\theta_r(r_2-x)}, & \text{if } x \in (\frac{r_1+r_2}{2}, r_2]; \\ \frac{(-1+\theta_r)x-\theta_r r_2+r_3}{r_3-r_2+\theta_r(x-r_2)}, & \text{if } x \in (r_2, \frac{r_2+r_3}{2}]; \\ \frac{(1+\theta_r)(r_3-x)}{r_3-r_2+\theta_r(r_3-x)}, & \text{if } x \in (\frac{r_2+r_3}{2}, r_3]. \end{cases} \quad (56)$$

(ii) Using the pessimistic CV reduction method, the reduction $\xi_2$ of $\tilde{\xi}$ has the following possibility distribution

$$\mu_{\xi_2}(x) = \begin{cases} \frac{x-r_1}{r_2-r_1+\theta_l(x-r_1)}, & \text{if } x \in [r_1, \frac{r_1+r_2}{2}]; \\ \frac{x-r_1}{r_2-r_1+\theta_l(r_2-x)}, & \text{if } x \in (\frac{r_1+r_2}{2}, r_2]; \\ \frac{r_3-x}{r_3-r_2+\theta_l(x-r_2)}, & \text{if } x \in (r_2, \frac{r_2+r_3}{2}]; \\ \frac{r_3-x}{r_3-r_2+\theta_l(r_3-x)}, & \text{if } x \in (\frac{r_2+r_3}{2}, r_3]. \end{cases} \quad (57)$$

(iii) Using the CV reduction method, the reduction $\xi_3$ of $\tilde{\xi}$ has the following possibility distribution

$$\mu_{\xi_3}(x) = \begin{cases} \frac{(1+\theta_r)(x-r_1)}{r_2-r_1+2\theta_r(x-r_1)}, & \text{if } x \in [r_1, \frac{r_1+r_2}{2}]; \\ \frac{(1-\theta_l)x+\theta_l r_2-r_1}{r_2-r_1+2\theta_l(r_2-x)}, & \text{if } x \in (\frac{r_1+r_2}{2}, r_2]; \\ \frac{(-1+\theta_l)x-\theta_l r_2+r_3}{r_3-r_2+2\theta_l(x-r_2)}, & \text{if } x \in (r_2, \frac{r_2+r_3}{2}]; \\ \frac{(1+\theta_r)(r_3-x)}{r_3-r_2+2\theta_r(r_3-x)}, & \text{if } x \in (\frac{r_2+r_3}{2}, r_3]. \end{cases} \quad (58)$$

**Example 2.8.** Consider the type-2 triangular fuzzy variable $\tilde{\xi}=(2,3,4;0.5,0.8)$ whose FOU is depicted in Fig. 6.

Then its optimistic CV, pessimistic CV and CV reductions are shown in the Fig. 8.

**Note 2.1:** The reduced type-1 fuzzy variables from T2 FVs as obtained by CV-based reduction methods are not always normalized, i.e. are general fuzzy variables. For instance, from Example 2.4 (continued) we observe that the reductions of T2 FV $\tilde{\xi}$ are not normal. For such cases, generalized credibility measure $\tilde{C}r$ is used instead of the credibility measure.

The generalized credibility measure $\tilde{C}r$ of a fuzzy event $\{\tilde{\xi} \in B\}$, $B \subset \Re$ is defined as

$$\tilde{C}r\{\tilde{\xi} \in B\} = \frac{1}{2}(\sup_{x\in\Re} \mu_{\tilde{\xi}}(x) + \sup_{x\in B} \mu_{\tilde{\xi}}(x) - \sup_{x\in B^c} \mu_{\tilde{\xi}}(x)).$$

It is obvious that if $\tilde{\xi}$ is normalized (i.e. $\sup_{x\in\Re} \mu_{\tilde{\xi}}(x) = 1$), then $\tilde{C}r$ coincides with usual credibility measure $Cr$.

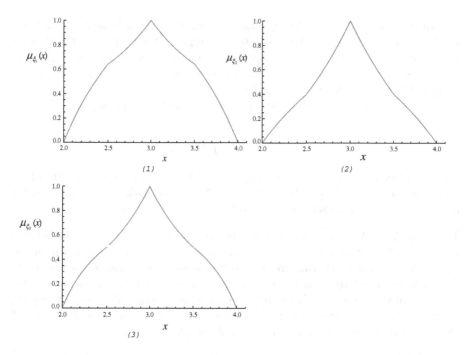

**Fig. 8.** (1) optimistic CV, (2) pessimistic CV, (3) CV reductions of $\tilde{\xi}$.

## 2.4 Rough Set

Here we introduced some basic idea of approximation of a subset of a certain universe by means of lower and upper approximation and the rough set theory. Suppose $U$ is a non-empty finite set of objects called the universe and $A$ is a non-empty finite set of attributes, then the pair $S = (U, A)$ is called information system. For any $B \subseteq A$ there is associated an equivalence relation $I(B)$ defined as $I(B) = \{(x, y) \in U \times U \mid \forall a \in B, a(x) = a(y)\}$, where $a(x)$ denotes the value of attribute $a$ for element $x$. $I(B)$ is called the $B$-indiscernibility relation. The equivalence classes of the $B$-indiscernibility relation are denoted by $[x]_B$.

For an information system $S = (U, A)$ and $B \subseteq A$, $X \subseteq U$ can be approximated using only the information contained in $B$ by constructing the $B$-lower and $B$-upper approximations (Pawlak [122]) of $X$, denoted $\underline{B}X$ and $\overline{B}X$ respectively, where

$$\underline{B}X = \{x \mid [x]_B \subseteq X\} \ and \ \overline{B}X = \{x \mid [x]_B \cap X \neq \phi\}.$$

Clearly, lower approximation $\underline{B}X$ is the definable (exact) set contained in $X$ so that the objects in $\underline{B}X$ can be with certainty classified as members of $X$ on the basis of knowledge in $B$, while the objects in $\overline{B}X$ can be only classified as possible members of $X$ on the basis of knowledge in $B$. The $B$-boundary region of $X$ is defined as

$$BN_B = \overline{B}X - \underline{B}X$$

and thus consists of those objects that we cannot decisively classify into $X$ on the basis of knowledge in $B$. The boundary region of a crisp (exact) set is an empty set as the lower and upper approximation of crisp set are equal. A set is said to be rough if the boundary region is non-empty, i.e., if $BN_B \neq \phi$ then $X$ is referred to as rough with respect to $B$.

Rough set can be also characterized numerically by the following coefficient

$$\alpha_B(X) = \frac{\mid \underline{B}X \mid}{\mid \overline{B}X \mid}$$

called the accuracy of approximation, where $|X|$ denotes the cardinality of X. Obviously $0 \leq \alpha_B(X) \leq 1$. If $\alpha_B(X) = 1$, X is crisp with respect to B and if $\alpha_B(X) < 1$, X is rough with respect to B.

**Example 2.9.** A simple information system (also known as attribute-value tables or information table) is shown in Table 1. This table contains information about patients suffering from a certain disease and objects in this table are patients, attributes can be, for example, headache, body temperature etc. Columns of the table are labeled by attributes (symptoms) and rows by objects (patients), whereas entries of the table are attribute values. Thus each row of the table can be seen as information about specific patient.

**Table 1.** An example of information system

| Patient | Headache | Muscle-pain | Temperature | Flu |
|---------|----------|-------------|-------------|-----|
| p1 | no | yes | high | yes |
| p2 | yes | no | high | yes |
| p3 | yes | yes | very high | yes |
| p4 | no | yes | normal | no |
| p5 | yes | no | high | no |
| p6 | no | yes | very high | yes |

From the table it is observed that patients p2, p3 and p5 have the same conditions with respect to the attribute Headache. So patients p2, p3 and p5 are indiscernible with respect to the attribute Headache. Similarly patients p3 and p6 are indiscernible with respect to attributes Muscle-pain and Flu, and patients p2 and p5 are indiscernible with respect to attributes Headache, Muscle-pain and Temperature. Hence, the attribute Headache generates two elementary sets $\{p2, p3, p5\}$ and $\{p1, p4, p6\}$, i.e., $I(Headache) = \{\{p2, p3, p5\}, \{p1, p4, p6\}\}$. Similarly the attributes Headache and Muscle-pain form the following elementary sets: $\{p1, p4, p6\}$, $\{p2, p5\}$ and $\{p3\}$.

Now we observe that patient p2 and p5 indiscernible with respect to the attributes Headache, Muscle-pain and Temperature, but patient p2 has flu, whereas patient p5 does not, hence flu cannot be characterized in terms of attributes Headache, Muscle-pain and Temperature. Hence p2 and p5 are the boundary-line cases, which cannot be properly classified in view of the available knowledge. The remaining patients p1, p3 and p6 display symptoms (Muscle-pain and at least high temperature, which are must in a patient having flu as seen from the table) which enable us to classify them with certainty as having flu and patient p4 for sure does not have flu, in view of the displayed symptoms. Thus the lower approximation of the set of patients having flu is the set $\{p1, p3, p6\}$ and the upper approximation of this set is the set $\{p1, p2, p3, p5, p6\}$, whereas the boundary-line cases are patients p2 and p5. Now consider the concept "flu", i.e., the set $X = \{p1, p2, p3, p6\}$ and the set of attributes B = Headache, Muscle-pain, Temperature. Then $\underline{B}X = \{p1, p3, p6\}$ and $\overline{B}X = \{p1, p2, p3, p5, p6\}$ and $BN_B = \overline{B}X - \underline{B}X = \{p2, p5\} \neq \phi$. So here $X$ can be referred to as rough with respect to $B$. Also in this case we get $\alpha_B(X) = 3/5$. It means that the concept "flu" can be characterized partially employing symptoms Headache, Muscle-pain and Temperature.

**Rough Variable:** The concept of rough variable is introduced by Liu [86]. The following definitions are based on Liu [86,88].

**Definition 2.11.** Let $\Lambda$ be a nonempty set, $\mathcal{A}$ be a $\sigma$-algebra of subsets of $\Lambda$, $\Delta$ be an element in $\mathcal{A}$, and $\pi$ be a nonnegative, real-valued, additive set function on $\mathcal{A}$. Then $(\Lambda, \Delta, \mathcal{A}, \pi)$ is called a rough space.

**Definition 2.12 (Rough Variable).** A rough variable $\xi$ on the rough space $(\Lambda, \Delta, \mathcal{A}, \pi)$ is a measurable function from $\Lambda$ to the set of real numbers $\Re$ such that for every Borel set $B$ of $\Re$, we have $\{\lambda \in \Lambda \mid \xi(\lambda) \in B\} \in \mathcal{A}$.

Then the lower and upper approximations of the rough variable $\xi$ are defined as follows:

$$\underline{\xi} = \{\xi(\lambda) \mid \lambda \in \Delta\} \text{ and } \overline{\xi} = \{\xi(\lambda) \mid \lambda \in \Lambda\}.$$

**Definition 2.13.** Let $\xi$ be a rough vector on the rough space $(\Lambda, \Delta, \mathcal{A}, \pi)$, and $f_j : \Re^n \to \Re$ be continuous functions, $j = 1, 2, ..., m$. Then the upper trust of the rough event characterized by $f_j(\xi) \leq 0; j = 1, 2, ..., m$ is defined by

$$T\bar{r}\{f_j(\xi) \leq 0_{,j=1,2,...,m}\} = \frac{\pi\{\lambda \in \Lambda | f_j(\xi(\lambda)) \leq 0, \ j = 1, 2, ..., m\}}{\pi(\Lambda)},$$

and the lower trust of the rough event characterized by $f_j(\xi) \leq 0; j = 1, 2, ..., m$ is defined by

$$T\underline{r}\{f_j(\xi) \leq 0_{,j=1,2,...,m}\} = \frac{\pi\{\lambda \in \Delta | f_j(\xi(\lambda)) \leq 0, \ j = 1, 2, ..., m\}}{\pi(\Delta)}.$$

If $\pi(\Delta) = 0$, then the upper trust and lower trust of the rough event are assumed to be equivalent, i.e., $T\bar{r}\{f_j(\xi) \leq 0_{,j=1,2,...,m}\} \equiv T\underline{r}\{f_j(\xi) \leq 0_{,j=1,2,...,m}\}$.
The trust of the rough event is defined as the average value of the lower and upper trusts, i.e.,

$$Tr\{f_j(\xi) \leq 0_{,j=1,2,...,m}\} = \frac{1}{2}(T\bar{r}\{f_j(\xi) \leq 0_{,j=1,2,...,m}\} + T\underline{r}\{f_j(\xi) \leq 0_{,j=1,2,...,m}\}).$$

**Definition 2.14.** Let $\xi$ be a rough variable on the rough space $(\Lambda, \Delta, \mathcal{A}, \pi)$ and $\alpha \in (0, 1]$, then

$$\xi_{sup}(\alpha) = \sup\{r | Tr\{\xi \geq r\} \geq \alpha\}$$

is called $\alpha$-optimistic value to $\xi$; and

$$\xi_{inf}(\alpha) = \inf\{r | Tr\{\xi \leq r\} \geq \alpha\}$$

is called $\alpha$-pessimistic value to $\xi$.

**Definition 2.15.** Let $\xi$ be a rough variable on the rough space $(\Lambda, \Delta, \mathcal{A}, \pi)$. The expected value of $\xi$ is defined by

$$E[\xi] = \int_0^\infty Tr\{\xi \geq r\}dr - \int_{-\infty}^0 Tr\{\xi \leq r\}dr.$$

**Theorem 2.5. (Liu [88]).** Let $\xi_{inf}(\alpha)$ and $\xi_{sup}(\alpha)$ be the $\alpha$-pessimistic and $\alpha$-optimistic values of the rough variable $\xi$, respectively.Then we have

(a) $Tr\{\xi \leq \xi_{inf}(\alpha) \geq \alpha\}$ and $Tr\{\xi \geq \xi_{sup}(\alpha) \geq \alpha\}$;
(b) $\xi_{inf}(\alpha)$ is an increasing and left-continuous function of $\alpha$;
(c) $\xi_{sup}(\alpha)$ is a decreasing and left-continuous function of $\alpha$;
(d) if $0 < \alpha \leq 1$, then $\xi_{inf}(\alpha) = \xi_{sup}(1 - \alpha)$ and $\xi_{sup}(\alpha) = \xi_{inf}(1 - \alpha)$;
(e) if $0 < \alpha \leq 0.5$, then $\xi_{inf}(\alpha) \leq \xi_{sup}(\alpha)$;
(f) if $0.5 < \alpha \leq 1$,then $\xi_{inf}(\alpha) \geq \xi_{sup}(\alpha)$.

**Example 2.10.** Consider that $\xi = ([a, b], [c, d])$ be a rough variable with $c \leq a < b \leq d$, where $[a, b]$ is the lower approximation and $[c, d]$ is the upper approximation. This means the elements in $[a, b]$ are certainly members of the variable and that of $[c, d]$ are possible members of the variable. Here $\Delta = \{\lambda | a \leq \lambda \leq b\}$ and $\Lambda = \{\lambda | c \leq \lambda \leq d\}$, $\xi(x) = x$ for all $x \in \Lambda$, $\mathcal{A}$ is the Borel algebra on $\Lambda$ and $\pi$ is the Lebesgue measure.

As an practical example consider the possible transportation cost of unit product to be transported from a source $i$ to certain destination $j$ through a conveyance $k$ for a certain time period. But as transportation cost depends upon fuel price, labor charges, tax charges, etc. and each of which is fluctuate time to time, so it is not always possible to determine its exact value. Suppose four experts give the possible unit transportation cost for $i - j$ route via conveyance $k$, determined in a certain time period as intervals [3,5], [3.5,6], [4,5] and [4,6] respectively. Denotes $c_{ijk}$ as - 'the possible value of the unit transportation cost for $i - j$ route through conveyance $k$ according to the all experts'. Then $c_{ijk}$ is not exact and can be approximated by means of lower and upper approximation. Here [4,5] can be taken as the lower approximation of $c_{ijk}$ as it is the greatest definable (exact) set that $c_{ijk}$ contain, i.e., every member of [4,5] is certainly a value of $c_{ijk}$ according to all experts. Here [3,6] is the upper approximation, as members of [3,6] may or may not be possible transportation cost according to all experts. So $c_{ijk}$ can be represented as the rough variable ([4,5],[3,6]).

For a given value $r$ and $\xi = ([a, b], [c, d])$, trust of rough events characterized by $\xi \leq r$ and $\xi \geq r$ (Liu [86,88]) are given by

$$Tr\{\xi \leq r\} = \begin{cases} 0, & \text{if } r \leq c; \\ \frac{r-c}{2(d-c)}, & \text{if } c \leq r \leq a; \\ \frac{1}{2}(\frac{r-a}{b-a} + \frac{r-c}{d-c}), & \text{if } a \leq r \leq b; \\ \frac{1}{2}(\frac{r-c}{d-c} + 1), & \text{if } b \leq r \leq d; \\ 1, & \text{if } r \geq d. \end{cases} \tag{59}$$

$$Tr\{\xi \geq r\} = \begin{cases} 0, & \text{if } r \geq d; \\ \frac{d-r}{2(d-c)}, & \text{if } b \leq r \leq d; \\ \frac{1}{2}(\frac{d-r}{d-c} + \frac{b-r}{b-a}), & \text{if } a \leq r \leq b; \\ \frac{1}{2}(\frac{d-r}{d-c} + 1), & \text{if } c \leq r \leq a; \\ 1, & \text{if } r \leq c. \end{cases} \tag{60}$$

For rough variable $\xi = ([4, 5], [2, 7])$, $Tr\{\xi \leq r\}$ and $Tr\{\xi \geq r\}$ are depicted in Fig. 9.

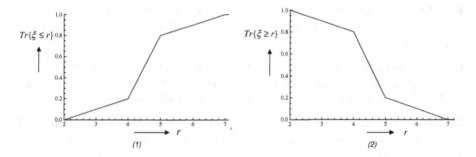

**Fig. 9.** The trust of the rough event characterized by (1) $\xi \leq r$ and (2) $\xi \geq r$.

$\alpha$-optimistic value to $\xi = ([a, b], [c, d])$ is

$$\xi_{sup}(\alpha) = \begin{cases} (1 - 2\alpha)d + 2\alpha c, & \text{if } \alpha \leq ((d - b)/2(d - c)); \\ 2(1 - \alpha)d + (2\alpha - 1)c, & \text{if } \alpha \geq ((2d - a - c)/2(d - c)); \\ \frac{d(b-a)+b(d-c)-2\alpha(b-a)(d-c)}{(b-a)+(d-c)}, & \text{otherwise.} \end{cases} \tag{61}$$

$\alpha$-pessimistic value to $\xi$ is

$$\xi_{inf}(\alpha) = \begin{cases} (1 - 2\alpha)c + 2\alpha d, & \text{if } \alpha \leq ((a - c)/2(d - c)); \\ 2(1 - \alpha)c + (2\alpha - 1)d, & \text{if } \alpha \geq ((b + d - 2c)/2(d - c)); \\ \frac{c(b-a)+a(d-c)+2\alpha(b-a)(d-c)}{(b-a)+(d-c)}, & \text{otherwise.} \end{cases} \tag{62}$$

The expected value of $\xi$ is $E(\xi) = \frac{1}{4}(a + b + c + d)$.

## 2.5    Single-Objective Optimization

**Single-Objective Linear Programming (SOLP)/Non-linear Programming (SONLP) Problem:** If an optimization problem consists of only one objective function, then problem is called a single-objective mathematical programming (SOMP) problem. The minimization of a constrained SOMP problem can be formulated as:

$$
\begin{cases}
Find & x = (x_1, x_2, ...., x_n)^T \\
which\ minimizes & f(x) \\
subject\ to & x \in X \\
where\ X = & x : \begin{cases} g_j(x) \le 0, j = 1, 2, ..., m; \\ x_i \ge 0, \quad i = 1, 2, ..., n. \end{cases}
\end{cases}
\tag{63}
$$

where, objective function $f(x)$ and constraints $g_j(x), j = 1, 2, ...., m$ are functions defined on n-dimensional set.

When both the objective function and the constraints are linear, the above SOMP problem becomes a single-objective linear programming problem (SOLP). Otherwise, it is a single-objective non-linear programming problem (SONLP).

**Feasible Solution:** A decision variable vector $x$ satisfying all the constraints is called a feasible solution to the problem. The collection of all such solutions forms the feasible region.

**Optimal Solution:** If a feasible solution $x^*$ of (63) be such that for each feasible point $x$, $f(x) \le f(x^*)$ for maximization problem and $f(x) \ge f(x^*)$ for minimization problem then $x^*$ is called an optimal solution of the problem.

**Local Optimum:** A feasible solution $x^*$ of (63) is said be local optimum if there exists an $\epsilon > 0$ such that $f(x) \ge f(x^*)$ for minimization problem and $f(x) \le f(x^*)$ for maximization problem, $\forall x \in X : \parallel x - x^* \parallel < \epsilon$.

**Global Optimum:** A feasible solution $x^*$ of (63) is said be global optimum if $f(x) \ge f(x^*)$ for minimization problem and $f(x) \le f(x^*)$ for maximization problem $\forall x \in X$.

**Necessary Condition for Optimality:** The necessary condition for a feasible solution $x^* \in X$ of (63) to be optimal is that all the partial derivatives $\frac{\partial f(x)}{\partial x_r}$ are exist at $x = x^*$ and $\frac{\partial f(x)}{\partial x_r} = 0$ for $r = 1, 2, ..., n$.

**Example 2.11.** As an example of a single-objective problem we consider a simple transportation problem with 3 sources $(i = 1, 2, 3)$ and 3 destinations $(j = 1, 2, 3)$ as follows:

$$
\begin{cases}
Minimize\ Z = \sum_{i=1}^{3} \sum_{j=1}^{3} c_{ij}\, x_{ij}, \\
\\
subject\ to \quad \sum_{j=1}^{3} x_{ij} \le a_i, \quad i = 1, 2, 3, \\
\qquad\qquad\quad \sum_{i=1}^{3} x_{ij} \ge b_j, \quad j = 1, 2, 3, \\
x_{ij} \ge 0,\ \forall\ i, j,\ \sum_{i=1}^{3} a_i \ge \sum_{j=1}^{3} b_j,
\end{cases}
\tag{64}
$$

where $c_{ij}$ is the cost for transportation of unit product from source $i$ to destination $j$, $x_{ij}$ is decision variable which represents the quantity to be transported from $i$-th origin to $j$-th destination and so that the objective function $Z$ represents the total transportation cost. The unit transportation costs are given as follows: $c_{11} = 5$, $c_{12} = 6$, $c_{13} = 8$, $c_{21} = 7$, $c_{22} = 9$, $c_{23} = 5$, $c_{31} = 8$, $c_{32} = 9$ and $c_{33} = 7$. The availabilities at each sources and demands of each destinations are as follows: $a_1 = 36$, $a_2 = 20.8$, $a_3 = 28.6$, $b_1 = 22.8$, $b_2 = 31$, $b_3 = 30$.

## 2.6 Solution Techniques for Single-Objective Linear/Non-linear Problem

**Generalized Reduced Gradient (GRG):** The GRG technique is a method for solving NLP problems with equality as well as inequality constraints. Consider the NLP problem as:

$$\begin{cases} Find & x = (x_1, x_2, ...., x_n)^T \\ which\ minimizes\ f(x) \\ subject\ to & x \in X \\ where\ X = & x : \begin{cases} g_j(x) \le 0,\ j = 1, 2, ..., m; \\ h_r(x) = 0,\ r = 1, 2, ..., p; \\ x_i \ge 0, \qquad i = 1, 2, ..., n. \end{cases} \end{cases} \quad (65)$$

By adding a non-negative slack variable $s_j (\ge 0)$, $j = 1, 2, ..., m$ to each of the above inequality constraints, the problem (65) can be stated as,

$$\begin{cases} Minimize\ \ f(x) \\ subject\ to\ \ \ x = (x_1, x_2, ...., x_n)^T \in X \\ where\ X = x : \begin{cases} g_j(x) + s_j = 0,\ j = 1, 2, ..., m; \\ h_r(x) = 0, \qquad r = 1, 2, ..., p; \\ x_i \ge 0, \qquad\quad i = 1, 2, ..., n. \\ s_j \ge 0, \qquad\quad j = 1, 2, ..., m \end{cases} \end{cases} \quad (66)$$

where the lower and upper bounds on the slack variables, $s_j$, $j = 1, 2, ..., m$ are taken as a zero and a large number (infinity) respectively.

Denote $s_j$ by $x_{n+j}$, $g_j(x) + s_j$ by $\xi_j$, $j = 1, 2, ..., m$ and $h_r(x)$ by $\xi_{m+r}$, $r = 1, 2, ..., p$. Then the above problem (66) becomes

$$\begin{cases} Minimize\ \ f(x) \\ subject\ to\ \ \ x = (x_1, x_2, ...., x_{n+m})^T \in X \\ where\ X = x : \begin{cases} \xi_j = 0,\ j = 1, 2, ..., m + p; \\ x_i \ge 0,\ i = 1, 2, ..., n + m. \end{cases} \end{cases} \quad (67)$$

This GRG technique is based on the idea of elimination of variables using the equality constraints. Theoretically, $(m + p)$ variables (dependent variables) can be expressed in terms of remaining $(n - p)$ variables (independent variables). Thus one can divide the $(n + m)$ decision variables arbitrarily into two sets as

$$x = (y, z)^T$$

where, $y$ is $(n-p)$ design or independent variables and $z$ is $(m+p)$ state or dependent variables and

$$y = (y_1, y_2, ...., y_{n-p})^T, \ z = (z_1, z_2, ...., z_{m+p})^T.$$

Here, the design variables are completely independent and the state variables are dependent on the design variables used to satisfy the constraints $\xi_j(x) = 0$, $(j = 1, 2, ..., m+p)$. Consider the first variations of the objective and constraint functions:

$$df(x) = \sum_{i=1}^{n-p} \frac{\partial f}{\partial y_i} dy_i + \sum_{i=1}^{m+p} \frac{\partial f}{\partial z_i} dz_i = \nabla_y^T f dy + \nabla_z^T f dz \tag{68}$$

$$d\xi_j(x) = \sum_{i=1}^{n-p} \frac{\partial \xi_j}{\partial y_i} dy_i + \sum_{i=1}^{m+p} \frac{\partial \xi_j}{\partial z_i} dz_i \tag{69}$$

$$or \ d\xi = C \ dy + D \ dz \tag{70}$$

$$where \ \nabla_y^T f = \left( \frac{\partial f}{\partial y_1}, \frac{\partial f}{\partial y_2}, ...., \frac{\partial f}{\partial y_{n-p}} \right)$$

$$\nabla_z^T f = \left( \frac{\partial f}{\partial z_1}, \frac{\partial f}{\partial z_2}, ...., \frac{\partial f}{\partial z_{m+p}} \right)$$

$$C = \begin{bmatrix} \frac{\partial \xi_1}{\partial y_1} & \cdots & \cdots & \frac{\partial \xi_1}{\partial y_{n-p}} \\ \frac{\partial \xi_2}{\partial y_1} & \cdots & \cdots & \frac{\partial \xi_2}{\partial y_{n-p}} \\ \cdots & \cdots & \cdots & \cdots \\ \cdots & \cdots & \cdots & \cdots \\ \frac{\partial \xi_{m+p}}{\partial y_1} & \cdots & \cdots & \frac{\partial \xi_{m+p}}{\partial y_{n-p}} \end{bmatrix}, \ D = \begin{bmatrix} \frac{\partial \xi_1}{\partial z_1} & \cdots & \cdots & \frac{\partial \xi_1}{\partial z_{m+p}} \\ \frac{\partial \xi_2}{\partial z_1} & \cdots & \cdots & \frac{\partial \xi_2}{\partial z_{m+p}} \\ \cdots & \cdots & \cdots & \cdots \\ \cdots & \cdots & \cdots & \cdots \\ \frac{\partial \xi_{m+p}}{\partial z_1} & \cdots & \cdots & \frac{\partial \xi_{m+p}}{\partial z_{m+p}} \end{bmatrix}$$

$$dy = (dy_1, dy_2, ..., dy_{n-p})^T$$

$$and \ dz = (dz_1, dz_2, ..., dz_{m+p})^T$$

Assuming that the constraints are originally satisfied at the vector $x$ $(\xi(x) = 0)$, any change in the vector $dx$ must correspond to $d\xi = 0$ to maintain feasibility at $x + dx$. Thus, Eq. (70) can be solved as

$$C \ dy + D \ dz = 0$$

$$or \ dz = -D^{-1} \ Cdy \tag{71}$$

The change in the objective function due to the change in $x$ is given by the Eq. (68), which can be expressed, using Eq. (71) as

$$df(x) = (\nabla_y^T f dy - \nabla_z^T f D^{-1} \ C)dy$$

$$or \ \frac{df(x)}{dy} = G_R \tag{72}$$

$$where \ G_R = \nabla_y^T f dy - \nabla_z^T f D^{-1} \ C \tag{73}$$

is called the generalized reduced gradient. Geometrically, the reduced gradient can be described as a projection of the original $n$-dimensional gradient into the $(n - m)$ dimensional feasible region described by the design variables.

A necessary condition for the existence of minimum of an unconstrained function is that the components of the gradient vanish. Similarly, a constrained function assumes its minimum value when the appropriate components of the reduced gradient are zero. In fact, the reduced gradient $G_R$ can be used to generate a search direction S to reduce the value of the constrained objective function. Similarly, to the gradient $\nabla f$ that can be used to generate a search direction S for an unconstrained function. A suitable step length $\lambda$ is to be chosen to minimize the value of $f(x)$ along the search direction. For any specific value of $\lambda$, the dependent variable vector $z$ is updated using Eq. (70). Noting that Eq. (68) is based on using a linear approximation to the original non-linear problem, so the constraints may not be exactly equal to zero at $\lambda$, i.e., $d\xi \neq 0$. Hence, when y is held fixed, in order to have

$$\xi_j(x) + d\xi_j(x) = 0, \ j = 1, 2, ..., m + p \tag{74}$$

following must be satisfied.

$$\xi(x) + d\xi(x) = 0. \tag{75}$$

Using Eq. (70) for $d\xi$ in Eq. (75), following is obtained

$$dz = D^{-1}(-\xi(x) - Cdy). \tag{76}$$

The value dz given by Eq. (76) is used to update the value of $z$ as

$$z_{update} = z_{current} + dz. \tag{77}$$

The constraints evaluated at the updated vector $x$, and the procedure of finding $dz$ using Eq. (76) is repeated until $dz$ is sufficiently small.

## 2.7  Single-Objective Problem in Fuzzy Environment

When in a single-objective optimization problem, some of the or all the parameters in objective function and constraints are not precisely defined or completely known, then if such parameters are represented by fuzzy numbers(/variables), the problem is termed as single-objective optimization problem in fuzzy environment. Consider the following fuzzy programming problem,

$$\begin{cases} Min & f(x, \tilde{\xi}) \\ subject\ to\ g_j(x, \tilde{\xi}) \leq \tilde{b}_j, \ j = 1, 2, ..., m \\ \quad\quad\quad x_i \geq 0, \ i = 1, 2, ..., n \end{cases} \tag{78}$$

where, $x = (x_1, x_2, ...., x_n)^T$ is a decision vector, $\tilde{\xi}$ is a fuzzy vector, $f(x, \tilde{\xi})$ is a return function, and $g_j(x, \tilde{\xi}) \leq \tilde{b}_j$ are constraints, $j = 1, 2, ..., m$. It is not possible to minimize directly a fuzzy quantity $f(x, \tilde{\xi})$ and also the constraints $g_j(x, \tilde{\xi}) \leq \tilde{b}_j$, $j = 1, 2, ..., m$ do not produce a crisp feasible set. In order to solve the above fuzzy model several researchers proposed a number of different methods. Fuzzy expected value model (EVM) (Liu and Liu [94]), chance-constrained programming model (CCP) (Liu and Iwamura [92]), dependent-chance programming (DCP) (Liu [85]) are some of the such available techniques. We now provide short description of these techniques.

**Expected Value Model:** In order to obtain a solution (here a decision with minimum expected return) of (78), Liu and Liu [94] provided a spectrum of fuzzy expected value model (EVM) as follows:

$$\begin{cases} Min & E[f(x, \tilde{\xi})] \\ subject\ to\ E[g_j(x, \tilde{\xi}) - \tilde{b}_j] \leq 0,\ j = 1, 2, ..., m \\ & x_i \geq 0,\ i = 1, 2, ..., n \end{cases} \tag{79}$$

For detail explanation, crisp equivalent form of the fuzzy EVM please see Liu and Liu [94], Liu [90], Yang and Liu [153].

**Chance-Constrained Programming Model:** This method is used to solve the problems with chance-constraints. In this method, the uncertain constraints are allowed to be violated such that constraints must be satisfied at some chance (/confidence) level. For example, since the fuzzy constraints $g_j(x, \tilde{\xi}) \leq \tilde{b}_j$, $j = 1, 2, ..., m$ do not define a deterministic feasible set, a natural idea is to provide a confidence level $\alpha$ at which it is desired that the fuzzy constraints hold. A chance constrained programming for the minimization problem (78) with fuzzy parameters using possibility measure may be constructed as follows:

$$\begin{cases} Min_x & (Min_{\bar{f}}\ \bar{f}) \\ subject\ to\ Pos\{f(x, \tilde{\xi}) \leq \bar{f}\} \geq \alpha, \\ & Pos\{\tilde{\xi} \mid g_j(x, \tilde{\xi}) \leq \tilde{b}_j\} \geq \alpha_j,\ j = 1, 2, ..., m \\ & x_i \geq 0,\ i = 1, 2, ..., n \end{cases} \tag{80}$$

where, $\alpha$ is a predetermined confidence level so that Min $\bar{f}$ indicates the minimum value that the objective function achieves with possibility at least $\alpha$ ($0 < \alpha \leq 1$). In other words we want to minimize the $\alpha$-optimistic return. $\alpha_j$ indicates the predetermined confidence level of satisfaction of the constraint.

A chance constrained programming for the minimization problem (78) with fuzzy parameters using credibility measure may be constructed as follows:

$$\begin{cases} Min_x & (Min_{\bar{f}}\ \bar{f}) \\ subject\ to\ Cr\{f(x, \tilde{\xi}) \leq \bar{f}\} \geq \alpha, \\ & Cr\{g_j(x, \tilde{\xi}) \leq \tilde{b}_j\} \geq \alpha_j,\ j = 1, 2, ..., m \\ & x_i \geq 0,\ i = 1, 2, ..., n \end{cases} \tag{81}$$

where, $\alpha$ is a predetermined confidence (credibility) level so that Min $\bar{f}$ indicates the minimum value that the objective function achieves with credibility degree(/level) at least $\alpha$ $(0 < \alpha \leq 1)$. In other words we want to minimize the $\alpha$-optimistic return. $\alpha_j$ indicates the predetermined credibility level of satisfaction of the constraint.

For detail explanation regarding the, crisp equivalent forms of the fuzzy CCP problem please see Liu and Iwamura [92], Mula et al. [116], Liu [84,90], Yang and Liu [153] and Kundu et al. [74].

**Dependent-Chance Programming:** The idea of dependent-chance programming (DCP) is to optimize the chance of an uncertain event. Suppose for the minimization problem like (78), a decision maker will satisfy with a solution (decision vector $x$) for which the objective value is not exceed a certain value. So a decision maker may fixed a satisfying predetermined maximal objective value and maximize the credibility level that objective value is not exceed the predetermined value. Then with respect to a given predetermined maximal objective value $\bar{f}$ the dependent chance-constrained programming model for the problem (78) is formulated as follows:

$$\begin{cases} Max & Cr\{f(x,\tilde{\xi}) \leq \bar{f}\} \\ subject\ to\ g_j(x,\tilde{\xi}) \leq \tilde{b}_j,\ j = 1, 2, ..., m \\ x_i \geq 0,\ i = 1, 2, ..., n \end{cases} \tag{82}$$

For detail explanation, crisp equivalent forms of the fuzzy CCP problem please see Liu [85], Liu [90], Yang and Liu [153].

## 2.8   Multi-objective Optimization

A general multi-objective programming problem (minimization problem) is of the following form:

$$\begin{cases} Find & x = (x_1, x_2, ...., x_n)^T \\ which\ minimizes\ F(x) = (f_1(x), f_2(x), ..., f_k(x))^T \\ subject\ to & x \in X \\ where\ X = & x : \begin{cases} g_j(x) \leq 0, j = 1, 2, ..., m; \\ x_i \geq 0, \quad i = 1, 2, ..., n. \end{cases} \end{cases} \tag{83}$$

where, $f_1(x), f_2(x), ..., f_k(x)$ are $k \geq 2$) objectives.

**Complete Optimal Solution:** A feasible solution $x^*$ is said to be a complete optimal solution to the multi-objective problem in (83) iff there exists $x^* \in X$ such that $f_i(x^*) \leq f_i(x)$, $i = 1, 2, ..., k$ for all $x \in X$. In general, the objective functions of the multi-objective problem conflict with each other, a complete optimal solution does not always exist and so Pareto (or non dominated) optimality concept is introduced.

**Pareto Optimal Solution:** A feasible solution $x^*$ is said to be a Pareto optimal solution to the (83) iff there does not exist another $x \in X$ such that $f_i(x) \leq f_i(x^*)$ for all $i$, $i = 1, 2, ..., k$ and $f_j(x) < f_j(x^*)$ for at least one index

$j$, $j = 1, 2, ...., k$. An objective vector $F^*$ is Pareto-optimal if there does not exist another objective vector $F(x)$ such that $f_i \leq f_i^*$, for all $i$, $i = 1, 2, ..., k$ and $f_j < f_j^*$ for at least one index $j$. Therefore, $F^*$ is Pareto optimal if the decision vector corresponding to it is Pareto optimal. Unless an optimization problem is convex, only locally optimal solution is guaranteed using standard mathematical programming techniques. Therefore, the concept of Pareto-optimality needs to be modified to introduce the notion of a locally Pareto-optimal solution for a non-convex problem as defined by Geoffrion [49].

**Locally Pareto Optimal Solution:** A feasible solution $x^*$ is said to be a locally Pareto optimal solution to the multi-objective problem (83) if and only if there exists an $r > 0$ such that $x^*$ is Pareto optimal in $X \cap N(x^*, r)$, where $N(x^*, r)$ is a $r$-neighborhood of $x^*$, i.e., there does not exist another $x \in X \cap N(x^*, r)$ such that $f_i(x) \leq f_i(x^*)$.

**Concept of Domination:** Most evolutionary multi-objective optimization algorithms use the concept of domination. In these algorithms, two solutions are compared on the basis of whether one dominates the other solution or not. Let us use the operator $\sqsupseteq$ between two solutions $i$ and $j$ as $i \sqsupseteq j$ denotes that solution $i$ is better than solution $j$ on a particular objective. Similarly $i \sqsubseteq j$ for a particular objective implies that solution $i$ is worse than solution $j$ on this objective. With this assumption a solution $x$ is said to dominate the other solution $y$, if both the following conditions hold.

The solution $x$ is not worse than $y$ in all the objectives.

The solution $x$ is strictly better than $y$ in at least one objective, i.e., $f_j(x) \sqsupseteq f_j(y)$ for at least one $j$, $j = 1, 2, ..., k$. Now, let us introduce some linear/non-linear programming techniques which are used to achieve at least local Pareto optimal solutions of multi-objective optimization problem.

**Example 2.12.** In the single-objective problem (64) as presented in Sect. 2.5, the objective function is minimization of total transportation cost. Now in case of transportation of highly breakable items (e.g. glass-goods, toys, ceramic goods, etc.), the breakability issue also should be considered. Breaking of items may be due to bad condition of road, long distance of a certain route, etc. Then an additional objective function which represents minimization of total breaking items is imposed in the problem and the problem becomes multi-objective. Suppose $r_{ij}$ be the rate of breakability (/percentage of breakability) for transportation of goods from source $i$ to destination $j$. Also suppose customer at destination compromises on receiving less amount than the demanded amount due to breaking of items. Then the problem becomes

$$\begin{cases} Minimize \ Z = \sum_{i=1}^{3}\sum_{j=1}^{3} c_{ij} \ x_{ij}, \\[2mm] Minimize \ Z' = \sum_{i=1}^{3}\sum_{j=1}^{3} r_{ij} \ x_{ij}, \\[2mm] subject \ to \quad \sum_{j=1}^{3} x_{ij} \leq a_i, \quad i = 1, 2, 3, \\ \qquad\qquad\quad \sum_{i=1}^{3} x_{ij} \geq b_j, \quad j = 1, 2, 3, \\ x_{ij} \geq 0, \ \forall \ i, j, \quad \sum_{i=1}^{3} a_i \geq \sum_{j=1}^{3} b_j, \end{cases} \qquad (84)$$

where values of $r_{ij}$ are given by $r_{11} = 2, r_{12} = 1.5, r_{13} = 1.2, r_{21} = 1, r_{22} = 1.2,$ $r_{23} = 1.5, r_{31} = 2.5, r_{32} = 2, r_{33} = 1.5$. The values of the other parameters are same as the problem (64). The solution of such multi-objective problem is discussed in the next section and also in Sect. 3 with numerical example.

## 2.9   Solution Techniques for Multi-objective Linear/Non-linear Problem

**Fuzzy Programming Technique:** Zimmarmann [159] introduced fuzzy linear programming approach for solving problem with multiple objectives and he showed that fuzzy linear programming always gives efficient solutions and an optimal compromise solution. The steps to solve the multi-objective models using fuzzy programming technique are as follows:

Step 1: Solve the multi-objective problem (83) as a single objective problem using, each time, only one objective $f_p (p = 1, 2, ..., k)$ (ignore all other objectives) to obtain the optimal solution $X^{p*} = x_i^p$ of $k$ different single objective solid transportation problem.

Step 2: Calculate the values of all the $k$ objective functions at all these $k$ optimal solutions $X^{p*}$ $(p = 1, 2, ..., k)$ and find the upper and lower bound for each objective given by $U_p = Max\{f_p(X^{1*}), f_p(X^{2*}), ..., f_p(X^{k*})\}$ and $L_p = f_p(X^{p*})$, $p = 1, 2, ..., k$ respectively.

Step 3: Then an initial fuzzy model is given by

$$\begin{cases} Find & x = (x_1, x_2, ...., x_n)^T \\ subject\ to & f_p(x) \leq L_p \\ x \in X,\ X = x : \begin{cases} g_j(x) \leq 0, j = 1, 2, ..., m; \\ x_i \geq 0, & i = 1, 2, ..., n. \end{cases} \end{cases} \qquad (85)$$

where $x = x_i, i = 1, 2, ..., n;\ p = 1, 2, ..., k$

Step 4: case(I). Construct linear membership function $\mu_p(f_p)$ corresponding to $p$-th objective as

$$\mu_p(f_p) = \begin{cases} 1, & if\ f_p \leq L_p; \\ \frac{U_p - f_p}{U_p - L_p}, & if\ L_p < f_p < U_p; \\ 0, & if\ f_p \geq L_p, \end{cases} \qquad \forall\ p.$$

or,

case(II): Construct hyperbolic membership function $\mu_p^H(f_p)$ corresponding to p-th objective as

$$\mu_p^H(f_p) = 1/2 + 1/2\ \tanh[(\frac{U_p + L_p}{2} - f_p)\alpha_p], \alpha_p = \frac{3}{(U_p - L_p)/2} = \frac{6}{U_p - L_p}.$$

Step 5: For case(I), formulate fuzzy linear programming problem using max-min operator as

$$\begin{cases} Max & \lambda \\ subject\ to & \lambda \leq \mu_p(f_p) = (U_p - f_p)/(U_p - L_p), \forall p \\ x \in X,\ X = x : \begin{cases} g_j(x) \leq 0, j = 1, 2, ..., m; \\ x_i \geq 0, & i = 1, 2, ..., n. \end{cases} \end{cases} \qquad (86)$$

$$\lambda \geq 0 \ and \ \lambda = min_p\{\mu_p(Z_p)\}.$$

For case(II), formulate fuzzy programming problem with hyperbolic membership function as

$$
\begin{cases}
Max & \lambda \\
subject\ to & 2\lambda - 1 \leq \tanh[(\frac{U_p+L_p}{2} - f_p)\alpha_p], \ \forall p \\
x \in X, \ X = x : \begin{cases} g_j(x) \leq 0, \ j = 1, 2, ..., m; \\ x_i \geq 0, \quad i = 1, 2, ..., n. \end{cases} \\
\lambda \geq 0.
\end{cases}
\tag{87}
$$

Let $\lambda' = \tanh^{-1}(2\lambda - 1)$, then above problem becomes

$$
\begin{cases}
Max & \lambda \\
subject\ to & \lambda' \leq (\frac{U_p+L_p}{2} - f_p)\alpha_p, \ \forall p \\
x \in X, \ X = x : \begin{cases} g_j(x) \leq 0, \ j = 1, 2, ..., m; \\ x_i \geq 0, \quad i = 1, 2, ..., n. \end{cases} \\
\lambda \geq 0.
\end{cases}
\tag{88}
$$

Then since tanh and $\tanh^{-1}$ are strictly increasing functions, the above problem equivalently becomes,

$$
\begin{cases}
Max & \lambda' \\
subject\ to & \lambda' + Z_p \, \alpha_p \leq (\frac{U_p+L_p}{2}) \, \alpha_p, \ \forall p \\
x \in X, \ X = x : \begin{cases} g_j(x) \leq 0, \ j = 1, 2, ..., m; \\ x_i \geq 0, \quad i = 1, 2, ..., n. \end{cases} \\
\lambda' \geq 0.
\end{cases}
\tag{89}
$$

Step-6: Now the reduced problems under case-(I) and case-(II) are solved by a linear optimization technique and in each case the optimum compromise solutions are obtained.

In case-(II), maximum overall satisfactory level of compromise is Max $\lambda = \lambda^* = 1/2 + (\tanh \lambda'^*)/2$.

**Global Criteria Method:** Global criteria method gives a compromise solution for a multi-objective problem. Actually this method is a way of achieving compromise in minimizing the sum in deviations of the ideal solutions (minimum value of the each objectives in case of minimization problem) from the respective objective functions. The steps of this method to solve the multi-objective model (83) are as follows:

Step-1: Solve the multi-objective problem as a single objective problem using, each time, only one objective $f_p$ ($p = 1, 2, ..., k$) ignoring all other objectives.

Step-2: From the results of step-1, determine the ideal objective vector, say $(f_1^{min}, f_2^{min}, ..., f_k^{min})$ and corresponding values of $(f_1^{max}, f_2^{max}, ..., f_k^{max})$.

Step-3: Formulate the following auxiliary problem

$$
\begin{cases}
Find & x = (x_1, x_2, ...., x_n)^T \\
which\ minimizes\ GC \\
subject\ to & x \in X \\
where\ X = & x : \begin{cases} g_j(x) \leq 0, \ j = 1, 2, ..., m; \\ x_i \geq 0, \quad i = 1, 2, ..., n. \end{cases}
\end{cases}
\tag{90}
$$

where

$$GC = Min\left\{ \sum_{t=1}^{k} \left( \frac{f_t(x) - f_t^{min}}{f_t^{min}} \right)^q \right\}^{\frac{1}{q}}, \tag{91}$$

$$or, \ GC = Min\left\{ \sum_{t=1}^{k} \left( \frac{f_t(x) - f_t^{min}}{f_t^{max} - f_t^{min}} \right)^q \right\}^{\frac{1}{q}}, \tag{92}$$

where $1 \leq q \leq \infty$. An usual value of $q$ is 2. This method is then called global criterion method in $L_2$ norm.

**Weighted Sum Method:** The weighted sum method scalarizes a set of objectives into a single objective by multiplying each objective with users supplied weights. The weights of an objective are usually chosen in proportion to the objectives relative importance in the problem. However setting up an appropriate weight vector depends on the scaling of each objective function. It is likely that different objectives take different orders of magnitude. When such objectives are weighted to form a composite objective function, it would be better to scale them appropriately so that each objective possesses more or less the same order of magnitude. This process is called normalization of objectives. After the objectives are normalized, a composite objective function $F(x)$ can be formed by summing the weighted normalized objectives and the multi-objective problem given in Eq. (83) is then converted to a single-objective optimization problem as follows:

$$\begin{cases} Find & x = (x_1, x_2, ...., x_n)^T \\ which \ minimizes & \sum_{i=1}^{k} w_i f_i(x), \ w_i \in [0, 1] \\ subject \ to & x \in X \\ where \ X = & x : \begin{cases} g_j(x) \leq 0, j = 1, 2, ..., m; \\ x_i \geq 0, \quad i = 1, 2, ..., n. \end{cases} \end{cases} \tag{93}$$

Here, $w_i$ is the weight of the $i$-th objective function. Since the minimum of the above problem does not change if all the weights are multiplied by a constant, it is the usual practice to choose weights such that their sum is one, i.e., $\sum_{i=1}^{k} w_i = 1$. Miettinen [109] proved that the solution of the weighted sum problem (2.71) is Pareto optimal if the weighting coefficients are positive, that is $w_i > 0$, $i = 1, 2, ..., k$.

# 3    Some Transportation Models with Fuzzy (Type-1) Parameters

If more than one objective is to be considered and optimized at the same time in an STP, then the problem is called multi-objective solid transportation problem (MOSTP). If more than one type of item/product is to be transported through the conveyances in an STP, then the problem is called multi-item solid transportation problem (MISTP). Also in a solid transportation system it may happen that several objectives are present and several types of items are to be transported, then we call this problem multi-objective multi-item solid transportation problem (MOMISTP). Besides source, destination and conveyance capacity constraints in an STP, there may exist several other types of constraints. For example, budget constraints may arise due to limited budget, space constraints may arise due to limited space in warehouses, stores, etc.

Due to insufficient information, lack of evidence, fluctuating financial market, the available data of a transportation system such as transportation costs, resources, demands, conveyance capacities are not always crisp or precise. For example, transportation cost depends upon fuel price, labor charges, tax charges, etc., each of which are fluctuate time to time. So for a future transportation planning it is not always easy to predict surely the possible unit transportation cost of a route in a certain time period. Similarly supply of a source can not be always exact, because it depends upon the availability of manpower, raw-materials, market competition, product demands, etc. Also it may not always possible to get relevant precise data/random data with a known distribution. So such a TP becomes more realistic if these parameters are assumed to be flexible/imprise i.e. fuzzy nature. For example if value of certain parameter of a decision making problem is given in an interval, then practically each of the point in the interval may not have the same importance or possibility. So it will be more realistic if those parameters are expressed by fuzzy numbers like triangular, trapezoidal, etc.

To solve constrained/unconstrained optimization problem with fuzzy parameters, several researchers developed many methodologies. Liu and Iwamura [92] presented chance-constrained programming with fuzzy parameters. Liu and Liu [94] presented expected value model for fuzzy programming. Yang and Liu [153] applied expected value model, chance-constrained programming model and dependent-chance programming to a fixed charge STP in fuzzy environment. Liang [82] presented a fuzzy goal programming approach for solving integrated production/transportation planning decision problems with fuzzy multiple goals. Mula et al. [116] applied possibilistic programming approach to a material requirement planing problem with fuzzy constraints and fuzzy coefficients.

In this chapter, we have investigated two solid transportation models namely, a multi-objective solid transportation problem with budget constraint and a multi-objective multi-item solid transportation problem both in fuzzy environment.

## 3.1 Related Results

**Theorem 3.1 (Yang and Liu [153]).** Suppose that $\tilde{\xi}$ is a fuzzy number with continuous membership function $\mu_{\tilde{\xi}}(x)$, and $r_0 = \sup\{r : \mu_{\tilde{\xi}}(r) = 1\}$, $g(x, \tilde{\xi}) = h(x) - \tilde{\xi}$. Then we have $Cr\{g(x, \tilde{\xi}) \geq 0\} \geq \alpha$ if and only if $h(x) \geq F_\alpha$, where

$$F_\alpha = \begin{cases} \inf\{F | F = \mu_{\tilde{\xi}}^{-1}(2\alpha)\}, & \text{if } \alpha \leq 0.5; \\ \inf\{F | F = \mu_{\tilde{\xi}}^{-1}(2(1-\alpha)), F > r_0\}, & \text{if } \alpha > 0.5. \end{cases}$$

**Theorem 3.2 (Yang and Liu [153]).** Suppose that $\tilde{\xi}$ is a fuzzy number with continuous membership function $\mu_{\tilde{\xi}}(x)$, and $r_0 = \inf\{r : \mu_{\tilde{\xi}}(r) = 1\}$, $g(x, \tilde{\xi}) = h(x) - \tilde{\xi}$. Then we have $Cr\{g(x, \tilde{\xi}) \le 0\} \ge \alpha$ if and only if $h(x) \le F_\alpha$, where

$$F_\alpha = \begin{cases} \sup\{F|F = \mu_{\tilde{\xi}}^{-1}(2\alpha)\}, & \text{if } \alpha \le 0.5; \\ \sup\{F|F = \mu_{\tilde{\xi}}^{-1}(2(1-\alpha)), F < r_0\}, & \text{if } \alpha > 0.5. \end{cases}$$

**Theorem 3.3 (Liu [90]).** Assume that the function $g(x, \xi)$ can be written as ,

$$g(x, \xi) = h_1(x)\tilde{\xi}_1 + h_1(x)\tilde{\xi}_2 + ... + h_t(x)\tilde{\xi}_t + h_0(x)$$

where $\tilde{\xi}_k$ are trapezoidal fuzzy variables $(r_{k1}, r_{k2}, r_{k3}, r_{k4})$, $k = 1, 2, ..., t$, respectively. We define two functions $h_k^+(x) = h_k(x) \vee 0$ and $h_k^-(x) = -(h_k(x) \wedge 0)$ for k=1,2,...,t. Then we have
(a) when $\alpha \le 1/2$, $Cr\{g(x, \xi) \le 0\} \ge \alpha$ if and only if

$$(1 - 2\alpha) \sum_{k=1}^{t}[r_{k1}h_k^+(x) - r_{k4}h_k^-(x)] + 2\alpha \sum_{k=1}^{t}[r_{k2}h_k^+(x) - r_{k3}h_k^-(x)] + h_0(x) \le 0;$$

(b) when $\alpha > 1/2$, $Cr\{g(x, \xi) \le 0\} \ge \alpha$ if and only if

$$(2 - 2\alpha) \sum_{k=1}^{t}[r_{k3}h_k^+(x) - r_{k2}h_k^-(x)] + (2\alpha - 1) \sum_{k=1}^{t}[r_{k4}h_k^+(x) - r_{k1}h_k^-(x)] + h_0(x) \le 0;$$

From the above theorem following corollaries are obtained.

**Corollary 1:** If $\tilde{\xi} = (r_1, r_2, r_3, r_4)$ is a trapezoidal fuzzy variable and $h(x)$ is a function of $x$, then $Cr\{h(x) \le \tilde{\xi}\} \ge \alpha$ if and only if $h(x) \le F_{\tilde{\xi}}$, where

$$F_{\tilde{\xi}} = \begin{cases} (1 - 2\alpha)r_4 + 2\alpha r_3, & \text{if } \alpha \le \frac{1}{2}; \\ 2(1 - \alpha)r_2 + (2\alpha - 1)r_1, & \text{if } \alpha > \frac{1}{2}. \end{cases}$$

**Proof:** $Cr\{h(x) \le \tilde{\xi}\} \ge \alpha \Leftrightarrow Cr\{-\tilde{\xi} + h(x) \le 0\} \ge \alpha \Leftrightarrow Cr\{\tilde{\xi}' + h(x) \le 0\} \ge \alpha$, where $\tilde{\xi}' = -\tilde{\xi} = (-r_4, -r_3, -r_2, -r_1)$.
Then from the above theorem it follows that this inequality holds if and only if

$$(a)(1 - 2\alpha)(-r_4) + 2\alpha(-r_3) + h(x) \le 0, \ if \ \alpha \le \frac{1}{2}$$

$$(b) \ 2(1 - \alpha)(-r_2) + (2\alpha - 1)(-r_1) + h(x) \le 0, \ if \ \alpha > \frac{1}{2}.$$

and hence the corollary follows.

**Corollary 2:** If $\tilde{\xi} = (r_1, r_2, r_3, r_4)$ is a trapezoidal fuzzy variable and $h(x)$ is a function of $x$, then $Cr\{h(x) \ge \tilde{\xi}\} \ge \alpha$ if and only if $h(x) \ge F_{\tilde{\xi}}$, where

$$F_{\tilde{\xi}} = \begin{cases} (1 - 2\alpha)r_1 + 2\alpha r_2, & \text{if } \alpha \le \frac{1}{2}; \\ 2(1 - \alpha)r_3 + (2\alpha - 1)r_4, & \text{if } \alpha > \frac{1}{2}. \end{cases}$$

**Proof:** $Cr\{h(x) \geq \tilde{\xi}\} \geq \alpha \Leftrightarrow Cr\{\tilde{\xi} - h(x) \leq 0\} \geq \alpha.$
Then from the above theorem it follows that this inequality holds if and only if

$$(a)(1 - 2\alpha)r_1 + 2\alpha r_2 - h(x) \leq 0, \ if \ \alpha \leq \frac{1}{2}$$

$$(b) \ 2(1 - \alpha)r_3 + (2\alpha - 1)r_4 - h(x) \leq 0, \ if \ \alpha > \frac{1}{2}.$$

and hence the corollary follows.

**Corollary 3:** If $\tilde{\xi} = (r_1, r_2, r_3, r_4)$ and $\tilde{\eta} = (t_1, t_2, t_3, t_4)$ are trapezoidal fuzzy variables and $h(x) \geq 0 \ \forall x$, then $Cr\{h(x)\tilde{\xi} \leq \tilde{\eta}\} \geq \alpha$ if and only if

$$(a)(1 - 2\alpha)h(x)r_1 + 2\alpha h(x)r_2 \leq (1 - 2\alpha)t_4 + 2\alpha t_3, \ when \ \alpha \leq \frac{1}{2}$$

$$(b) \ 2(1 - \alpha)h(x)r_3 + (2\alpha - 1)h(x)r_4 \leq 2(1 - \alpha)t_2 + (2\alpha - 1)t_1, \ when \ \alpha > \frac{1}{2}.$$

**Proof:** $Cr\{h(x)\tilde{\xi} \leq \tilde{\eta}\} \geq \alpha \Leftrightarrow Cr\{(h(x)\tilde{\xi} - \tilde{\eta}) \leq 0\} \geq \alpha \Leftrightarrow Cr\{(h(x)\tilde{\xi} + \tilde{\eta}') \leq 0\} \geq \alpha$, where $\tilde{\eta}' = -\tilde{\eta} = (-t_4, -t_3, -t_2, -t_1)$.

Then from the above theorem it follows that this inequality holds if and only if

$$(a)(1 - 2\alpha)(h(x)r_1 - t_4) + 2\alpha(h(x)r_2 - t_3) \leq 0, \ when \ \alpha \leq \frac{1}{2}$$

$$(b) \ 2(1 - \alpha)(h(x)r_3 - t_2) + (2\alpha - 1)(h(x)r_4 - t_1) + h(x) \leq 0, \ if \ \alpha > \frac{1}{2}.$$

and hence the corollary follows.
This is obvious that these three corollaries help us to determine crisp equivalences of various inequalities with fuzzy parameters.

## 3.2  A Defuzzification Method

Kikuchi [66] proposed a defuzzification method to find the most appropriate set of crisp numbers for a set of fuzzy numbers which satisfy a set of rigid relationships among them. The main idea of the method is to find the best set of crisp values satisfying the relationships those maximizes the minimum degree of membership that one of those values takes. Fuzzy linear programming is applied in this method. The method is summarized as follows.

Let $\tilde{X}_1, \tilde{X}_2,...,\tilde{X}_n$ are fuzzy numbers with membership functions $\mu_{\tilde{X}_1}$, $\mu_{\tilde{X}_2},...,\mu_{\tilde{X}_n}$ respectively. Suppose we have to find corresponding crisp values $x_1$, $x_2,...,x_n$ those satisfy some relationships $R_j(x)$, $j \in \mathbb{N}$ among them. Then the following linear programming based on fuzzy linear programming is formulated.

$$Max \ \lambda$$

$$s.t. \ \mu_{\tilde{X}_i}(x_i) \geq \lambda, \ i = 1, 2, ..., n$$

and the relationships $R_j(x)$, $j \in \mathbb{N}$

$$x_i,\ \lambda \geq 0,\ i = 1, 2, ..., n.$$

where $\lambda$ is the minimum degree of membership that one of the values $x_1, x_2,...,x_n$ takes, i.e. $\lambda^* = Max\lambda = Max\ Min[\mu_{\tilde{X}_1}(x_1), \mu_{\tilde{X}_2}(x_2), ..., \mu_{\tilde{X}_n}(x_n)]$.
Kikuchi [66] applied this method to a traffic volume consistency problem taking all observed values as triangular fuzzy numbers. Dey and Yadav [36] modified this method with trapezoidal fuzzy numbers.

## 3.3    Model 3.1: Multi-objective Solid Transportation Problem Having Budget Constraint with Fuzzy Parameters

Here a multi-objective solid transportation problem (MOSTP) is formulated with unit transportation penalties/costs, supplies, demands and conveyance capacities as fuzzy numbers (variables). Here the several objectives may be minimization of total transportation costs, minimization of total deterioration of goods, etc. Also, apart from source, demand and capacity constraints, an extra constraint on the total transportation budget at each destination is imposed. Obviously these budget constraints are performed for the objective function which represents minimization of the total transportation cost. The following notations are used to formulate the model.

**Notations:**

(i) $\tilde{c}^p_{ijk}$: Fuzzy unit transportation penalties from $i$-th source to $j$-th destination via $k$-th conveyance for the $p$-th objective.

(ii) $x^p_{ijk}$: The decision variable which represents amount of product to be transported from $i$-th origin to $j$-th destination via $k$-th conveyance for the $p$-th objective.

(iii) $\tilde{a}_i$: The fuzzy amount of the product available at the $i$-th origin.

(iv) $\tilde{b}_j$: The fuzzy demand of the product of $j$-th destination.

(v) $\tilde{e}_k$: Fuzzy transportation capacity of conveyance $k$.

(vi) $Z_p$: The $p$-th objective.

(vii) $\tilde{B}^l_j$: Available fuzzy budget amount for $j$-th destination for objective $Z_l$.

**Mathematical Model:** Mathematically the MOSTP with budget constraints having $P$ objectives, $m$ origins, $n$ destinations and $K$ conveyances is formulated as follows.

$$Min\ Z_p = \sum_{i=1}^{m}\sum_{j=1}^{n}\sum_{k=1}^{K} \tilde{c}^p_{ijk}\ x_{ijk}\ ,p = 1, 2, ..., P \tag{94}$$

$$s.t.\ \sum_{j=1}^{n}\sum_{k=1}^{K} x_{ijk} \leq \tilde{a}_i,\ i = 1, 2, ..., m, \tag{95}$$

$$\sum_{i=1}^{m}\sum_{k=1}^{K} x_{ijk} \geq \tilde{b}_j, \quad j=1,2,...,n, \tag{96}$$

$$\sum_{i=1}^{m}\sum_{j=1}^{n} x_{ijk} \leq \tilde{e}_k, \quad k=1,2,...,K, \tag{97}$$

$$\sum_{i=1}^{m}\sum_{k=1}^{K} \tilde{c}_{ijk}^{l} x_{ijk} \leq \tilde{B}_j^l, \quad j=1,2,...,n \ , \ l \in \{1,...,P\}, \tag{98}$$

$$x_{ijk} \geq 0, \forall \ i,j,k.$$

Here for $p = l$ (say), $l \in \{1,2,...,P\}$, $\tilde{c}_{ijk}^{l}$ represent unit transportation cost so that available fuzzy budget amount for $j$-th destination, i.e. $\tilde{B}_j^l$ is imposed for objective $Z_l$.

**Solution Methodology: Chance-Constrained Programming:** We apply the chance constrained programming (CCP) technique using credibility measure to the above model and then it is formulated as

$$Min[\bar{Z}_1, \bar{Z}_2, ..., \bar{Z}_P] \tag{99}$$

$$s.t. \ Cr\{\sum_{i=1}^{m}\sum_{j=1}^{n}\sum_{k=1}^{K} c_{ijk}^{\tilde{p}} x_{ijk} \leq \bar{Z}_p\} \geq \eta_p, \ p=1,2,...,P, \tag{100}$$

$$Cr\{\sum_{j=1}^{n}\sum_{k=1}^{K} x_{ijk} \leq \tilde{a}_i\} \geq \alpha_i, \quad i=1,2,...,m, \tag{101}$$

$$Cr\{\sum_{i=1}^{m}\sum_{k=1}^{K} x_{ijk} \geq \tilde{b}_j\} \geq \beta_j, \quad j=1,2,...,n, \tag{102}$$

$$Cr\{\sum_{i=1}^{m}\sum_{j=1}^{n} x_{ijk} \leq \tilde{e}_k\} \geq \gamma_k, \quad k=1,2,...,K, \tag{103}$$

$$Cr\{\sum_{i=1}^{m}\sum_{k=1}^{K} \tilde{c}_{ijk}^{l} x_{ijk} \leq \tilde{B}_j^l\} \geq \delta_j^l, \quad j=1,2,...,n \ , \ l \in \{1,...,P\}, \tag{104}$$

$$x_{ijk} \geq 0, \forall \ i,j,k,$$

where $\eta_p$ indicates that we are going to optimize the $\eta_p$-critical value of the objective $Z_p(p=1,2,...,P)$, and $\alpha_i$, $\beta_j$, $\gamma_k$ and $\delta_j^l$ are predetermined credibility levels of satisfaction of the above constraints (101), (102), (103) and (104) respectively. In other words, the constraint (101) indicates that total amount transported from source $i$ must be less than or equal to its supply capacity $\tilde{a}_i$ at the credibility level at least $\alpha_i$; the constraint (102) indicates that total amount transported to destination $j$ must satisfy its requirement $\tilde{b}_j$ at the credibility at least $\beta_j$, the constraint (103) indicates that total amount transported through conveyance $k$ must not be more than its capacity $\tilde{e}_k$ at the credibility at least $\gamma_k$ and the constraints (104) indicates that for the specific objective $Z_l$, total transportation costs for $j$-th destination must not exceed the available budget amount $\tilde{B}_j^l$ at the credibility at least $\delta_j^l$.

**Crisp Equivalences:** Let $c_{ijk}^{\tilde{p}} = (c_{ijk}^{p1}, c_{ijk}^{p2}, c_{ijk}^{p3}, c_{ijk}^{p4})$, $\tilde{a}_i = (a_i^1, a_i^2, a_i^3, a_i^4)$, $\tilde{b}_j = (b_j^1, b_j^2, b_j^3, b_j^4)$, $\tilde{e}_k = (e_k^1, e_k^2, e_k^3, e_k^4)$, $\tilde{B}_j^l = (B_j^{l1}, B_j^{l2}, B_j^{l3}, B_j^{l4})$ are trapezoidal fuzzy numbers for all $p, i, j$ and $k$.

Now since $c_{ijk}^{\tilde{p}}$ are trapezoidal fuzzy numbers and $x_{ijk} \geq 0$ for all $i, j, k$, so $Z_p(x) = \sum_{i=1}^m \sum_{j=1}^n \sum_{k=1}^K c_{ijk}^{\tilde{p}} x_{ijk}$ are also trapezoidal fuzzy numbers for any feasible solution $x$ and given by $Z_p(x) = (r_1^p(x), r_2^p(x), r_3^p(x), r_4^p(x))$, where

$$r_1^p(x) = \sum_{i=1}^m \sum_{j=1}^n \sum_{k=1}^K c_{ijk}^{p1} x_{ijk} \, , \quad r_2^p(x) = \sum_{i=1}^m \sum_{j=1}^n \sum_{k=1}^K c_{ijk}^{p2} x_{ijk}, \qquad (105)$$

$$r_3^p(x) = \sum_{i=1}^m \sum_{j=1}^n \sum_{k=1}^K c_{ijk}^{p3} x_{ijk} \, , \quad r_4^p(x) = \sum_{i=1}^m \sum_{j=1}^n \sum_{k=1}^K c_{ijk}^{p4} x_{ijk}, \qquad (106)$$

$$p = 1, 2, ..., P.$$

Then the objective $\bar{Z}_p$ in (99), i.e. $Min \ \bar{Z}_P$, s.t. $Cr\{Z_p(x) \leq \bar{Z}_p\} \geq \eta_p$ is equivalently computed as $\bar{Z}_p = \inf\{r : Cr\{Z_P(x) \leq r\} \geq \eta_p\}$ which is nothing but $\eta_p$ - pessimistic value to $Z_p$ (i.e. $Z_{p_{inf}}(\eta_p)$) and so is equal to $Z_p'(x)$, where

$$Z_p'(x) = \begin{cases} (1 - 2\eta_p)r_1^p(x) + 2\eta_p \ r_2^p(x), & \text{if } \alpha \leq 0.5; \\ 2(1 - \eta_p)r_3^p(x) + (2\eta_p - 1)r_4^p(x), & \text{if } \alpha > 0.5. \end{cases}$$

Now from corollaries 1 and 2 of the Theorem 3.3, the constraint (101) and (102) and from corollary 1, the constraint (103) can be written respectively in equivalent crisp forms as

$$\sum_{j=1}^n \sum_{k=1}^K x_{ijk} \leq F_{\alpha_i}, \ i = 1, 2, ..., m \qquad (107)$$

$$\sum_{i=1}^m \sum_{k=1}^K x_{ijk} \geq F_{\beta_j}, \ j = 1, 2, ..., n \qquad (108)$$

$$\sum_{i=1}^m \sum_{j=1}^n x_{ijk} \leq F_{\gamma_k}, \ k = 1, 2, ..., K \qquad (109)$$

where, $F_{\alpha_i} = \begin{cases} (1 - 2\alpha_i)a_i^4 + 2\alpha_i a_i^3, & \text{if } \alpha_i \leq 0.5; \\ 2(1 - \alpha_i)a_i^2 + (2\alpha_i - 1)a_i^1, & \text{if } \alpha_i > 0.5. \end{cases}$

$F_{\beta_j} = \begin{cases} (1 - 2\beta_j)b_j^1 + 2\beta_j b_j^2, & \text{if } \beta_j \leq 0.5; \\ 2(1 - \beta_j)b_j^3 + (2\beta_j - 1)b_j^4, & \text{if } \beta_j > 0.5. \end{cases}$

$$F_{\gamma_k} = \begin{cases} (1 - 2\gamma_k)e_k^4 + 2\gamma_k e_k^3, & \text{if } \gamma_k \leq 0.5; \\ 2(1 - \gamma_k)e_k^2 + (2\gamma_k - 1)e_k^1, & \text{if } \gamma_k > 0.5. \end{cases}$$

Now the budget constraint (104) is in the form

$$Cr\{\sum_{i=1}^{m}\sum_{k=1}^{K} x_{ijk} \ (c_{ijk}^{l1}, c_{ijk}^{l2}, c_{ijk}^{l3}, c_{ijk}^{l4}) + (-1)(B_j^{l1}, B_j^{l2}, B_j^{l3}, B_j^{l4}) \leq 0\} \geq \delta_j^l$$

Since $x_{ijk} \geq 0$ for all $i, j, k$, from Corollary 2 of the Theorem 3.3, it is obvious that this constraint will be active if and only if $g_j^l \leq 0$, where

$$g_j^l = \begin{cases} (1 - 2\delta_j^l)(\sum_{i=1}^{m}\sum_{k=1}^{K} c_{ijk}^{l1} \ x_{ijk} - B_j^{l4}) + \\ \hspace{4cm} \text{if } \delta_j \leq 0.5; \\ +2\delta_j^l(\sum_{i=1}^{m}\sum_{k=1}^{K} c_{ijk}^{l2} \ x_{ijk} - B_j^{l3}), \\ 2(1 - \delta_j^l)(\sum_{i=1}^{m}\sum_{k=1}^{K} c_{ijk}^{l3} \ x_{ijk} - B_j^{l2}) + \\ \hspace{4cm} \text{if } \delta_j > 0.5. \\ +(2\delta_j^l - 1)(\sum_{i=1}^{m}\sum_{k=1}^{K} c_{ijk}^{l4} \ x_{ijk} - B_j^{l1}), \end{cases}$$

So finally an equivalent crisp form of the above CCP model (99)–(104) formulated for the model (94)–(98) can be written as

$$Min[Z_1'(x), Z_2'(x), ..., Z_P'(x)]$$

$$s.t. \quad g_j^l \leq 0, \ j = 1, 2, ..., n \ , \tag{110}$$

with the constraints (107), (108), (109),

$$x_{ijk} \geq 0 \, for \, all \ i, j, k.$$

Now the problem (110) is a multi-objective problem and so can be solved by fuzzy programming technique, Global criteria method (cf. Sect. 2.9), etc.

**Numerical Experiment:** To illustrate the Model 3.1 ((94)–(98)), we consider an example where from the past record of a transport company, the possible values of the parameters such as the unit transportation costs, the supplies, the demands, the available conveyance capacities can not be precisely determined. For instance, unit transportation cost for a route is "about 6", the supply of a source is "around 28–30", etc. These linguistic data can be transferred into triangular or trapezoidal fuzzy numbers. For example if it is seen from the past record that most possible value of unit transportation cost of a route is 13 and it vary from 12 to 14 with less possibility, then it is "about 13" and represented by the fuzzy number $(12,13,14)\sim(12,13,13,14)$. Similarly the most possible value of supply of a source ranges between 28 to 30 and

**Table 2.** Penalties (costs) $\tilde{c}^1_{ijk}$

| $i \setminus j$ | 1 | 2 | 1 | 2 |
|---|---|---|---|---|
| 1 | (10,11,13,14) | (7,10,11,12) | (11,13,13.5,14.5) | (15,16,18,19) |
| 2 | (13,14,16,17) | (8,10,10.5,11.5) | (16,17,17,18) | (12,13,15,17) |
| $k$ | 1 | | 2 | |

**Table 3.** Penalties $\tilde{c}^2_{ijk}$

| $i \setminus j$ | 1 | 2 | 1 | 2 |
|---|---|---|---|---|
| 1 | (13,14,16,17) | (7,8,10,11) | (10,11.5,13,13.5) | (12,13,15,16) |
| 2 | (12,13.5,14.5,16) | (13,14,15,16) | (12,13,13,14) | (9,12,13,14) |
| $k$ | 1 | | 2 | |

is not less than 27 and greater than 32, i.e. it is "around 28-30" and represented by TrFN (27,28,30,32). For the current model, two sources, two destinations, two conveyances and two objectives are considered, i.e. $i, j, k, p = 1, 2$.
The fuzzy penalties associated with the two objectives $Z_1$ and $Z_2$ are given in
Tables 2 and 3 respectively. The values of all the parameters associated with
two resources, two destinations, two conveyances and two objectives are given
below. Also budget constraint is imposed on the objective $Z_1$ (i.e. $l = 1$).
$\tilde{a}_1 = (37, 40, 46, 48)$, $\tilde{a}_2 = (28, 32, 35, 37)$, $\tilde{b}_1 = (28, 29, 30, 31)$, $\tilde{b}_2 = (31, 33, 34, 35)$, $\tilde{e}_1 = (27, 29, 32, 34)$, $\tilde{e}_2 = (39, 41, 44, 47)$, $\tilde{B}^1_1 = (480, 485, 491, 497)$, and $\tilde{B}^1_2 = (501, 505, 510, 515)$.

Now applying chance-constrained programming technique to solve the problem, we reconstruct the problem as model (99)–(104) and use corresponding
crisp equivalent form (110). For this purpose, let us consider the credibility
level $\alpha_i = 0.9$ for the source constraints, $\beta_j = 0.9$ for the demand constraints,
$\gamma_k = 0.95$ for the capacity constraints, $\delta^l_j = 0.9$ for the budget constraints, where
$i, j, k = 1, 2$, $l = 1$ and let $\eta_p = 0.9$ $(p = 1, 2)$, which implies that we want to
minimize 0.9-critical value of the objectives. Then using (110), the proposed
problem becomes

$$Min[Z'_p(x)] \quad p = 1, 2$$

$$Z'_p(x) = 0.2\, r^p_3(x) + 0.8\, r^p_4(x) \quad, p = 1, 2$$

$$\sum_{j=1}^{2}\sum_{k=1}^{2} x_{ijk} \leq F_{\alpha_i}, \; i = 1, 2$$

$$\sum_{i=1}^{2}\sum_{k=1}^{2} x_{ijk} \geq F_{\beta_j}, \; j = 1, 2 \tag{111}$$

$$\sum_{i=1}^{2}\sum_{j=1}^{2} x_{ijk} \le F_{\gamma_k}, \ k = 1, 2$$

$$0.2(\sum_{i=1}^{2}\sum_{k=1}^{2} c_{ijk}^{13} \ x_{ijk}) + 0.8(\sum_{i=1}^{2}\sum_{k=1}^{2} c_{ijk}^{14} \ x_{ijk}) \le (0.2B_{j}^{12} + 0.8B_{j}^{11}), \ j = 1, 2$$

where expression of $r_3^P(x)$ and $r_4^P(x)$ are same as in (106) for $i, j, k = 1, 2$.
With the given data, we have $F_{\alpha_1} = 37.6$, $F_{\alpha_2} = 28.8$, $F_{\beta_1} = 30.8$, $F_{\beta_2} = 34.6$, $F_{\gamma_1} = 27.4$, $F_{\gamma_1} = 39.4$

Solving this problem by fuzzy programming technique (with linear membership function) (cf. Sect. 2.9), the obtained optimum compromise solution is presented in Table 4. The solution is obtained by using the standard optimization solver - LINGO.

Table 4. Optimum results for Model (94)–(98)

| |
|---|
| $x_{121} = 14.99$, $x_{211} = 12.40$, $x_{112} = 2.0$, $x_{122} = 19.60$, $x_{212} = 16.40$, |
| $B_1^1 = 478.19$, $B_2^1 = 370.80$, $\lambda = 0.55$, $Z_1' = 901.30$, $Z_2' = 895.58$. |

$B_1^1$ and $B_2^1$ represent the budget values for $j=1,2$ respectively for objective $Z_1$ and $\lambda$ represents maximum overall satisfactory level of compromise.

## 3.4  Model 3.2: Multi-objective Multi-item Solid Transportation Problem with Fuzzy Parameters

A multi-objective multi-item solid transportation problem (MOMISTP) with fuzzy parameters is formulated in which several objectives (e.g., minimization of transportation costs, minimization of total deterioration of goods, etc.) are involved and also several types of items/goods are to be transported from sources to destinations through the conveyances. The following notations are used to formulate the model.

**Notations:**

(i) $\tilde{c}_{ijk}^{tp}$: for the objective $Z_t$, fuzzy unit transportation penalty from $i$-th origin to $j$-th destination by $k$-th conveyance for $p$-th item.

(ii) $x_{ijk}^p$: the decision variable that represents the amount of $p$-th item to be transported from $i$-th source to $j$-th destination by $k$-th conveyance.

(iii) $\tilde{a}_i^p$: amount of $p$-th item represented by fuzzy number available at $i$-th origin

(iv) $\tilde{b}_j^p$: fuzzy demand of $j$-th destination for $p$-th item

(v) $\tilde{e}_k$: total fuzzy capacity of $k$-th conveyance.

**Mathematical Model:** An MOMISTP with $R$ objectives, $l$ different items, $m$ origins, $n$ destinations and $K$ types of conveyances is formulated as follows:

$$Min\ Z_t = \sum_{p=1}^{l}\sum_{i=1}^{m}\sum_{j=1}^{n}\sum_{k=1}^{K} c_{ijk}^{tp}\ x_{ijk}^{p}\ , t = 1, 2, ..., R$$

$$s.t.\ \sum_{j=1}^{n}\sum_{k=1}^{K} x_{ijk}^{p} \leq \tilde{a}_i^p,\ \ i = 1, 2, ..., m; p = 1, 2, ...l,$$

$$\sum_{i=1}^{m}\sum_{k=1}^{K} x_{ijk}^{p} \geq \tilde{b}_j^p,\ \ j = 1, 2, ..., n; p = 1, 2, ..., l, \tag{112}$$

$$\sum_{p=1}^{l}\sum_{i=1}^{m}\sum_{j=1}^{n} x_{ijk}^{p} \leq \tilde{e}_k,\ \ k = 1, 2, ..., K,$$

$$x_{ijk}^{p} \geq 0,\ \forall\ i, j, k, p.$$

**Defuzzification Process:** Consider $\tilde{a}_i^p$, $\tilde{b}_j^p$ and $\tilde{e}_k$ ($\forall\ i, j, k, p$) as trapezoidal fuzzy numbers defined by $\tilde{a}_i^p = (a_i^{p1}, a_i^{p2}, a_i^{p3}, a_i^{p4})$, $\tilde{b}_j^p = (b_j^{p1}, b_j^{p2}, b_j^{p3}, b_j^{p4})$ and $\tilde{e}_k = (e_k^1, e_k^2, e_k^3, e_k^4)$ and their membership functions are $\mu_{\tilde{a}_i^p}$, $\mu_{\tilde{b}_j^p}$ and $\mu_{\tilde{e}_k}$ respectively. Now to solve the above problem, we first find corresponding defuzzified (crisp) values, say, $a_{ic}^p$, $b_{jc}^p$ and $e_{kc}$ ($\forall\ i, j, k, p$) so that for each item, total available resources greater than or equal to the total demands and also total conveyance capacities greater than or equal to the total demands for all items, i.e.

$$\sum_{i=1}^{m} a_{ic}^p \geq \sum_{j=1}^{n} b_{jc}^p\ ,\ p = 1, 2, ..., l\ and\ \sum_{k=1}^{K} e_{kc} \geq \sum_{p=1}^{l}\sum_{j=1}^{n} b_{jc}^p.$$

Because defuzzified values of availabilities, demands and conveyance capacities must have to be satisfy these conditions to have a feasible solution of the above problem. For this purpose we apply the defuzzification method (cf Sect. 3.2) based on fuzzy linear programming.

The method is to introduce an auxiliary variable $\lambda$ and formulate the following linear programming.

$$Max\ \lambda$$

$$s.t.\ \mu_{\tilde{a}_i^p}(a_{ic}^p) \geq \lambda,\ \mu_{\tilde{b}_j^p}(b_{jc}^p) \geq \lambda,\ \mu_{\tilde{e}_k}(e_{kc}) \geq \lambda,$$

$$\sum_{i=1}^{m} a_{ic}^p \geq \sum_{j=1}^{n} b_{jc}^p,\ \sum_{k=1}^{K} e_{kc} \geq \sum_{p=1}^{l}\sum_{j=1}^{n} b_{jc}^p,$$

$$\forall\ i,\ j,\ k,\ p.$$

where $\lambda$ is the minimum degree of membership that one of the values of the variables $a_{ic}^p$, $b_{jc}^p$, $e_{kc}$ takes, i.e. Max $\lambda = \lambda^* = $ Max Min $[\mu_{\tilde{a}_i^p}(a_{ic}^p),\ \mu_{\tilde{b}_j^p}(b_{jc}^p),\ \mu_{\tilde{e}_k}(e_{kc})]$, where

$$\mu_{\tilde{a}_i^p}(a_{ic}^p) = \begin{cases} \frac{a_{ic}^p - a_i^{p1}}{a_i^{p2} - a_i^{p1}}, & \text{if } a_i^{p1} \le a_{ic}^p \le a_i^{p2}; \\ 1, & \text{if } a_i^{p2} \le a_{ic}^p \le a_{ic}^{p3}; \\ \frac{a_i^{p4} - a_{ic}^p}{a_i^{p4} - a_i^{p3}}, & \text{if } a_i^{p3} \le a_{ic}^p \le a_i^{p4}. \end{cases}$$

and similarly for $\mu_{\tilde{b}_j^p}(b_{jc}^p)$ and $\mu_{\tilde{e}_k}(e_{kc})$.

Now if we denote left and right sides of the membership function $\mu_{\tilde{a}_i^p}(a_{ic}^p)$ by $\mu_{\tilde{a}_i^p}^l(a_{ic}^p)$ and $\mu_{\tilde{a}_i^p}^r(a_{ic}^p)$ respectively and so on for $\mu_{\tilde{b}_j^p}(b_{jc}^p)$ and $\mu_{\tilde{e}_k}(e_{kc})$, then the above programming becomes

$$Max\ \lambda$$

$$s.t.\ \ \mu_{\tilde{a}_i^p}^l(a_{ic}^p) \ge \lambda,\ \mu_{\tilde{a}_i^p}^r(a_{ic}^p) \ge \lambda$$

$$\mu_{\tilde{b}_j^p}^l(b_{jc}^p) \ge \lambda,\ \mu_{\tilde{b}_j^p}^r(b_{jc}^p) \ge \lambda$$

$$\mu_{\tilde{e}_k}^l(e_{kc}) \ge \lambda,\ \mu_{\tilde{e}_k}^r(e_{kc}) \ge \lambda \tag{113}$$

$$\sum_{i=1}^m a_{ic}^p \ge \sum_{j=1}^n b_{jc}^p,\ \ \sum_{k=1}^K e_{kc} \ge \sum_{p=1}^l \sum_{j=1}^n b_{jc}^p,$$

$\forall\ i,\ j,\ k,\ p$.

**Solution Methodology:** Consider that $c_{ijk}^{\tilde{t}p}$ are all independent trapezoidal fuzzy numbers represented as $(c_{ijk}^{tp1}, c_{ijk}^{tp2}, c_{ijk}^{tp3}, c_{ijk}^{tp4})$. Now after obtaining the defuzzified values $a_{ic}^p$, $b_{jc}^p$ and $e_{kc}$ ($\forall\ i, j, k, p$) by above procedure (i.e. using (113)), the problem (112) becomes,

$$Min\ Z_t = \sum_{p=1}^l \sum_{i=1}^m \sum_{j=1}^n \sum_{k=1}^K c_{ijk}^{\tilde{t}p}\, x_{ijk}^p\ \ , t = 1, 2, ..., R \tag{114}$$

$$s.t.\ \sum_{j=1}^n \sum_{k=1}^K x_{ijk}^p \le a_{ic}^p,\ \ i = 1, 2, ..., m; p = 1, 2, ...l \tag{115}$$

$$\sum_{i=1}^m \sum_{k=1}^K x_{ijk}^p \ge b_{jc}^p,\ \ j = 1, 2, ..., n; p = 1, 2, ..., l \tag{116}$$

$$\sum_{p=1}^l \sum_{i=1}^m \sum_{j=1}^n x_{ijk}^p \le e_{kc},\ \ k = 1, 2, ..., K \tag{117}$$

$$x_{ijk}^p \ge 0,\ \forall\ i, j, k, p.$$

Now, we use following methods to solve this problem.

**Method-1: Using the Concept-Minimum of Fuzzy Number:** The objective functions in (114) are $Z_t = \sum_{p=1}^{l} \sum_{i=1}^{m} \sum_{j=1}^{n} \sum_{k=1}^{K} \tilde{c}_{ijk}^{tp} x_{ijk}^{p}$ , $t = 1, 2, ..., R$. Since $\tilde{c}_{ijk}^{tp}$ are trapezoidal fuzzy numbers and $x_{ijk}^{p} \geq 0$ for all $i, j, k$ and $p$, so each $Z_t$ for $t = 1, 2, ..., R$ is also a trapezoidal fuzzy number for any feasible solution and is given by $Z_t = (Z_t^1, Z_t^2, Z_t^3, Z_t^4)$ where

$$Z_t^r = \sum_{p=1}^{l} \sum_{i=1}^{m} \sum_{j=1}^{n} \sum_{k=1}^{K} c_{ijk}^{tpr} x_{ijk}^{p} \text{ for } r = 1, 2, 3, 4.$$

As it is not possible to minimize directly a fuzzy number $\tilde{Z}$, here we use a method proposed by Buckly et al. [14]. They applied this method to a fuzzy inventory control problem. The method is to convert min $\tilde{Z}$ into a multi-objective problem

$$Min\tilde{Z} = (Max\ A_L(\tilde{Z}), Min\ C(\tilde{Z}), Min\ A_R(\tilde{Z})),$$

where $C(\tilde{Z})$ is the center of the core of the fuzzy number and $A_L(\tilde{Z})$, $A_R(\tilde{Z})$ are the area under graph of the membership function of $\tilde{Z}$ to the left and right of $C(Z)$ (minimization of a TrFN $\tilde{Z}$ is shown in Fig. 10). If the support of $\tilde{Z}$ be $[u_1, u_3]$ and the center of the core of $\tilde{Z}$ be at $u_2$, then

$$A_L(\tilde{Z}) = \int_{u_1}^{u_2} \mu_{\tilde{Z}}(x)\ dx \text{ and } A_R(\tilde{Z}) = \int_{u_2}^{u_3} \mu_{\tilde{Z}}(x)\ dx .$$

Then this multi-objective problem is converted to a single objective problem as follows

$$Min\tilde{Z} = Min\{\lambda_1[M - A_L(\tilde{Z})] + \lambda_2 C(\tilde{Z}) + \lambda_3 A_R(\tilde{Z})\}, \quad (118)$$

where $\lambda_l > 0$, for $l = 1, 2, 3$, $\lambda_1 + \lambda_2 + \lambda_3 = 1$ and M is a large positive number so that Max $A_L(\tilde{Z})$ is equivalent to Min $[M - A_L(\tilde{Z})]$. The values of $\lambda_l$ are taken

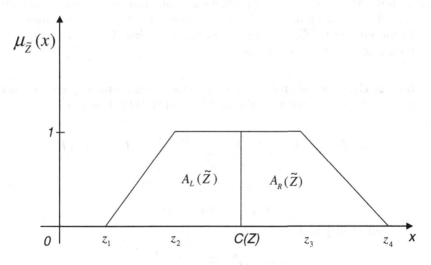

**Fig. 10.** Minimization of TrFN $\tilde{Z}$

by decision maker or a sensitivity analysis can be done taking different values of $\lambda_l$ to choose appropriate values of $\lambda_l$.

Now for a trapezoidal fuzzy number $\tilde{Z} = (z^1, z^2, z^3, z^4)$, the membership function is

$$\mu_{\tilde{Z}}(x) = \begin{cases} \frac{x-z^1}{z^2-z^1}, & \text{if } z^1 \le x \le z^2; \\ 1, & \text{if } z^2 \le x \le z^3; \\ \frac{z^4-x}{z^4-z^3}, & \text{if } z^3 \le x \le z^4; \\ 0, & \text{otherwise.} \end{cases}$$

So the core of $\tilde{Z}$ is $[z^2, z^3]$ and $C(\tilde{Z}) = \frac{z^2+z^3}{2}$.

$$A_L(\tilde{Z}) = \int_{z^1}^{\frac{z^2+z^3}{2}} \mu_{\tilde{Z}}(x) \, dx = \int_{z^1}^{z^2} \frac{x-z^1}{z^2-z^1} \, dx + \int_{z^2}^{\frac{z^2+z^3}{2}} 1 \, dx = \frac{1}{2}(z^3 - z^1),$$

$$A_R(\tilde{Z}) = \int_{\frac{z^2+z^3}{2}}^{z^3} 1 \, dx + \int_{z^3}^{z^4} \frac{z^4-x}{z^4-z^3} \, dx = \frac{1}{2}(z^4 - z^2).$$

Applying this method to the objective function (114) of the problem (114)–(117), the corresponding crisp form of the problem becomes

$$Min \bar{Z}_t = \lambda_1[M - A_L(Z_t)] + \lambda_2 C(Z_t) + \lambda_3 A_R(Z_t) \,, t = 1, 2, ...R,$$

$$A_L(Z_t) = \frac{Z_t^3 - Z_t^1}{2}, \quad C(Z_t) = \frac{Z_t^2 + Z_t^3}{2}, \quad A_R(Z_t) = \frac{Z_t^4 - Z_t^2}{2}, \tag{119}$$

subject to the constraints (115)–(117),

$$x_{ijk}^p \ge 0, \ \forall \ i,j,k,p, \quad \lambda_1 + \lambda_2 + \lambda_3 = 1, \ \lambda_l > 0, \ l = 1, 2, 3.$$

Though the choice of values of $\lambda_l$ depends upon decision maker(s), it should be kept in mind that, as the above problem is a minimization problem, our aim should be more in maximizing $A_L(Z_t)$ (i.e. possibility of getting less values than $C(Z_t)$) and minimizing $C(Z_t)$ rather than in minimizing $A_R(Z_t)$ (i.e. possibility of getting more values than $C(Z_t)$).

**Method-2: Using Expected Value:** Here we minimize the expected value of the objective functions and then the problem (114)–(117) becomes

$$Min \ E[Z_t] = E[\sum_{p=1}^{l} \sum_{i=1}^{m} \sum_{j=1}^{n} \sum_{k=1}^{K} \tilde{c}_{ijk}^{tp} \, x_{ijk}^p] \,, t = 1, 2, ..., R$$

$$s.t. \ the \ constraints \ (115) - (117), \tag{120}$$

$$x_{ijk}^p \ge 0, \ \forall \ i,j,k,p$$

which is equivalently written as

$$Min \ E[Z_t] = \bar{Z}_t = \sum_{p=1}^{l} \sum_{i=1}^{m} \sum_{j=1}^{n} \sum_{k=1}^{K} E[\tilde{c}_{ijk}^{tp}] \, x_{ijk}^p \,, t = 1, 2, ..., R$$

$$s.t. \ the \ constraints \ (115) - (117), \qquad (121)$$

$$x^p_{ijk} \geq 0, \ \forall \ i, j, k, p.$$

Now the deterministic models (119) and (121) are multi-objective problems and so can be solved by fuzzy programming technique, Global criteria method (cf. Sect. 2.9), etc.

**Note:** Deterministic forms obtained using expected value model (EVM), chance-constrained programming model for an optimization problem with fuzzy parameters having interrelated constraints like in STP may not always possesses any feasible solution. For example if we construct the EVM for the problem (112) by using expected value to both the objective functions and the constraints then it becomes:

$$Min \ E[Z_t] = E\Big[\sum_{p=1}^{l}\sum_{i=1}^{m}\sum_{j=1}^{n}\sum_{k=1}^{K} \tilde{c}^{tp}_{ijk} \ x^p_{ijk}\Big] \ , t = 1, 2, ..., R$$

$$s.t. \ E\Big[\sum_{j=1}^{n}\sum_{k=1}^{K} x^p_{ijk} - \tilde{a}^p_i\Big] \leq 0, \quad i = 1, 2, ..., m; p = 1, 2, ...l,$$

$$E\Big[\sum_{i=1}^{m}\sum_{k=1}^{K} x^p_{ijk} - \tilde{b}^p_j\Big] \geq 0, \quad j = 1, 2, ..., n; p = 1, 2, ..., l, \qquad (122)$$

$$E\Big[\sum_{p=1}^{l}\sum_{i=1}^{m}\sum_{j=1}^{n} x^p_{ijk} - \tilde{e}_k\Big] \leq 0, \quad k = 1, 2, ..., K,$$

$$x^p_{ijk} \geq 0, \ \forall \ i, j, k, p.$$

Then by the linearity property of expected value operator, the crisp equivalence form of this model becomes

$$Min \ E[Z_t] = \bar{Z}_t = \sum_{p=1}^{l}\sum_{i=1}^{m}\sum_{j=1}^{n}\sum_{k=1}^{K} E[\tilde{c}^{tp}_{ijk}] \ x^p_{ijk} \ , t = 1, 2, ..., R$$

$$s.t. \ \sum_{j=1}^{n}\sum_{k=1}^{K} x^p_{ijk} \leq E[\tilde{a}^p_i], \quad i = 1, 2, ..., m; p = 1, 2, ...l,$$

$$\sum_{i=1}^{m}\sum_{k=1}^{K} x^p_{ijk} \geq E[\tilde{b}^p_j], \quad j = 1, 2, ..., n; p = 1, 2, ..., l, \qquad (123)$$

$$\sum_{p=1}^{l}\sum_{i=1}^{m}\sum_{j=1}^{n} x^p_{ijk} \leq E[\tilde{e}_k], \quad k = 1, 2, ..., K,$$

$$x^p_{ijk} \geq 0, \ \forall \ i, j, k, p.$$

**Table 5.** Penalties/costs $c_{ijk}^{\tilde{1}1}$

| $i \backslash j$ | 1 | 2 | 3 | 1 | 2 | 3 |
|---|---|---|---|---|---|---|
| 1 | (5,8,9,11) | (4,6,9,11) | (10,12,14,16) | (9,11,13,15) | (6,8,10,12) | (7,9,12,14) |
| 2 | (8,10,13,15) | (6,7,8,9) | (11,13,15,17) | (10,11,13,15) | (6,8,10,12) | (14,16,18,20) |
| $k$ | | 1 | | | 2 | |

**Table 6.** Penalties/costs $c_{ijk}^{\tilde{1}2}$

| $i \backslash j$ | 1 | 2 | 3 | 1 | 2 | 3 |
|---|---|---|---|---|---|---|
| 1 | (9,10,12,13) | (5,8,10,12) | (10,11,12,13) | (11,13,14,15) | (6,7,9,11) | (8,10,11,13) |
| 2 | (11,13,14,16) | (7,9,12,14) | (12,14,16,18) | (14,16,18,20) | (9,11,13,14) | (13,14,15,16) |
| $k$ | | 1 | | | 2 | |

But in the above crisp equivalence form, the deterministic values of supplies, demands and conveyance capacities, i.e. $E[\tilde{a}_i^p]$, $E[\tilde{b}_j^p]$ and $E[\tilde{e}_k]$ respectively may not satisfy the required conditions for feasible solution, i.e. for each item, total supplies greater than or equal to the total demands and also total conveyance capacities greater than or equal to the total demands for all items. So this method gives a feasible solution only when the fuzzy supplies, demands and conveyance capacities are so that their respective expected values automatically satisfy those conditions.

**Numerical Experiment:** To illustrate numerically the Model 3.2 (112), we consider an example with $p = 1, 2 = i, k$; $j = 1, 2, 3$ and the following data. The unit transportation penalties are given in Tables 5, 6, 7 and 8. $\tilde{a}_1^1 = (21, 24, 26, 28)$, $\tilde{a}_2^1 = (28, 32, 35, 37)$, $\tilde{b}_1^1 = (14, 16, 19, 22)$, $\tilde{b}_2^1 = (17, 20, 22, 25)$, $\tilde{b}_3^1 = (12, 15, 18, 21)$, $\tilde{a}_1^2 = (32, 34, 37, 39)$, $\tilde{a}_2^2 = (25, 28, 30, 33)$, $\tilde{b}_1^2 = (20, 23, 25, 28)$, $\tilde{b}_2^2 = (16, 18, 19, 22)$, $\tilde{b}_3^2 = (15, 17, 19, 21)$, $\tilde{e}_1 = (46, 49, 51, 53)$, $\tilde{e}_2 = (51, 53, 56, 59)$.

Now to get the corresponding defuzzified values $a_{ic}^p$, $b_{jc}^p$, $e_{kc}$, $i = 1, 2$, $j = 1, 2, 3$, $k = 1, 2$, $p = 1, 2$, we apply the fuzzy programming (113) and the obtained values are $a_{1c}^1 = 23.7$, $a_{2c}^1 = 31.6$, $b_{1c}^1 = 15.8$, $b_{2c}^1 = 19.7$, $b_{3c}^1 = 14.7$, $a_{1c}^2 = 33.8$, $a_{2c}^2 = 27.7$, $b_{1c}^2 = 22.7$, $b_{2c}^2 = 17.8$, $b_{3c}^2 = 16.8$, $e_{1c} = 51.2$, $e_{2c} = 56.3$, with $\lambda = 0.9$.

**Results Using Minimum of Fuzzy Number (Method-1):** To solve the above considered problem we convert the problem as in (119) and take $\lambda_1 = \lambda_2 = 0.4$, $\lambda_3 = 0.2$ (as we concentrate more in maximizing $A_L(Z_t)$ and minimizing $C(Z_t)$ than in minimizing $A_R(Z_t)$) and M=500.

**Table 7.** Penalties/costs $c_{ijk}^{\tilde{2}1}$

| $i \backslash j$ | 1 | 2 | 3 | 1 | 2 | 3 |
|---|---|---|---|---|---|---|
| 1 | (4,5,7,8) | (3,5,6,8) | (7,9,10,12) | (6,7,8,9) | (4,6,7,9) | (5,7,9,11) |
| 2 | (6,8,9,11) | (5,6,7,8) | (6,7,9,10) | (4,6,8,10) | (7,9,11,13) | (9,10,11,12) |
| $k$ | | 1 | | | 2 | |

**Table 8.** Penalties/costs $c_{ijk}^{\tilde{2}2}$

| $i \backslash j$ | 1 | 2 | 3 | 1 | 2 | 3 |
|---|---|---|---|---|---|---|
| 1 | (5,7,9,10) | (4,6,7,9) | (9,11,12,13) | (7,8,9,10) | (4,5,7,8) | (8,10,11,12) |
| 2 | (10,11,13,14) | (6,7,8,9) | (7,9,11,12) | (6,8,10,12) | (5,7,9,11) | (9,10,12,14) |
| $k$ | | 1 | | | 2 | |

Applying fuzzy linear programming (i.e. fuzzy programming with linear membership function) (cf. Sect. 2.9) we get the following results
$L_1 = \min \bar{Z}_1 = 601.9$ $(A_L(Z_1) = 186, \ C(Z_1) = 1095.75, \ A_R(Z_1) = 190)$,
$U_1 = 656.73$ $(A_L(Z_1) = 196.05, \ C(Z_1) = 1231.45, \ A_R(Z_1) = 212.85)$,
$L_2 = \min \bar{Z}_2 = 483.2$ $(A_L(Z_2) = 190.9, \ C(Z_1) = 811.65, \ A_R(Z_1) = 174.5)$,
$U_2 = 541.71$ $(A_L(Z_2) = 166.15, \ C(Z_2) = 941.55, \ A_R(Z_2) = 157.75)$,
and the optimal compromise solution is
$x_{111}^1 = 9, \ x_{221}^1 = 19.7, \ x_{132}^1 = 14.7, \ x_{212}^1 = 6.8, \ x_{111}^2 = 18.38394, \ x_{231}^2 = 4.116056$,
$x_{122}^2 = 2.732113, \ x_{132}^2 = 12.68394, \ x_{212}^2 = 4.316056, \ x_{222}^2 = 15.06789, \ \lambda = 0.753$.
$\bar{Z}_1 = 615.4325$, in which $A_L(Z_1) = 182.35, \ C(Z_1) = 1133.39, \ A_R(Z_1) = 175.0821$, so that the core of the optimum value of objective function $Z_1$ is $[1039.232, 1227.548]$ and $1133.39$ is the center of the core.
$\bar{Z}_2 = 497.6407$, in which $A_L(Z_2) = 183.092, \ C(Z_2) = 844.4438, \ A_R(Z_2) = 165.5$, so that the core of the optimum value of objective function $Z_2$ is $[753.1357, 935.7518]$ and $844.4438$ is the center of the core. The optimum values of $Z_1$ and $Z_2$ are shown in Figs. 11 and 12 respectively. Applying global criterion method in $L_2$ norm (cf. Sect. 2.9) the following results are obtained.
$x_{111}^1 = 9, \ x_{221}^1 = 19.7, \ x_{132}^1 = 14.7, \ x_{212}^1 = 6.8, \ x_{111}^2 = 16.4902, \ x_{231}^2 = 6.0097$,
$x_{122}^2 = 6.5195, \ x_{132}^2 = 10.7902, \ x_{212}^2 = 6.2097, \ x_{222}^2 = 11.2804$.
$\bar{Z}_1 = 618.0837$, in which $A_L(Z_1) = 182.35, \ C(Z_1) = 1138.12, \ A_R(Z_1) = 178.8695$, so that the core of the optimum value of objective function $Z_1$ is the interval $[1043.019, 1233.229]$ and $1138.12$ is the center of the core.
$\bar{Z}_2 = 495.3683$, in which $A_L(Z_2) = 182.1451, \ C(Z_2) = 837.8158, \ A_R(Z_2) = 165.5$, so that the core of the optimum value of objective function $Z_2$ is $[745.5609, 930.0706]$ and $837.8158$ is the center of the core.

It is observed from the optimal solutions of the objective functions that the decision makers have more information in hand about the objective function values.

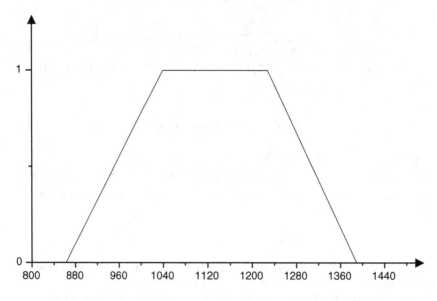

**Fig. 11.** Optimum values of $Z_1$

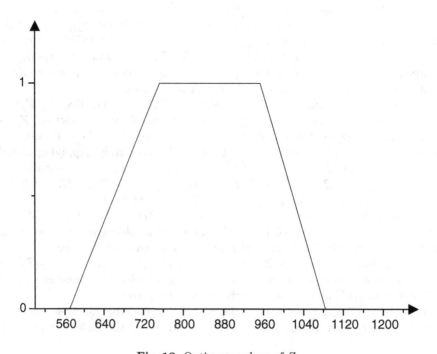

**Fig. 12.** Optimum values of $Z_2$

**Results Using Expected Value of the Objective Function (Method-2):**
To minimize the expected value of the objective functions of the above considered problem, we reconstruct the problem as (120) and transform it to corresponding crisp equivalence form as (121).

Then applying fuzzy linear programming (cf. Sect. 2.9) we get the following result.

$L_1 = \min \bar{Z}_1 = 1092.45$, $U_1 = 1248.85$, $L_2 = \min \bar{Z}_2 = 800.45$, $U_2 = 923.45$ and the optimal compromise solution is

$x^1_{111} = 3.1686$, $x^1_{121} = 5.8313$, $x^1_{221} = 13.8686$, $x^1_{132} = 14.7$, $x^1_{212} = 12.6314$, $x^2_{111} = 22.7$, $x^2_{221} = 5.6314$, $x^2_{122} = 11.1$, $x^2_{222} = 1.0686$, $x^2_{232} = 16.8$, $\lambda = 0.6989$, $\bar{Z}_1 = 1139.536$, $\bar{Z}_2 = 837.4808$.

Applying global criterion method in $L_2$ (cf. Sect. 2.9) norm we get the following result.

$\bar{Z}_1{}^{\min} = 1092.45$ and $\bar{Z}_2{}^{\min} = 800.45$ and the compromise optimum solution is

$x^1_{111} = 1.5624$, $x^1_{121} = 7.4375$, $x^1_{221} = 12.2625$, $x^1_{132} = 14.7$, $x^1_{212} = 14.2375$, $x^2_{111} = 22.7$, $x^2_{221} = 6.7$, $x^2_{231} = 0.5375$, $x^2_{122} = 11.1$, $x^2_{232} = 16.2625$, $\bar{Z}_1 = 1144.894$ and $\bar{Z}_2 = 832.1250$.

Using the crisp equivalence form (123) of the expected value model (122) we can not find any feasible solution for this numerical example. However for any other example with suitable data set, this method can gives feasible solution.

**Overview of the Results by Two Methods:** We note that the optimal expected values of the objective functions $Z_1(= 1139.536)$ and $Z_2(= 837.4808)$ as obtained using expected value (method-2) and fuzzy programming technique, lie within the core $[1039.232, 1227.548]$ and $[753.1357, 935.7518]$ of $Z_1$ and $Z_2$ respectively as obtained using method-1 (method based on minimum of fuzzy number). Also the optimal expected values $Z_1$ and $Z_2$ that obtained by method-2 are close to the center of core 1133.39 and 844.4438 of $Z_1$ and $Z_2$ respectively that are obtained by method-1.

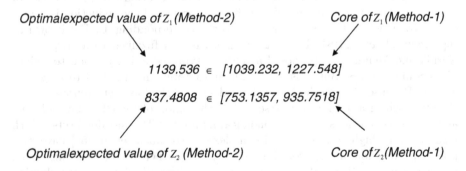

*Optimalexpected value of $z_1$ (Method-2)*          *Core of $z_1$ (Method-1)*

1139.536 ∈ [1039.232, 1227.548]

837.4808 ∈ [753.1357, 935.7518]

*Optimalexpected value of $z_2$ (Method-2)*          *Core of $z_2$ (Method-1)*

**Fig. 13.** Discussion of results obtained by method-1 and method-2.

This discussion is given pictorially in Fig. 13. This is also true for the results obtained by using global criterion method.

## 3.5    Overall Conclusion

In Model 3.1, a MOSTP with fuzzy penalties, resources, demands, conveyance capacities and budget constraints has been formulated. Budget constraints with fuzzy budget amounts are imposed in the problem. The presented problem is solved using chance-constrained programming with credibility measure.

For the first time, a multi-objective multi-item solid transportation problem (Model 3.2) with fuzzy penalties, sources, demands and conveyance capacities is formulated and solved. For defuzzification, two methods (cf. Sects. 3.2 and 3.4 (Method-1), available in the literature and not applied so far in STP have been successfully applied in MOMISTP. Multi-objective problems are solved by two methods and results are compared.

The presented models can be extended to include space constraints, price discount on the basis of amount of transported units, etc. The methods, used for solution here are quite general in nature and these can be applied to other similar uncertain/impricise models in other areas such as inventory control, ecology, sustainable farm management, etc.

## 4    Some Transportation Models with Type-2 Fuzzy Parameters

In many real world problems, due to lack of input information, noises in data, linguistic information, multiple sources of the collected data, bad statistical analysis etc., it is sometimes difficult to determine exact membership grades to represent an uncertain parameter by ordinary fuzzy set and as a result type-2 fuzzy set (T2 FS) appear. Due to fuzziness in membership function, the computational complexity is very high to deal with T2 FS. For the high computational complexity of general T2 FS, till now the most widely used T2 FS is interval T2 FS (IT2 FS), which is a special case of general T2 FS.

In case of a T2 FS, generally complete defuzzification process consists of two parts - type reduction and defuzzification proper. Type reduction is a procedure by which a T2 FS is converted to the corresponding T1 FS, known as type reduced set (TRS). The TRS is then easily defuzzified to a crisp value. Karnik and Mendel [63] proposed a centroid type reduction method to reduce IT2 FS into T1 FS. But it is very difficult to apply this method to a general T2 FS. Because this method was derived from embedded sets representation theory in which a T2 FS is represented as the union of its T2 embedded sets, and this union consists of an astronomical number of T2 embedded sets, which results a very high computational complexity. Greenfield et al. [51] have proposed an extension of this procedure to discretised generalized type-2 fuzzy sets. Other researchers (Liu, [91]; Wu and Tan, [147]) have developed type reduction strategies for continuous generalized T2 FS. Coupland and John [31] proposed a geometric defuzzification method for T2 FSs by converting a T2 FS into a geometric T2 FS. In terms of possibility theory, Liu and Liu [99] defined a type-2 fuzzy variable as a map from a fuzzy possibility space to the set of real numbers.

Qin et al. [127] introduced three kinds of reduction methods called optimistic CV, pessimistic CV and CV reduction methods for type-2 fuzzy variables (T2 FVs) based on CVs (critical values) of regular fuzzy variables.

At present, type-2 fuzzy set theories are being developed and applied in various fields such as group decision making system (Chen and Lee [23,24], Chen et al. [21], Chen et al. [26]), neural network (Aliev et al. [5]), Portfolio selection problem (Hasuike and Ishi [54]), Pattern recognition (Mitchell [112]), data envelopment analysis (Qin et al. [127]), Ad hoc networks (Yuste et al. [155]), etc. Figueroa-Garca and Hernndez [43] first considered a transportation problem with interval type-2 fuzzy demands and supplies. At the best of our knowledge, we are the first to consider the transportation problems with type-2 fuzzy parameters. Thus this is an emerging area and is yet to be developed. In this chapter, we have proposed a defuzzification method of type-2 fuzzy variables. We have also introduced an interval approximation method of continuous type-2 fuzzy variables. We have formulated and solved two fixed charge transportation problems and a multi-item solid transportation problem with type-2 fuzzy parameters.

## 4.1   Related Theorem

The following theorem approaches to find crisp equivalent forms of constraints involving type-2 triangular fuzzy variables. This theorem is established using generalized credibility measure for the reduced fuzzy variable from type-2 triangular fuzzy variable by CV (critical value) reduction method.

**Theorem 4.1** (Qin et al. [127]). Let $\xi_i$ be the reduction of the type-2 triangular fuzzy variable $\tilde{\xi}_i = (r_1^i, r_2^i, r_3^i; \theta_{l,i}, \theta_{r,i})$ obtained by the CV reduction method for $i = 1, 2, ..., n$. Suppose $\xi_1, \xi_2, ..., \xi_n$ are mutually independent, and $k_i \geq 0$ for $i = 1, 2, ..., n$.

(i) Given the generalized credibility level $\alpha \in (0, 0.5]$, if $\alpha \in (0, 0.25]$, then $\tilde{C}r\{\sum_{i=1}^n k_i\xi_i \leq t\} \geq \alpha$ is equivalent to

$$\sum_{i=1}^n \frac{(1 - 2\alpha + (1 - 4\alpha)\theta_{r,i})k_i r_1^i + 2\alpha k_i r_2^i}{1 + (1 - 4\alpha)\theta_{r,i}} \leq t, \qquad (124)$$

and if $\alpha \in (0.25, 0.5]$, then $\tilde{C}r\{\sum_{i=1}^n k_i\xi_i \leq t\} \geq \alpha$ is equivalent to

$$\sum_{i=1}^n \frac{(1 - 2\alpha)k_i r_1^i + (2\alpha + (4\alpha - 1)\theta_{l,i})k_i r_2^i}{1 + (4\alpha - 1)\theta_{l,i}} \leq t. \qquad (125)$$

(ii) Given the generalized credibility level $\alpha \in (0.5, 1]$, if $\alpha \in (0.5, 0.75]$, then $\tilde{C}r\{\sum_{i=1}^n k_i\xi_i \leq t\} \geq \alpha$ is equivalent to

$$\sum_{i=1}^n \frac{(2\alpha - 1)k_i r_3^i + (2(1 - \alpha) + (3 - 4\alpha)\theta_{l,i})k_i r_2^i}{1 + (3 - 4\alpha)\theta_{l,i}} \leq t, \qquad (126)$$

and if $\alpha \in (0.75, 1]$, then $\tilde{Cr}\{\sum_{i=1}^{n} k_i \xi_i \leq t\} \geq \alpha$ is equivalent to

$$\sum_{i=1}^{n} \frac{(2\alpha - 1 + (4\alpha - 3)\theta_{r,i})k_i r_3^i + 2(1 - \alpha)k_i r_2^i}{1 + (4\alpha - 3)\theta_{r,i}} \leq t. \tag{127}$$

**Corollary 4.1:** From the above theorem, equivalent expressions of $\tilde{Cr}\{\sum_{i=1}^{n} k_i \xi_i \geq t\} \geq \alpha$ are easily obtained, since

$$\tilde{Cr}\{\sum_{i=1}^{n} k_i \xi_i \geq t\} \geq \alpha \Rightarrow \tilde{Cr}\{\sum_{i=1}^{n} - k_i \xi_i \leq -t\} \geq \alpha$$

$$\Rightarrow \tilde{Cr}\{\sum_{i=1}^{n} k_i \xi_i' \leq t'\} \geq \alpha,$$

where $\xi_i' = -\xi_i$ is the reduction of $-\tilde{\xi}_i = (-r_3^i, -r_2^i, -r_1^i; \theta_{r,i}, \theta_{l,i})$ and $t' = -t$.

So from (i) of the above theorem, given the generalized credibility level $\alpha \in (0, 0.5]$, if $\alpha \in (0, 0.25]$, then $\tilde{Cr}\{\sum_{i=1}^{n} k_i \xi_i \geq t\} \geq \alpha$, i.e. $\tilde{Cr}\{\sum_{i=1}^{n} k_i \xi_i' \leq t'\} \geq \alpha$ is equivalent to

$$\sum_{i=1}^{n} \frac{(1 - 2\alpha + (1 - 4\alpha)\theta_{l,i})k_i(-r_3^i) + 2\alpha k_i(-r_2^i)}{1 + (1 - 4\alpha)\theta_{l,i}} \leq t' = -t, \tag{128}$$

which implies

$$\sum_{i=1}^{n} \frac{(1 - 2\alpha + (1 - 4\alpha)\theta_{l,i})k_i r_3^i + 2\alpha k_i r_2^i}{1 + (1 - 4\alpha)\theta_{l,i}} \geq t, \tag{129}$$

and if $\alpha \in (0.25, 0.5]$, then $\tilde{Cr}\{\sum_{i=1}^{n} k_i \xi_i \geq t\} \geq \alpha$ is equivalent to

$$\sum_{i=1}^{n} \frac{(1 - 2\alpha)k_i(-r_3^i) + (2\alpha + (4\alpha - 1)\theta_{r,i})k_i(-r_2^i)}{1 + (4\alpha - 1)\theta_{r,i}} \leq -t \tag{130}$$

which implies

$$\sum_{i=1}^{n} \frac{(1 - 2\alpha)k_i r_3^i + (2\alpha + (4\alpha - 1)\theta_{r,i})k_i r_2^i}{1 + (4\alpha - 1)\theta_{r,i}} \geq t. \tag{131}$$

The equivalent expressions for other values of $\alpha$ are similarly obtained.

## 4.2   Theoretical Developments

**Defuzzification of Type-2 Fuzzy Variables** ([75]): Here we have introduced a defuzzification process of type-2 fuzzy variables. This method consists of two parts. First CV-based reduction method (Sect. 2.3) is applied to transform the type-2 fuzzy variables into corresponding type-1 fuzzy variables. Then, to get corresponding defuzzified (crisp) values, centroid method described in Sect. 2.2 is applied to these reduced type-1 fuzzy variables. For continuous case the formula $\int_{-\infty}^{\infty} x\mu_{\tilde{A}}(x)dx / \int_{-\infty}^{\infty} \mu_{\tilde{A}}(x)dx$ is used while $\sum_x x\mu_{\tilde{A}}(x) / \sum_x \mu_{\tilde{A}}(x)$ ia applied for discrete case.

The entire defuzzification process is shown in Fig. 14 and illustrated with the following two examples.

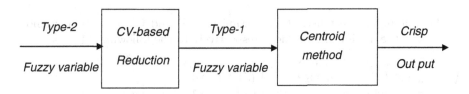

**Fig. 14.** Defuzzification of a type-2 fuzzy variable

**Example 4.1.** Let us consider $\tilde{A} = \{(x, \tilde{\mu}_{\tilde{A}}(x)) : x \in X\}$ where $X = \{4, 5, 6\}$ and the primary memberships (possibilities) of the points of X are, respectively, $J_4 = \{0.3, 0.4, 0.6\}$, $J_5 = \{0.6, 0.8, 0.9\}$, and $J_6 = \{0.5, 0.6, 0.7, 0.8\}$. The secondary possibility of the point 4 is

$$\tilde{\mu}_{\tilde{A}}(4) = \mu_{\tilde{A}}(4, u) = (0.6/0.3) + (1.0/0.4) + (0.7/0.6) \sim \begin{pmatrix} 0.3\ 0.4\ 0.6 \\ 0.6\ 1.0\ 0.7 \end{pmatrix},$$

which represents a regular fuzzy variable (RFV). Similarly

$$\tilde{\mu}_{\tilde{A}}(5) = \mu_{\tilde{A}}(5, u) = (0.7/0.6) + (1.0/0.8) + (0.8/0.9) \sim \begin{pmatrix} 0.6\ 0.8\ 0.9 \\ 0.7\ 1.0\ 0.8 \end{pmatrix},$$

$$\tilde{\mu}_{\tilde{A}}(6) = \mu_{\tilde{A}}(6, u) = (0.3/0.5) + (0.4/0.6) + (1.0/0.7) + (0.5/0.8) \sim \begin{pmatrix} 0.5\ 0.6\ 0.7\ 0.8 \\ 0.3\ 0.4\ 1.0\ 0.5 \end{pmatrix}.$$

So discrete type-2 fuzzy variable $\tilde{A}$ is given by

$$\tilde{A} = \begin{cases} 4, \text{ with membership } \tilde{\mu}_{\tilde{A}}(4); \\ 5, \text{ with membership } \tilde{\mu}_{\tilde{A}}(5); \\ 6, \text{ with membership } \tilde{\mu}_{\tilde{A}}(6). \end{cases} \tag{132}$$

For this T2 FV $\tilde{A}$, $\tilde{\mu}_{\tilde{A}}(4)$, $\tilde{\mu}_{\tilde{A}}(5)$ and $\tilde{\mu}_{\tilde{A}}(6)$ are discrete RFVs. So the CVs of these RFVs can be obtained by using the formula (47), (48) and (49) (cf. Sect. 2.3) as illustrated in Example 2.7, e.g., from (47), we have $CV^*[\tilde{\mu}_{\tilde{A}}(4)] = \sup_{\alpha \in [0,1]} [\alpha \wedge Pos\{\tilde{\mu}_{\tilde{A}}(4) \geq \alpha\}]$, where

$$Pos\{\tilde{\mu}_{\tilde{A}}(4) \geq \alpha\} = \begin{cases} 1, & \text{if } \alpha \leq 0.4; \\ 0.7, & \text{if } 0.4 < \alpha \leq 0.6; \\ 0, & \text{if } 0.6 < \alpha \leq 1. \end{cases} \tag{133}$$

so that

$$CV^*[\tilde{\mu}_{\tilde{A}}(4)] = \sup_{\alpha \in [0,0.4]} [\alpha \wedge 1] \ \vee \ \sup_{\alpha \in (0.4,0.6]} [\alpha \wedge 0.7] \ \vee \ \sup_{\alpha \in (0.6,1]} [\alpha \wedge 0]$$

$$= 0.4 \ \vee \ 0.6 \ \vee \ 0 = 0.6. \tag{134}$$

In this way, from (47), (48) and (49) (cf. Sect. 2.3) we obtain $\text{CV}^*[\tilde{\mu}_{\tilde{A}}(4)] = 0.6$, $\text{CV}^*[\tilde{\mu}_{\tilde{A}}(5)] = 0.8$, $\text{CV}^*[\tilde{\mu}_{\tilde{A}}(6)] = 0.6$,

$$\text{CV}_*[\tilde{\mu}_{\tilde{A}}(4)] = 0.4,\ \text{CV}_*[\tilde{\mu}_{\tilde{A}}(5)] = 0.6,\ \text{CV}_*[\tilde{\mu}_{\tilde{A}}(6)] = 0.6,$$

$$\text{CV}[\tilde{\mu}_{\tilde{A}}(4)] = 0.4,\ \text{CV}[\tilde{\mu}_{\tilde{A}}(5)] = 0.65,\ \text{CV}[\tilde{\mu}_{\tilde{A}}(6)] = 0.6.$$

Then applying optimistic CV, pessimistic CV and CV reduction methods (Sect. 2.3), the T2 FV $\tilde{A}$ is reduced respectively to the following T1 FVs

$$\begin{pmatrix} 4 & 5 & 6 \\ 0.6 & 0.8 & 0.6 \end{pmatrix},\ \begin{pmatrix} 4 & 5 & 6 \\ 0.4 & 0.6 & 0.6 \end{pmatrix}\ \text{and}\ \begin{pmatrix} 4 & 5 & 6 \\ 0.4 & 0.65 & 0.6 \end{pmatrix}.$$

Then applying centroid method to these T1 FVs we get the corresponding complete defuzzified (crisp) values 5, 5.125 and 5.121 respectively. For practical use, which of the defuzzified values should be considered, it is up to the decision maker. However we recommend to take the defuzzified value (e.g., 5.121 for this example) obtained by applying the centroid method to the reduced T1 FV as derived using CV reduction method. This is because optimistic CV and pessimistic CV reduction methods are developed using possibility and necessity measure respectively, while CV reduction method uses credibility measure which is the average of possibility and necessity measures.

**Example 4.2.** Consider the T2 FV $\tilde{\xi}$ presented in Example 2.4 in Sect. 2.3, Chap. 2. Also in Example 2.4 (continued), applying optimistic CV, pessimistic CV and CV reduction method to $\tilde{\xi}$ we already obtain the corresponding reduced type-1 fuzzy variables as $\begin{pmatrix} 5 & 6 & 7 \\ \frac{1}{2} & \frac{2}{3} & \frac{7}{12} \end{pmatrix}$, $\begin{pmatrix} 5 & 6 & 7 \\ \frac{1}{3} & \frac{1}{2} & \frac{1}{4} \end{pmatrix}$ and $\begin{pmatrix} 5 & 6 & 7 \\ \frac{3}{7} & \frac{4}{7} & \frac{1}{2} \end{pmatrix}$ respectively.

Then applying centroid method to these T1 FVs we get the corresponding complete defuzzified (crisp) values 6.0476, 5.923 and 6.0476 respectively.
Comparison with geometric defuzzification method (Coupland and John [31]):
Applying the geometric defuzzification method (cf. Sect. 2.3) to Example 4.1 we find the defuzzified value of the discrete type-2 fuzzy variable $\tilde{A}$ as 5.158 as compared to 5.121, obtained by the above proposed method.

Since the domains of the secondary possibilities (memberships) of all the points of the type-2 fuzzy variable $\tilde{\xi}$ of Example 4.2 are continuous over [0,1], so to apply geometric defuzzification method we have to dicretize the continuous domains. We discretize the continuous domains of the secondary possibilities of the points of $\tilde{\xi}$ with equidistant 0.05 and applying geometric defuzzification method we obtain defuzzified value 6.1403 of $\tilde{\xi}$, compared to earlier result 6.0476.

**Nearest Interval Approximation of Continuous Type-2 Fuzzy Variables** ([76]). Here we have proposed a method of approximation of continuous type-2 fuzzy variable by crisp interval. For this purpose we first find the CV-based reductions of the type-2 fuzzy variable. Then we derive the corresponding $\alpha$-cuts of these CV-based reductions. Finally applying interval approximation method (Grzegorzewski [52]) to the $\alpha$-cuts we find approximate crisp intervals. The entire method is shown in the Fig. 15.

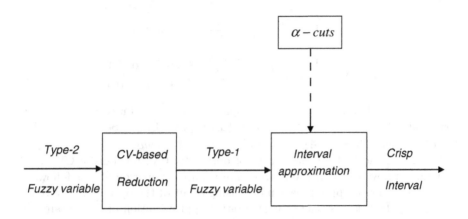

**Fig. 15.** Nearest interval approximation of continuous T2 fuzzy variable.

**Example 4.3.** Now we illustrate the above method with type-2 triangular fuzzy variable. Let $\tilde{\xi}$ be a type-2 triangular fuzzy variable defined as $\tilde{\xi} = (r_1, r_2, r_3; \theta_l, \theta_r)$. Then from Theorem 2.4 (cf. Sect. 2.3) we have the optimistic CV reduction, pessimistic CV reduction and CV reduction of $\tilde{\xi}$ as $\xi_1$, $\xi_2$ and $\xi_3$ respectively with the possibility distributions given by (56), (57) and (58) (cf. Sect. 2.3). Now using the definition of $\alpha$-cut of a fuzzy number we find $\alpha$-cuts of the reductions of $\tilde{\xi}$.

$\alpha$-cut of the optimistic CV reduction $\xi_1$ of $\tilde{\xi}$: Applying the definition of $\alpha$-cut of a fuzzy variable we find the $\alpha$-cut of the reduction $\xi_1$ as $[\xi_{1L}(\alpha), \xi_{1R}(\alpha)]$, where,

$$\xi_{1L}(\alpha) = \begin{cases} \frac{(1+\theta_r)r_1 + (r_2 - r_1 - \theta_r r_1)\alpha}{(1+\theta_r) - \theta_r \alpha}, & 0 \le \alpha \le 0.5; \\ \frac{(r_1 - \theta_r r_2) + (r_2 - r_1 + \theta_r r_2)\alpha}{(1-\theta_r) + \theta_r \alpha}, & 0.5 < \alpha \le 1. \end{cases} \tag{135}$$

$$\xi_{1R}(\alpha) = \begin{cases} \frac{(r_3 - \theta_r r_2) - (r_3 - r_2 - \theta_r r_2)\alpha}{(1-\theta_r) + \theta_r \alpha}, & 0.5 \le \alpha \le 1; \\ \frac{(1+\theta_r)r_3 - (r_3 - r_2 + \theta_r r_3)\alpha}{(1+\theta_r) - \theta_r \alpha}, & 0 \le \alpha < 0.5. \end{cases} \tag{136}$$

$\alpha$-cut of the pessimistic CV reduction $\xi_2$ of $\tilde{\xi}$: $\alpha$-cut of the reduction $\xi_2$ is obtained as $[\xi_{2L}(\alpha), \xi_{2R}(\alpha)]$, where,

$$\xi_{2L}(\alpha) = \begin{cases} \frac{r_1 + (r_2 - r_1 - \theta_l r_1)\alpha}{1 - \theta_l \alpha}, & 0 \le \alpha \le 0.5; \\ \frac{r_1 + (r_2 - r_1 + \theta_l r_2)\alpha}{1 + \theta_l \alpha}, & 0.5 < \alpha \le 1. \end{cases} \tag{137}$$

$$\xi_{2R}(\alpha) = \begin{cases} \frac{r_3 - (r_3 - r_2 - \theta_l r_2)\alpha}{1 + \theta_l \alpha}, & 0.5 \le \alpha < 1; \\ \frac{r_3 - (r_3 - r_2 + \theta_l r_3)\alpha}{1 - \theta_l \alpha}, & 0 \le \alpha < 0.5. \end{cases} \tag{138}$$

$\alpha$-cut of the CV reduction $\xi_3$ of $\tilde{\xi}$: $\alpha$-cut of the reduction $\xi_3$ is obtained as $[\xi_{3L}(\alpha), \xi_{3R}(\alpha)]$, where,

$$\xi_{3L}(\alpha) = \begin{cases} \frac{(1+\theta_r)r_1+(r_2-r_1-2\theta_r r_1)\alpha}{(1+\theta_r)-2\theta_r\alpha}, & 0 \le \alpha \le 0.5; \\ \frac{(r_1-\theta_l r_2)+(r_2-r_1+2\theta_l r_2)\alpha}{(1-\theta_l)+2\theta_l\alpha}, & 0.5 < \alpha \le 1. \end{cases} \qquad (139)$$

$$\xi_{3R}(\alpha) = \begin{cases} \frac{(r_3-\theta_l r_2)-(r_3-r_2-2\theta_l r_2)\alpha}{(1-\theta_l)+2\theta_l\alpha}, & 0.5 \le \alpha \le 1; \\ \frac{(1+\theta_r)r_3-(r_3-r_2+2\theta_r r_3)\alpha}{(1+\theta_r)-2\theta_r\alpha}, & 0 \le \alpha < 0.5. \end{cases} \qquad (140)$$

Now we know that nearest interval approximation of a fuzzy number (Grzegorzewski [52]) $\tilde{A}$ with $\alpha$-cut $[A_L(\alpha), A_R(\alpha)]$ is given by (cf. Sect. 2.2) $C_d(\tilde{A}) = [C_L, C_R]$, where $C_L = \int_0^1 A_L(\alpha)d\alpha$ and $C_R = \int_0^1 A_R(\alpha)d\alpha$.

Using this method for the $\alpha$-cuts of optimistic CV, pessimistic CV or CV reduction of $\tilde{\xi}$ we can find the nearest interval approximation of $\tilde{\xi}$ as follows.

Nearest interval approximation of $\tilde{\xi}$ using $\alpha$-cut of the optimistic CV reduction $\xi_1$ of $\tilde{\xi}$: In this case the nearest interval approximation of $\tilde{\xi}$ is obtained as $[C_L, C_R]$ where,

$$\begin{aligned} C_L &= \int_0^1 \xi_{1L}(\alpha)d\alpha \\ &= \int_0^{0.5} \frac{(1+\theta_r)r_1 + (r_2 - r_1 - \theta_r r_1)\alpha}{(1+\theta_r) - \theta_r\alpha}d\alpha \\ &\quad + \int_{0.5}^1 \frac{(r_1 - \theta_r r_2) + (r_2 - r_1 + \theta_r r_2)\alpha}{(1-\theta_r) + \theta_r\alpha}d\alpha = C_{L1} + C_{L2}, \quad (141) \end{aligned}$$

$$C_{L1} = \frac{(1+\theta_r)r_1}{\theta_r}\ln\left(\frac{1+\theta_r}{1+0.5\theta_r}\right) - \frac{r_2 - r_1 - \theta_r r_1}{\theta_r^2}\left[0.5\theta_r - (1+\theta_r)\ln\left(\frac{1+\theta_r}{1+0.5\theta_r}\right)\right],$$

$$C_{L2} = -\frac{r_1 - \theta_r r_2}{\theta_r}\ln(1-0.5\theta_r) + \frac{r_2 - r_1 + \theta_r r_2}{\theta_r^2}\left[0.5\theta_r + (1-\theta_r)\ln(1-0.5\theta_r)\right].$$

$$\begin{aligned} C_R &= \int_0^1 \xi_{1R}(\alpha)d\alpha \\ &= \int_0^{0.5} \frac{(1+\theta_r)r_3 - (r_3 - r_2 + \theta_r r_3)\alpha}{(1+\theta_r) - \theta_r\alpha}d\alpha \\ &\quad + \int_{0.5}^1 \frac{(r_3 - \theta_r r_2) - (r_3 - r_2 - \theta_r r_2)\alpha}{(1-\theta_r) + \theta_r\alpha}d\alpha \\ &= C_{R1} + C_{R2}, \quad (142) \end{aligned}$$

$$C_{R1} = \frac{(1+\theta_r)r_3}{\theta_r}\ln\left(\frac{1+\theta_r}{1+0.5\theta_r}\right) + \frac{r_3 - r_2 + \theta_r r_3}{\theta_r^2}\left[0.5\theta_r - (1+\theta_r)\ln\left(\frac{1+\theta_r}{1+0.5\theta_r}\right)\right],$$

$$C_{R2} = -\frac{r_3 - \theta_r r_2}{\theta_r}\ln(1-0.5\theta_r) - \frac{r_3 - r_2 - \theta_r r_2}{\theta_r^2}\left[0.5\theta_r + (1-\theta_r)\ln(1-0.5\theta_r)\right].$$

We call this interval as optimistic interval approximation of $\tilde{\xi}$.

Nearest interval approximation of $\tilde{\xi}$ using $\alpha$-cut of the pessimistic CV reduction $\xi_2$ of $\tilde{\xi}$: In this case the nearest interval approximation of $\tilde{\xi}$ is obtained as $[C_L, C_R]$ where,

$$
\begin{aligned}
C_L &= \int_0^1 \xi_{2L}(\alpha)d\alpha \\
&= \int_0^{0.5} \frac{r_1 + (r_2 - r_1 - \theta_l r_1)\alpha}{1 - \theta_l \alpha} d\alpha + \int_{0.5}^1 \frac{r_1 + (r_2 - r_1 + \theta_l r_2)\alpha}{1 + \theta_l \alpha} d\alpha \\
&= C_{L1} + C_{L2},
\end{aligned}
\tag{143}
$$

$$
C_{L1} = -\frac{r_1}{\theta_l} \ln(1 - 0.5\theta_l) - \frac{r_2 - r_1 - \theta_l r_1}{\theta_l^2} \left[0.5\theta_l + \ln(1 - 0.5\theta_l)\right],
$$

$$
C_{L2} = \frac{r_1}{\theta_l} \ln\left(\frac{1 + \theta_l}{1 + 0.5\theta_l}\right) + \frac{r_2 - r_1 + \theta_l r_2}{\theta_l^2} \left[0.5\theta_l - \ln\left(\frac{1 + \theta_l}{1 + 0.5\theta_l}\right)\right].
$$

$$
\begin{aligned}
C_R &= \int_0^1 \xi_{2R}(\alpha)d\alpha \\
&= \int_0^{0.5} \frac{r_3 - (r_3 - r_2 + \theta_l r_3)\alpha}{1 - \theta_l \alpha} d\alpha + \int_{0.5}^1 \frac{r_3 - (r_3 - r_2 - \theta_l r_2)\alpha}{1 + \theta_l \alpha} d\alpha \\
&= C_{R1} + C_{R2},
\end{aligned}
\tag{144}
$$

$$
C_{R1} = -\frac{r_3}{\theta_l} \ln(1 - 0.5\theta_l) + \frac{r_3 - r_2 + \theta_l r_3}{\theta_l^2} \left[0.5\theta_l + \ln(1 - 0.5\theta_l)\right],
$$

$$
C_{R2} = \frac{r_3}{\theta_l} \ln\left(\frac{1 + \theta_l}{1 + 0.5\theta_l}\right) - \frac{r_3 - (r_3 - r_2 - \theta_l r_2)}{\theta_l^2} \left[0.5\theta_l - \ln\left(\frac{1 + \theta_l}{1 + 0.5\theta_l}\right)\right].
$$

We call this interval as pessimistic interval approximation of $\tilde{\xi}$.

Nearest interval approximation of $\tilde{\xi}$ using $\alpha$-cut of the CV reduction $\xi_3$ of $\tilde{\xi}$: In this case the nearest interval approximation of $\tilde{\xi}$ is obtained as $[C_L, C_R]$ where,

$$
\begin{aligned}
C_L &= \int_0^1 \xi_{3L}(\alpha)d\alpha \\
&= \int_0^{0.5} \frac{(1 + \theta_r)r_1 + (r_2 - r_1 - 2\theta_r r_1)\alpha}{(1 + \theta_r) - 2\theta_r \alpha} d\alpha \\
&\quad + \int_{0.5}^1 \frac{(r_1 - \theta_l r_2) + (r_2 - r_1 + 2\theta_l r_2)\alpha}{(1 - \theta_l) + 2\theta_l \alpha} \\
&= C_{L1} + C_{L2},
\end{aligned}
\tag{145}
$$

$$
C_{L1} = \frac{(1 + \theta_r)r_1}{2\theta_r} \ln(1 + \theta_r) - \frac{r_2 - r_1 - 2\theta_r r_1}{4\theta_r^2} \left[\theta_r - (1 + \theta_r)\ln(1 + \theta_r)\right],
$$

$$C_{L2} = \frac{r_1 - \theta_l r_2}{2\theta_l} \ln(1 + \theta_l) + \frac{r_2 - r_1 + 2\theta_l r_2}{4\theta_l^2} \left[\theta_l - (1 - \theta_l)\ln(1 + \theta_l)\right].$$

$$\begin{aligned}
C_R &= \int_0^1 \xi_{3R}(\alpha)d\alpha \\
&= \int_0^{0.5} \frac{(1 + \theta_r)r_3 - (r_3 - r_2 + 2\theta_r r_3)\alpha}{(1 + \theta_r) - 2\theta_r \alpha} d\alpha \\
&\quad + \int_{0.5}^1 \frac{(r_3 - \theta_l r_2) - (r_3 - r_2 - 2\theta_l r_2)\alpha}{(1 - \theta_l) + 2\theta_l \alpha} d\alpha \\
&= C_{R1} + C_{R2},
\end{aligned} \tag{146}$$

$$C_{R1} = \frac{(1 + \theta_r)r_3}{2\theta_r} \ln(1 + \theta_r) + \frac{r_3 - r_2 + 2\theta_r r_3}{4\theta_r^2} \left[\theta_r - (1 + \theta_r)\ln(1 + \theta_r)\right],$$

$$C_{R2} = \frac{r_3 - \theta_l r_2}{2\theta_l} \ln(1 + \theta_l) - \frac{r_3 - r_2 - 2\theta_l r_2}{4\theta_l^2} \left[\theta_l - (1 - \theta_l)\ln(1 + \theta_l)\right].$$

We call this interval as credibilistic interval approximation of $\tilde{\tilde{\xi}}$.

For example consider the type-2 triangular fuzzy variable $\tilde{\tilde{\xi}} = (2, 3, 4; 0.5, 0.8)$ whose FOU is depicted in Fig. 6 and its optimistic CV, pessimistic CV and CV reductions are shown in the Fig. 8 (Sect. 2.3). We find nearest interval approximation of $\tilde{\tilde{\xi}}$. From Eqs. (141), (142), (143), (144), (145) and (146), the optimistic, pessimistic and credibilistic interval approximations of $\tilde{\tilde{\xi}}$ are obtained as

$$[2.4086, 3.5913], \quad [2.5567, 3.4432] \quad and \quad [2.4925, 3.5074]$$

respectively. These results are shown in the Fig. 16.

## 4.3    Model 4.1: Fixed Charge Transportation Problem with Type-2 Fuzzy Cost Parameters

Here a fixed charge transportation problem (FCTP) with unit transportation costs and fixed(/additional) costs as type-2 fuzzy variables is formulated.

**Notations:**

(i) $\tilde{\tilde{c}}_{ij}$: The unit transportation cost from $i$-th source to $j$-th destination represented by type-2 fuzzy variable.

(ii) $\tilde{\tilde{d}}_{ij}$: Fixed(/additional) cost associated with route $(i, j)$ represented by type-2 fuzzy variable.

(iii) $x_{ij}$: The decision variable which represents amount of product to be transported from $i$-th origin to $j$-th destination.

(iv) $Z$: The objective function.

(v) $a_i$: The amount of the product available at the $i$-th origin.

(vi) $b_j$: The demand of the product at $j$-th destination.

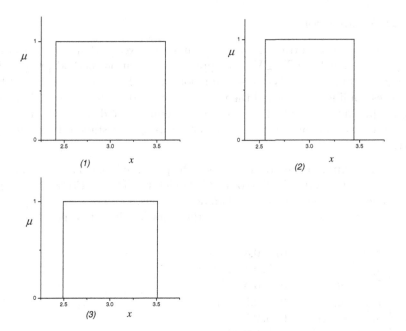

**Fig. 16.** Interval approximation of $\tilde{\xi}$ using (1) optimistic CV, (2) pessimistic CV, (3) CV reductions.

**Mathematical Model:** A FCTP with $m$ sources and $n$ destinations and direct costs and fixed cost parameters as T2 FVs is as follows:

$$Min\ Z = \sum_{i=1}^{m}\sum_{j=1}^{n}(\tilde{c}_{ij}\ x_{ij} + \tilde{d}_{ij}y_{ij}), \tag{147}$$

$$subject\ to\ \sum_{j=1}^{n}x_{ij}\ \leq a_i,\quad i = 1, 2, ..., m, \tag{148}$$

$$\sum_{i=1}^{m}x_{ij}\ \geq b_j,\quad j = 1, 2, ..., n, \tag{149}$$

$$x_{ij} \geq 0,\ y_{ij} = \begin{cases} 1, & \text{if } x_{ij} > 0; \\ 0, & \text{otherwise.} \end{cases}\quad \forall\ i, j, \tag{150}$$

$$\sum_{i=1}^{m}a_i \geq \sum_{j=1}^{n}b_j \tag{151}$$

It is obvious that the fixed charge $\tilde{d}_{ij}$ will be costed for a route $(i, j)$ only if any transportation activity is assigned to that route. So $y_{ij}$ is defined such that if $x_{ij} > 0$ then $y_{ij} = 1$, otherwise it will be 0.

## 4.4　Defuzzifiaction

Since the unit transportation costs $\tilde{\tilde{c}}_{ij}$s and the fixed(/additional) costs $\tilde{\tilde{d}}_{ij}$s in the above model are T2 FVs, we apply defuzzification method presented in Sect. 4.2. We first apply CV-based reduction method to transform the T2 FVs into corresponding T1 FVs and then centroid method to the reduced T1 FVs to get corresponding defuzzified (crisp) values. Taking these defuzzified (crisp) cost values, the problem can be then easily solved using any standard optimization solver.

**Numerical Experiment:** In this section the presented model and methods are illustrated numerically. To illustrate the Model 4.1 ((147)–(151)), we consider an example with three sources and two destinations, i.e., $i = 1, 2, 3$ and $j = 1, 2$.

The unit transportation costs $\tilde{\tilde{c}}_{ij}$ are the following discrete type-2 fuzzy variables.

$$\tilde{\tilde{c}}_{11} = \begin{cases} 2, \text{with} \tilde{\mu}_{c\tilde{1}1}(2) = (0.2, 0.4, 0.6, 0.8); \\ 4, \text{with} \tilde{\mu}_{c\tilde{1}1}(4) = (0.5, 0.7, 0.9); \\ 5, \text{with} \tilde{\mu}_{c\tilde{1}1}(5) = (0.3, 0.5, 0.7). \end{cases}$$

$$\tilde{\tilde{c}}_{12} = \begin{cases} 7, \text{with} \tilde{\mu}_{c\tilde{1}2}(7) = (0.4, 0.6, 0.7); \\ 8, \text{with} \tilde{\mu}_{c\tilde{1}2}(8) = (0.5, 0.7, 0.8); \\ 9, \text{with} \tilde{\mu}_{c\tilde{1}2}(9) = (0.7, 0.9, 1). \end{cases}$$

$$\tilde{\tilde{c}}_{21} = \begin{cases} 4, \text{with} \tilde{\mu}_{c\tilde{2}1}(4) = \begin{pmatrix} 0.3 & 0.5 & 0.7 \\ 0.4 & 1 & 0.7 \end{pmatrix}; \\ 5, \text{with} \tilde{\mu}_{c\tilde{2}1}(5) = \begin{pmatrix} 0.6 & 0.8 & 0.9 \\ 0.5 & 0.9 & 1 \end{pmatrix}; \\ 7, \text{with} \tilde{\mu}_{c\tilde{2}1}(7) = \begin{pmatrix} 0.5 & 0.7 & 0.8 \\ 0.4 & 1 & 0.7 \end{pmatrix}. \end{cases}$$

$$\tilde{\tilde{c}}_{22} = \begin{cases} 6, \text{with} \tilde{\mu}_{c\tilde{2}2}(6) = (0.4, 0.5, 0.7, 0.8); \\ 7, \text{with} \tilde{\mu}_{c\tilde{2}2}(7) = (0.6, 0.8, 0.9); \\ 9, \text{with} \tilde{\mu}_{c\tilde{2}2}(9) = (0.4, 0.6, 0.7). \end{cases}$$

$$\tilde{\tilde{c}}_{31} = \begin{cases} 3, \text{with} \tilde{\mu}_{c\tilde{3}1}(3) = (0.3, 0.4, 0.6); \\ 5, \text{with} \tilde{\mu}_{c\tilde{3}1}(5) = (0.7, 0.9, 1); \\ 6, \text{with} \tilde{\mu}_{c\tilde{3}1}(6) = (0.4, 0.6, 0.7). \end{cases}$$

$$\tilde{\tilde{c}}_{32} = \begin{cases} 8, \;\; \text{with} \tilde{\mu}_{c\tilde{3}2}(8) = (0.3, 0.5, 0.6); \\ 9, \;\; \text{with} \tilde{\mu}_{c\tilde{3}2}(9) = (0.5, 0.7, 0.8, 0.9); \\ 10, \text{with} \tilde{\mu}_{c\tilde{3}2}(10) = (0.5, 0.6, 0.8). \end{cases}$$

The supplies $a_i$ and demands $b_j$ are as follows:
$a_1 = 20$, $a_2 = 14$, $a_3 = 18$, $b_1 = 28$, $b_2 = 21$.
For convenience of computing we suppose that the fixed charge $\tilde{\tilde{d}}_{ij} = 0.5 \tilde{\tilde{c}}_{ij}$.

**Solution Using Proposed Defuzzification Method (cf. Sect. 4.2):** To solve the above problem we first find corresponding defuzzified (crisp) values of the type-2 fuzzy cost parameters $\tilde{\tilde{c}}_{ij}$. For this purpose we first apply CV reduction method to reduce type-2 fuzzy variables $\tilde{\tilde{c}}_{ij}$ to type-1 fuzzy variables, then applying centroid method we get the corresponding crisp values. We denote these crisp values as $c_{ij}^c$ which are obtained as

$c_{11}^c = 3.6956$, $c_{12}^c = 8.1071$, $c_{21}^c = 5.4615$, $c_{22}^c = 7.36$, $c_{31}^c = 4.8523$ and $c_{32}^c = 9.0482$.

Now using these crisp costs values, the optimum solution of the problem is obtained by the standard optimization solver - LINGO and given in Table 9.

**Solution Using Geometric Defuzzification (cf. Sect. 2.3):** Using geometric defuzzification method we obtain the defuzzified values of the type-2 fuzzy cost parameters $\tilde{c}_{ij}$ as follows.

$c_{11}^c = 3.6896$, $c_{12}^c = 8.219$, $c_{21}^c = 5.6355$, $c_{22}^c = 7.5651$, $c_{31}^c = 4.65$ and $c_{32}^c = 9.1932$.

Using these defuzzified cost values, the optimum solution of the problem is obtained and presented in Table 9.

**Table 9.** Optimum results for model-4.1

| Method | Defuzzified cost parameters | Optimum costs Min Z | Optimum transported amounts |
|---|---|---|---|
| Proposed defuzzification method | $c_{11}^c = 3.6956$, $c_{12}^c = 8.1071$, $c_{21}^c = 5.4615$, $c_{22}^c = 7.36$, $c_{31}^c = 4.8523$, $c_{32}^c = 9.0482$ | 283.3245 | $x_{11} = 13$, $x_{12} = 7$ $x_{22} = 14$, $x_{31} = 15$ |
| Geometric defuzzification method | $c_{11}^c = 3.6896$, $c_{12}^c = 8.219$, $c_{21}^c = 5.6355$, $c_{22}^c = 7.5651$, $c_{31}^c = 4.65$, $c_{32}^c = 9.1932$ | 293.2211 | $x_{11} = 13$, $x_{12} = 7$ $x_{22} = 14$, $x_{31} = 15$ |

So from the above two results, we see that the optimum allocations (i.e., values of $x_{ij}s$) as obtained by the two approaches are the same. However the optimum objective value (minimum transportation cost) as obtained using the geometric defuzzification method is something more than that of using proposed defuzzification method.

### 4.5 Model 4.2: Fixed Charge Transportation Problem with Type-2 Fuzzy Costs, Supplies and Demands

**Notations:**

(i) $\tilde{c}_{ij}$: The unit transportation costs from $i$-th source to $j$-th destination represented by type-2 fuzzy variable.

(ii) $\tilde{d}_{ij}$: Fixed(/additional) cost associated with route $(i, j)$ represented by type-2 fuzzy variable.

(iii) $x_{ij}$: The decision variable which represents amount of product to be transported from $i$-th origin to $j$-th destination.

(iv) $Z$: The objective function.

(v) $\tilde{a}_i$: The amount of the product available at the $i$-th origin represented by type-2 fuzzy variable.

(vi) $\tilde{b}_j$: The demand of the product at $j$-th destination represented by type-2 fuzzy variable.

**Mathematical Model:** A FCTP with $m$ sources, $n$ destinations and unit transportation costs, fixed costs, supplies and demands as T2 FVs is formulated as follows:

$$Min \ Z = \sum_{i=1}^{m} \sum_{j=1}^{n} (\tilde{\tilde{c}}_{ij} \ x_{ij} + \tilde{\tilde{d}}_{ij} y_{ij}), \tag{152}$$

$$subject \ to \ \sum_{j=1}^{n} x_{ij} \leq \tilde{\tilde{a}}_i, \quad i = 1, 2, ..., m, \tag{153}$$

$$\sum_{i=1}^{m} x_{ij} \geq \tilde{\tilde{b}}_j, \quad j = 1, 2, ..., n, \tag{154}$$

$$x_{ij} \geq 0, \ y_{ij} = \begin{cases} 1, & \text{if } x_{ij} > 0; \\ 0, & \text{otherwise.} \end{cases} \quad \forall \ i, j \tag{155}$$

**Solution Methodology: Chance-Constrained Programming Using Generalized Credibility:** Suppose that $\tilde{c}'_{ij}$, $\tilde{d}'_{ij}$, $\tilde{a}'_i$ and $\tilde{b}'_j$ are the reduced T1 FVs (may not be normalized) of the T2 FVs $\tilde{\tilde{c}}_{ij}$, $\tilde{\tilde{d}}_{ij}$, $\tilde{\tilde{a}}_i$ and $\tilde{\tilde{b}}_j$ respectively according to CV-based reduction method. Now to solve the above problem we formulate a chance-constrained programming model with these reduced fuzzy parameters. Chance-constrained programming with fuzzy (type-1) parameters was introduced by Liu and Iwamura [92] using possibility measure. Latter it is developed (Liu [90], Yang and Liu [153], Kundu et al. [74]) by using credibility measure. But since the reduced fuzzy parameters $\tilde{c}'_{ij}$, $\tilde{d}'_{ij}$, $\tilde{a}'_i$ and $\tilde{b}'_j$ may not be normalized, so usual credibility measure can not be used and hence using generalized credibility (Note 2.1, Sect. 2.3), as the problem is minimization problem, the following chance-constrained programming model is formulated:

$$Min_x \ (Min_{\bar{f}} \ \bar{f}) \tag{156}$$

$$subject \ to \ \tilde{Cr}\{\sum_{i=1}^{m} \sum_{j=1}^{n} (\tilde{c}'_{ij} \ x_{ij} + \tilde{d}'_{ij} y_{ij}) \leq \bar{f}\} \geq \alpha \tag{157}$$

$$\tilde{Cr}\{\sum_{j=1}^{n} x_{ij} \leq \tilde{a}'_i\} \geq \alpha_i, \quad i = 1, 2, ..., m \tag{158}$$

$$\tilde{Cr}\{\sum_{i=1}^{m} x_{ij} \geq \tilde{b}'_j\} \geq \beta_j, \quad j = 1, 2, ..., n \tag{159}$$

$$x_{ij} \geq 0, \ y_{ij} = \begin{cases} 1, & \text{if } x_{ij} > 0; \\ 0, & \text{otherwise.} \end{cases} \quad \forall \ i, j \tag{160}$$

where Min $\bar{f}$ indicates the minimum possible value that the objective function less or equal to it with generalized credibility at least $\alpha$ ($0 < \alpha \leq 1$). $\alpha_i, \beta_j$ ($0 < \alpha_i, \beta_j \leq 1$) are the predetermined generalized credibility levels of satisfaction of the source and destination constraints respectively for all $i, j$.

**Crisp Equivalences:** Suppose that the $\tilde{c}_{ij}$, $\tilde{d}_{ij}$, $\tilde{a}_i$ and $\tilde{b}_j$ are all mutually independent type-2 triangular fuzzy variables defined by $\tilde{c}_{ij} = (c_{ij}^1, c_{ij}^2, c_{ij}^3; \theta_{l,ij}, \theta_{r,ij})$, $\tilde{d}_{ij} = (d_{ij}^1, d_{ij}^2, d_{ij}^3; \theta'_{l,ij}, \theta'_{r,ij})$, $\tilde{a}_i = (a_i^1, a_i^2, a_i^3; \theta_{l,i}, \theta_{r,i})$ and $\tilde{b}_j = (b_j^1, b_j^2, b_j^3; \theta_{l,j}, \theta_{r,j})$. Also let $\tilde{c}'_{ij}$, $\tilde{d}'_{ij}$, $\tilde{a}'_i$ and $\tilde{b}'_j$ are the corresponding reductions by the CV reduction method.

Then from Theorem 4.1 and its corollary, the chance-constrained model formulation (156)–(160) of Model-4.2 (i.e., (152)–(155)) can be turned into the following crisp equivalent (for proof see the Appendix) parametric programming problems:

Case-I: $0 < \alpha \leq 0.25$: Then the equivalent parametric programming problem for the model representation (156)–(160) is

$$Min \sum_{i=1}^{m} \sum_{j=1}^{n} \left[ \frac{(1 - 2\alpha + (1 - 4\alpha)\theta_{r,ij})c_{ij}^1 x_{ij} + 2\alpha c_{ij}^2 x_{ij}}{1 + (1 - 4\alpha)\theta_{r,ij}} + \right.$$
$$\left. \frac{(1 - 2\alpha + (1 - 4\alpha)\theta'_{r,ij})d_{ij}^1 y_{ij} + 2\alpha d_{ij}^2 y_{ij}}{1 + (1 - 4\alpha)\theta'_{r,ij}} \right] \qquad (161)$$

$$subject\ to\ \sum_{j=1}^{n} x_{ij} \leq F_{a_i},\ i = 1, 2, ..., m \qquad (162)$$

$$\sum_{i=1}^{m} x_{ij} \geq F_{b_j},\ j = 1, 2, ..., n \qquad (163)$$

$$x_{ij} \geq 0,\ y_{ij} = \begin{cases} 1, \text{if } x_{ij} > 0; \\ 0, \text{otherwise.} \end{cases} \forall\ i, j, \qquad (164)$$

where $F_{a_i}$ and $F_{b_j}$ are given by (177) and (178) respectively.

Case-II: $0.25 < \alpha \leq 0.5$: Then the equivalent parametric programming problem for the model (156)–(160) is

$$Min \sum_{i=1}^{m} \sum_{j=1}^{n} \left[ \frac{(1 - 2\alpha)c_{ij}^1 x_{ij} + (2\alpha + (4\alpha - 1)\theta_{l,ij})c_{ij}^2 x_{ij}}{1 + (4\alpha - 1)\theta_{l,ij}} + \right.$$
$$\left. \frac{(1 - 2\alpha)d_{ij}^1 y_{ij} + (2\alpha + (4\alpha - 1)\theta'_{l,ij})d_{ij}^2 y_{ij}}{1 + (4\alpha - 1)\theta'_{l,ij}} \right] \qquad (165)$$

$$subject\ to\ \sum_{j=1}^{n} x_{ij} \leq F_{a_i},\ i = 1, 2, ..., m \qquad (166)$$

$$\sum_{i=1}^{m} x_{ij} \geq F_{b_j},\ j = 1, 2, ..., n \qquad (167)$$

$$x_{ij} \geq 0,\ y_{ij} = \begin{cases} 1, \text{if } x_{ij} > 0; \\ 0, \text{otherwise.} \end{cases} \forall\ i, j \qquad (168)$$

Case-III: $0.5 < \alpha \leq 0.75$: Then the equivalent parametric programming problem for the model (156)–(160) is

$$
Min \sum_{i=1}^{m} \sum_{j=1}^{n} \left[ \frac{(2\alpha - 1)c_{ij}^3 x_{ij} + (2(1-\alpha) + (3 - 4\alpha)\theta_{l,ij})c_{ij}^2 x_{ij}}{1 + (3 - 4\alpha)\theta_{l,ij}} + \right.
$$
$$
\left. \frac{(2\alpha - 1)d_{ij}^3 y_{ij} + (2(1-\alpha) + (3 - 4\alpha)\theta'_{l,ij})d_{ij}^2 y_{ij}}{1 + (3 - 4\alpha)\theta'_{l,ij}} \right] \tag{169}
$$

$$
subject\ to\ \sum_{j=1}^{n} x_{ij} \leq F_{a_i},\ i = 1, 2, ..., m \tag{170}
$$

$$
\sum_{i=1}^{m} x_{ij} \geq F_{b_j},\ j = 1, 2, ..., n \tag{171}
$$

$$
x_{ij} \geq 0,\ y_{ij} = \begin{cases} 1, \text{ if } x_{ij} > 0; \\ 0, \text{ otherwise.} \end{cases} \forall\, i, j \tag{172}
$$

Case-IV: $0.75 < \alpha \leq 1$: Then the equivalent parametric programming problem for the model (156)–(160) is

$$
Min \sum_{i=1}^{m} \sum_{j=1}^{n} \left[ \frac{(2\alpha - 1 + (4\alpha - 3)\theta_{r,ij})c_{ij}^3 x_{ij} + 2(1 - \alpha)c_{ij}^2 x_{ij}}{1 + (4\alpha - 3)\theta_{r,ij}} + \right.
$$
$$
\left. \frac{(2\alpha - 1 + (4\alpha - 3)\theta'_{r,ij})d_{ij}^3 y_{ij} + 2(1 - \alpha)d_{ij}^2 y_{ij}}{1 + (4\alpha - 3)\theta'_{r,ij}} \right] \tag{173}
$$

$$
subject\ to\ \sum_{j=1}^{n} x_{ij} \leq F_{a_i},\ i = 1, 2, ..., m \tag{174}
$$

$$
\sum_{i=1}^{m} x_{ij} \geq F_{b_j},\ j = 1, 2, ..., n \tag{175}
$$

$$
x_{ij} \geq 0,\ y_{ij} = \begin{cases} 1, \text{ if } x_{ij} > 0; \\ 0, \text{ otherwise.} \end{cases} \forall\, i, j, \tag{176}
$$

where,

$$
F_{a_i} = \begin{cases} \frac{(1 - 2\alpha_i + (1 - 4\alpha_i)\theta_{l,i})a_i^3 + 2\alpha_i a_i^2}{1 + (1 - 4\alpha_i)\theta_{l,i}}, & \text{if } 0 < \alpha_i \leq 0.25; \\ \frac{(1 - 2\alpha_i)a_i^3 + (2\alpha_i + (4\alpha_i - 1)\theta_{r,i})a_i^2}{1 + (4\alpha_i - 1)\theta_{r,i}}, & \text{if } 0.25 < \alpha_i \leq 0.5; \\ \frac{(2\alpha_i - 1)a_i^3 + (3 - \alpha_i) + (3 - 4\alpha_i)\theta_{r,i})a_i^2}{1 + (3 - 4\alpha_i)\theta_{r,i}}, & \text{if } 0.5 < \alpha_i \leq 0.75; \\ \frac{(2\alpha_i - 1 + (4\alpha_i - 3)\theta_{l,i})a_i^1 + 2(1 - \alpha_i)a_i^2}{1 + (4\alpha_i - 3)\theta_{l,i}}, & \text{if } 0.75 < \alpha_i \leq 1. \end{cases} \tag{177}
$$

$$F_{b_j} = \begin{cases} \dfrac{(1-2\beta_j+(1-4\beta_j)\theta_{r,j})b_j^1+2\beta_j b_j^2}{1+(1-4\beta_j)\beta_{r,j}}, & \text{if } 0 < \beta_j \le 0.25; \\[2ex] \dfrac{(1-2\beta_j)b_j^1+(2\beta_j+(4\beta_j-1)\theta_{l,j})b_j^2}{1+(4\beta_j-1)\theta_{l,j}}, & \text{if } 0.25 < \beta_j \le 0.5; \\[2ex] \dfrac{(2\beta_j-1)b_j^3+(2(1-\beta_j)+(3-4\beta_j)\theta_{l,j})b_j^2}{1+(3-4\beta_j)\theta_{l,j}}, & \text{if } 0.5 < \beta_j \le 0.75; \\[2ex] \dfrac{(2\beta_j-1+(4\beta_j-3)\theta_{r,j})b_j^3+2(1-\beta_j)b_j^2}{1+(4\beta_j-3)\theta_{r,j}}, & \text{if } 0.75 < \beta_j \le 1. \end{cases} \quad (178)$$

**Numerical Experiment:** To illustrate the Model 4.2 ((152)–(155)) numerically, consider a problem having three sources and two destinations with the following type-2 fuzzy data.

$\tilde{a}_1 = (16, 18, 20; 0.5, 0.8)$, $\tilde{a}_2 = (15, 16, 18; 0.4, 0.6)$, $\tilde{a}_3 = (14, 15, 16; 0.6, 0.9)$
$\tilde{b}_1 = (20, 22, 24; 0.5, 0.5)$, $\tilde{b}_2 = (18, 19, 20; 0.6, 0.8)$.

The direct and fixed costs for this problem are given in Table 10.

**Table 10.** Direct costs and fixed costs

| $i$ | $j$ | | $j$ | |
|---|---|---|---|---|
| | 1 | 2 | 1 | 2 |
| 1 | (2,3,4;0.4,0.7) | (3.5,4,4.2;0.5,0.8) | (3.5,4,4.5;0.5,0.5) | (3,4,5;0.6,0.4) |
| 2 | (4,5,6;0.7,0.8) | (4,6,7;0.6,0.4) | (4.5,5,5.4;0.4,0.7) | (4.6,5.2,5.6;0.6,0.8) |
| 3 | (4.4,4.6,5;0.5,0.5) | (5,5.5,6;0.5,0.7) | (5,6,6.4;0.6,0.6) | (4,4.2,4.8;0.4,0.6) |
| | $c_{ij}$ | | $d_{ij}$ | |

The predetermined general credibility levels for the chance-constrained programming model (156)–(160) as formulated to solve the Model-4.2 are taken as $\alpha = 0.9$, $\alpha_i = 0.9$, $\beta_j = 0.9$, $i = 1, 2, 3$; $j = 1, 2$.

Now using (173)–(176), the equivalent parametric programming problem becomes

$$Min \sum_{i=1}^{3} \sum_{j=1}^{2} \Big[ \frac{(0.8 + 0.6\theta_{r,ij})c_{ij}^3 x_{ij} + 0.2c_{ij}^2 x_{ij}}{1 + 0.6\theta_{r,ij}} +$$

$$\frac{(0.8 + 0.6\theta'_{r,ij})d_{ij}^3 y_{ij} + 0.2d_{ij}^2 y_{ij}}{1 + 0.6\theta'_{r,ij}} \Big] \quad (179)$$

$$subject \ to \ \sum_{j=1}^{2} x_{ij} \le F_{a_i}, \ i = 1, 2, 3 \quad (180)$$

$$\sum_{i=1}^{3} x_{ij} \ge F_{b_j}, \ j = 1, 2 \quad (181)$$

$$x_{ij} \ge 0, \ y_{ij} = \begin{cases} 1, & \text{if } x_{ij} > 0; \\ 0, & \text{otherwise.} \end{cases} \quad \forall \, i, j \quad (182)$$

where $F_{a_1} = 16.30$, $F_{a_2} = 15.16$, $F_{a_3} = 14.14$, $F_{b_1} = 23.69$ and $F_{b_2} = 19.86$ are obtained from Eqs. (177) and (178).

Solving this, the optimum results are $x_{12} = 16.3$, $x_{21} = 13.11$, $x_{31} = 10.58$, $x_{32} = 3.56$ and the objective function value (minimum transportation cost)= 239.5014.

## 4.6    Sensitivity Analysis for the Numerical Experiment Of Model 4.2

A sensitivity analysis for the numerical experiment of Model-4.2 is presented to show the efficiency and logically correctness of the crisp equivalent form and solution approaches of the presented problem. For some different generalized credibility levels for the objective function, source constraints and destination constraints of the model representation (156)–(160) formulated to solve the Model 4.2, the changes in the objective function value (minimum transportation cost) are presented in the Table 11.

It is observed from the Table 11 that for fixed credibility levels of the objective function and the source constraints, i.e., for fixed $\alpha$ and $\alpha_i (i = 1, 2, 3)$, minimum transportation cost increases with the increased credibility levels $(\beta_j, j = 1, 2)$ of the destination constraints. The reason of this fact is that as the credibility levels $\beta_j$ increase, the defuzzified amount of the demands $(F_{b_j}, j = 1, 2)$ are also increased (e.g., for $\beta_j = 0.7$, $F_{b_1} = 22.27$, $F_{b_2} = 19.35$ and for $\beta_j = 0.8$, $F_{b_1} = 23.27$, $F_{b_2} = 19.65$) and as a result total transported amount also increases.

Now for fixed $\alpha$ and $\beta_j (j = 1, 2)$, minimum transportation cost increases with the increased credibility levels $(\alpha_i, i = 1, 2, 3)$ of the source constraints. The interesting fact is that in this case, though total transported amounts are the same but still transportation cost increases as the credibility levels $\alpha_i (i = 1, 2, 3)$ increase. The reason is that, as the credibility levels $\alpha_i (i = 1, 2, 3)$ increase, the defuzzified amount of the availabilities $(F_{a_i}, i = 1, 2, 3)$ decrease (e.g., for $\alpha_i = 0.7$, $F_{a_1} = 17.31$, $F_{a_2} = 15.64$, $F_{a_3} = 14.66$ and for $\alpha_i = 0.8$, $F_{a_1} = 16.72$, $F_{a_2} = 15.37$, $F_{a_3} = 14.35$) and as a result for the fixed demand, the allocation options of the product (to the less cost path) decrease.

Also we observe from the Table 11 that for fixed $\alpha_i (i = 1, 2, 3)$ and $\beta_j (j = 1, 2)$ minimum transportation cost increases as the credibility level $\alpha$ increases. This is because in this case defuzzified cost coefficients of the objective function increase with increased credibility level $\alpha$.

## 4.7    Model 4.3: Multi-item Solid Transportation Problem Having Restriction on Conveyances with Type-2 Fuzzy Parameters

Here, we formulate a multi-item solid transportation problem (MISTP) with restriction on some items and conveyances in the sense that some specific items prohibited to be transported through some particular conveyances. In this problem the transportation parameters, e.g., unit transportation costs, supplies, demands, conveyance capacities are type-2 triangular fuzzy variables.

**Table 11.** Changes in transportation cost for different credibility levels

| $\alpha$ | $\alpha_i$ | $\beta_j$ | Transported amount | Transportation cost |
|---|---|---|---|---|
|  |  | 0.7 | 42.07 | 230.3159 |
| 0.9 | 0.9 | 0.8 | 42.92 | 235.5984 |
|  |  | 0.9 | 43.55 | 239.5014 |
|  |  | 0.95 | 43.78 | 240.9197 |
|  | 0.7 |  |  | 236.3098 |
| 0.9 | 0.8 | 0.9 | 43.55 | 238.1800 |
|  | 0.9 |  |  | 239.5014 |
|  | 0.95 |  |  | 240.0048 |
| 0.7 |  |  |  | 227.0400 |
| 0.8 | 0.9 | 0.9 | 43.55 | 234.3038 |
| 0.9 |  |  |  | 239.5014 |
| 0.95 |  |  |  | 241.3965 |

**Notations:**

(i) $\tilde{c}^p_{ijk}$: The unit transportation costs from $i$-th source to $j$-th destination by $k$-th conveyance for $p$-th item, represented by type-2 fuzzy variable.

(ii) $x^p_{ijk}$: The decision variable which represents amount of $p$-th item to be transported from $i$-th origin to $j$-th destination by $k$-th conveyance.

(iii) $Z$: The objective function.

(iv) $\tilde{a}^p_i$: The amount of the $p$-th item available at the $i$-th origin, represented by type-2 fuzzy variable.

(v) $\tilde{b}^p_j$: The demand of the $p$-th item at $j$-th destination, represented by type-2 fuzzy variable.

(vi) $\tilde{e}_k$: Total transportation capacity of conveyance $k$, represented by type-2 fuzzy variable.

**Mathematical Model:** Let $l$ items are to be transported from $m$ origins (or sources) $O_i$ ($i = 1, 2, ..., m$) to $n$ destinations $D_j$ ($j = 1, 2, ..., n$) by means of $K$ different modes of transportation (conveyance). Also there are some restrictions on some specific items and conveyances such a way that some items can not be transported through some conveyances. Let us denote $I_k$ as the set of items which can be transported through conveyance $k$ ($k = 1, 2, ..., K$). We use notation $p$ ($= 1, 2, ..., l$) to denote the items. Then the mathematical formulation of the problem is as follows:

$$Min\ Z = \sum_{p \in I_1} \sum_{i=1}^{m} \sum_{j=1}^{n} \tilde{c}^p_{ij1}\, x^p_{ij1} + \sum_{p \in I_2} \sum_{i=1}^{m} \sum_{j=1}^{n} \tilde{c}^p_{ij2}\, x^p_{ij2} + ... + \sum_{p \in I_K} \sum_{i=1}^{m} \sum_{j=1}^{n} \tilde{c}^p_{ijK}\, x^p_{ijK},$$

where $|I_1 \bigcup I_2 \bigcup ... \bigcup I_K| = l$

$$= \sum_{p=1}^{l}\sum_{i=1}^{m}\sum_{j=1}^{n}\sum_{k=1}^{K} d_{ijk}^{p}.(c_{ijk}^{\tilde{p}}\, x_{ijk}^{p}), \tag{183}$$

$$subject\ to\ \sum_{j=1}^{n}\sum_{k=1}^{K} d_{ijk}^{p}\, x_{ijk}^{p}\ \le \tilde{a}_{i}^{p},\quad i=1,2,...,m; p=1,2,...l, \tag{184}$$

$$\sum_{i=1}^{m}\sum_{k=1}^{K} d_{ijk}^{p}\, x_{ijk}^{p}\ \ge \tilde{b}_{j}^{p},\quad j=1,2,...,n; p=1,2,...,l, \tag{185}$$

$$\sum_{p=1}^{l}\sum_{i=1}^{m}\sum_{j=1}^{n} d_{ijk}^{p}\, x_{ijk}^{p}\ \le \tilde{e}_{k},\quad k=1,2,...,K, \tag{186}$$

$$x_{ijk}^{p}\ge 0,\ \forall\ i,j,k,p \tag{187}$$

where $d_{ijk}^{p}$ is defined as $d_{ijk}^{p}= \begin{cases} 1, & \text{if } p\in I_k; \\ 0, & \text{otherwise.} \end{cases}\ \forall\ i,j,k,p.$

**Solution Methodology 1: Chance-Constrained Programming Using Generalized Credibility:** Suppose that $c_{ijk}^{\tilde{p}'}$, $a_{i}^{\tilde{p}'}$, $b_{j}^{\tilde{p}'}$ and $\tilde{e}_{k}'$ are the reduced fuzzy (type-1) variables from type-2 fuzzy variables $c_{ijk}^{\tilde{p}}$, $a_{i}^{\tilde{p}}$, $b_{j}^{\tilde{p}}$ and $\tilde{e}_{k}$ respectively based on CV-based reduction method. Now to solve the above problem we formulate a chance-constrained programming model with these reduced fuzzy parameters. Since the reduced fuzzy parameters $c_{ijk}^{\tilde{p}'}$, $a_{i}^{\tilde{p}'}$, $b_{j}^{\tilde{p}'}$ and $\tilde{e}_{k}'$ may not be normalized, so using generalized credibility for the objective function as well as for the constraints the following chance constrained programming model is formulated for the above problem (183)–(187).

$$Min_{x}\ (Min_{\bar{f}}\ \bar{f}) \tag{188}$$

$$\tilde{C}r\{\sum_{p=1}^{l}\sum_{i=1}^{m}\sum_{j=1}^{n}\sum_{k=1}^{K} d_{ijk}^{p}.(c_{ijk}^{\tilde{p}'}\, x_{ijk}^{p})\le \bar{f}\}\ge \alpha, \tag{189}$$

$$s.t.\ \tilde{C}r\{\sum_{j=1}^{n}\sum_{k=1}^{K} d_{ijk}^{p}\, x_{ijk}^{p}\ \le a_{i}^{\tilde{p}'}\}\ge \alpha_{i}^{p},\quad i=1,2,...,m; p=1,2,...l, \tag{190}$$

$$\tilde{C}r\{\sum_{i=1}^{m}\sum_{k=1}^{K} d_{ijk}^{p}\, x_{ijk}^{p}\ \ge b_{j}^{\tilde{p}'}\}\ge \beta_{j}^{p},\quad j=1,2,...,n; p=1,2,...,l, \tag{191}$$

$$\tilde{C}r\{\sum_{p=1}^{l}\sum_{i=1}^{m}\sum_{j=1}^{n} d_{ijk}^{p}\, x_{ijk}^{p}\ \le \tilde{e}_{k}'\}\ge \gamma_{k},\quad k=1,2,...,K, \tag{192}$$

$$x_{ijk}^p \geq 0, \ d_{ijk}^p = \begin{cases} 1, \text{ if } p \in I_k; \\ 0, \text{ otherwise.} \end{cases} \forall \ i,j,k,p \qquad (193)$$

where Min $\bar{f}$ indicates the minimum value that the objective function achieves with generalized credibility at least $\alpha$ $(0 < \alpha \leq 1)$. $\alpha_i^p$, $\beta_j^p$ and $\gamma_k$ are predetermined generalized credibility levels of satisfaction of the respective constraints for all $i,j,k,p$.

**Crisp Equivalence:** We consider $\tilde{c}_{ijk}^p$, $\tilde{a}_i^p$, $\tilde{b}_j^p$ and $\tilde{e}_k$ are all mutually independent type-2 triangular fuzzy variables as $\tilde{c}_{ijk}^p = (c_{ijk}^{p1}, c_{ijk}^{p2}, c_{ijk}^{p3}; \theta_{l,ijk}^p, \theta_{r,ijk}^p)$, $\tilde{a}_i^p = (a_i^{p1}, a_i^{p2}, a_i^{p3}; \theta_{l,i}^p, \theta_{r,i}^p)$, $\tilde{b}_j^p = (b_j^{p1}, b_j^{p2}, b_j^{p3}; \theta_{l,j}^p, \theta_{r,j}^p)$ and $\tilde{e}_k = (e_k^1, e_k^2, e_k^3; \theta_{l,k}, \theta_{r,k})$. Then from Theorem-4.1 and its corollary, the chance-constrained model formulation (188)–(193) is turned into the following crisp equivalent parametric programming problems:

Case-I: $0 < \alpha \leq 0.25$: The equivalent parametric programming problem for model (188)–(193) is

$$Min \ \sum_{p=1}^l \sum_{i=1}^m \sum_{j=1}^n \sum_{k=1}^K d_{ijk}^p \Big[ \frac{(1 - 2\alpha + (1 - 4\alpha)\theta_{r,ijk}^p)c_{ijk}^{p1} x_{ijk}^p + 2\alpha c_{ijk}^{p2} x_{ijk}^p}{1 + (1 - 4\alpha)\theta_{r,ijk}^p} \Big],$$

$$(194)$$

$$\text{sub. to} \ \sum_{j=1}^n \sum_{k=1}^K d_{ijk}^p \ x_{ijk}^p \leq F_{a_i^p}, \ i = 1, 2, ..., m; \ p = 1, 2, ...l, \qquad (195)$$

$$\sum_{i=1}^m \sum_{k=1}^K d_{ijk}^p \ x_{ijk}^p \geq F_{b_j^p}, \ j = 1, 2, ..., n; \ p = 1, 2, ...l, \qquad (196)$$

$$\sum_{p=1}^l \sum_{i=1}^m \sum_{j=1}^n d_{ijk}^p \ x_{ijk}^p \leq F_{e_k}, \ k = 1, 2, ..., K, \qquad (197)$$

$$x_{ijk}^p \geq 0, \ d_{ijk}^p = \begin{cases} 1, \text{ if } p \in I_k; \\ 0, \text{ otherwise.} \end{cases} \forall \ i,j,k,p, \qquad (198)$$

where $F_{a_i^p}$, $F_{b_j^p}$ and $F_{e_k}$ are given by (214), (215) and (216) respectively.
Case-II: $0.25 < \alpha \leq 0.5$: Then the equivalent parametric programming problem for model (188)–(193) is

$$Min \ \sum_{p=1}^l \sum_{i=1}^m \sum_{j=1}^n \sum_{k=1}^K d_{ijk}^p \Big[ \frac{(1 - 2\alpha)c_{ijk}^{p1} x_{ijk}^p + (2\alpha + (4\alpha - 1)\theta_{l,ijk}^p)c_{ijk}^{p2} x_{ijk}^p}{1 + (4\alpha - 1)\theta_{l,ijk}^p} \Big]$$

$$(199)$$

$$\text{sub. to} \ \sum_{j=1}^n \sum_{k=1}^K d_{ijk}^p \ x_{ijk}^p \leq F_{a_i^p}, \ i = 1, 2, ..., m; \ p = 1, 2, ...l, \qquad (200)$$

$$\sum_{i=1}^{m}\sum_{k=1}^{K}d_{ijk}^{p}\,x_{ijk}^{p}\geq F_{b_{j}^{p}},\ j=1,2,...,n;\ p=1,2,...l, \tag{201}$$

$$\sum_{p=1}^{l}\sum_{i=1}^{m}\sum_{j=1}^{n}d_{ijk}^{p}\,x_{ijk}^{p}\leq F_{e_{k}},\ k=1,2,...,K, \tag{202}$$

$$x_{ijk}^{p}\geq 0,\ d_{ijk}^{p}=\begin{cases}1,\ \mathrm{if}\,p\in I_{k};\\0,\ \mathrm{otherwise}.\end{cases}\ \forall\,i,j,k,p. \tag{203}$$

Case-III: $0.5 < \alpha \leq 0.75$: Then the equivalent parametric programming problem for model (188)–(193) is

$$Min\ \sum_{p=1}^{l}\sum_{i=1}^{m}\sum_{j=1}^{n}\sum_{k=1}^{K}d_{ijk}^{p}\Big[\frac{(2\alpha-1)c_{ijk}^{p3}x_{ijk}^{p}+(2(1-\alpha)+(3-4\alpha)\theta_{l,ijk}^{p})c_{ijk}^{p2}x_{ijk}^{p}}{1+(3-4\alpha)\theta_{l,ijk}^{p}}\Big] \tag{204}$$

$$sub.\ to\ \sum_{j=1}^{n}\sum_{k=1}^{K}d_{ijk}^{p}\,x_{ijk}^{p}\leq F_{a_{i}^{p}},\ i=1,2,...,m;\ p=1,2,...l, \tag{205}$$

$$\sum_{i=1}^{m}\sum_{k=1}^{K}d_{ijk}^{p}\,x_{ijk}^{p}\geq F_{b_{j}^{p}},\ j=1,2,...,n;\ p=1,2,...l, \tag{206}$$

$$\sum_{p=1}^{l}\sum_{i=1}^{m}\sum_{j=1}^{n}d_{ijk}^{p}\,x_{ijk}^{p}\leq F_{e_{k}},\ k=1,2,...,K, \tag{207}$$

$$x_{ijk}^{p}\geq 0,\ d_{ijk}^{p}=\begin{cases}1,\ \mathrm{if}\,p\in I_{k};\\0,\ \mathrm{otherwise}.\end{cases}\ \forall\,i,j,k,p. \tag{208}$$

Case-IV: $0.75 < \alpha \leq 1$: Then the equivalent parametric programming problem for model (188)–(193) is

$$Min\ \sum_{p=1}^{l}\sum_{i=1}^{m}\sum_{j=1}^{n}\sum_{k=1}^{K}d_{ijk}^{p}\Big[\frac{(2\alpha-1+(4\alpha-3)\theta_{r,ijk}^{p})c_{ijk}^{p3}x_{ijk}^{p}+2(1-\alpha)c_{ijk}^{p2}x_{ijk}^{p}}{1+(4\alpha-3)\theta_{r,ijk}^{p}}\Big] \tag{209}$$

$$sub.\ to\ \sum_{j=1}^{n}\sum_{k=1}^{K}d_{ijk}^{p}\,x_{ijk}^{p}\leq F_{a_{i}^{p}},\ i=1,2,...,m;\ p=1,2,...l, \tag{210}$$

$$\sum_{i=1}^{m}\sum_{k=1}^{K}d_{ijk}^{p}\,x_{ijk}^{p}\geq F_{b_{j}^{p}},\ j=1,2,...,n;\ p=1,2,...l, \tag{211}$$

$$\sum_{p=1}^{l}\sum_{i=1}^{m}\sum_{j=1}^{n}d_{ijk}^{p}\,x_{ijk}^{p}\leq F_{e_{k}},\ k=1,2,...,K, \tag{212}$$

$$x_{ijk}^p \geq 0, \; d_{ijk}^p = \begin{cases} 1, & \text{if } p \in I_k; \\ 0, & \text{otherwise.} \end{cases} \; \forall \, i,j,k,p \tag{213}$$

where,

$$F_{a_i^p} = \begin{cases} \dfrac{(1-2\alpha_i^p+(1-4\alpha_i^p)\theta_{l,i}^p)a_i^{p3}+2\alpha_i^p a_i^{p2}}{1+(1-4\alpha_i^p)\theta_{l,i}^p}, & \text{if } 0 < \alpha_i^p \leq 0.25; \\[2mm] \dfrac{(1-2\alpha_i^p)a_i^{p3}+(2\alpha_i^p+(4\alpha_i^p-1)\theta_{r,i}^p)a_i^{p2}}{1+(4\alpha_i^p-1)\theta_{r,i}^p}, & \text{if } 0.25 < \alpha_i^p \leq 0.5; \\[2mm] \dfrac{(2\alpha_i^p-1)a_i^{p1}+(2(1-\alpha_i^p)+(3-4\alpha_i^p)\theta_{r,i}^p)a_i^{p2}}{1+(3-4\alpha_i^p)\theta_{r,i}^p}, & \text{if } 0.5 < \alpha_i^p \leq 0.75; \\[2mm] \dfrac{(2\alpha_i^p-1+(4\alpha_i^p-3)\theta_{l,i}^p)a_i^{p1}+2(1-\alpha_i^p)a_i^{p2}}{1+(4\alpha_i^p-3)\theta_{l,i}^p}, & \text{if } 0.75 < \alpha_i^p \leq 1. \end{cases} \tag{214}$$

$$F_{b_j^p} = \begin{cases} \dfrac{(1-2\beta_j^p+(1-4\beta_j^p)\theta_{r,j}^p)b_j^{p1}+2\beta_j^p b_j^{p2}}{1+(1-4\beta_j^p)\beta_{r,j}^p}, & \text{if } 0 < \beta_j^p \leq 0.25; \\[2mm] \dfrac{(1-2\beta_j^p)b_j^{p1}+(2\beta_j^p+(4\beta_j^p-1)\theta_{l,j}^p)b_j^{p2}}{1+(4\beta_j^p-1)\theta_{l,j}^p}, & \text{if } 0.25 < \beta_j^p \leq 0.5; \\[2mm] \dfrac{(2\beta_j p-1)b_j^{p3}+(2(1-\beta_j^p)+(3-4\beta_j^p)\theta_{l,j}^p)b_j^{p2}}{1+(3-4\beta_j^p)\theta_{l,j}^p}, & \text{if } 0.5 < \beta_j^p \leq 0.75; \\[2mm] \dfrac{(2\beta_j^p-1+(4\beta_j^p-3)\theta_{r,j}^p)b_j^{p3}+2(1-\beta_j^p)b_j^{p2}}{1+(4\beta_j^p-3)\theta_{r,j}^p}, & \text{if } 0.75 < \beta_j^p \leq 1. \end{cases} \tag{215}$$

$$F_{e_k} = \begin{cases} \dfrac{(1-2\gamma_k+(1-4\gamma_k)\theta_{l,k})e_k^3+2\gamma_k e_k^2}{1+(1-4\gamma_k)\theta_{l,k}}, & \text{if } 0 < \gamma_k \leq 0.25; \\[2mm] \dfrac{(1-2\gamma_k)e_k^3+(2\gamma_k+(4\gamma_k-1)\theta_{r,k})e_k^2}{1+(4\gamma_k-1)\theta_{r,k}}, & \text{if } 0.25 < \gamma_k \leq 0.5; \\[2mm] \dfrac{(2\gamma_k-1)e_k^1+(2(1-\gamma_k)+(3-4\gamma_k)\theta_{r,k})e_k^2}{1+(3-4\gamma_k)\theta_{r,k}}, & \text{if } 0.5 < \gamma_k \leq 0.75; \\[2mm] \dfrac{(2\gamma_k-1+(4\gamma_k-3)\theta_{l,k})e_k^1+2(1-\gamma_k)e_k^2}{1+(4\gamma_k-3)\theta_{l,k}}, & \text{if } 0.75 < \gamma_k \leq 1. \end{cases} \tag{216}$$

**Solution Methodology 2: Using Nearest Interval Approximation:** Consider costs $c_{ijk}^{\tilde{p}}$, supplies $a_i^{\tilde{p}}$, demands $b_j^{\tilde{p}}$ and conveyance capacities $\tilde{e}_k$ are all mutually independent type-2 triangular fuzzy variables defined by $c_{ijk}^{\tilde{p}} = (c_{ijk}^{p1}, c_{ijk}^{p2}, c_{ijk}^{p3}; \theta_{l,ijk}^p, \theta_{r,ijk}^p)$, $a_i^{\tilde{p}} = (a_i^{p1}, a_i^{p2}, a_i^{p3}; \theta_{l,i}^p, \theta_{r,i}^p)$, $b_j^{\tilde{p}} = (b_j^{p1}, b_j^{p2}, b_j^{p3}; \theta_{l,j}^p, \theta_{r,j}^p)$ and $\tilde{e}_k = (e_k^1, e_k^2, e_k^3; \theta_{l,k}, \theta_{r,k})$. Then we find nearest interval approximations (credibilistic interval approximation, cf. Sect. 4.2) of $c_{ijk}^{\tilde{p}}$, $a_i^{\tilde{p}}$, $b_j^{\tilde{p}}$ and $\tilde{e}_k$, suppose these are $[c_{ijkL}^p, c_{ijkR}^p]$, $[a_{iL}^p, a_{iR}^p]$, $[b_{jL}^p, b_{jR}^p]$ and $[e_{kL}, e_{kR}]$ respectively. Then with these nearest interval approximations, the Model (183)–(183) becomes

$$Min \; Z = \sum_{p=1}^l \sum_{i=1}^m \sum_{j=1}^n \sum_{k=1}^K d_{ijk}^p \cdot ([c_{ijkL}^p, c_{ijkR}^p] \, x_{ijk}^p) \tag{217}$$

$$subject \; to \; \sum_{j=1}^n \sum_{k=1}^K d_{ijk}^p \, x_{ijk}^p \leq [a_{iL}^p, a_{iR}^p], \quad i=1,2,...,m; p=1,2,...l, \tag{218}$$

$$\sum_{i=1}^m \sum_{k=1}^K d_{ijk}^p \, x_{ijk}^p \geq [b_{jL}^p, b_{jR}^p], \quad j=1,2,...,n; p=1,2,...,l, \tag{219}$$

$$\sum_{p=1}^{l}\sum_{i=1}^{m}\sum_{j=1}^{n} d_{ijk}^{p}\, x_{ijk}^{p} \leq [e_{kL}, e_{kR}], \quad k = 1, 2, ..., K, \tag{220}$$

$$x_{ijk}^{p} \geq 0, \; d_{ijk}^{p} = \begin{cases} 1, \text{ if } p \in I_k; \\ 0, \text{ otherwise.} \end{cases} \forall\, i, j, k, p. \tag{221}$$

**Deterministic Form:** We first obtain deterministic forms of the uncertain constraints using the idea of possibility degree of interval number (Zhang et al. [161]) representing certain degree by which one interval is larger or smaller than another. Now we denote the left hand side expressions of the source, destination and conveyance capacity constraints, i.e. (218), (219) and (220) respectively of the model (217)–(221) by $S_i^p$, $D_j^p$ and $E_k$ respectively. Here the right hand sides of these constraints are interval numbers and left sides are crisp, then the possibility degree of satisfaction of these constraints are defined as

$$P_{S_i^p \leq [a_{iL}^p, a_{iR}^p]} = \begin{cases} 1, & S_i^p \leq a_{iL}^p; \\ \frac{a_{iR}^p - S_i^p}{a_{iR}^p - a_{iL}^p}, & a_{iL}^p < S_i^p \leq a_{iR}^p; \\ 0, & S_i^p > a_{iR}^p. \end{cases}$$

$$P_{D_j^p \geq [b_{jL}^p, b_{jR}^p]} = \begin{cases} 0, & D_j^p < b_{jL}^p; \\ \frac{D_j^p - b_{jL}^p}{b_{jR}^p - b_{jL}^p}, & b_{jL}^p \leq D_j^p < b_{jR}^p; \\ 1, & D_j^p > b_{jR}^p. \end{cases}$$

$$P_{E_k \leq [e_{kL}, e_{kR}]} = \begin{cases} 1, & E_k \leq e_{kL}; \\ \frac{e_{kR} - E_k}{e_{kR} - e_{kL}}, & e_{kL} < E_k \leq e_{kR}; \\ 0, & E_k > e_{kR}. \end{cases}$$

Now if the constraints are allowed to be satisfied with some predetermined possibility degree level $\alpha_i^p$, $\beta_j^p$ and $\gamma_k$ ($0 \leq \alpha_i^p, \beta_j^p, \gamma_k \leq 1$) respectively, i.e. $P_{S_i^p \leq [a_{iL}^p, a_{iR}^p]} \geq \alpha_i^p$, $P_{D_j^p \geq [b_{jL}^p, b_{jR}^p]} \geq \beta_j^p$ and $P_{E_k \leq [e_{kL}, e_{kR}]} \geq \gamma_k \; \forall\, i, j, k, p$, then the equivalent deterministic inequalities of the respective constraints are obtained as follows:

$$S_i^p \leq a_{iR}^p - \alpha_i^p(a_{iR}^p - a_{iL}^p), \; i = 1, 2, ..., m; \; p = 1, 2, ..., l, \tag{222}$$

$$D_j^p \geq b_{jL}^p + \beta_j^p(b_{jR}^p - b_{jL}^p), \; j = 1, 2, ..., n; \; p = 1, 2, ..., l, \tag{223}$$

$$E_k \leq e_{kR} - \gamma_k(e_{kR} - e_{kL}), \; k = 1, 2, ..., K. \tag{224}$$

Now to deal with objective function we find minimum possible objective function value (say $\underline{Z}$) and maximum possible objective function value (say $\overline{Z}$) for the interval costs $[c_{ijkL}^p, c_{ijkR}^p]$, by solving the following two problems:

$$\underline{Z} = Min_{c_{ijkL}^p \leq c_{ijk}^p \leq c_{ijkR}^p} [Min \sum_{p=1}^{l}\sum_{i=1}^{m}\sum_{j=1}^{n}\sum_{k=1}^{K} d_{ijk}^p(c_{ijk}^p\, x_{ijk}^p)] \tag{225}$$

$$\overline{Z} = Max_{c_{ijkL}^p \leq c_{ijk}^p \leq c_{ijkR}^p} [Min \sum_{p=1}^{l}\sum_{i=1}^{m}\sum_{j=1}^{n}\sum_{k=1}^{K} d_{ijk}^p(c_{ijk}^p\, x_{ijk}^p)] \tag{226}$$

subject to the above constraints (4.99)–(4.101) for both cases.

So we get the range of the optimal value of the objective function of the problem (217)–(221) as $[\underline{Z}, \overline{Z}]$. Assume that the solution of the problem (225) is $x' = \{x_{ijk}^{p'}\}$ with corresponding costs $c' = \{c_{ijk}^{p'}\}$ and the solution of the problem (226) is $x'' = \{x_{ijk}^{p''}\}$ with corresponding costs $c'' = \{c_{ijk}^{p''}\}$, $\forall\, i, j, k, p$.

Now we find compromise optimal solution by treating above problems (225) and (226) together as bi-objective problem and applying fuzzy linear programming (Zimmermann [159]) as follows:

Let us denote
$Z_1 = \sum_{p=1}^{l} \sum_{i=1}^{m} \sum_{j=1}^{n} \sum_{k=1}^{K} d_{ijk}^{p}(c_{ijk}^{p'}\, x_{ijk}^{p})$ and
$Z_2 = \sum_{p=1}^{l} \sum_{i=1}^{m} \sum_{j=1}^{n} \sum_{k=1}^{K} d_{ijk}^{p}(c_{ijk}^{p''}\, x_{ijk}^{p})$, so that $Z_1(x_{ijk}^{p'}) = \underline{Z}$ and $Z_2(x_{ijk}^{p''}) = \overline{Z}$.

Now we find lower and upper bound for both the objective as $L_1 = Z_1(x_{ijk}^{p'})$, $U_1 = Z_1(x_{ijk}^{p''})$ and $L_2 = Z_2(x_{ijk}^{p''})$, $U_2 = Z_2(x_{ijk}^{p'})$ respectively.

Then construct the following two membership function for the objective functions respectively as

$$\mu_1(Z_1) = \begin{cases} 1, & \text{if } Z_1 \leq L_1; \\ \frac{U_1 - Z_1}{U_1 - L_1}, & \text{if } L_1 < Z_1 < U_1; \\ 0, & \text{if } Z_1 \geq U_1. \end{cases} \quad and \quad \mu_2(Z_2) = \begin{cases} 1, & \text{if } Z_2 \leq L_2; \\ \frac{U_2 - Z_2}{U_2 - L_2}, & \text{if } L_2 < Z_2 < U_2; \\ 0, & \text{if } Z_2 \geq U_2. \end{cases}$$

Finally solve the following problem

$$Max\ \lambda$$

$$subject\ to\ \mu_1(Z_1) \geq \lambda,\ \mu_2(Z_2) \geq \lambda \qquad (227)$$

$$and\ the\ constraints\ (222)–(224)$$

$$0 \leq \lambda \leq 1.$$

Solving this we get the optimal solution, say $x_{ijk}^{p*}$, $\forall\, i, j, k, p$ which minimizes both the objectives $Z_1, Z_2$ with certain degree $\lambda = \lambda^*$ (say) and values of the objectives $Z_1, Z_2$ at $x_{ijk}^{p*}$ give the range of the objective value, say $[\underline{Z}^*, \overline{Z}^*]$.

**Numerical Experiment:** Consider the Model 4.3 ((183)–(187)) with 3 ($p = 1, 2, 3$) items, 4 ($k = 1, 2, 3, 4$) conveyances, sources $i = 1, 2$ and destinations $j = 1, 2, 3$. Also $I_1 = \{1, 2\}$, $I_2 = \{1, 2, 3\}$, $I_3 = \{3\}$, $I_4 = \{1, 2\}$.

The transportation costs are given in the Tables 12, 13 and 14.

The supplies, demands and conveyance capacities are as follows:
$a_1^1 = (21, 24, 25; 0.5, 0.5)$, $a_2^1 = (26, 28, 30; 0.6, 0.8)$, $b_1^1 = (10, 12, 14; 0.7, 0.9)$, $b_2^1 = (12, 13, 15; 0.4, 0.7)$, $b_3^1 = (9, 12, 15; 0.4, 0.6)$,
$a_1^2 = (26, 28, 31; 0.5, 1)$, $a_2^2 = (20, 24, 26; 0.6, 0.8)$, $b_1^2 = (14, 16, 17; 0.4, 0.6)$, $b_2^2 = (11, 13, 15; 0.8, 0.5)$, $b_3^2 = (10, 11, 12; 0.5, 0.5)$,
$a_1^3 = (24, 26, 28; 0.6, 0.9)$, $a_2^3 = (32, 35, 37; 0.8, 0.5)$, $b_1^3 = (16, 18, 20; 0.4, 0.6)$, $b_2^3 = (12, 14, 16; 0.6, 1)$, $b_3^3 = (12, 15, 17; 0.5, 0.5)$,
$e_1 = (34, 36, 38; 0.5, 1)$, $e_2 = (46, 49, 51; 0.6, 0.8)$, $e_3 = (28, 30, 32; 0.7, 0.9)$, $e_4 = (40, 43, 44; 0.5, 0.5)$.

**Table 12.** Costs $c^{\tilde{1}}_{ijk}$

| $i\backslash j$ | 1 | 2 | 3 | k |
|---|---|---|---|---|
| 1 | (3,5,6;0.6,0.8) | (4,6,7;0.5,0.5) | (3,5,8;0.4,0.7) | 1 |
| 2 | (4,5,7;0.7,0.9) | (5,7,8;0.7,0.9) | (5,7,8;0.6,1) | |
| 1 | (4,5,7;0.6,0.9) | (3,5,6;0.5,0.5) | (4,6,8;0.8,1) | 2 |
| 2 | (6,8,9;0.4,0.6) | (4,6,8;0.5,0.7) | (5,7,8;0.7,0.8) | |
| 1 | (5,6,8;0.6,0.8) | (5,6,7;0.5,0.7) | (6,8,9;0.8,0.7) | 4 |
| 2 | (5,6,8;0.5,0.9) | (7,8,9;0.8,0.6) | (6,8,10;0.4,0.8) | |

**Table 13.** Costs $c^{\tilde{2}}_{ijk}$

| $i\backslash j$ | 1 | 2 | 3 | k |
|---|---|---|---|---|
| 1 | (5,7,9;0.5,0.6) | (4,6,8;0.4,0.8) | (6,8,10;0.5,0.5) | 1 |
| 2 | (6,7,8;0.8,0.6) | (7,9,10;0.5,0.8) | (4,5,7;0.6,0.9) | |
| 1 | (6,8,10;0.8,0.9) | (3,4,6;0.6,0.8) | (7,8,9;0.5,0.8) | 2 |
| 2 | (7,9,10;0.5,1) | (6,7,9;0.4,0.8) | (5,7,8;0.8,0.6) | |
| 1 | (4,6,8;0.5,0.5) | (7,9,10;0.7,0.9) | (5,7,8;0.6,0.7) | 4 |
| 2 | (6,8,9;1,0.6) | (4,5,7;0.8,0.6) | (6,7,9;0.7,0.9) | |

**Table 14.** Costs $c^{\tilde{3}}_{ijk}$

| $i\backslash j$ | 1 | 2 | 3 | k |
|---|---|---|---|---|
| 1 | (10,11,13;0.7,0.9) | (8,10,11;0.5,0.5) | (6,8,9;0.4,0.6) | 2 |
| 2 | (9,11,14;0.8,0.6) | (12,13,15;0.5,1) | (7,9,11;0.6,0.9) | |
| 1 | (12,14,15;0.4,0.7) | (7,9,11;0.5,0.1) | (8,10,12;0.8,0.9) | 4 |
| 2 | (6,8,9;0.6,0.8) | (13,14,16;0.8,0.5) | (7,10,12;0.9,0.6) | |

**Solution Using Chance-Constrained Programming (c.f. Sect. 4.7):** The predetermined general credibility levels for objective function and constraints are taken as $\alpha = 0.9$, $\alpha^p_i = 0.9$, $\beta^p_j = 0.9$, $\gamma_k = 0.9$, $p = 1, 2, 3$, $i = 1, 2$, $j = 1, 2, 3$, $k = 1, 2, 3, 4$. Then using (209)–(213) the equivalent deterministic form of the problem becomes

$$Min \sum_{p=1}^{l}\sum_{i=1}^{m}\sum_{j=1}^{n}\sum_{k=1}^{K} d^p_{ijk}\left[\frac{(0.8 + 0.6\theta^p_{r,ijk})c^{p3}_{ijk}x^p_{ijk} + 0.2c^{p2}_{ijk}x^p_{ijk}}{1 + 0.6\theta^p_{r,ijk}}\right]$$

$$sub.to \ \sum_{j=1}^{n}\sum_{k=1}^{K} d^p_{ijk}\, x^p_{ijk} \le F_{a^p_i}, \ i = 1, 2, ..., m; \ p = 1, 2, ...l, \qquad (228)$$

$$\sum_{i=1}^{m}\sum_{k=1}^{K} d_{ijk}^{p}\, x_{ijk}^{p} \geq F_{b_{j}^{p}}, \; j=1,2,...,n; \; p=1,2,...l,$$

$$\sum_{p=1}^{l}\sum_{i=1}^{m}\sum_{j=1}^{n} d_{ijk}^{p}\, x_{ijk}^{p} \leq F_{e_{k}}, \; k=1,2,...,K,$$

$$x_{ijk}^{p} \geq 0, \; d_{ijk}^{p} = \begin{cases} 1, \text{if } p \in I_{k}; \\ 0, \text{otherwise.} \end{cases} \forall\, i,j,k,p,$$

where, $F_{a_1^1} = 21.46$, $F_{a_2^1} = 26.29$, $F_{a_1^2} = 26.30$, $F_{a_2^2} = 20.58$, $F_{a_1^3} = 24.29$, $F_{a_2^3} = 32.40$, $F_{b_1^1} = 13.74$, $F_{b_2^1} = 14.71$, $F_{b_3^1} = 14.55$, $F_{b_1^2} = 16.85$, $F_{b_2^2} = 14.69$, $F_{b_3^2} = 11.84$, $F_{b_1^3} = 19.70$, $F_{b_2^3} = 15.75$, $F_{b_3^3} = 16.69$, $F_{e_1} = 34.30$, $F_{e_2} = 46.44$, $F_{e_3} = 28.28$, $F_{e_4} = 40.46$.

Solving this problem using LINGO solver, based upon Generalized Reduced Gradient (GRG) algorithm, we get the optimum solution as follows:
$x_{111}^1 = 6.75$, $x_{211}^1 = 6.99$, $x_{231}^1 = 5.07$, $x_{232}^1 = 9.48$, $x_{124}^2 = 14.71$, $x_{211}^2 = 3.65$, $x_{231}^2 = 11.84$, $x_{122}^2 = 13.1$, $x_{114}^3 = 13.2$, $x_{224}^2 = 1.59$, $x_{122}^3 = 7.17$, $x_{132}^3 = 8.54$, $x_{232}^3 = 8.15$, $x_{123}^3 = 8.58$, $x_{213}^3 = 19.7$ and minimum transportation cost (objective value)= 1093.482.

**Solution Using Nearest Interval Approximation (c.f. Sect. 4.7):** The nearest interval approximations (credibilistic) of the given triangular type-2 fuzzy parameters are calculated using the formula (145) and (146).

**Table 15.** Costs $\tilde{c}_{ijk}^{l}$

| $i\backslash j$ | 1 | 2 | 3 | k |
|---|---|---|---|---|
| 1 | [3.9904,5.5047] | [5.0,6.50] | [3.9841,6.5238] | 1 |
| 2 | [4.4955,6.0089] | [6.0109,7.4945] | [4.9820,7.0179] | |
| 1 | [4.4930,6.0138] | [4.0,5.50] | [4.9915,7.0084] | 2 |
| 2 | [6.9890,8.5054] | [4.9897,7.0102] | [5.9953,7.5023] | |
| 1 | [5.4952,7.0095] | [5.4948,6.5051] | [7.0095,8.4952] | 4 |
| 2 | [5.4904,7.0191] | [7.5047,8.4952] | [6.9795,9.0204] | |

The corresponding unit transportation costs as obtained are presented in Tables 15, 16 and 17 and supplies, demands, capacities are as follows:
$a_1^1 = [22.50, 24.50]$, $a_2^1 = [26.9904, 29.0095]$, $b_1^1 = [10.9910, 13.0089]$,
$b_2^1 = [12.4920, 14.0158]$, $b_3^1 = [10.4835, 13.5164]$,
$a_1^2 = [26.9767, 29.5348]$, $a_2^2 = [21.9809, 25.0095]$, $b_1^2 = [14.9890, 16.5054]$, $b_2^2 = [12.0148, 13.9851]$, $b_3^2 = [10.50, 11.50]$,

**Table 16.** Costs $c_{ijk}^{\tilde{2}}$

| $i\backslash j$ | 1 | 2 | 3 | k |
|---|---|---|---|---|
| 1 | [5.9947,8.0052] | [4.9795,7.0204] | [7.0,9.0] | 1 |
| 2 | [6.5047,7.4952] | [7.9851,9.5074] | [4.4930,6.0138] | |
| 1 | [6.9956,9.0043] | [3.4952,5.0095] | [7.4925,8.5074] | 2 |
| 2 | [7.9767,9.5116] | [6.4897,8.0204] | [6.0095,7.4952] | |
| 1 | [5.0,7.0] | [7.9910,9.5044] | [5.9950,7.5024] | 4 |
| 2 | [7.0179,8.4910] | [4.5047,5.9904] | [6.4955,8.0089] | |

**Table 17.** Costs $c_{ijk}^{\tilde{3}}$

| $i\backslash j$ | 1 | 2 | 3 | k |
|---|---|---|---|---|
| 1 | [10.4955,12.0089] | [9.0,10.50] | [6.9890,8.5054] | 2 |
| 2 | [10.0095,12.4856] | [12.4833,14.0232] | [7.9861,10.0138] | |
| 1 | [12.9841,14.5079] | [7.9767,10.0232] | [8.9956,11.0043] | 4 |
| 2 | [6.9904,8.5047] | [13.5074,14.9851] | [8.5208,10.9861] | |

$a_1^3 = [24.9861, 27.0138]$, $a_2^3 = [33.5222, 35.9851]$, $b_1^3 = [16.9890, 19.0109]$, $b_2^3 = [12.9820, 15.0179]$, $b_3^3 = [13.50, 16.0]$,
$e_1 = [34.9767, 37.0232]$, $e_2 = [47.4856, 50.0095]$, $e_3 = [28.9910, 31.0089]$, $e_4 = [41.50, 43.50]$.

Consider that the possibility degree of satisfaction of each of the source, destination and conveyance capacity constraints with interval right hand sides is 0.9. Then the equivalent deterministic forms of all the constraints are obtained using (222)–(224). Now subject to these deterministic constraints we find minimum and maximum possible value of the objective function by solving (225) and (226) and corresponding optimal solutions are obtained as follows:
$\underline{Z} = 725.9498$; $x_{131}^1 = 13.2131$, $x_{211}^1 = 10.5682$, $x_{112}^1 = 2.2388$, $x_{122}^1 = 7.2480$, $x_{222}^1 = 6.6154$, $x_{231}^2 = 11.4$, $x_{122}^2 = 10.8787$, $x_{114}^2 = 16.3537$, $x_{224}^2 = 2.9093$, $x_{122}^3 = 4.4302$, $x_{132}^3 = 10.3745$, $x_{232}^3 = 5.3754$, $x_{123}^3 = 10.3840$, $x_{213}^3 = 18.8087$ and $\overline{Z} = 950.1511$; $x_{111}^1 = 8.8365$, $x_{211}^1 = 3.9705$, $x_{231}^1 = 10.6858$, $x_{122}^1 = 13.8634$, $x_{232}^1 = 2.5272$, $x_{211}^2 = 0.2883$, $x_{231}^2 = 11.4$, $x_{122}^2 = 11.1671$, $x_{114}^2 = 16.0654$, $x_{224}^2 = 2.6209$, $x_{122}^3 = 4.4302$, $x_{132}^3 = 10.3745$, $x_{232}^3 = 5.3754$, $x_{123}^3 = 10.3840$, $x_{111}^3 = 18.8087$.

We now apply fuzzy linear programming to obtain an unique optimum allocation. We get $L_1 = 725.9498$, $U_1 = 741.4106$, $L_2 = 950.1511$, $U_2 = 956.9979$ and hence compromise optimal solution as
$x_{111}^1 = 0.2095$, $x_{131}^1 = 8.6270$, $x_{211}^1 = 12.5975$, $x_{231}^1 = 2.3472$, $x_{122}^1 = 13.8634$, $x_{232}^1 = 2.2388$, $x_{231}^2 = 11.4$, $x_{122}^2 = 10.8787$, $x_{114}^2 = 16.3537$, $x_{224}^2 = 2.9093$, $x_{122}^3 = 4.4302$, $x_{132}^3 = 10.3745$, $x_{232}^3 = 5.3754$, $x_{123}^3 = 10.3840$, $x_{213}^3 = 18.8087$, $\lambda = 0.987$, $\underline{Z}^* = 726.1475$, $\overline{Z}^* = 950.2386$.

## 4.8   Overall Conclusion

In this section, a defuzzification method of general type-2 fuzzy variable is outlined and compared numerically with geometric defuzzification method. A nearest interval approximation for continuous T2 FV is also introduced. Interval approximation method has been illustrated with type-2 triangular fuzzy variable. For the first time, two FCTPs and a MISTP with type-2 fuzzy parameters have been formulated and solved. Chance-constrained programming problems for a FCTP and MISTP with type-2 triangular fuzzy variables are formulated and solved. The MISTP with type-2 triangular fuzzy parameters is also solved using interval approximations of type-2 triangular fuzzy variables. Now-a-days, the volume and complexity of the collected data in various fields is growing rapidly. In order to describe and extract the useful information hidden in uncertain data and to use this data properly in practical problems, many researchers have proposed a number of improved theories including type-2 fuzzy set. The methodologies used in this chapter are quite general and these can be applied to the decision making problems in different areas with type-2 fuzzy parameters. The presented models can be extended to different types of transportation problems including price discounts, transportation time constraints, breakable/deteriorating items, etc.

# 5   Transportation Mode Selection Problem with Linguistic Terms

## 5.1   Introduction

Solid transportation problem (STP) is a problem of transporting goods from some sources to some destinations through several types of conveyances (modes of transportation) and the objective may be minimization of cost, time, maximization of profit, etc. But cost or time may not be the only criteria for selecting modes. There may be several other criteria for which all modes may not be equally preferable in a transportation system. Generally the available modes of transportation are rail, road, water, air, pipeline etc. Choice of modes depends upon several parameters (criteria) such as transportation cost, time, distance, product characteristics (e.g. weight, volume, value, life cycle etc.), flexibility, safety factor, inventory cost, etc. The main difficulty to select best mode is the conflicting nature of the modes under different criteria, i.e., under certain criteria, a mode may be superior than another but may not be under another criteria. Also all the criteria related to a transportation system may not have equal priority. For example generally faster modes are preferable than slower modes for time saving, but for product having low value to weight ratio, slower modes are preferable for transportation cost saving and in this case time has less priority than transportation cost. The product having short life cycle need rapid transportation modes, because here the main priority is time saving. So the tusk is to select overall best transportation mode with respect to all the selection criteria in a transportation system. Obviously multi-criteria (/attribute) decision making

(MCDM/MADM), which is a procedure to determine best alternative among some feasible alternatives, can be an efficient method to solve transportation mode selection problem. In literature there are several articles available related to transportation mode selection problem (Kiesmüller et al. [65], Kumru and Kumru [71], Monahan and Berger [113], Tuzkaya and Önüt [134], Eskigun et al. [41], Wang and Lee [138]).

Multi-criteria (/attribute) decision making (MCDM/MADM) (Anand et al. [7], Baleentis and Zeng [10], Chen and Lee [23, 24], Chen et al. [26], Dalalah et al. [32], Ding and Liang [37], Fu [45], Wang and Lee [139], Wang and Parkan [32], Wu and Chen [148]), is a method to select most convenient alternative among some available alternatives with respect to some evaluation criteria provided by decision maker(s) for a particular problem. This type of problems are often called multi-criteria(/attribute) group decision making (MCGDM/MAGDM) problem in presence of several decision makers. The evaluation ratings of the alternatives with respect to the criteria and criteria weights as provided by the decision makers are generally linguistic terms (e.g., very high, medium, fair, good, etc.). Human judgements are not always precise and also a word does not have the same meaning to different people and is therefore uncertain. Zadeh [157, 158] first used a fuzzy set (Zadey [156]) to model a word. Many researchers (Anand et al. [7], Cheng and Lin [28], Dalalah et al. [32], Ding and Liang [37], Dursun et al. [39], Hatami-Marbini and Tavana [55], Tuzkaya and Önüt [134], Wang and Lee [139], Wang and Parkan [142]) developed MCDM problems where type-1 fuzzy sets(/numbers) are used to describe linguistic uncertainties rather than just single numeric value. Then the problem is called fuzzy multi-criteria decision making (FMCDM) problem in which evaluation ratings and criteria weights are fuzzy numbers.

Fuzzy analytical hierarchy process (FAHP) (Anand et al. [7], Chan, N. Kumar [16], Mikhailov and Tsvetinov [111]), Fuzzy analytical network process (FANP) (Ertay et al. [40], Mikhailov and Sing [110], Tuzkaya and Önüt [134]), fuzzy preference relation based decision making (Lee [78], Wang [137]), fuzzy TOPSIS method (Chen [20], Wang and Elhag [141], Wang et al. [140], Wang and Lee [139]) are some available methods for solving FMCDM problems. The main drawback of the FAHP and FANP methods is that these methods consist of large number of fuzzy pair-wise comparison which makes the methods difficult for computation. Lee's (Lee [78]) method using extended fuzzy preference relation is computationally efficient, but in case of two alternatives, whatever the ratings of the two alternatives are, this method always gives total performance index of one alternative as 1 and that of another 0. So it is difficult to compare the alternatives with each other in the sense how much one is preferable than another and to use the performance indices in any further requirements. In fuzzy TOPSIS method of Wang and Elhag [141], for each member of fuzzy decision matrix, different $\alpha$-level sets are to be evaluated and for each different $\alpha$-levels, two NLP models are to be solved. So to find accurate result, large number of $\alpha$-levels are to be set and then corresponding time complexity becomes high. Fuzzy TOPSIS of Wang et al. [140] is less complex, but the positive and negative ideal solutions as derived by Max and Min operations under fuzzy environment may not be founded on feasible alternatives.

Ranking of fuzzy number is an important issue in group decision making, specially for decision making with linguistic terms, which are generally represented by fuzzy numbers. There are several methods of fuzzy ranking (Abbasbandy and Asady [1], Cheng [27], Chu and Tsao [29], Fortems and Roubens [44], Liu [90], Lee [78], Liou and Wang [100]) available in the literature, each of which has some advantages and disadvantages. Also many ranking methods are based on defuzzification (e.g., expected value (Liu [90]), centroid (Wang et al. [143]), magnitude (Abbasbandy and Hajjari [2]) of fuzzy number) in which, from fuzzy numbers corresponding crisp quantities are obtained using some utility functions and fuzzy numbers are ranked according to these crisp values. Drawback of defuzzification is that it tends to loss some information and thus is unable to grasp the sense of uncertainty. For example, expected value (Liu and Liu [90]) of a trapezoidal fuzzy number $(r_1, r_2, r_3, r_4)$ is just the average $(r_1 + r_2 + r_3 + r_4)/4$, though each $r_i$ does not have the same membership (/possibility) degree. Some of the techniques (Abbasbandy and Asady [1], Chen [19]; Liou and Wang [100]) are case dependent and produce different results in different cases for certain fuzzy numbers. Also some methods are found to be logically incorrect. For example, Asady and Zendehnam's [8] distance minimization method and Chen's [19] method are seemed to be logically incorrect as shown by Abbasbandy and Hajjari [2] and Liou and Wang [100] respectively.

Mendel [102] explained using Popper's Falsificationism that modeling word using type-1 fuzzy set is not scientifically correct. Mendel [102, 103] also explained that a sensible way to model a word is to using interval type-2 fuzzy set (IT2 FS). There are some methodologies, such as the interval approach (Liu and Mendel [95], the person membership function approach (Mendel [104]) and the interval end-points approach (Mendel and Wu [108]) available to obtain mathematical models for IT2 FS for words. Chen and Lee [23, 24] developed fuzzy multiple attributes group decision-making methods (FMAGDM) based on ranking IT2 FSs and interval type-2 TOPSIS method respectively where linguistic weights are represented by IT2 FSs. Chen et al. [26], Chen and Wang [25] developed FMAGDM method based on ranking IT2 FSs.

In this section, a new ranking method of fuzzy numbers is developed using a ranking function which we define using credibility measure. This ranking function is bounded over [0,1] so that it is easy to compare two fuzzy numbers with each other. We also provide a method of ranking interval type-2 fuzzy variables (IT2 FVs) using a ranking function which we define with the help of generalized credibility measure. Then we propose two computationally efficient fuzzy MCGDM (FMCGDM) methods, first one based on proposed ranking method of fuzzy numbers and the second one based on the proposed ranking method of IT2 FVs. We discuss how to assign weights of modes if a decision maker wish, to a STP in addition to the main criteria. The proposed FMCGDM methods are applied to two transportation mode selection problems where evaluation ratings and criteria weights are expressed by linguistic terms.

## 5.2   Theoretical Developments

In the following Subsects. 5.4 and 5.6 we have developed two fuzzy multi-criteria group decision making process, first one based on ranking fuzzy numbers and second one based on ranking interval type-2 fuzzy variables. For the construction of the methods we need some results which are given below.

**Theorem 5.1:** For any two trapezoidal fuzzy numbers $\tilde{A} = (a_1, a_2, a_3, a_4)$ and $\tilde{B} = (b_1, b_2, b_3, b_4)$, $Cr\{\tilde{A} \le \tilde{B}\} \ge \alpha$ if and only if

$$(1 - 2\alpha)(a_1 - b_4) + 2\alpha(a_2 - b_3) \le 0 \quad for \ \alpha \le 0.5$$

$$2(1 - \alpha)(a_3 - b_2) + (2\alpha - 1)(a_4 - b_1) \le 0 \quad for \ \alpha > 0.5.$$

Proof: $Cr\{\tilde{A} \le \tilde{B}\} \ge \alpha \Leftrightarrow Cr\{(\tilde{A} - \tilde{B}) \le 0\} \ge \alpha \Leftrightarrow Cr\{(\tilde{A} + \tilde{B}') \le 0\} \ge \alpha$, where $\tilde{B}' = -\tilde{B} = (-b_4, -b_3, -b_2, -b_1)$.

Then the theorem follows from Theorem 3.3 (Sect. 3).

**Corollary 5.1.** For any two trapezoidal fuzzy numbers $\tilde{A} = (a_1, a_2, a_3, a_4)$ and $\tilde{B} = (b_1, b_2, b_3, b_4)$, $Cr\{\tilde{A} \ge \tilde{B}\} \ge \alpha$ if and only if

$$(1 - 2\alpha)(a_4 - b_1) + 2\alpha(a_3 - b_2) \ge 0 \quad for \ \alpha \le 0.5$$

$$2(1 - \alpha)(a_2 - b_3) + (2\alpha - 1)(a_1 - b_4) \ge 0 \quad for \ \alpha > 0.5.$$

**Theorem 5.2:** For any two triangular fuzzy numbers $\tilde{A} = (a_1, a_2, a_3)$ and $\tilde{B} = (b_1, b_2, b_3)$, $Cr\{\tilde{A} \le \tilde{B}\} \ge \alpha$ if and only if

$$(1 - 2\alpha)(a_1 - b_3) + 2\alpha(a_2 - b_2) \le 0 \quad for \ \alpha \le 0.5$$

$$2(1 - \alpha)(a_2 - b_2) + (2\alpha - 1)(a_3 - b_1) \le 0 \quad for \ \alpha > 0.5.$$

**Corollary 5.2.** For any two triangular fuzzy numbers $\tilde{A} = (a_1, a_2, a_3)$ and $\tilde{B} = (b_1, b_2, b_3)$, $Cr\{\tilde{A} \ge \tilde{B}\} \ge \alpha$ if and only if

$$(1 - 2\alpha)(a_3 - b_1) + 2\alpha(a_2 - b_2) \ge 0 \quad for \ \alpha \le 0.5$$

$$2(1 - \alpha)(a_2 - b_2) + (2\alpha - 1)(a_1 - b_3) \ge 0 \quad for \ \alpha > 0.5.$$

## 5.3   A New Approach for Ranking of Fuzzy Numbers

To rank fuzzy numbers $\tilde{A}$ and $\tilde{B}$, we propose to find the possible credibility degree to which $\tilde{A} \ge \tilde{B}$ or $\tilde{A} \le \tilde{B}$. For this purpose we find the maximum satisfied credibility degree that $\tilde{A} \ge \tilde{B}$ or maximum satisfied credibility degree that $\tilde{A} \le \tilde{B}$, i.e.

$$Max_{\alpha \in [0,1]}[Cr\{\tilde{A} \ge \tilde{B}\} = \alpha] \qquad (229)$$

$$or, \ Max_{\alpha \in [0,1]}[Cr\{\tilde{A} \le \tilde{B}\} = \alpha]. \qquad (230)$$

We denote (229) by $M_\alpha(\tilde{A} \ge \tilde{B})$ and (230) by $M_\alpha(\tilde{A} \le \tilde{B})$.
Suppose for two fuzzy numbers $\tilde{A}$ and $\tilde{B}$, $M_\alpha(\tilde{A} \ge \tilde{B}) = \alpha'$, then we say that $\tilde{A} \ge \tilde{B}$ with credibility $\alpha'$.

**Example 5.1.** Suppose $\tilde{A}$ and $\tilde{B}$ be two trapezoidal fuzzy numbers defined by $\tilde{A} = (a_1, a_2, a_3, a_4)$ and $\tilde{B} = (b_1, b_2, b_3, b_4)$. Then from Corollary 5.1 it follows that $M_\alpha(\tilde{A} \geq \tilde{B}) = Max_{\alpha \in [0,1]}[Cr\{\tilde{A} \geq \tilde{B}\} = \alpha]$ is obtained by solving

$$Max\ \alpha$$

$$(1 - 2\alpha)(a_4 - b_1) + 2\alpha(a_3 - b_2) \geq 0 \quad for\ 0 \leq \alpha \leq 0.5 \qquad (231)$$
$$2(1 - \alpha)(a_2 - b_3) + (2\alpha - 1)(a_1 - b_4) \geq 0 \quad for\ 0.5 < \alpha \leq 1$$

$$0 \leq \alpha \leq 1.$$

As our object is to find maximum possible credibility degree $\alpha$, an easy way of solving (231) is that first solve Max $\alpha$ with respect to the second constraint, if such $\alpha$ $(0.5 < \alpha \leq 1)$ exist then this is the required solution and if it does not exist then solve Max $\alpha$ with respect to the first constraint.

If we consider $\tilde{A} = (4, 6, 7, 9)$ and $\tilde{B} = (2, 3, 5, 7)$, then $M_\alpha(\tilde{A} \geq \tilde{B}) = Max_{\alpha \in [0,1]}[Cr\{\tilde{A} \geq \tilde{B}\} = \alpha] = 0.625$.

**Example 5.2.** Suppose $\tilde{A}$ and $\tilde{B}$ be two triangular fuzzy numbers defined by $\tilde{A} = (a_1, a_2, a_3)$ and $\tilde{B} = (b_1, b_2, b_3)$. Then from Corollary 5.2 it follows that $M_\alpha(\tilde{A} \geq \tilde{B}) = Max_{\alpha \in [0,1]}[Cr\{\tilde{A} \geq \tilde{B}\} = \alpha]$ is obtained by solving

$$Max\ \alpha$$

$$(1 - 2\alpha)(a_3 - b_1) + 2\alpha(a_2 - b_2) \geq 0 \quad for\ 0 \leq \alpha \leq 0.5 \qquad (232)$$
$$2(1 - \alpha)(a_2 - b_2) + (2\alpha - 1)(a_1 - b_3) \geq 0 \quad for\ 0.5 < \alpha \leq 1$$

$$0 \leq \alpha \leq 1.$$

For example if $\tilde{A} = (2, 4, 6)$ and $\tilde{B} = (3, 5, 6)$ then $M_\alpha(\tilde{A} \geq \tilde{B}) = Max_{\alpha \in [0,1]}[Cr\{\tilde{A} \geq \tilde{B}\} = \alpha] = 0.375$.

**Ranking Function:** We define ranking function R to rank one fuzzy number $\tilde{A}$ upon another fuzzy number $\tilde{B}$ as follows:

$$R(\tilde{A}, \tilde{B}) = \begin{cases} M_\alpha(\tilde{A} \geq \tilde{B}), & \text{if it exist;} \\ 0, & \text{otherwise.} \end{cases} \qquad (233)$$

Obviously

$$R(\tilde{B}, \tilde{A}) = \begin{cases} M_\alpha(\tilde{A} \leq \tilde{B}), & \text{if it exist;} \\ 0, & \text{otherwise.} \end{cases} \qquad (234)$$

It follows from the definition of $R(\tilde{A}, \tilde{B})$ and self-duality property of the credibility measure that R is reciprocal, i.e., $R(\tilde{A}, \tilde{B}) = 1 - R(\tilde{B}, \tilde{A})$. Also from (231) and (232) it is clear that R is transitive for trapezoidal or triangular fuzzy numbers, i.e. $R(\tilde{A}, \tilde{B}) \geq 1/2$ and $R(\tilde{B}, \tilde{C}) \geq 1/2 \Rightarrow R(\tilde{A}, \tilde{C}) \geq 1/2$ for any trapezoidal or triangular fuzzy numbers $\tilde{A}$, $\tilde{B}$, $\tilde{C}$. So R is total ordering and satisfies all the criteria proposed by Yuan [154]. For any two fuzzy numbers $\tilde{A}$ and $\tilde{B}$, the ranking of $\tilde{A}$, $\tilde{B}$ is done as follows:

Table 18. Comparative results of Example 5.3

| Methods | Evaluation | Set 1 | Set 2 | Set 3 |
|---------|-----------|-------|-------|-------|
| Proposed method | $R(\tilde{A}, \tilde{B})$ | 0.625 | 0.545 | 0.5 |
| Order relation | | $\tilde{A} \succ \tilde{B}$ | $\tilde{A} \succ \tilde{B}$ | $\tilde{A} \sim \tilde{B}$ |
| Expected value (Liu [90]) | $E(\tilde{A})$ | 6.5 | 3.5 | 4.25 |
| | $E(\tilde{B})$ | 4.25 | 3.4 | 4.25 |
| Order relation | | $\tilde{A} \succ \tilde{B}$ | $\tilde{A} \succ \tilde{B}$ | $\tilde{A} \sim \tilde{B}$ |
| Sign distance for $p = 1$ | $d_p(\tilde{A}, 0)$ | 13 | 7 | 8.5 |
| (Abbasbandy and Asady [1]) | $d_p(\tilde{B}, 0)$ | 8.5 | 6.8 | 8.5 |
| Order relation | | $\tilde{A} \succ \tilde{B}$ | $\tilde{A} \succ \tilde{B}$ | $\tilde{A} \sim \tilde{B}$ |
| Sign distance for $p = 2$ | $d_p(\tilde{A}, 0)$ | 9.469 | 4.9665 | 6.298 |
| (Abbasbandy and Asady [1]) | $d_p(\tilde{B}, 0)$ | 6.531 | 4.9625 | 6.531 |
| Order relation | | $\tilde{A} \succ \tilde{B}$ | $\tilde{A} \succ \tilde{B}$ | $\tilde{A} \prec \tilde{B}$ |
| Lee [78] | $\mu_F(\tilde{A}, \tilde{B})$ | 4.5 | 0.2 | 0 |
| Order relation | | $\tilde{A} \succ \tilde{B}$ | $\tilde{A} \succ \tilde{B}$ | $\tilde{A} \sim \tilde{B}$ |
| Liou and Wang [100] ($\alpha = 1/2$) | $I_T^\alpha(\tilde{A})$ | 6.5 | 3.5 | 4.25 |
| | $I_T^\alpha(\tilde{B})$ | 4.25 | 3.4 | 4.25 |
| Order relation | | $\tilde{A} \succ \tilde{B}$ | $\tilde{A} \succ \tilde{B}$ | $\tilde{A} \sim \tilde{B}$ |
| Fortems and Roubens [44] | $C(\tilde{A} \geq \tilde{B})$ | 2.25 | 0.1 | 0 |
| Order relation | | $\tilde{A} \succ \tilde{B}$ | $\tilde{A} \succ \tilde{B}$ | $\tilde{A} \sim \tilde{B}$ |
| Cheng distance (Cheng [27]) | $R(\tilde{A})$ | 6.519 | 3.535 | 4.231 |
| | $R(\tilde{B})$ | 4.314 | 3.467 | 4.313 |
| Order relation | | $\tilde{A} \succ \tilde{B}$ | $\tilde{A} \succ \tilde{B}$ | $\tilde{A} \prec \tilde{B}$ |
| Chu and Tsao [29] | $R(\tilde{A})$ | 3.25 | 1.75 | 2.141 |
| | $R(\tilde{B})$ | 2.101 | 1.699 | 2.101 |
| Order relation | | $\tilde{A} \succ \tilde{B}$ | $\tilde{A} \succ \tilde{B}$ | $\tilde{A} \succ \tilde{B}$ |
| Abbasbandy and Hajjari [2] | $Mag(\tilde{A})$ | 6.5 | 3.5 | 4.416 |
| | $Mag(\tilde{B})$ | 4.08 | 2.73 | 4.083 |
| Order relation | | $\tilde{A} \succ \tilde{B}$ | $\tilde{A} \succ \tilde{B}$ | $\tilde{A} \succ \tilde{B}$ |

(i) $\tilde{A} \succ \tilde{B}$ iff $R(\tilde{A}, \tilde{B}) > 1/2$.

(ii) $\tilde{A} \prec \tilde{B}$ iff $R(\tilde{A}, \tilde{B}) < 1/2$.

(iii) But if $R(\tilde{A}, \tilde{B}) = 1/2$, then it is difficult to determine which is larger and which is smaller. In this case we may conclude $\tilde{A} \sim \tilde{B}$.

**Example 5.3.** Consider the following sets.

Set 1: $\tilde{A} = (4, 6, 7, 9)$, $\tilde{B} = (2, 3, 5, 7)$;

Set 2: $\tilde{A} = (3, 3.5, 4)$, $\tilde{B} = (2, 3.3, 5)$

Set 3: $\tilde{A} = (2, 4, 5, 6)$, $\tilde{B} = (2, 3, 5, 7)$

A comparative results of our proposed method and several other methods are presented in Table 18.

**Remark:** From the Table 18 we observe that when value of $R(\tilde{A}, \tilde{B})$ (i.e. the credibility that $\tilde{A} \geq \tilde{B}$) in our proposed method is far from 0.5 (larger or smaller than 0.5), then all the methods give the same result. For example for Set 1, $R(\tilde{A}, \tilde{B}) = 0.625$ and all methods give the same result that $\tilde{A} \succ \tilde{B}$. But as the credibility becomes close to 0.5, all methods do not give the same result. For example for the Set 3, results of our proposed method, Expected value (Liu [90]), Sign distance for $p = 1$ (Abbasbandy and Asady [1]), Lee's [78], Liou and Wang's [100], Fortems and Roubens's [44] methods are $\tilde{A} \sim \tilde{B}$, but Sign distance for $p = 2$ (Abbasbandy and Asady [1]), Cheng distance (Cheng [27]) methods give $\tilde{A} \prec \tilde{B}$ and Chu and Tsao's [29], Abbasbandy and Hajjari's [2] methods furnish $\tilde{A} \succ \tilde{B}$.

## 5.4  The Proposed FMCGDM Method Based on Ranking Fuzzy Numbers

Suppose $A_1, A_2, ..., A_m$ are $m$ alternatives and these alternatives are evaluated on basis of the criteria $C_1, C_2, ..., C_n$ by the decision makers $D_l, l = 1, 2, ..., p$. Suppose rating of $A_i$ based on criteria $C_j$ according to the decision maker $D_l$ is $\tilde{A}_{ij}^l$ which is represented by fuzzy number, where $i = 1, 2, ..., m$, $j = 1, 2, ..., n$ and $l = 1, 2, ..., p$. Let $\tilde{w}_j^l$ be the fuzzy weight of the criteria $C_j$ indicating its importance given by the decision maker $D_l$ for all $j$ and $l$. The proposed fuzzy MCGDM method to rank the alternatives is as follows:

**Step-1:** Construct the decision matrix $\tilde{D} = [\tilde{A}_{ij}]_{m \times n}$ where each $\tilde{A}_{ij}$ is the average of the ratings of alternative $A_i$ given by the decision makers $D_l, l = 1, 2, ..., p$ based on criteria $C_j$, i.e.,

$$
\begin{array}{c}
\phantom{\tilde{D} =} \quad C_1 \; C_2 \; .... \; C_n \\[4pt]
\tilde{D} = \begin{array}{c} A_1 \\ A_2 \\ \vdots \\ A_m \end{array}
\begin{bmatrix}
\tilde{A}_{11} & \tilde{A}_{12} & ... & \tilde{A}_{1n} \\
\tilde{A}_{21} & \tilde{A}_{22} & ... & \tilde{A}_{2n} \\
\vdots & \vdots & \vdots & \vdots \\
\tilde{A}_{m1} & \tilde{A}_{m2} & ... & \tilde{A}_{mn}
\end{bmatrix},
\end{array}
$$

where

$$
\tilde{A}_{ij} = \frac{\tilde{A}_{ij}^1 \oplus \tilde{A}_{ij}^2 \oplus ... \oplus \tilde{A}_{ij}^p}{p}. \tag{235}
$$

Calculate the average weights $w_j$ of the each criteria $C_j$ by averaging their weights given by the decision makers $D_l, l = 1, 2, ..., p$, i.e.

$$
\tilde{w}_j = (\tilde{w}_j^1 \oplus \tilde{w}_j^2 \oplus .... \oplus \tilde{w}_j^p)/p. \tag{236}
$$

**Step-2:** Normalize the decision matrix $\tilde{D}$. Suppose $\tilde{D}' = [\tilde{A}'_{ij}]_{m \times n}$ be the normalized decision matrix (normalizing process is shown in the end of this method).

**Step-3:** Derive the relative preference (/performance) matrix $P = [r_{ij}]_{m \times n}$, where

$$r_{ij} = \sum_{k \neq i} R(\tilde{A}'_{ij}, \tilde{A}'_{kj}), \tag{237}$$

$R$ is the ranking function as defined in (233). $r_{ij}$ is called the relative preference index of the alternative $A_i$ with respective to all the remaining alternatives for the criteria $C_j$.

**Step-4:** Calculate the fuzzy weighted relative preference of the each alternative by

$$A_i^* = \sum_{j=1}^{n} r_{ij} \otimes \tilde{w}_j, \ i = 1, 2, ..., m. \tag{238}$$

**Step-5:** Find the total preference index of each alternative by

$$r_i = \sum_{k \neq i} R(\tilde{A}_i^*, \tilde{A}_k^*), \ i = 1, 2, ..., m. \tag{239}$$

**Step-6:** Normalize the preference indices $r_i$ to obtain preference weights of the alternatives that sum to 1 by

$$w_i^P = \frac{r_i}{\sum_j^m r_j}, \ i = 1, 2, .., m. \tag{240}$$

**Step-7:** Rank alternatives according to their weights $w_i^P$, $i = 1, 2, ..., m$.
The process of normalization of $\tilde{D} = [\tilde{A}_{ij}]_{m \times n}$ is shown below in case when $\tilde{A}_{ij}$ are triangular fuzzy numbers:
Suppose each $\tilde{A}_{ij}$ is a triangular fuzzy number defined by $\tilde{A}_{ij} = (a_{ij}^L, a_{ij}^M, a_{ij}^U)$, then

$$\tilde{A}'_{ij} = \left( \frac{a_{ij}^L}{a_j^*}, \frac{a_{ij}^M}{a_j^*}, \frac{a_{ij}^U}{a_j^*} \right), \ where \ a_j^* = \max_i \{a_{ij}^U\},$$

where $j$ is the benefit criteria or the ratings are given in favor of the criteria (i.e. in positive sense).

But if $j$ is a cost criteria and $\tilde{A}_{ij}$ is given as amount of cost but not as rating, then normalization is done as follows:

$$\tilde{A}'_{ij} = \left( \frac{a_j^*}{a_{ij}^U}, \frac{a_j^*}{a_{ij}^M}, \frac{a_j^*}{a_{ij}^L} \right), \ where \ a_j^* = \min_i \{a_{ij}^L\}.$$

The flow-chart of the above method is presented in Fig. 17.

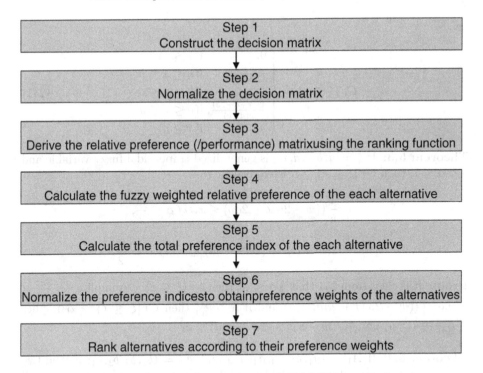

**Fig. 17.** Flow-chart of the proposed FMCGDM method

## 5.5    A Method of Ranking Trapezoidal Interval Type-2 Fuzzy Variables

**Some Results on Generalized Trapezoidal Fuzzy Variables:** In Sect. 2, we already mentioned that if a fuzzy variable is not normal, i.e. for generalized fuzzy variable, generalized credibility measure $\tilde{Cr}$ has to be used instead of the usual credibility measure. For a generalized trapezoidal fuzzy variable $\tilde{\xi} = (a, b, c, d; w)$, $\tilde{Cr}\{\tilde{\xi} \leq x\}$ is obtained as follows:

$$
\begin{aligned}
\tilde{Cr}\{\tilde{\xi} \leq x\} &= \frac{1}{2}(w + \sup_{r \leq x} \mu_{\tilde{\xi}}(x) - \sup_{r > x} \mu_{\tilde{\xi}}(x)) \\
&= \frac{1}{2}\{w + 0 - w\} = 0, \; if \; x \leq a \\
&= \frac{1}{2}\{w + \frac{w(x - a)}{b - a} - w\} = \frac{w(x - a)}{2(b - a)}, \; if \; a \leq x \leq b \\
&= \frac{1}{2}\{w + w - w\} = \frac{w}{2}, \; if \; b \leq x \leq c \\
&= \frac{1}{2}\{w + w - \frac{w(d - x)}{d - c}\} = \frac{w(x + d - 2c)}{2(d - c)}, \; if \; c \leq x \leq d \\
&= \frac{1}{2}\{w + w - 0\} = w, \; if \; x \geq d,
\end{aligned}
$$

i.e.,

$$\tilde{Cr}\{\tilde{\xi} \le x\} = \begin{cases} 0, & \text{if } x \le a; \\ \frac{w(x-a)}{2(b-a)}, & \text{if } a \le x \le b; \\ \frac{w}{2}, & \text{if } b \le x \le c; \\ \frac{w(x+d-2c)}{2(d-c)}, & \text{if } c \le x \le d; \\ w, & \text{if } x \ge d. \end{cases} \tag{241}$$

**Theorem 5.3:** If $\tilde{\xi} = (a, b, c, d; w)$ is generalized trapezoidal fuzzy variable and $0 < \alpha \le 1$, then $Cr\{\tilde{\xi} \le x\} \ge \alpha$ is equivalent to

$$(i) \ \frac{1}{w}((w - 2\alpha)a + 2\alpha b) \le x, \ if \ \alpha \le \frac{w}{2},$$

$$(ii) \ \frac{1}{w}(2(w - \alpha)c + (2\alpha - w)d) \le x, \ if \ \alpha > \frac{w}{2}.$$

**Proof:** It is clear from 241 that if $\alpha \le \frac{w}{2}$, then $Cr\{\tilde{\xi} \le x\} \ge \alpha$ implies $\frac{w(x-a)}{2(b-a)} \ge \alpha$, i.e. $\frac{1}{w}((w - 2\alpha)a + 2\alpha b) \le x$ and if $\alpha > \frac{w}{2}$, then $Cr\{\tilde{\xi} \le x\} \ge \alpha$ implies $\frac{w(x+d-2c)}{2(d-c)} \ge \alpha$, i.e. $\frac{1}{w}(2(w - \alpha)c + (2\alpha - w)d) \le x$.

**Theorem 5.4:** If $\tilde{A}_1 = (a_1, a_2, a_3, a_4; w_1)$ and $\tilde{A}_2 = (b_1, b_2, b_3, b_4; w_2)$ be two generalized trapezoidal fuzzy variables and $0 < \alpha \le 1$, then $\tilde{Cr}\{\tilde{A}_1 \le \tilde{A}_2\} \ge \alpha$ if

$$(w - 2\alpha)(a_1 - b_4) + 2\alpha(a_2 - b_3) \le 0 \ \ for \ \alpha \le w/2$$

$$2(w - \alpha)(a_3 - b_2) + (2\alpha - w)(a_4 - b_1) \le 0 \ \ for \ \alpha > w/2,$$

where $w = min(w_1, w_2)$.

**Proof:** $-\tilde{A}_2 = (-b_4, -b_3, -b_2, -b_1; w_2)$ and so $\tilde{A}_1 - \tilde{A}_2 = (a_1 - b_4, a_2 - b_3, a_3 - b_2, a_4 - b_1; min(w_1, w_2))$. Now $Cr\{\tilde{A}_1 \le \tilde{A}_2\} \ge \alpha \Leftrightarrow Cr\{(\tilde{A}_1 - \tilde{A}_2) \le 0\} \ge \alpha$ and hence the theorem follows from Theorem 5.3.

**Corollary 5.3:** If $\tilde{A}_1 = (a_1, a_2, a_3, a_4; w_1)$ and $\tilde{A}_2 = (b_1, b_2, b_3, b_4; w_2)$ be two generalized trapezoidal fuzzy variables and $0 < \alpha \le 1$, then $\tilde{Cr}\{\tilde{A}_1 \ge \tilde{A}_2\} \ge \alpha$ if

$$(w - 2\alpha)(a_4 - b_1) + 2\alpha(a_3 - b_2) \ge 0 \ \ for \ \alpha \le w/2$$

$$2(w - \alpha)(a_2 - b_3) + (2\alpha - w)(a_1 - b_4) \ge 0 \ \ for \ \alpha > w/2,$$

where $w = min(w_1, w_2)$.

**Ranking Function and Relative Preference Index:** Here we define a ranking function which can be used to rank two generalized fuzzy variables, say $\tilde{A}_1$ and $\tilde{A}_2$. For this purpose we find the possible credibility (generalized) degree to which $\tilde{A}_1 \geq \tilde{A}_2$ or $\tilde{A}_1 \leq \tilde{A}_2$, i.e. we find the maximum satisfied credibility degree that $\tilde{A}_1 \geq \tilde{A}_2$ or maximum satisfied credibility degree that $\tilde{A}_1 \leq \tilde{A}_2$, i.e.

$$Max_{\alpha \in [0,w]}[\tilde{C}r\{\tilde{A}_1 \geq \tilde{A}_2\} = \alpha] \tag{242}$$

$$or, \ Max_{\alpha \in [0,w]}[\tilde{C}r\{\tilde{A}_1 \leq \tilde{A}_2\} = \alpha], \tag{243}$$

where $w$ is minimum of heights of $\tilde{A}_1$ and $\tilde{A}_2$, so $0 < w \leq 1$.

We denote (242) by $M_\alpha(\tilde{A}_1 \geq \tilde{A}_2)$ and (242) by $M_\alpha(\tilde{A}_1 \leq \tilde{A}_2)$. Suppose for two generalized fuzzy numbers $\tilde{A}_1$ and $\tilde{A}_2$, $M_\alpha(\tilde{A}_1 \geq \tilde{A}_2) = \alpha'$, then we say that $\tilde{A}_1 \geq \tilde{A}_2$ with credibility $\alpha'$.

Suppose $\tilde{A}_1 = (a_1, a_2, a_3, a_4; w_1)$ and $\tilde{A}_2 = (b_1, b_2, b_3, b_4; w_2)$ be two generalized trapezoidal fuzzy variables and $w = min(w_1, w_2)$. Then from Corollary 5.3 it follows that $M_\alpha(\tilde{A}_1 \geq \tilde{A}_2) = Max_{\alpha \in [0,w]}[\tilde{C}r\{\tilde{A}_1 \geq \tilde{A}_2\} = \alpha]$ is obtained by solving

$$Max \ \alpha$$

$$s.t. \ (w - 2\alpha)(a_4 - b_1) + 2\alpha(a_3 - b_2) \geq 0 \ \ for \ 0 \leq \alpha \leq w/2 \tag{244}$$
$$2(w - \alpha)(a_2 - b_3) + (2\alpha - w)(a_1 - b_4) \geq 0 \ \ for \ w/2 < \alpha \leq w$$

$$0 \leq \alpha \leq w.$$

As the objective is to find maximum possible credibility degree $\alpha$, so to solve 244, one may first find Max $\alpha$ with respect to the second constraint, if such $\alpha$ ($w/2 < \alpha \leq w$) exist then this is the required solution and if it does not exist then find Max $\alpha$ satisfying the first constraint.

**Example 5.4:** As an example consider $\tilde{A}_1 = (5, 7, 8, 9; 1)$ and $\tilde{A}_2 = (4, 5, 6, 9; 0.8)$, then $w = 0.8$ and solving (5.16) for this example we have $M_\alpha(\tilde{A}_1 \geq \tilde{A}_2) = Max_{\alpha \in [0,w]}[Cr\{\tilde{A}_1 \geq \tilde{A}_2\} = \alpha] = 0.48$.

**Ranking Function:** We define a ranking function that can be used to rank two generalized fuzzy variable $\tilde{A}_1$ and $\tilde{A}_2$ as follows:

$$R(\tilde{A}_1 \geq \tilde{A}_2) = \begin{cases} M_\alpha(\tilde{A}_1 \geq \tilde{A}_2), & \text{if it exist;} \\ 0, & \text{otherwise.} \end{cases} \tag{245}$$

Obviously

$$R(\tilde{A}_1 \leq \tilde{A}_2) = \begin{cases} M_\alpha(\tilde{A}_1 \leq \tilde{A}_1), & \text{if it exist;} \\ 0, & \text{otherwise.} \end{cases} \tag{246}$$

It follows from the definition of $R(\tilde{A}_1, \tilde{A}_2)$ and self-duality property of the credibility measure that $R(\tilde{A}_1 \geq \tilde{A}_2) = w - R(\tilde{A}_1 \leq \tilde{A}_2)$. In particular if $\tilde{A}_1$ and $\tilde{A}_2$ are normalized, i.e. $w = 1$, then R is reciprocal. Also from (244) it is clear that R is transitive for generalized trapezoidal fuzzy variables, i.e. $R(\tilde{A}_1, \tilde{A}_2) \geq w/2$

and $R(\tilde{A}_2, \tilde{A}_3) \geq w/2 \Rightarrow R(\tilde{A}_1, \tilde{A}_3) \geq w/2$ for any trapezoidal fuzzy variables $\tilde{A}_1, \tilde{A}_1, \tilde{A}_1$. So R is total ordering and satisfies all the criteria of fuzzy ranking proposed by Yuan [154]. For any two generalized fuzzy variables $\tilde{A}_1$ and $\tilde{A}_2$, the ranking of $\tilde{A}_1, \tilde{A}_2$ is done as follows:

(i) $\tilde{A}_1 \succ \tilde{A}_2$ iff $R(\tilde{A}_1 \geq \tilde{A}_2) > w/2$.
(ii) $\tilde{A}_1 \prec \tilde{A}_2$ iff $R(\tilde{A}_1 \geq \tilde{A}_2) < w/2$.
(iii) But if $R(\tilde{A}_1 \geq \tilde{A}_2) = w/2$, then rank them including their heights, i.e., $\tilde{A}_1 \sim \tilde{A}_2$ also if $w_1 = w_2$, $\tilde{A}_1 < \tilde{A}_2$ also if $w_1 < w_2$ and $\tilde{A}_1 > \tilde{A}_2$ also if $w_1 > w_2$.

Now to include the three subcases of the equality case $R(\tilde{A}_1 \geq \tilde{A}_2) = w/2$ and for any further use of ranking values, we define relative preference index $r_i$ of each $\tilde{A}_i$ by adding an additional value based on their heights to the ranking value $R(\tilde{A}_i \geq \tilde{A}_j)$ as follows:

$$r_i = R(\tilde{A}_i \geq \tilde{A}_j) + \frac{w_i - w}{2}, \ i \neq j, \ i, j = 1, 2,$$

$w = min(w_1, w_2)$. Then the fuzzy variables $\tilde{A}_i$ are ranked based on relative preference indices $r_i$.

**The Method of Ranking Trapezoidal Interval Type-2 Fuzzy Variables:**
Suppose $\tilde{A}_i, i = 1, 2, ..., n$ are $n$ trapezoidal interval type-2 fuzzy variables, where $\tilde{A}_i = (\tilde{A}_i^U, \tilde{A}_i^L) = ((a_{i1}^U, a_{i2}^U, a_{i3}^U, a_{i4}^U; w_i^U), (a_{i1}^L, a_{i2}^L, a_{i3}^L, a_{i4}^L; w_i^L))$. Denote $w_M^U = min_i\{w_i^U\}$ and $w_M^L = min_i\{w_i^L\}$. The proposed procedure of ranking $\tilde{A}_i, i = 1, 2, ..., n$ is as follows:
First find upper relative preference index $r_i^U$ of each $\tilde{A}_i$ by

$$r_i^U = \sum_{k \neq i} R(\tilde{A}_i^U \geq \tilde{A}_k^U) + \frac{w_i^U - w_M^U}{2}, \ i = 1, 2, ..., n$$

and lower relative preference index $r_i^L$ of each $\tilde{A}_i$ by

$$r_i^L = \sum_{k \neq i} R(\tilde{A}_i^L \geq \tilde{A}_k^L) + \frac{w_i^L - w_M^L}{2}, \ i = 1, 2, ..., n.$$

Then the relative preference index $r_i$ of each $\tilde{A}_i$ is calculated by

$$r_i = \frac{r_i^U + r_i^L}{2}, \ i = 1, 2, ..., n.$$

Rank $\tilde{A}_i$ according to the value of $r_i$, i.e. the larger the value of $r_i$, the better the ranking order of $\tilde{A}_i$.

**Example 5.5:** Consider trapezoidal interval type-2 fuzzy variables
$A_1 = ((0.4, 0.7, 0.9, 1.2; 1), (0.5, 0.7, 0.9, 1.1; 0.9))$,

$A_2 = ((0.3, 0.5, 0.6, 0.9; 1), (0.4, 0.55, 0.65, 0.8; 0.9))$

and $A_3 = ((0.6, 0.8, 1.1, 1.4; 1), (0.7, 0.9, 1, 1.2; 0.9))$. Then $w_i^U = 1$, $w_i^L = 0.9$, $i = 1, 2, 3$ and so $w_M^U = 1$ and $w_M^L = 0.9$. Now,

$r_1^U = R(\tilde{A}_1^U \geq \tilde{A}_2^U) + R(\tilde{A}_1^U \geq \tilde{A}_3^U) + \frac{w_1^U - w_M^U}{2} = 0.583 + 0.5 + 0 = 1.083$,

$r_1^L = R(\tilde{A}_1^L \geq \tilde{A}_2^L) + R(\tilde{A}_1^L \geq \tilde{A}_3^L) + \frac{w_1^L - w_M^L}{2} = 0.514 + 0.45 + 0 = 0.964$ and so $r_1 = 1.023$

Similarly we obtain $r_2^U = 0.717$, $r_2^L = 0.515$, $r_2 = 0.616$ and $r_3^U = 1.2$, $r_3^L = 1.221$, $r_3 = 1.2105$. Hence $A_3 \succ A_1 \succ A_2$.

**A Comparison of the Above Ranking Result:** We now compare the above ranking result of the interval type-2 fuzzy variables $A_1$, $A_2$ and $A_3$ with few existing methods as given in Table 19.

**Table 19.** Comparative results of Example 5.5

| Methods | Evaluation | $A_1$ | $A_2$ | $A_3$ | Order relation |
|---|---|---|---|---|---|
| Chen and Lee [23] | $Rank(A_i)$ | 0.385 | 0.2103 | 0.437 | $A_3 \succ A_1 \succ A_2$ |
| Chen et al. [26] | $RV(\tilde{A}_i)$ | 1.4 | 0.9106 | 1.876 | $A_3 \succ A_1 \succ A_2$ |
| Chen and Wang [25] | $Score(\tilde{A}_i)$ | | | | |
| | $\alpha = 0$ | 0.3495 | 0.1788 | 0.5068 | $A_3 \succ A_1 \succ A_2$ |
| | $\alpha = 0.5$ | 0.4021 | 0.201 | 0.5784 | $A_3 \succ A_1 \succ A_2$ |
| | $\alpha = 1$ | 0.4546 | 0.2232 | 0.6501 | $A_3 \succ A_1 \succ A_2$ |
| Proposed method | $r_i$ | 1.023 | 0.616 | 1.2105 | $A_3 \succ A_1 \succ A_2$ |

## 5.6 Proposed Fuzzy MCGDM Based on Ranking Interval Type-2 Fuzzy Variables

Suppose $A_1, A_2, ..., A_m$ are $m$ alternatives and these alternatives are evaluated on basis of the criteria $C_1, C_2, ..., C_n$ by the decision makers $D_l, l = 1, 2, ..., p$. Suppose rating of $A_i$ based on criteria $C_j$ according to the decision maker $D_l$ is $\tilde{A}_{ij}^l$ which is represented by trapezoidal interval type-2 fuzzy variable, where $i = 1, 2, ..., m$, $j = 1, 2, ..., n$ and $l = 1, 2, ..., p$. Let $\tilde{w}_j^l$ be the weight of the criteria $C_j$ indicating its importance given by the decision maker $D_l$, where $\tilde{w}_j^l$ is represented by interval type-2 fuzzy variable for all $j$ and $l$. The proposed fuzzy MCGDM method to rank the alternatives is as follows:

**Step 1:** Construct the decision matrix $\tilde{D} = [\tilde{A}_{ij}]_{m \times n}$ where each $\tilde{A}_{ij}$ is the average of the ratings of alternative $A_i$ given by the decision makers $D_l, l = 1, 2, ..., p$ based on criteria $C_j$, i.e.,

$$C_1 \; C_2 \; .... \; C_n$$

$$\tilde{D} = \begin{matrix} A_1 \\ A_2 \\ \vdots \\ A_m \end{matrix} \begin{bmatrix} \tilde{A}_{11} & \tilde{A}_{12} & ... & \tilde{A}_{1n} \\ \tilde{A}_{21} & \tilde{A}_{22} & ... & \tilde{A}_{2n} \\ \vdots & \vdots & \vdots & \vdots \\ \tilde{A}_{m1} & \tilde{A}_{m2} & ... & \tilde{A}_{mn} \end{bmatrix},$$

where

$$\tilde{A}_{ij} = \frac{\tilde{A}_{ij}^1 \oplus \tilde{A}_{ij}^2 \oplus ... \oplus \tilde{A}_{ij}^p}{p}. \tag{247}$$

Suppose each $\tilde{A}_{ij}$ is represented by $\tilde{A}_{ij} = (\tilde{A}_{ij}^U, \tilde{A}_{ij}^L)$ with heights of $\tilde{A}_{ij}^U$ and $\tilde{A}_{ij}^L$ as $w_{ij}^U$ and $w_{ij}^L$ respectively.

Calculate the average weights $w_j$ of the each criteria $C_j$ by averaging their weights given by the decision makers $D_l, l = 1, 2, ..., p$, i.e.

$$\tilde{w}_j = (\tilde{w}_j^1 \oplus \tilde{w}_j^2 \oplus .... \oplus \tilde{w}_j^p)/p. \tag{248}$$

**Step 2:** Derive the upper relative preference matrix $RP^U = [r_{ij}^U]_{m \times n}$, where $r_{ij}^U$ are the upper relative preference indices of alternatives $A_i$ based on criteria $C_j$, i.e.,

$$r_{ij}^U = \sum_{k \neq i} R(\tilde{A}_{ij}^U \geq \tilde{A}_{kj}^U) + \frac{w_{ij}^U - w_{Mj}^U}{2}, \quad 1 \leq i \leq m, \ 1 \leq j \leq n \tag{249}$$

and similarly the lower relative preference matrix $RP^L = [r_{ij}^L]_{m \times n}$, where

$$r_{ij}^L = \sum_{k \neq i} R(\tilde{A}_{ij}^L \geq \tilde{A}_{kj}^L) + \frac{w_{ij}^L - w_{Mj}^L}{2}, \quad 1 \leq i \leq m, \ 1 \leq j \leq n, \tag{250}$$

$w_{Mj}^U = min_i\{w_{ij}^U\}$ and $w_{Mj}^L = min_i\{w_{ij}^L\}$, $w_{ij}^U$ and $w_{ij}^L$ are the heights of the upper and lower membership of $\tilde{A}_{ij}^l$ respectively.

Finally derive the relative preference matrix $RP = [r_{ij}]_{m \times n}$, where $r_{ij} = \frac{r_{ij}^U + r_{ij}^L}{2}, 1 \leq i \leq m, \ 1 \leq j \leq n$.

**Step 3:** Calculate the fuzzy weighted relative preference of each alternative by employing the importance weights of the criteria as follows:

$$\tilde{A}_i^* = r_{11} \cdot \tilde{w}_1 \oplus r_{12} \cdot \tilde{w}_2 \oplus ... \oplus r_{1n} \cdot \tilde{w}_n = (A_i^{*U}, A_i^{*L}) \ (say), \quad i = 1, 2, ..., n. \tag{251}$$

**Step 4:** Find the final upper preference index of each alternative by

$$r_i^U = \sum_{k \neq i} R(A_i^{*U} \geq A_k^{*U}) + \frac{w_i^U - w_M^U}{2}, \quad i = 1, 2, ..., m. \tag{252}$$

and the final lower preference index

$$r_i^L = \sum_{k \neq i} R(A_i^{*L} \geq A_k^{*L}) + \frac{w_i^L - w_M^L}{2}, \quad i = 1, 2, ..., m, \tag{253}$$

where $w_i^U$, $w_i^L$ are the heights of $\tilde{A}_i^{*U}$, $\tilde{A}_i^{*L}$ respectively and $w_M^U = \min_i\{w_i^U\}$, $w_M^L = \min_i\{w_i^L\}$. Then final preference index $r_i$ of each alternative is obtained by $r_i = \frac{r_i^U + r_i^L}{2}$, $i = 1, 2, ..., m$.

Now the alternatives $A_i$ can be ranked according to their ranking values $r_i$. However for better comparison we find preference weights of the alternatives that sum to 1 as in the following step.

**Step 5:** Obtain preference weights $W_i$ of the alternatives that sum to 1 by normalizing the preference indices $r_i$ as

$$W_i = \frac{r_i}{\sum_{j=1}^m r_j}, \quad i = 1, 2, ..., m, \tag{254}$$

where $0 \leq W_i \leq 1$ and $\sum_{i=1}^m W_i = 1$.
Rank the alternatives $A_i$ according to their preference weights $W_i$, $i = 1, 2, ..., m$.

### 5.7 Problem 5.1: A Transportation Mode Selection Problem with Linguistic Weights and Ratings Generated by Fuzzy Numbers and Its Application to STP

Suppose in a solid transportation problem (STP) there are two modes of transportation (conveyances) available - rail and road. Along with the main criteria (transportation cost), decision makers want to rate the two modes with respect to some other criteria, which are also very important for a transportation policy. Three decision makers $D_1$, $D_2$, $D_3$ select five main criteria- cost (C), speed/time (S), product characteristics (P), flexibility (F) and safety factor (SF). Also there are subcriteria associated with each main criteria as follows:

(1) Cost (C): This main criterion contains cost factors that are involved in transportation system.
   - $C_1$: Transportation cost for shipment of goods from source to destination.
   - $C_2$: Cost of damages to freight incurred at the transportation or transshipment stages.
   - $C_3$: Fixed cost (transport equipment, maintenance, terminal facilities, etc.).
(2) Speed/time (S): This criterion contains time related subcriteria.
   - $S_1$: The average speed that the conveyance can provide.
   - $S_2$: The time consumed for loading, storing and unloading process.
   - $S_3$: The ratio of the distance between supply and demand points to the transportation time.
   - $S_4$: Time reliability, i.e. the past record for delivering in time.
(3) Product characteristics (P): This criteria involved with product related features.
   - $P_1$: The weight of the freight permissible.
   - $P_2$: The volume of the freight permissible.
   - $P_3$: Value to weight of the freight.

(4) Flexibility (F): This criterion contains subcriteria involving capacity, route, time schedule flexibility.
   - $F_1$: The ability to change the transportation route for unexpected cause during transportation.
   - $F_2$: The ability to change the volume and weight capacity of the vehicles.
   - $F_3$: The ability to change the predetermined time schedule.
(5) Safety factor (SF): This criterion contains safety problem related features.
   - $SF_1$: The accidental rate in a determined time period.
   - $SF_2$: The rate of product being damaged during transportation.

The decision makers compare the criteria with each other and gives the importance weights for each criteria. The linguistic terms and related fuzzy numbers (Lee [78], Wang and Elhag [141]) for criteria weights and evaluation ratings are shown in Table 20. The linguistic importance weights of the main and subcriteria are given in Tables 21 and 23 respectively. The average fuzzy weights of the main criteria are obtained using Tables 20 and 21 by averaging their weights given by the three decision makers and presented in Table 22. Similarly using Tables 20 and 23, the average fuzzy weights of the subcriteria are obtained and presented in Table 24. The average weight of the each subcriterion is then multiplied by the corresponding main criterion weight and presented in Table 25 and thus effective weight of the each subcriterion is obtained. Denotes these effective weights of the subcriteria $j$ by $\tilde{w}_j$, $j = 1, 2, ..., 15$, where $j = 1$ indicates the criterion $C_1$, $j = 2$ indicates the criterion $C_2$ and in this way $j = 15$ indicates the criterion $SF_2$. The evaluation ratings of the transportation modes - rail and road as determined by the decision makers based on the selection criteria are given in Table 26. Based on the Table 26, the group fuzzy decision matrix is derived by averaging the ratings of the decision makers and is presented in Table 27. So this decision matrix is denoted by $\tilde{D} = [\tilde{A}_{ij}]_{2 \times 15}$, where $i = 1, 2$ indicate the alternatives rail and road respectively, $j = 1$ indicates the criterion $C_1$, $j = 2$ indicates the criterion $C_2$ and in this way $j = 15$ indicates the criterion $SF_2$. Now we apply our proposed FMCGDM method (cf. Sect.5.4) based on ranking fuzzy numbers step by step as follows:

Step-1: The fuzzy group decision matrix (Table 27) is normalized and shown in Table 28.

Step-2: Based on Table 28, the relative preference matrix $[r_{ij}]_{2 \times 15}$ is derived using Eq. (237) and shown in Table 29. For example, $r_{11} = R(\tilde{A}'_{11}, \tilde{A}'_{21}) = M_\alpha(\tilde{A}'_{11} \geq \tilde{A}'_{21}) = 0.833$, obtained by solving (232) where $\tilde{A}'_{11} = (0.83, 0.97, 1)$, $\tilde{A}'_{21} = (0.57, 0.77, 0.93)$ and $r_{21} = R(\tilde{A}'_{21}, \tilde{A}'_{11}) = 1 - R(\tilde{A}'_{11}, \tilde{A}'_{21}) = 1 - r_{11} = 0.167$.

Step-3: Fuzzy weighted relative preferences of the two alternatives are calculated through Tables 29 and 25 using Eq. (238) and shown in Table 30.

Step-4: From Table 30, using Eq. (239) total preference indices of the two alternatives are obtained and shown in Table 31.

Step-5: Normalizing the preference indices according to Eq. (240) the preference weights of the two alternatives are obtained as $w_1^P = 0.586$ for rail and $w_2^P = 0.414$ for road.

**Table 20.** Linguistic terms and related fuzzy numbers of criteria weights and evaluation ratings

| Linguistic terms | Fuzzy weights | Linguistic terms | Fuzzy ratings |
|---|---|---|---|
| Very low (VL) | (0,0,0.1) | Very poor (VP) | (0,0,1) |
| Low (L) | (0,0.1,0.3) | Poor (P) | (0,1,3) |
| Medium low (ML) | (0.1,0.3,0.5) | Medium poor (MP) | (1,3,5) |
| Medium (M) | (0.3,0.5,0.7) | Fair (F) | (3,5,7) |
| Medium high (MH) | (0.5,0.7,0.9) | Medium good (MG) | (5,7,9) |
| High (H) | (0.7,0.9,1.0) | Good (G) | (7,9,10) |
| Very high (VH) | (0.9,1.0,1.0) | Very good (VG) | (9,10,10) |

**Table 21.** Linguistic importance weights of the main criteria

| Main criteria | $D_1$ | $D_2$ | $D_3$ |
|---|---|---|---|
| Cost (C) | VH | H | VH |
| Speed (S) | H | MH | H |
| Product characteristics (P) | MH | MH | MH |
| Flexibility (F) | M | MH | M |
| Safety factors (SF) | MH | M | MH |

**Table 22.** Average weights of the main criteria

| C | S | P | F | SF |
|---|---|---|---|---|
| (0.83,0.97,1) | (0.63,0.83,0.97) | (0.5,0.7,0.9) | (0.37,0.57,0.77) | (0.43,0.63,0.83) |

**Table 23.** Linguistic importance weights of the subcriteria

| | $D_1$ | $D_2$ | $D_3$ | | $D_1$ | $D_2$ | $D_3$ | | $D_1$ | $D_2$ | $D_3$ |
|---|---|---|---|---|---|---|---|---|---|---|---|
| $C_1$ | VH | H | VH | $S_3$ | H | H | MH | $F_1$ | M | MH | M |
| $C_2$ | H | H | MH | $S_4$ | MH | M | MH | $F_2$ | M | M | M |
| $C_3$ | H | MH | H | $P_1$ | MH | M | M | $F_3$ | MH | M | M |
| $S_1$ | H | MH | MH | $P_2$ | MH | M | M | $SF_1$ | M | M | M |
| $S_2$ | MH | MH | MH | $P_3$ | MH | MH | MH | $SF_2$ | MH | M | MH |

From the above results we observe that for the current problem rail mode is preferred than road and weights of rail and road are 0.586 and 0.414 respectively. So sum of the weights is 1 and these weights can be used for any further requirements.

**Table 24.** Average weights of the subcriteria

|          | $C_1$ | $C_2$ | $C_3$ | $S_1$ |
|----------|-------|-------|-------|-------|
| Weights | (0.83,0.97,1) | (0.63,0.83,0.97) | (0.63,0.83,0.97) | (0.57,0.77,0.93) |
|          | $S_2$ | $S_3$ | $S_4$ | $P_1$ |
| Weights | (0.5,0.7,0.9) | (0.63,0.83,0.97) | (0.43,0.63,0.83) | (0.37,0.57,0.77) |
|          | $P_2$ | $P_3$ | $F_1$ | $F_2$ |
| Weights | (0.37,0.57,0.77) | (0.3,0.5,0.7) | (0.37,0.57,0.77) | (0.3,0.5,0.7) |
|          | $F_3$ | $SF_1$ | $SF_2$ | |
| Weights | (0.37,0.57,0.77) | (0.3,0.5,0.7) | (0.43,0.63,0.83) | |

**Table 25.** Average weights of the subcriteria multiplied by the corresponding main criteria weights

|          | $C_1$ | $C_2$ | $C_3$ | $S_1$ |
|----------|-------|-------|-------|-------|
| Weights | (0.69,0.94,1) | (0.52,0.8,0.97) | (0.52,0.8,0.97) | (0.34,0.64,0.9) |
|          | $S_2$ | $S_3$ | $S_4$ | $P_1$ |
| Weights | (0.32,0.58,0.87) | (0.4,0.69,0.94) | (0.27,0.52,0.81) | (0.19,0.4,0.7) |
|          | $P_2$ | $P_3$ | $F_1$ | $F_2$ |
| Weights | (0.19,0.4,0.7) | (0.25,0.49,0.81) | (0.14,0.32,0.59) | (0.11,0.29,0.54) |
|          | $F_3$ | $SF_1$ | $SF_2$ | |
| Weights | (0.14,0.32,0.59) | (0.13,0.32,0.58) | (0.18,0.4,0.69) | |

**Table 26.** Linguistic ratings of the alternatives with respect to each criteria

|      | $D_1$ | $D_2$ | $D_3$ | $D_1$ | $D_2$ | $D_3$ | $D_1$ | $D_2$ | $D_3$ | $D_1$ | $D_2$ | $D_3$ |
|------|-------|-------|-------|-------|-------|-------|-------|-------|-------|-------|-------|-------|
|      | $C_1$ | | | $C_2$ | | | $C_3$ | | | $S_1$ | | |
| Rail | VG | VG | G | VG | G | G | MG | G | F | VG | G | VG |
| Road | MG | G | MG | G | G | MG | G | G | G | MG | MG | G |
|      | $S_2$ | | | $S_3$ | | | $S_4$ | | | $P_1$ | | |
| Rail | MG | MG | G | MG | MG | MG | G | G | MG | VG | VG | G |
| Road | G | G | VG | G | G | G | MG | MG | G | MG | G | MG |
|      | $P_2$ | | | $P_3$ | | | $F_1$ | | | $F_2$ | | |
| Rail | VG | VG | G | MG | G | MG | MP | F | F | VG | VG | G |
| Road | MG | G | MG | G | G | G | G | MG | G | F | MP | F |
|      | $F_3$ | | | $SF_1$ | | | $SF_2$ | | | | | |
| Rail | F | F | MG | G | G | VG | VG | G | G | | | |
| Road | G | VG | G | MG | MG | G | MG | G | G | | | |

**Table 27.** Fuzzy group decision matrix

|      | $C_1$ | $C_2$ | $C_3$ | $S_1$ |
|------|-------|-------|-------|-------|
| Rail | (8.3,9.7,10) | (7.7,9.3,10) | (5,7,8.7) | (8.3,9.7,10) |
| Road | (5.7,7.7,9.3) | (6.3,8.3,9.7) | (7,9,10) | (5.7,7.7,9.3) |
|      | $S_2$ | $S_3$ | $S_4$ | $P_1$ |
| Rail | (5.7,7.7,9.3) | (5,7,9) | (6.3,8.3,9.7) | (8.3,9.7,10) |
| Road | (7.7,9.3,10) | (7,9,10) | (5.7,7.7,9.3) | (5.7,7.7,9.3) |
|      | $P_2$ | $P_3$ | $F_1$ | $F_2$ |
| Rail | (8.3,9.7,10) | (5.7,7.7,9.3) | (2.3,4.3,6.3) | (8.3,9.7,10) |
| Road | (5.7,7.7,9.3) | (7,9,10) | (6.3,8.3,9.7) | (2.3,4.3,6.3) |
|      | $F_3$ | $SF_1$ | $SF_2$ | |
| Rail | (3.7,5.7,7.7) | (7.7,9.3,10) | (7.7,9.3,10) | |
| Road | (7.7,9.3,10) | (5.7,7.7,9.3) | (6.3,8.3,9.7) | |

**Table 28.** Normalized group fuzzy decision matrix

|      | $C_1$ | $C_2$ | $C_3$ | $S_1$ |
|------|-------|-------|-------|-------|
| Rail | (0.83,0.97,1) | (0.77,0.93,1) | (0.5,0.7,0.87) | (0.83,0.97,1) |
| Road | (0.57,0.77,0.93) | (0.63,0.83,0.97) | (0.7,0.9,1) | (0.57,0.77,0.93) |
|      | $S_2$ | $S_3$ | $S_4$ | $P_1$ |
| Rail | (0.57,0.77,0.93) | (0.5,0.7,0.9) | (0.65,0.85,1) | (0.83,0.97,1) |
| Road | (0.77,0.93,1) | (0.7,0.9,1) | (0.59,0.79,0.96) | (0.57,0.77,0.93) |
|      | $P_2$ | $P_3$ | $F_1$ | $F_2$ |
| Rail | (0.83,0.97,1) | (0.57,0.77,0.93) | (0.24,0.44,0.65) | (0.83,0.97,1) |
| Road | (0.57,0.77,0.93) | (0.7,0.9,1) | (0.65,0.85,1) | (0.23,0.43,0.63) |
|      | $F_3$ | $SF_1$ | $SF_2$ | |
| Rail | (0.37,0.57,0.77) | (0.77,0.93,1) | (0.77,0.93,1) | |
| Road | (0.77,0.93,1) | (0.57,0.77,0.93) | (0.63,0.83,0.97) | |

**Table 29.** Relative preference matrix

|      | $C_1$ | $C_2$ | $C_3$ | $S_1$ | $S_2$ | $S_3$ | $S_4$ | $P_1$ |
|------|-------|-------|-------|-------|-------|-------|-------|-------|
| Rail | 0.833 | 0.667 | 0.23 | 0.833 | 0.25 | 0.25 | 0.581 | 0.833 |
| Road | 0.167 | 0.333 | 0.77 | 0.167 | 0.75 | 0.75 | 0.419 | 0.167 |
|      | $P_2$ | $P_3$ | $F_1$ | $F_2$ | $F_3$ | $SF_1$ | $SF_2$ | |
| Rail | 0.833 | 0.319 | 0 | 1 | 0 | 0.75 | 0.667 | |
| Road | 0.167 | 0.681 | 1 | 0 | 1 | 0.25 | 0.333 | |

**Table 30.** Fuzzy weighted relative preferences of the alternatives

| Rail | (2.385,4.272,6.236) |
|------|---------------------|
| Road | (2.005,3.637,5.424) |

**Table 31.** Total preference indices

| Rail | Road |
|-------|-------|
| 0.586 | 0.414 |

**Table 32.** Comparative results of presented problem 5.1.

| Methods | Evaluation | Rail | Road | Preferable mode |
|---------|-----------|------|------|-----------------|
| Proposed method | Preference weights | 0.586 | 0.414 | rail |
| Lee [78] | Performance index | 1 | 0 | rail |
| Wang and Lee [139] | Closeness coefficient | 1 | 0 | rail |
| Cheng and Lin [28] | defuzzified evaluations | 6.854 | 6.791 | rail |

**Table 33.** Penalties (costs) $c_{ijk}$

| $i \backslash j$ | 2 | 3 | 1 | 2 | 3 |
|------------------|---|---|---|---|---|
| 1 | 4 | 6 | 8 | 7 | 8 | 4 |
| 2 | 7 | 9 | 7 | 5 | 6 | 9 |
| 3 | 6 | 8 | 6 | 4 | 10 | 5 |
| $k$ |   | 1 |   |   | 2 |   |

**Comparison with Some Other Methods:** We solve the above problem by three existing methods- Lee's [78] method based on extended fuzzy preference relation, fuzzy TOPSIS of Wang and Lee [139] and method of Cheng and Lin [28] based on fuzzy Delphi method and the results are presented in Table 32. Lee's method gives total performance index of rail mode as 1 and that of road 0 so that rail mode is preferable than road for this problem. Wang and Lee's method gives the closeness coefficient of rail as 1 and that of road 0 so that by this method rail mode is preferable than road. Actually in case of two alternatives, whatever the ratings of the alternatives and criteria weights are, these two methods always give total performance index or closeness coefficient of one alternative as 1 and that of another 0. Cheng and Lin's method gives the defuzzified values of the aggregate fuzzy evaluations (here aggregate triangular fuzzy numbers) for rail and road as 6.854 and 6.791 respectively. Hence both the three methods give same preference as obtained by our proposed method, i.e. rail mode is preferred than road for the current problem.

## 5.8 Assigning the Preference Weights of the Different Modes into The STP

**How to Assign:** Suppose in a STP, K types of modes of transportation (conveyances) available for transportation. If objective function of a STP is minimization of transportation cost, then obviously transportation cost be the main criterion of choosing conveyances for certain route. However, if it is observed that besides main criterion there are also some other criteria such as speed/time, flexibility, safety factor of conveyances, etc. those are also vary important for a particular problem, then the decision maker may seek to find overall importance weights of the modes with respect all the criteria. Then assign the weights of the modes to the main objective function of the problem so that optimal transportation policy is according to the main criterion in addition to the other selected important criteria. Suppose $w_k^P$ is the weights of the conveyance $k (= 1, 2, ..., K)$ as obtained by the FMCGDM method under some predetermined criteria. These transportation mode weights are assigned to the STP so that the amounts of goods transported through conveyances are according to their weights in addition to the main criterion such as cost or time etc. Actually the main aim is to transport the goods through the best mode as maximum as possible. The objective function of the STP is

$$Max/Min \ Z = \sum_{i=1}^{m} \sum_{j=1}^{n} \sum_{k=1}^{K} c_{ijk} \ x_{ijk}.$$

(i) If the problem is a maximization problem, i.e. $c_{ijk}$ represents profit, amount etc., then to find optimum result (values of $x_{ijk}$'s), assign $w_k^P$ in the objective function as follows:

$$Max \ Z' = \sum_{i=1}^{m} \sum_{j=1}^{n} \sum_{k=1}^{K} w_k^P (c_{ijk} \ x_{ijk}).$$

and then find the actual value of $Z$ (total profit, amount, etc.) using the values of obtained $x_{ijk}$'s and corresponding $c_{ijk}$'s.

(ii) If the problem is a minimization problem, i.e. $c_{ijk}$ represents transportation cost, etc., then to find optimum result (values of $x_{ijk}$'s), assign $w_k^P$ in the objective function as follows:

$$Min \ Z' = \sum_{i=1}^{m} \sum_{j=1}^{n} \sum_{k=1}^{K} \frac{1}{w_k^P} (c_{ijk} \ x_{ijk}),$$

because higher value of $w_k^P$ (i.e., lower value of $1/w_k^P$) ensures the possibility of increasing the amount of goods transported through the conveyance $k$. Now the actual value of $Z$ (total transportation cost) is derived using the values of obtained $x_{ijk}$'s and corresponding $c_{ijk}$'s.

**Numerical Illustration:** Consider a STP with three sources $(i = 1, 2, 3)$, three destinations $(j = 1, 2, 3)$ and two conveyances $(k = 1, 2)$. Here conveyance $k = 1$

indicates rail and $k = 2$ indicates road. The unit transportation costs $(c_{ijk})$ are presented in Table 33 and the availabilities $(a_i)$, demands $(b_j)$ are given below. $a_1 = 35$, $a_2 = 30$, $a_3 = 42$, $b_1 = 32$, $b_2 = 36$, $b_3 = 35$, $e_1 = 60$, $e_2 = 52$.

So mathematically the problem becomes

$$Min\ Z = \sum_{i=1}^{3} \sum_{j=1}^{3} \sum_{k=1}^{2} (c_{ijk}\ x_{ijk}),$$

$$s.t.\ \sum_{j=1}^{3} \sum_{k=1}^{2} x_{ijk} \leq a_i,\ \ i = 1, 2, 3,$$

$$\sum_{i=1}^{3} \sum_{k=1}^{2} x_{ijk} \geq b_j,\ \ j = 1, 2, 3, \tag{255}$$

$$\sum_{i=1}^{3} \sum_{j=1}^{3} x_{ijk} \leq e_k,\ \ k = 1, 2,$$

$$x_{ijk} \geq 0,\ \forall\ i, j, k.$$

Now introducing weights of the transportation modes in the objective function, the problem becomes

$$Min\ Z' = \sum_{i=1}^{3} \sum_{j=1}^{3} \sum_{k=1}^{2} \frac{1}{w_k^P} (c_{ijk}\ x_{ijk}),$$

$$s.t.\ \sum_{j=1}^{3} \sum_{k=1}^{2} x_{ijk} \leq a_i,\ \ i = 1, 2, 3,$$

$$\sum_{i=1}^{3} \sum_{k=1}^{2} x_{ijk} \geq b_j,\ \ j = 1, 2, 3, \tag{256}$$

$$\sum_{i=1}^{3} \sum_{j=1}^{3} x_{ijk} \leq e_k,\ \ k = 1, 2,$$

$$x_{ijk} \geq 0,\ \forall\ i, j, k,$$

where $w_1^P = 0.586$ (for rail) and $w_2^P = 0.414$ (for road).
Solving the problem (256) we have
$x_{111} = 20$, $x_{121} = 10$, $x_{331} = 30$, $x_{132} = 5$, $x_{222} = 26$, $x_{312} = 12$ and $MinZ = 4 \cdot 20 + 6 \cdot 10 + 6 \cdot 30 + 4 \cdot 5 + 6 \cdot 26 + 4 \cdot 12 = 544$.

Now solving the problem without mode weights (problem (255)), we have
$x_{111} = 19.5$, $x_{121} = 6$, $x_{331} = 25.5$, $x_{132} = 9.5$, $x_{222} = 30$, $x_{312} = 12.5$ and
$MinZ = 4 \cdot 19.5 + 6 \cdot 6 + 6 \cdot 25.5 + 4 \cdot 9.5 + 6 \cdot 30 + 4 \cdot 12.5 = 535$.

**Table 34.** Linguistic terms and related fuzzy variables of criteria weights

| Linguistic terms | Fuzzy weights |
|---|---|
| Very low (VL) | ((0,0,0,0.1;1),(0,0,0,0.05;0.9)) |
| Low (L) | ((0,0.1,0.1,0.3;1),(0.05,0.1,0.1,0.2;0.9)) |
| Medium low (ML) | ((0.1,0.3,0.3,0.5;1),(0.2,0.3,0.3,0.4;0.9)) |
| Medium (M) | ((0.3,0.5,0.5,0.7;1),(0.4,0.5,0.5,0.6;0.9)) |
| Medium high (MH) | ((0.5,0.7,0.7,0.9;1),(0.6,0.7,0.7,0.8;0.9)) |
| High (H) | ((0.7,0.9,0.9,1;1),(0.8,0.9,0.9,0.95;0.9)) |
| Very high (VH) | ((0.9,1,1,1;1),(0.95,1,1,1;0.9)) |

**Table 35.** Linguistic terms and related fuzzy variables of evaluation ratings

| Linguistic terms | Fuzzy ratings |
|---|---|
| Very poor (VP) | ((0,0,0,1;1),(0,0,0,0.5;0.9)) |
| Poor (P) | ((0,1,1,3;1),(0.5,1,1,2;0.9)) |
| Medium poor (MP) | ((1,3,3,5;1),(2,3,3,4;0.9)) |
| Fair (F) | ((3,5,5,7;1),(4,5,5,6;0.9)) |
| Medium good (MG) | ((5,7,7,9;1),(6,7,7,8;0.9)) |
| Good (G) | ((7,9,9,10;1),(8,9,9,9.5;0.9)) |
| Very good (VG) | ((9,10,10,10;1),(9.5,10,10,10;0.9)) |

**Remark:** We see that in case of the problem without mode weights, total transported amount through rail is 51 and through road is 52. Where as for the problem with mode weights total transported amount through rail is 60 and through road is 43. This is as per expectation because here rail mode has higher preference weight than road. Also we observe that the problem without mode weights provides less transportation cost. So it is up to the decision makers whether they decide to determine transportation policy only according to the main criterion (i.e. cost) or according to all other criteria including the main criterion.

## 5.9    Problem 5.2: A Transportation Mode Selection Problem with Linguistic Weights and Ratings Generated by IT2 FVs

Suppose in a transportation system there are two modes of transportation (conveyances) available - rail and road. Besides the main criterion (transportation cost), decision makers want to rate the two modes with respect to some other criteria, which are also very important for a transportation policy. The selection criteria are already presented in Sect. 5.3 for the Problem 5.1.

The decision makers compare the criteria with each other and gives the importance weights for each criteria. The linguistic terms and related fuzzy variables (Anand et al. [7], Chen and Lee [23,24]) for criteria weights and evaluation ratings are shown in Tables 34 and 35 respectively. The linguistic importance weights of the main and subcriteria as given by the decision makers are same as presented in Tables 21 and 23 respectively. Average fuzzy weights of the main criteria and subcriteria are obtained by averaging the related IT2 fuzzy variables (based on Eq. (248)) of the criteria weights. For example, average weight of the cost criteria C is found $((0.83,0.97,0.97,1;1),(0.9,0.97,0.97,0.98;0.9))$, obtained by averaging the related fuzzy variables of the linguistic weights VH, H and VH. Similarly the average weight of the subcriteria $C_1$ is found $((0.83,0.97,0.97,1;1),$ $(0.9,0.97,0.97,0.98;0.9))$. The average weight of the each subcriterion is then multiplied by the corresponding main criteria weight and thus effective weight of the each subcriterion is obtained. For example, effective weight of the subcriterion $C_1$ is obtained $((0.69,0.94,0.94,1;1),(0.81,0.94,0.94,0.96;0.9))$ by multiplying the average weight of the subcriterion $C_1$ with the weight of the corresponding main criteria C. In this way we find the effective weights of all the subcriteria as follows.

$C_1$: $((0.69,0.94,0.94,1;1),(0.81,0.94,0.94,0.96;0.9))$,
$C_2$: $((0.52,0.8,0.8,0.97;1),(0.53,0.8,0.8,0.88;0.9))$,
$C_3$: $((0.52,0.8,0.8,0.97;1),(0.53,0.8,0.8,0.88;0.9))$,
$S_1$: $((0.34,0.64,0.64,0.9;1),(0.49,0.64,0.64,0.71;0.9))$,
$S_2$: $((0.32,0.58,0.58,0.87;1),(0.44,0.58,0.58,0.72))$,
$S_3$: $((0.4,0.69,0.69,0.94),(0.53,0.69,0.69,0.81;0.9))$,
$S_4$: $((0.27,0.52,0.52,0.81;1),(0.39,0.52,0.52,0.66;0.9))$,
$P_1$: $((0.19,0.4,0.4,0.7;1),(0.28,0.4,0.4,0.54;0.9))$,
$P_2$: $((0.19,0.4,0.4,0.7;1),(0.28,0.4,0.4,0.54;0.9))$,
$P_3$: $((0.25,0.49,0.49,0.81;1),(0.36,0.49,0.49,0.64;0.9))$,
$F_1$: $((0.14,0.32,0.32,0.59;1),(0.22,0.32,0.32,0.45;0.9))$,
$F_2$: $((0.11,0.29,0.29,0.54;1),(0.22,0.29,0.29,0.4;0.9))$,
$F_3$: $((0.14,0.32,0.32,0.59),(0.22,0.32,0.32,0.45;0.9))$,
$SF_1$: $((0.13,0.31,0.31,0.58),(0.21,0.31,0.31,0.44;0.9))$,
$SF_2$: $((0.18,0.4,0.4,0.69),(0.28,0.4,0.4,0.53;0.9))$.

The evaluation ratings of the transportation modes - rail and road as determined by the decision makers based on the selection criteria are same as presented in Table 26.

Now we apply our proposed FMCGDM method based on ranking interval type-2 fuzzy variables step by step (cf. Sect. 5.6) as follows:

**Step-1:** Based on the Eq. (247), the group fuzzy decision matrix is derived by averaging the linguistic ratings of the decision makers and is presented in Table 36. The average effective weights of all the subcriteria are already obtained in the above.

**Step-2:** Upper and lower relative preference indices, i.e., $r_{ij}^U$ and $r_{ij}^L$ of the alternatives (rail and road) with respect to each subcriteria are obtained from Table 36 using Eqs. (249) and (250) respectively and presented in Table 37. Then the relative preference indices $r_{ij}$ of the alternatives are obtained by averaging their upper and lower relative preference indices as presented in Table 37.

**Table 36.** Fuzzy group decision matrix

| | $C_1$ | $C_2$ |
|---|---|---|
| Rail | ((8.3,9.7,9.7,10;1),(9,9.7,9.7,9.8;0.9)) | ((7.7,9.3,9.3,10;1),(8.5,9.3,9.3,9.7;0.9)) |
| Road | ((6.3,8.3,8.3,9.7;1),(7.3,8.3,8.3,9;0.9)) | ((6.3,8.3,8.3,9.7;1),(7.3,8.3,8.3,9;0.9)) |
| | $C_3$ | $S_1$ |
| Rail | ((5,7,7,8.7;1),(6,7,7,7.8;0.9)) | ((8.3,9.7,9.7,10;1),(9,9.7,9.7,9.8;0.9)) |
| Road | ((7,9,9,10;1),(8,9,9,9.5;0.9)) | ((5.7,7.7,7.7,9.3;1),(6.7,7.7,7.7,8.5;0.9)) |
| | $S_2$ | $S_3$ |
| Rail | ((5.7,7.7,7.7,9.3;1),(6.7,7.7,7.7,8.5;0.9)) | ((5,7,7,9;1),(6,7,7,8;0.9)) |
| Road | ((7.7,9.3,9.3,10;1),(8.5,9.3,9.3,9.7;0.9)) | ((7,9,9,10;1),(8,9,9,9.5;0.9)) |
| | $S_4$ | $P_1$ |
| Rail | ((6.3,8.3,8.3,9.7;1),(7.3,8.3,8.3,9;0.9)) | ((8.3,9.7,9.7,10;1),(9,9.7,9.7,9.8;0.9)) |
| Road | ((5.7,7.7,7.7,9.3;1),(6.7,7.7,7.7,8.5;0.9)) | ((5.7,7.7,7.7,9.3;1),(6.7,7.7,7.7,8.5;0.9)) |
| | $P_2$ | $P_3$ |
| Rail | ((8.3,9.7,9.7,10;1),(9,9.7,9.7,9.8;0.9)) | ((5.7,7.7,7.7,9.3;1),(6.7,7.7,7.7,8.5;0.9)) |
| Road | ((5.7,7.7,7.7,9.3;1),(6.7,7.7,7.7,8.5;0.9)) | ((7,9,9,10;1),(8,9,9,9.5;0.9)) |
| | $F_1$ | $F_2$ |
| Rail | ((2.3,4.3,4.3,6.3),(3.3,4.3,4.3,5.3;0.9)) | ((8.3,9.7,9.7,10;1),(9,9.7,9.7,9.8;0.9)) |
| Road | ((8.3,9.7,9.7,10;1),(9,9.7,9.7,9.8;0.9)) | ((4.3,6.3,6.3,8.3;1),(5.3,6.3,6.3,7.3;0.9)) |
| | $F_3$ | $SF_1$ |
| Rail | ((1.7,3.7,3.7,5.7;1),(2.7,3.7,3.7,4.7;0.9)) | ((7.7,9.3,9.3,10;1),(8.5,9.3,9.3,9.7;0.9)) |
| Road | ((7.7,9.3,9.3,10;1),(8.5,9.3,9.3,9.7;0.9)) | ((5.7,7.7,7.7,9.3;1),(6.7,7.7,7.7,8.5;0.9)) |
| | $SF_2$ | |
| Rail | ((7.7,9.3,9.3,10;1),(8.5,9.3,9.3,9.7;0.9)) | |
| Road | ((6.3,8.3,8.3,9.7;1),(7.3,8.3,8.3,9;0.9)) | |

**Step-3:** Then the fuzzy weighted relative preferences of the alternatives rail and road are calculated by employing the effective weights of the subcriteria as Eq. (251) and presented in Table 38.

**Step-4:** From Table 38, the final upper preference indices $r_i^U$ and lower preference indices $r_i^L$ of the alternatives are obtained using Eqs. (252) and (253) respectively and final preference indices $r_i$ are calculated by averaging them. These results are presented in Table 39.

**Step-5:** Preference weights $(W_i)$ of the alternatives that sum to 1 are obtained by normalizing the preference indices $r_i$ and shown in Table 39.

From Table 39 we observe that weights of rail and road are 0.63 and 0.37 respectively and so for the current problem rail mode is preferred than road.

Table 37. Relative preference matrix

|  | $r_{ij}^U$ | $r_{ij}^L$ | $r_{ij}$ | $r_{ij}^U$ | $r_{ij}^L$ | $r_{ij}$ | $r_{ij}^U$ | $r_{ij}^L$ | $r_{ij}$ |
|---|---|---|---|---|---|---|---|---|---|
|  | $C_1$ |  |  | $C_2$ |  |  | $C_3$ |  |  |
| Rail | 0.75 | 0.9 | 0.825 | 0.667 | 0.75 | 0.708 | 0.23 | 0 | 0.115 |
| Road | 0.25 | 0 | 0.125 | 0.333 | 0.15 | 0.241 | 0.77 | 0.9 | 0.835 |
|  | $S_1$ |  |  | $S_2$ |  |  | $S_3$ |  |  |
| Rail | 0.833 | 0.9 | 0.866 | 0.25 | 0 | 0.125 | 0.25 | 0 | 0.125 |
| Road | 0.167 | 0 | 0.083 | 0.75 | 0.9 | 0.825 | 0.75 | 0.9 | 0.825 |
|  | $S_4$ |  |  | $P_1$ |  |  | $P_2$ |  |  |
| Rail | 0.583 | 0.6 | 0.591 | 0.833 | 0.9 | 0.866 | 0.833 | 0.9 | 0.866 |
| Road | 0.417 | 0.3 | 0.358 | 0.167 | 0 | 0.083 | 0.167 | 0 | 0.083 |
|  | $P_3$ |  |  | $F_1$ |  |  | $F_2$ |  |  |
| Rail | 0.319 | 0.125 | 0.222 | 0 | 0 | 0 | 1 | 0.9 | 0.95 |
| Road | 0.681 | 0.775 | 0.782 | 1 | 0.9 | 0.95 | 0 | 0 | 0 |
|  | $F_3$ |  |  | $SF_1$ |  |  | $SF_2$ |  |  |
| Rail | 0 | 0 | 0 | 0.75 | 0.9 | 0.825 | 0.667 | 0.75 | 0.708 |
| Road | 1 | 0.9 | 0.95 | 0.25 | 0 | 0.125 | 0.333 | 0.15 | 0.241 |

Table 38. Fuzzy weighted relative preferences of the alternatives

| Rail | ((2.265,4.07,4.07,5.961;1),(3.026,4.07,4.07,4.908;0.9)) |
|---|---|
| Road | ((1.917,3.458,3.458,4.879;1),(2.492,3.458,3.458,4.252;0.9)) |

Table 39. Final preference indices and preference weights

|  | $r_i^U$ | $r_i^L$ | $r_i$ | $W_i$ |
|---|---|---|---|---|
| Rail | 0.5949 | 0.5998 | 0.597 | 0.63 |
| Road | 0.4051 | 0.3002 | 0.352 | 0.37 |

## 5.10    Overall Conclusion

Selection of suitable transportation modes is a major issue in transportation systems. There may exist large number of conflicting criteria for selecting convenient modes. Also human judgments are usually imprecise (i.e., linguistic, interval etc.) rather than precise numeric values.

In this section, we have proposed a computationally efficient fuzzy multicriteria group decision making (FMCGDM) method (cf. Sect. 5.4) based on ranking fuzzy numbers. For this purpose we have defined a ranking function (cf. Sect. 5.3) based on credibility measure to rank a fuzzy number over another fuzzy number. The proposed fuzzy MCGDM method is applied (cf. Sect. 5.7) to find most convenient transportation mode alternatives in which the evaluation ratings and criteria weights are expressed in linguistic terms generated by fuzzy numbers. Also this method gives the weights of the alternatives which can be used for further requirements. The mode weights as founded by the method are assigned to a STP so that best mode can be used as maximum as possible.

In Sect. 5.5, a new method of ranking IT2 FVs based on generalized credibility measure is proposed. In Sect. 5.6, we have presented a new FMCGDM method based on the proposed ranking method of IT2 FVs. The proposed FMCGDM method is applied to a transportation mode selection problem (cf. Sect. 5.9) in which the evaluation ratings and criteria weights are expressed in linguistic terms generated by trapezoidal IT2 FVs.

The proposed methods are computationally efficient and we expect that these methods may have potential applications in many industry based FMCGDM problems in the future.

# 6 Solid Transportation Models with Transportation Cost Parameters as Rough Variables

## 6.1 Introduction

Traditionally the solid transportation problem (STP) (Haley [53], Gen et al. [48], Jiménez and Verdegay [60], Li et al. [81]) is modeled taking total supply capacity of all the conveyances and it is assumed that this total capacity is available for utilization for all source to destination routs whatever be the amount of product allocated in the routs for transportation. But in many practical situations this may not always happen. Practically most of time full vehicles, e.g., trucks, rail coaches are to be booked and the availability of each type of conveyance at each source may not be the same and vehicles available at one source may not be utilized at another source due to long distance between them or some other problems. Also fulfillment of capacity of a vehicle effects the optimal transportation policy. These practical situations motivated us to formulate some useful solid transportation models.

Rough set theory is one of the most convenient and accepted tool to deal with uncertainty. Though transportation problems in various types of uncertain environments such as fuzzy, random are studied by many researchers, there are very few research papers about TP in rough uncertain environment. Since rough set theory is proposed by Pawlak [121], it is developed by many researchers (Pawlak [122], Pawlak and Skowron [124], Polkowski [125], Liu and Zhu [96]) in theoretical aspect and applied into many practical fields such as data envelopment analysis (DEA) (Shafiee and Shams-e-alam [131], Xu et al. [151]), data mining (Lin et al. [83]), multi-criteria decision analysis (Dembczynski et al. [35], Pawlak and Slowinski [123]), medical diagnosis (Hirano and Tsumoto [56], Tsumoto [133], Zhang et al. [163]), neural network (Azadeh et al. [9], Zhang et al. [162]), etc. Liu [86] proposed the concept rough variable which is a measurable function from rough space to the set of real numbers. Liu [87] discussed some inequalities of rough variables and convergence concept of sequence of rough variables. Liu and Zhu [97] introduced rough variable with values in measurable spaces. Liu [86,88] studied some rough programming models with rough variables as parameters. Xu and Yao [150] studied a two-person zero-sum matrix games with payoffs as rough variables. Tao and Xu [132] developed a rough multi-objective programming for dealing with multi-objective solid transportation problem assuming that the feasible region is not fixed but flexible due to imprecise parameters. Xu et al. [151] proposed a rough DEA model to solve a supply chain performance evaluation problem with rough parameters. Xiao and Lai [149] considered power-aware VLIW instruction scheduling problem with power consumption parameters as rough variables. Mondal et al. [114] considered a production-repairing inventory model with fuzzy rough variables. But at the best of our knowledge none studied STPs with any of the parameters as rough variables before Kundu et al. [73].

In this section, we formulate solid transportation model with vehicle capacity and an additional cost which is incurred due to not fulfilling the vehicle capacity. The unit transportation costs and unit additional costs in the models are taken as rough variables. To solve the said models with transportation costs as rough variables we have presented rough chance-constrained programming, rough expected value and rough dependent-chance programming models.

## 6.2   Model 6.1: New Solid Transportation Model with Vehicle Capacity

We first describe and formulate the model deterministically and then consider the model with rough cost parameters.

### Notations:

(i) $c_{ijk}$: The unit transportation costs from $i$-th source to $j$-th destination via $k$-th conveyance according to full utilization of the vehicle capacity.

(ii) $x_{ijk}$: The decision variable which represents amount of product to be transported from $i$-th origin to $j$-th destination via $k$-th conveyance.

(iii) $Z$: The objective function.

(iv) $a_i$: The amount of the product available at the $i$-th origin.

(v) $b_j$: The demand of the product at $j$-th destination.

(vi) $q_k$: The capacity of singe vehicle of $k$-th type conveyance.

(vii) $z_{ijk}$: The frequency (number of required vehicles) of conveyance $k$ for transporting goods from source $i$ to destination $j$ via conveyance $k$.

(viii) $\epsilon_{ijk}$: Total additional (penalty) cost for $i - j - k$ route due to not fulfilling the vehicle capacity.

**Description of the Problem and Model Formulations:** In traditional STP, total transportation capacity of conveyances is taken and the problem is solved assuming that this total capacity can be utilized for all routes whatever the allocation of products is in the routes. But in many real transportation systems, full vehicles (e.g. trucks for road transportation, coaches for rail transportation, etc.) are to be booked and number of vehicles required are according to amount of product to be transported through a particular route. The difficulty in this case arises when the amount of allocated product is not sufficient to fill up the capacity of the vehicle, because then extra cost is incurred despite the unit transportation cost due to not fulfilling the vehicle capacity. Here we formulate some solid transportation models with vehicle capacity to deal with such situations.

Suppose $q_k$ be the capacity of singe vehicle of $k$-th type conveyance. Let $z_{ijk}$ be the frequency (number of required vehicles) of conveyance $k$ for transporting goods from source $i$ to destination $j$ via conveyance $k$ and $x_{ijk}$ (decision variable) be the corresponding amount of goods. Then $z_{ijk}$ is a decision variable which takes only positive integer or zero. Also we have

$$x_{ijk} \leq z_{ijk} \cdot q_k.$$

Now in such vehicle transportation system obviously calculation of unit transportation cost is according to the full utilization of the capacity of the vehicle. That is for a particular route $i-j-k$ if the unit transportation cost $c_{ijk}$ is according to full utilization of the vehicle capacity $q_k$ then an extra cost (penalty) will be added if the capacity $q_k$ is not fully utilized. Determination of additional cost for deficit amount depends upon the relevant transportation authority. Two cases may arise, either authority do not want to compromise for deficit amount and so direct cost $c_{ijk}$ is also represent the additional cost for unit deficit amount, or they agree to compromise and fixed an additional cost for unit deficit amount. For calculating additional cost first deficit amount of goods is to be calculated for each route. This can be done by two ways - calculating deficit amount for $i-j-k$ route directly as $(z_{ijk} \cdot q_k - x_{ijk})$ or by calculating the empty ratio (Yang et al. [152]) of each vehicle of $k$-th type conveyance for transporting goods from source $i$ to destination $j$ as

$$d_{ijk} = \begin{cases} 0, & \text{if } \frac{x_{ijk}}{q_k} = \lceil \frac{x_{ijk}}{q_k} \rceil; \\ 1 - (\frac{x_{ijk}}{q_k} - \lceil \frac{x_{ijk}}{q_k} \rceil), & \text{otherwise.} \end{cases}$$

Then the amount of deficit amount for $i - j - k$ route is given by $q_k \cdot d_{ijk}$. Now if $u_{ijk}$ represents additional cost for unit amount of deficit from source $i$ to destination $j$ via conveyance $k$, then additional cost for this route is given by

$$\epsilon_{ijk} = u_{ijk}(z_{ijk} \cdot q_k - x_{ijk}) \text{ or } \epsilon_{ijk} = u_{ijk} \cdot q_k \cdot d_{ijk}.$$

The total additional (penalty) cost for the problem is

$$C(x) = \sum_{i=1}^{m} \sum_{j=1}^{n} \sum_{k=1}^{K} \epsilon_{ijk}.$$

So the STP model becomes

$$Min \ Z = \sum_{i=1}^{m} \sum_{j=1}^{n} \sum_{k=1}^{K} (c_{ijk} \ x_{ijk} + \epsilon_{ijk})$$

$$s.t. \quad \sum_{j=1}^{n}\sum_{k=1}^{K} x_{ijk} \le a_i, \quad i = 1, 2, ..., m,$$

$$\sum_{i=1}^{m}\sum_{k=1}^{K} x_{ijk} \ge b_j, \quad j = 1, 2, ..., n, \tag{257}$$

$$x_{ijk} \le z_{ijk} \cdot q_k, \quad i = 1, 2, ..., m; j = 1, 2, ..., n; k = 1, 2, ..., K,$$

$$\sum_{i=1}^{m} a_i \ge \sum_{j=1}^{n} b_j, \quad x_{ijk} \ge 0, \quad z_{ijk} \in Z^+, \forall \, i, j, k.$$

In the above model it is assumed that there are sufficient number of vehicles of each type of conveyance available to transport the required amount of goods (i.e., there is no restriction on number of available vehicles of each type of conveyances). If number of vehicles of conveyances limited to certain number, suppose $Q_k$ for $k$-th type conveyance then an another constraint

$$\sum_{i=1}^{m}\sum_{j=1}^{n} z_{ijk} \le Q_k, \quad k = 1, 2, ..., K$$

is added to the model (257), then the above model becomes

$$Min \; Z = \sum_{i=1}^{m}\sum_{j=1}^{n}\sum_{k=1}^{K} (c_{ijk} \, x_{ijk} + \epsilon_{ijk})$$

$$s.t. \quad \sum_{j=1}^{n}\sum_{k=1}^{K} x_{ijk} \le a_i, \quad i = 1, 2, ..., m,$$

$$\sum_{i=1}^{m}\sum_{k=1}^{K} x_{ijk} \ge b_j, \quad j = 1, 2, ..., n, \tag{258}$$

$$x_{ijk} \le z_{ijk} \cdot q_k, \quad i = 1, 2, ..., m; j = 1, 2, ..., n; k = 1, 2, ..., K,$$

$$\sum_{i=1}^{m}\sum_{j=1}^{n} z_{ijk} \le Q_k, \quad k = 1, 2, ..., K,$$

$$\sum_{i=1}^{m} a_i \ge \sum_{j=1}^{n} b_j, \quad x_{ijk} \ge 0, \quad z_{ijk} \in Z^+, \forall \, i, j, k.$$

This limitation of number of vehicles can effect the optimal transportation policy. For example unavailability of sufficient number of vehicles of certain type of conveyance may force to use another type of conveyance which costs higher than the previous.

The hierarchical structures of the model (258) is shown in the Fig. 18.

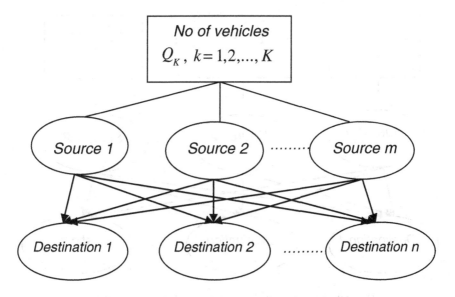

**Fig. 18.** The hierarchical structures of the model (258).

In the above two models it is assumed that total available vehicles can be utilized in each source as they required. But in reality in each source, the availability of different vehicles may not be the same and the vehicles available at one source may not be utilized for another source due to long distance between them. So there may be a situation arises that in a certain source there are more than sufficient number of particular vehicles available to transport product to destinations but at the same time in an another source there are less number of that vehicles available than the requirement. As a result it may happen that vehicle having less transportation cost leaving from certain source to destination without being fully loaded, while vehicle having comparably high transportation cost leaving from another source to destination with fully loaded. So it is realistic to include a constraint defining source-wise vehicle availability. Suppose at source $i$, the number of available vehicles of $k$-th type conveyance is $V_i^k$ and vehicles at each source can not be shared to other sources.

Then the constraints

$$\sum_{j=1}^{n} z_{ijk} \leq V_i^k, \ i = 1, 2, ..., m; k = 1, 2, ..., K$$

is added to the model (257) and so the model becomes

$$Min \ Z = \sum_{i=1}^{m} \sum_{j=1}^{n} \sum_{k=1}^{K} \left( c_{ijk} \ x_{ijk} + \epsilon_{ijk} \right)$$

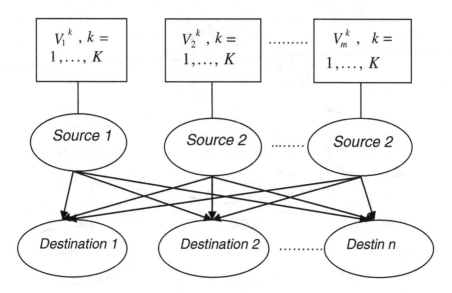

**Fig. 19.** The hierarchical structures of the model (259).

$$s.t. \ \sum_{j=1}^{n}\sum_{k=1}^{K} x_{ijk} \leq a_i, \ \ i = 1, 2, ..., m,$$

$$\sum_{i=1}^{m}\sum_{k=1}^{K} x_{ijk} \geq b_j, \ \ j = 1, 2, ..., n, \qquad (259)$$

$$x_{ijk} \leq z_{ijk} \cdot q_k, \ \ i = 1, 2, ..., m; j = 1, 2, ..., n; k = 1, 2, ..., K,$$

$$\sum_{j=1}^{n} z_{ijk} \leq V_i^k, \ \ i = 1, 2, ..., m; k = 1, 2, ..., K,$$

$$\sum_{i=1}^{m} a_i \geq \sum_{j=1}^{n} b_j, \ x_{ijk} \geq 0, \ z_{ijk} \in Z^+, \forall \ i, j, k.$$

The hierarchical structures of the model (259) is shown in the Fig. 19.

## 6.3   The Model with Unit Transportation and Additional Costs (Penalty) as Rough Variables

Consider the unit transportation costs $c_{ijk}$ and as well as unit additional costs $u_{ijk}$ for the model (257) are rough variables represented by $c_{ijk} = ([c_{ijk}^2, c_{ijk}^3],$ $[c_{ijk}^1, c_{ijk}^4])$, $c_{ijk}^1 \leq c_{ijk}^2 < c_{ijk}^3 \leq c_{ijk}^4$ and $u_{ijk} = ([u_{ijk}^2, u_{ijk}^3], [u_{ijk}^1, u_{ijk}^4])$, $u_{ijk}^1 \leq u_{ijk}^2 < u_{ijk}^3 \leq u_{ijk}^4$. Then, the objective function of the model (257), given by

$Z = \sum_{i=1}^{m} \sum_{j=1}^{n} \sum_{k=1}^{K} (c_{ijk} \, x_{ijk} + \epsilon_{ijk})$, $\epsilon_{ijk} = u_{ijk}(z_{ijk} \cdot q_k - x_{ijk})$ becomes a rough variable defined as $Z = ([Z^2, Z^3], [Z^1, Z^4])$, where

$$Z^r = \sum_{i=1}^{m} \sum_{j=1}^{n} \sum_{k=1}^{K} (c_{ijk}^r \, x_{ijk} + \epsilon_{ijk}^r), \quad r = 1, 2, 3, 4, \tag{260}$$

$\epsilon_{ijk}^r = u_{ijk}^r(z_{ijk} \cdot q_k - x_{ijk})$, $r = 1, 2, 3, 4$.

**Rough Chance-Constrained Programming Model:** For the above problem with rough objective function one can not directly minimize it. The main idea of chance-constrained method is that a uncertain constraint is allowed to violet ensuring that it must be hold at some chance/confidence level. We apply the idea of chance-constrained programming (CCP) to the objective function of the model (257) with rough costs (i.e. rough objective function) with the help of trust measure. Since the problem is a minimization problem, we minimize the smallest objective $\bar{Z}$ satisfying $Tr\{Z \leq \bar{Z}\} \geq \alpha$, where $\alpha \in (0, 1]$ is a specified trust (confidence) level, i.e., we minimize the $\alpha$-pessimistic value $Z_{inf}(\alpha)$ of $Z$. This implies that the optimum objective value will below the $\bar{Z}$ with a trust level at least $\alpha$. So the rough CCP becomes

$$Min \; (Min \; \bar{Z}) \tag{261}$$

$$s.t. \; Tr\{Z \leq \bar{Z}\} \geq \alpha,$$

$$\sum_{j=1}^{n} \sum_{k=1}^{K} x_{ijk} \leq a_i, \quad i = 1, 2, ..., m, \tag{262}$$

$$\sum_{i=1}^{m} \sum_{k=1}^{K} x_{ijk} \geq b_j, \quad j = 1, 2, ..., n, \tag{263}$$

$$x_{ijk} \leq z_{ijk} \cdot q_k, \quad i = 1, 2, ..., m; j = 1, 2, ..., n; k = 1, 2, ..., K, \tag{264}$$

$$\sum_{i=1}^{m} a_i \geq \sum_{j=1}^{n} b_j, \quad x_{ijk} \geq 0, \quad z_{ijk} \in Z^+, \forall \, i, j, k. \tag{265}$$

Now we also formulate another rough CCP for the model (257) with rough costs, to minimize the greatest objective $\underline{Z}$ satisfying $Tr\{Z \geq \underline{Z}\} \geq \alpha$, where $\alpha \in (0, 1]$ is a specified trust (confidence) level, i.e., we minimize the $\alpha$-optimistic value $Z_{sup}(\alpha)$ of $Z$. In other words, we minimize maximum $\underline{Z}$ so that the optimum objective value will greater or equal to the $\underline{Z}$ with a trust level at least $\alpha$. So the rough CCP becomes

$$Min \; (Max \; \underline{Z})$$

$$s.t. \; Tr\{Z \geq \underline{Z}\} \geq \alpha, \tag{266}$$

*and the constraints (262)–(265).*

**Deterministic Forms:** From the definition of $\alpha$-pessimistic value (Definition 2.14, Sect. 2.4), the above CCP (261)–(265) equivalently becomes

$$Min\ Z'$$

$$s.t.\ the\ constraints\ (262) - (265), \tag{267}$$

where

$Z' = Z_{inf}(\alpha)$

$$= \begin{cases} (1 - 2\alpha)Z^1 + 2\alpha Z^4, & if\ \alpha \le ((Z^2 - Z^1)/2(Z^4 - Z^1)); \\ 2(1 - \alpha)Z^1 + (2\alpha - 1)Z^4, & if\ \alpha \ge ((Z^3 + Z^4 - 2Z^1)/2(Z^4 - Z^1)); \\ \frac{Z^1(Z^3 - Z^2) + Z^2(Z^4 - Z^1) + 2\alpha(Z^3 - Z^2)(Z^4 - Z^1)}{(Z^3 - Z^2) + (Z^4 - Z^1)}, & otherwise. \end{cases}$$

From the definition of $\alpha$-optimistic value, the above CCP (266) equivalently becomes

$$Min\ Z''$$

$$s.t.\ the\ constraints\ (262) - (265), \tag{268}$$

where

$Z'' = Z_{sup}(\alpha)$

$$= \begin{cases} (1 - 2\alpha)Z^4 + 2\alpha Z^1, & if\ \alpha \le ((Z^4 - Z^3)/2(Z^4 - Z^1)); \\ 2(1 - \alpha)Z^4 + (2\alpha - 1)Z^1, & if\ \alpha \ge ((2Z^4 - Z^2 - Z^1)/2(Z^4 - Z^1)); \\ \frac{Z^4(Z^3 - Z^2) + Z^3(Z^4 - Z^1) - 2\alpha(Z^3 - Z^2)(Z^4 - Z^1)}{(Z^3 - Z^2) + (Z^4 - Z^1)}, & otherwise. \end{cases}$$

Since for $0.5 < \alpha \le 1$, $Z_{inf}(\alpha) \ge Z_{sup}(\alpha)$, so solving the problems (267) and (267) with trust level $\alpha$ ($0.5 < \alpha \le 1$) we conclude that optimum objective value lie within the range $[Z'', Z']$ with the trust level at least $\alpha$.

In case of models (258) and (259) with unit transportation and additional costs as rough variables, rough CCP can be developed same way as above.

**Rough Expected Value Model:** We find the expected value for the objective function of the model (257) with rough costs, so the problem becomes

$$Min\ E[Z] = E\left[\sum_{i=1}^{m}\sum_{j=1}^{n}\sum_{k=1}^{K} c_{ijk}\, x_{ijk} + \epsilon_{ijk}\right], \tag{269}$$

$$s.t.\ the\ constraints\ (262) - (265).$$

**Deterministic Forms:** From the expected value of a rough variable (Definition 2.15, Example 2.9, Sect. 2.4), the deterministic form of the above problem (269) becomes

$$Min\ E[Z] = (Z^1 + Z^2 + Z^3 + Z^4)/4, \tag{270}$$

$$s.t.\ the\ constraints\ (262) - (265),$$

$$where\ Z^r = \sum_{i=1}^{m}\sum_{j=1}^{n}\sum_{k=1}^{K} (c^r_{ijk}\, x_{ijk} + \epsilon^r_{ijk}),\ r = 1, 2, 3, 4,$$

**Rough Dependent-Chance Programming Model:** The idea of dependent-chance programming is to optimize the chance of an uncertain event. Suppose in view of previous experiment, a decision maker will satisfy with a transportation plan for which the total transportation cost is not exceed a certain value. So a decision maker may fixed a satisfying predetermined maximal objective value, i.e. total transportation cost and maximize the trust value that total transportation cost is not exceed the predetermined cost. So to obtain the most trastable transportation plan with respect to a given predetermined maximal cost $\bar{Z}$ the dependent chance-constrained programming model for the model (257) with rough objective function is formulated as follows:

$$Max \ Tr\{Z = \sum_{i=1}^{m}\sum_{j=1}^{n}\sum_{k=1}^{K} (c_{ijk} \ x_{ijk} + \epsilon_{ijk}) \leq \bar{Z}\}, \quad (271)$$

$$s.t. \ the \ constraints \ (262) - (265),$$

**Deterministic Forms:** The deterministic form of the objective function of (271) can be obtained by the trust of a rough event as discussed in Example 2.9, Sect. 2.4. $Tr\{Z = \sum_{i=1}^{m}\sum_{j=1}^{n}\sum_{k=1}^{K} (c_{ijk} \ x_{ijk} + \epsilon_{ijk}) \leq \bar{Z}\}$, where $Z = ([Z^2, Z^3], [Z^1, Z^4])$, $Z^r$ is given by (260), can be written as the following function:

$$Z' = \begin{cases} 0, & \text{if } \bar{Z} \leq Z^1; \\ \frac{\bar{Z}-Z^1}{2(Z^4-Z^1)}, & \text{if } Z^1 \leq \bar{Z} \leq Z^2; \\ \frac{1}{2}(\frac{\bar{Z}-Z^2}{Z^3-Z^2} + \frac{\bar{Z}-Z^1}{Z^4-Z^1}), & \text{if } Z^2 \leq \bar{Z} \leq Z^3; \\ \frac{1}{2}(\frac{\bar{Z}-Z^1}{Z^4-Z^1} + 1), & \text{if } Z^3 \leq \bar{Z} \leq Z^4; \\ 1, & \text{if } \bar{Z} \geq Z^4. \end{cases} \quad (272)$$

So deterministic form of the above problem (271) becomes Max $Z'$ with respect to the constraints (262)–(265).

## 6.4   Numerical Experiments

**Models with Unit Transportation and Additional Costs as Crisp Numbers:** Here we demonstrate the models with crisp cost parameters for better understanding and to show the efficiency of the models. Consider a problem with three sources ($i = 1, 2, 3$), three destinations ($j = 1, 2, 3$), two types of conveyances ($k = 1, 2$). The unit transportation costs are given in Table 40. The availabilities at each sources, demands of each destinations and capacity of single vehicle of each type of conveyances are given in Table 41.

For convenience suppose additional costs for unit deficit amount is $u_{ijk} = 0.8 \cdot c_{ijk}$.

Now if there are sufficient number of vehicles of each type conveyances available as required (i.e., there is no restriction on number of available vehicles of each type of conveyances), then for the above problem solving the model (257) we have the solution given in Table 42.

So total number of required vehicles of conveyance $k = 1$ is 10 and that of conveyance $k = 2$ is 11.

Now as we say earlier, it may happen that number of vehicles of certain type of conveyance is so limited that it is not sufficient to fulfill its requirement for a transportation system.

Suppose in the above example the number of available vehicles of conveyance $k = 1$ is 14 and that of conveyance $k = 2$ is 10, i.e., $Q_1 = 14$ and $Q_2 = 10$. Then with the same data as given in Tables 40 and 41, solving the model (258) we have the solution given in Table 43.

It should be mentioned that here in case of model (258), if number of available vehicles of each type of conveyances at each source are greater or equal to as required in model (257), i.e., if $Q_1 \geq 10$ and $Q_2 \geq 11$ then model (258) gives the same result as model (257).

Now to demonstrate model (259), consider the same data as given in Tables 40 and 41 and suppose availability of vehicles of each type conveyances at each sources are $V_1^1 = 5$, $V_1^2 = 3$, $V_2^1 = 4$, $V_2^2 = 6$, $V_3^1 = 4$, $V_3^2 = 5$.

Then solving the model (259) we have the solution as presented in Table 44.

**Models with Unit Transportation and Additional Costs as Rough Variables:** Consider the model (257) with three sources ($i = 1, 2, 3$), three destinations ($j = 1, 2, 3$), two types of conveyances ($k = 1, 2$). The unit transportation costs are rough variables as given in Tables 45 and 46.

The availabilities at each sources, demands of each destinations and capacity of single vehicle of each type of conveyances are same as in Table 41.

Table 40. Unit transportation costs $c_{ijk}$

| $i \backslash j$ | 1 | 2 | 3 | 1 | 2 | 3 |
|---|---|---|---|---|---|---|
| 1 | 8 | 11 | 12 | 12 | 9 | 13 |
| 2 | 8 | 10 | 7 | 11 | 8 | 10 |
| 3 | 9 | 14 | 9 | 12 | 10 | 9 |
| $k$ | | 1 | | | 2 | |

Table 41. Availabilities, demands and vehicle capacity.

$a_1 = 25.6$, $a_2 = 16.8$, $a_3 = 32.4$, $b_1 = 14.8$, $b_2 = 26.8$, $b_3 = 23.8$, $q_1 = 2.48$, $q_2 = 3.78$

Table 42. Optimum results for model (257)

$x_{111} = 14.8$, $x_{121} = 2.44$, $x_{221} = 1.68$, $x_{331} = 4.96$, $x_{122} = 7.56$, $x_{222} = 15.12$, $x_{332} = 18.84$, Min $Z = 572.936$, $z_{111} = 6$, $z_{121} = 1$, $z_{221} = 1$, $z_{331} = 2$, $z_{122} = 2$, $z_{222} = 4$, $z_{332} = 5$

**Table 43.** Optimum results for model (258)

---

$x_{111} = 14.8$, $x_{121} = 2.48$, $x_{221} = 1.64$, $x_{231} = 3.82$, $x_{331} = 4.86$, $x_{122} = 7.56$, $x_{222} = 11.34$, $x_{322} = 3.78$, $x_{332} = 15.12$, Min $Z = 579.536$, $z_{111} = 6$, $z_{121} = 1$, $z_{221} = 1$, $z_{231} = 2$, $z_{331} = 2$, $z_{122} = 2$, $z_{222} = 3$, $z_{322} = 1$, $z_{332} = 4$

---

**Table 44.** Optimum results for model (259)

---

$x_{111} = 9.92$, $x_{121} = 1.64$, $x_{221} = 2.48$, $x_{231} = 2.48$, $x_{311} = 4.88$, $x_{331} = 2.48$, $x_{122} = 11.34$, $x_{222} = 11.34$, $x_{332} = 18.9$, Min $Z = 576.54$, $z_{111} = 4$, $z_{121} = 1$, $z_{221} = 1$, $z_{231} = 1$, $z_{311} = 2$, $z_{331} = 1$, $z_{122} = 3$, $z_{222} = 3$, $z_{332} = 5$

---

**Table 45.** Unit transportation costs $c_{ij1}$

| $i \setminus j$ | 1 | 2 | 3 |
|---|---|---|---|
| 1 | ([7,9],[6,10]) | ([10,11],[8,12]) | ([11,13],[10,12]) |
| 2 | ([6,8],[5,9]) | ([9,10],[7,11]) | ([5,7],[4,8]) |
| 3 | ([8,10],[7,11]) | ([13,15],[12,16]) | ([8,10],[7,11]) |

**Table 46.** Unit transportation costs $c_{ij2}$

| $i \setminus j$ | 1 | 2 | 3 |
|---|---|---|---|
| 1 | ([10,12],[9,13]) | ([8,10],[7,11]) | ([12,14],[11,15]) |
| 2 | ([11,12],[9,13]) | ([6,8],[5,9]) | ([9,10],[7,11]) |
| 3 | ([11,12],[10,13]) | ([10,11],[9,12]) | ([8,9],[7,11]) |

For convenience suppose additional costs for unit deficit amount is $u_{ijk} = 0.8 \cdot c_{ijk}$.

**Solution Using Rough CCP:** Now constructing rough CCP as (261)–(265) with trust level $\alpha = 0.9$, we have corresponding deterministic form using (267) as follows:

$$Min \ Z'$$

$$s.t. \ \sum_{j=1}^{3}\sum_{k=1}^{2} x_{ijk} \leq a_i, \quad i = 1, 2, 3,$$

$$\sum_{i=1}^{3}\sum_{k=1}^{2} x_{ijk} \geq b_j, \quad j = 1, 2, 3, \tag{273}$$

$$x_{ijk} \leq z_{ijk} \cdot q_k, \quad i = 1, 2, 3; j = 1, 2, 3; k = 1, 2,$$

$$x_{ijk} \geq 0, \ z_{ijk} \in Z^+, \forall \, i, j, k.$$

where,

$$Z' = \begin{cases} -0.8Z^1 + 1.8Z^4, & \text{if } 0.9 \leq ((Z^2 - Z^1)/2(Z^4 - Z^1)); \\ 0.2Z^1 + 0.8Z^4, & \text{if } 0.9 \geq ((Z^3 + Z^4 - 2Z^1)/2(Z^4 - Z^1)); \\ \frac{Z^1(Z^3 - Z^2) + Z^2(Z^4 - Z^1) + 1.8(Z^3 - Z^2)(Z^4 - Z^1)}{(Z^3 - Z^2) + (Z^4 - Z^1)}, & \text{otherwise.} \end{cases}$$

$$Z^r = \sum_{i=1}^{3} \sum_{j=1}^{3} \sum_{k=1}^{2} (c_{ijk}^r \, x_{ijk} + \epsilon_{ijk}^r), \ r = 1, 2, 3, 4,$$

$\epsilon_{ijk}^r = 0.8 c_{ijk}^r (z_{ijk} \cdot q_k - x_{ijk}), \ r = 1, 2, 3, 4.$

Solving this problem we get the solution presented in Table 47.

**Table 47.** Optimum results for model (257) with transportation costs as rough variables using rough CCP

| |
|---|
| $x_{111} = 12.4$, $x_{121} = 4.87$, $x_{231} = 2.43$, $x_{311} = 2.4$, $x_{331} = 2.46$, $x_{122} = 7.56$, $x_{222} = 14.36$, $x_{332} = 18.9$, Min $Z' = 630.2688$, $z_{111} = 5$, $z_{121} = 2$, $z_{231} = 1$, $z_{311} = 1$, $z_{331} = 1$, $z_{122} = 2$, $z_{222} = 4$, $z_{332} = 5$ |

From this solution we conclude that the objective value will less or equal to 630.2688 with trust level at least 0.9.

We now construct rough CCP as (266) with trust level $\alpha = 0.9$ and then we have corresponding deterministic form using (268) as follows:

$$Min \ Z''$$

$$s.t. \ \sum_{j=1}^{3} \sum_{k=1}^{2} x_{ijk} \leq a_i, \ i = 1, 2, 3,$$

$$\sum_{i=1}^{3} \sum_{k=1}^{2} x_{ijk} \geq b_j, \ j = 1, 2, 3, \tag{274}$$

$$x_{ijk} \leq z_{ijk} \cdot q_k, \ i = 1, 2, 3; j = 1, 2, 3; k = 1, 2,$$

$$x_{ijk} \geq 0, \ z_{ijk} \in Z^+, \forall \, i, j, k.$$

where,

$$Z'' = \begin{cases} -0.8Z^4 + 1.8Z^1, & \text{if } 0.9 \leq ((Z^4 - Z^3)/2(Z^4 - Z^1)); \\ 0.2Z^4 + 0.8Z^1, & \text{if } 0.9 \geq ((2Z^4 - Z^2 - Z^1)/2(Z^4 - Z^1)); \\ \frac{Z^4(Z^3 - Z^2) + Z^3(Z^4 - Z^1) - 1.8(Z^3 - Z^2)(Z^4 - Z^1)}{(Z^3 - Z^2) + (Z^4 - Z^1)}, & \text{otherwise.} \end{cases}$$

Solving this we get $MinZ'' = 471.427$. So the objective value will greater or equal to 471.427 with trust level at least 0.9.

As we know for $0.5 < \alpha \leq 1$, $Z_{inf}(\alpha) \geq Z_{sup}(\alpha)$, here our results ($Z' > Z''$) shows this truth. Finally we can conclude that the optimum objective value lie within the range $[471.427, 630.2688]$ with trust level at least 0.9.

**Table 48.** Optimum results using rough expected value

| |
| --- |
| $x_{111} = 12.4$, $x_{121} = 4.86$, $x_{231} = 2.42$, $x_{311} = 2.4$, $x_{331} = 2.48$, $x_{122} = 7.56$, |
| $x_{222} = 14.38$, $x_{332} = 18.9$, Min $E[Z] = 547.358.$, $z_{111} = 5$, $z_{121} = 2$, $z_{231} = 1$, |
| $z_{311} = 1$, $z_{331} = 1$, $z_{122} = 2$, $z_{222} = 4$, $z_{332} = 5$ |

**Solution Using Rough Expected Value:** To solve the current problem using rough expected value, we use the rough expected value model (269). Then using its deterministic form (270) and solving it we get the solution presented in Table 48.

So we see that the expected objective value lie within the range of objective value as obtained by rough CCP.

**Solution Using Rough Dependent CCP:** The objective of this model is that for a predetermined maximal objective value find a solution with maximum satisfied trust level so that the optimum objective value is not more than that predetermined value. For the current problem, suppose the decision maker satisfied with a transportation plan for which the objective value is not exceed 600. So construct the problem as (271) with $\bar{Z} = 600$. Then using (272) we find the maximum trust level Max $\alpha = \alpha' = 0.843$ and the corresponding transportation planing is presented in Table 49.

**Table 49.** Optimum results using rough dependent CCP

| |
| --- |
| $x_{111} = 12.4$, $x_{121} = 4.895$, $x_{231} = 2.455$, $x_{311} = 2.4$, $x_{331} = 2.444$, $x_{122} = 7.56$, |
| $x_{222} = 14.344, x_{332} = 18.9$, $z_{111} = 5$, $z_{121} = 2$, $z_{231} = 1$, |
| $z_{311} = 1$, $z_{331} = 1$, $z_{122} = 2$, $z_{222} = 4$, $z_{332} = 5$ |

## 6.5    Overall Conclusion

This section presents solid transportation model for the transportation system where full vehicles are used for transportation so that unit transportation costs are determined according to full utilization of the vehicle capacity. To deal with different situations like availability of each type of conveyances, whether the available vehicles at one source can be utilized at another source or not, this presented model is extended to different models with different constraints. STP with different types of uncertain variables such as fuzzy, random, fuzzy random are discussed by many researchers, but STP with rough variables is not discussed before. In this paper we only assume the unit transportation costs as rough variables, the STP with all the parameters, i.e., costs, availabilities, demands, conveyance capacities as rough variables may be taken as a future work.

# 7    Overall Contribution and Future Extension

In this thesis, we have discussed several useful transportation models in different uncertain (e.g. fuzzy, type-2 fuzzy, rough, linguistic) environments. The thesis broadly addresses the following major sub-topics, namely:

- Transportation modeling with fuzzy parameters.
- Transportation modeling with type-2 fuzzy parameters.
- Transportation modeling with rough parameters.
- Transportation mode selection with linguistic information.

In Sect. 3 of this Article, a multi-objective solid transportation problem with type-1 fuzzy parameters is formulated and solved. In this problem, a fuzzy budget amount for each destination is imposed so that total transportation cost should not exceed that budget amount. In the budget constraint, both left and right sides have fuzzy quantity. To deal with such type of constraints, a deterministic form is derived by the idea of chance-constraint. Here, we have also formulated a general model (MOMISTP) to deal with transportation problem with multiple objectives and several types of goods to be transported. In this problem the corresponding parameters are taken as fuzzy numbers. A defuzzification process to find crisp values of corresponding fuzzy resources, demands and conveyance capacities is introduced so that the conditions that total available resources and total conveyance capacities are greater than or equal to the total demands must be satisfied. We have discussed that some well established methods like expected value model may not yield any feasible solution for the problem having constraints with such type of conditional relations. The idea of minimum of fuzzy numbers is also applied to the fuzzy objective function and we obtained fuzzy solution for the objective function with coefficients as fuzzy numbers.

For high computational complexity, there are very few methods available to deal with type-2 fuzzy set. In Sect. 4, we have proposed a defuzzification method of type-2 fuzzy variables. We have also introduced an interval approximation method of continuous type-2 fuzzy variables. For the first time, different transportation problems with type-2 fuzzy parameters are formulated and solved. Defuzzification method is applied to solve a FCTP with type-2 fuzzy cost parameters. A chance-constrained programming model is formulated using generalized credibility measure to solve a FCTP with type-2 fuzzy parameters. A

MISTP having restriction on conveyances is formulated with type-2 fuzzy parameters. A deterministic form for the problem is obtained by applying interval analysis using the interval approximations of continuous type-2 fuzzy variables.

In Sect. 5, we have proposed a computationally efficient fuzzy MCGDM method based on a ranking function which is defined based on credibility measure to rank a fuzzy number over another fuzzy number. The proposed FMCGDM method is successfully applied to transportation mode selection problem with linguistic terms generated by fuzzy numbers. We have also proposed a computationally efficient fuzzy MCGDM method based on a ranking interval type-2 fuzzy variables. This proposed FMCGDM method is applied to a transportation mode selection problem where linguistic ratings of the alternatives and criteria weights are represented by IT2 FVs.

In remaining part of the this Article (Sect. 6), a practical solid transportation model is formulated considering per trip capacity for each type of conveyances. This is applicable for the system in which full vehicles, e.g. trucks, rail coaches are to be booked for transportation of products so that transportation cost is determined on the basis of full conveyances. We have represented fluctuating cost parameters by rough variables. To solve the problem with rough cost parameters, we have used rough chance constrained programming model, rough expected value model and rough dependent-chance programming model developed on the basis of trust measure theory.

**Future Extension:** Improvement/development in existing transportation models is a major issue in transportation research. To overcome different types of increased complexities and new challenges model should be adaptively changed and solution strategies should be developed. The transportation models presented in the thesis also can be extended to form different types of realistic models. For example, for transportation of several types of items, optimal distribution of available vehicle capacity among the items is a very important issue. In such case, space constraints can be implemented considering amount of goods, preferability of goods to be transported, availability of vehicle capacity, etc. For transportation of highly breakable items (e.g. glass-goods, toys, ceramic goods, etc.), the breakability issue should be considered. Also safety of transportation of goods through a particular route (specially in roadways due to land slide, insurgency, robbery, bad road, etc.) is also very important in the transportation system. So consideration of safety factor of the routes may be taken into account as an additional objective or a constraint.

In Sect. 4, the interval approximation method of continuous type-2 fuzzy variables is illustrated with type-2 triangular fuzzy variable. This interval approximation method can be applied to other T2 FVs such as type-2 normal fuzzy variable, type-2 gamma fuzzy variable, etc.

In Sect. 6, the solid transportation model is formulated with only the unit transportation costs as rough variables, the STP with all the parameters, i.e., costs, availabilities, demands, conveyance capacities as rough variables may be taken as a future work.

The models formulated in this dissertation can be formulated and solved in fuzzy random, random fuzzy, fuzzy rough and bifuzzy environments with unit transportation costs, sources, demands, conveyance capacities, etc. as the corresponding imprecise parameters/variables.

**Acknowledgements.** I would like to acknowledge my deepest regards, sincere appreciation and heartfelt thanks to my supervisors Dr. Samarjit Kar and Prof. Manoranjan Maiti for their kind supervision and guidance. I would like to express my sincere thanks to the faculty members of Department of Mathematics, National Institute of Technology Durgapur, India.

I am thankful to Prof. Andrzej Skowron for his valuable suggestions and inspiring comments. I am especially thankful to my friends Dr. Partha Pratim Gop Mandal, Surajit Dan, Dr. Om Prakash and Dr. Anirban Saha for their enormous help and support.

Above all I would like to pay sincere respect and gratitude to my family members for their unconditional love and support throughout my life. Lastly, I pay deep respect to the Almighty who had made everything possible.

# References

1. Abbasbandy, S., Asady, B.: Ranking of fuzzy numbers by sign distance. Inf. Sci. **176**, 2405–2416 (2006)
2. Abbasbandy, S., Hajjari, T.: A new approach for ranking of trapezoidal fuzzy numbers. Comput. Math. Appl. **57**, 413–419 (2009)
3. Adlakha, V., Kowalski, K.: On the fixed-charge transportation problem. Omega **27**, 381–388 (1999)
4. Adlakha, V., Kowalski, K., Vemuganti, R.R., Lev, B.: More-for-less algorithm for fixed-charge transportation problems. Omega **35**, 116–127 (2007)
5. Aliev, R.A., Pedrycz, W., Guirimov, B., Aliev, R.R., Ilhan, U., Babagil, M., Mammadli, S.: Type-2 fuzzy neural networks with fuzzy clustering and differential evolution optimization. Inf. Sci. **181**(9), 1591–1608 (2011)
6. Ammar, E.E., Youness, E.A.: Study on multiobjective transportation problem with fuzzy numbers. Appl. Math. Comput. **166**, 241–253 (2005)
7. Anand, M.D., Kumanan, T.S.S., Johnny, M.A.: Application of multi-criteria decision making for selection of robotic system using fuzzy analytic hierarchy process. Int. J. Manage. Decis. Making **9**(1), 75–98 (2008)
8. Asady, B., Zendehnam, A.: Ranking fuzzy numbers by distance minimization. Appl. Math. Model. **31**, 2589–2598 (2007)
9. Azadeh, A., Saberi, M., Moghaddam, R.T., Javanmardi, L.: An integrated data envelopment analysis-artificial neural network-rough set algorithm for assessment of personnel efficiency. Expert Syst. Appl. **38**(3), 1364–1373 (2011)
10. Baleentis, T., Zeng, S.: Group multi-criteria decision making based upon interval-valued fuzzy numbers: an extension of the MULTIMOORA method. Expert Syst. Appl. **40**(2), 543–550 (2013)
11. Balinski, M.L.: Fixed cost transportation problems. Nav. Res. Logist. Q. **8**, 41–54 (1961)
12. Bector, C.R., Chandra, S.: Fuzzy Mathematical Programming and Fuzzy Matrix Games. Springer, Heidelberg (2005)

13. Bit, A.K., Biswal, M.P., Alam, S.S.: Fuzzy programming approach to multi-objective solid transportation problem. Fuzzy Sets Syst. **57**, 183–194 (1993)
14. Buckley, J.J., Feuring, T., Hayashi, Y.: Solving fuzzy problems in operations research: inventory control. Soft. Comput. **7**, 121–129 (2002)
15. Chakraborty, A., Chakraborty, M.: Cost-time minimization in a transportation problem with fuzzy parameters: a case study. J. Transp. Syst. Eng. Inf. Technol. **10**(6), 53–63 (2010)
16. Chan, F.T.S., Kumar, N.: Global supplier development considering risk factors using fuzzy extended AHP-based approach. Omega **35**(4), 417–431 (2007)
17. Chanas, S., Kolosziejczyj, W., Machaj, A.: A fuzzy approach to the transportation problem. Fuzzy Sets Syst. **13**, 211–221 (1984)
18. Chanas, S., Kuchta, D.: A concept of the optimal solution of the transportation problem with fuzzy cost coefficients. Fuzzy Sets Syst. **82**, 299–305 (1996)
19. Chen, S.: Ranking fuzzy numbers with maximizing set and minimizing set. Fuzzy Sets Syst. **17**, 113–129 (1985)
20. Chen, C.T.: Extensions to the TOPSIS for group decision-making under fuzzy environment. Fuzzy Sets Syst. **114**, 1–9 (2000)
21. Chen, T.Y., Chang, C.H., Lu, J.R.: The extended QUALIFLEX method for multiple criteria decision analysis based on interval type-2 fuzzy sets and applications to medical decision making. Eur. J. Oper. Res. **226**(3), 615–625 (2013)
22. Chen, S.H., Hsieh, C.H.: Representation, ranking, distance, and similarity of L-R type fuzzy number and application. Aust. J. Intell. Inf. Proc. Syst. **6**, 217–229 (2000)
23. Chen, S.M., Lee, L.W.: Fuzzy multiple attributes group decision-making based on the ranking values and the arithmetic operations of interval type-2 fuzzy sets. Expert Syst. Appl. **37**(1), 824–833 (2010)
24. Chen, S.M., Lee, L.W.: Fuzzy multiple attributes group decision-making based on interval type-2 TOPSIS method. Expert Syst. Appl. **37**(4), 2790–2798 (2010)
25. Chen, S.M., Wang, C.Y.: Fuzzy decision making systems based on interval type-2 fuzzy sets. Inf. Sci. **242**, 1–21 (2013)
26. Chen, S.M., Yang, M.Y., Lee, L.W., Yang, S.W.: Fuzzy multiple attributes group decision-making based on ranking interval type-2 fuzzy sets. Expert Syst. Appl. **39**, 5295–5308 (2012)
27. Cheng, C.H.: A new approach for ranking fuzzy numbers by distance method. Fuzzy Sets Syst. **95**, 307–317 (1998)
28. Cheng, C.H., Lin, Y.: Evaluating the best main battle tank using fuzzy decision theory with linguistic criteria evaluation. Eur. J. Oper. Res. **142**, 174–186 (2002)
29. Chu, T.C., Tsao, C.T.: Ranking fuzzy numbers with an area between the centroid point and original point. Comput. Math. Appl. **43**, 111–117 (2002)
30. Coupland S.: Type-2 fuzzy sets: geometric defuzzification and type reduction. In: Proceedings of the IEEE Symposium on Foundations of Computational Intelligence, Honolulu, HI, pp. 622–629 (2007)
31. Coupland, S., John, R.: A fast geometric method for defuzzification of type-2 fuzzy sets. IEEE Trans. Fuzzy Syst. **16**(4), 929–941 (2008)
32. Dalalah, D., Hayajneh, M., Batieha, F.: A fuzzy multi-criteria decision making model for supplier selection. Expert Syst. Appl. **38**(7), 8384–8391 (2011)
33. Dantzig, G.B.: Application of the simplex method to a transportation problem, Chapter XXII. In: Koopmans, T.C. (ed.) Active Analysis of Production and Allocation. Wiley, New York (2011)
34. Dantzig, G.B.: Linear Programming and Extensions. Princeton University Press, Princeton (1963)

35. Dembczynski, K., Greco, S., Slowinski, R.: Rough set approach to multiple criteria classification with imprecise evaluations and assignments. Eur. J. Oper. Res. **198**, 626–636 (2009)
36. Dey, P.K., Yadav, B.: Approuch to defuzzify the trapezoidal fuzzy number in transportation problem. Int. J. Comput. Cogn. **8**(4), 64–67 (2010)
37. Ding, J.F., Liang, G.S.: Using fuzzy MCDM to select pertners selection of strategic alliances for linear shipping. Inf. Sci. **173**, 197–225 (2005)
38. Dubois, D., Prade, H.: Possibility Theory: An Approach to Computerized Processing of Uncertainty. Plenum, New York (1998)
39. Dursun, M., Karsak, E.E., Karadayi, M.A.: A fuzzy multi-criteria group decision making framework for evaluating health-care waste disposal alternatives. Expert Syst. Appl. **38**(9), 11453–11462 (2011)
40. Ertay, T., Buyukozkan, G., Kahraman, C., Ruan, D.: Quality function deployment implementation based on analytical network process with linguistic data: an application in automotive industry. J. Intell. Fuzzy Syst. **16**(3), 221–232 (2005)
41. Eskigun, E., Uzsoy, R., Preckel, P.V., Beaujon, G., Krishnan, S., Tew, J.D.: Outbound supply chain network design with mode selection, lead times and capacitated vehical destribution centers. Eur. J. Oper. Res. **165**, 182–206 (2005)
42. Fegad, M.R., Jadhav, V.A., Muley, A.A.: Finding an optimal solution of transportation problem using interval and triangular membership functions. Eur. J. Sci. Res. **60**(3), 415–421 (2011)
43. Figueroa-García, J.C., Hernández, G.: A transportation model with interval type-2 fuzzy demands and supplies. In: Huang, D.-S., Jiang, C., Bevilacqua, V., Figueroa, J.C. (eds.) ICIC 2012. LNCS, vol. 7389, pp. 610–617. Springer, Heidelberg (2012)
44. Fortemps, P., Roubens, M.: Ranking and defuzzification methods based on area compensation. Fuzzy Sets Syst. **82**, 319–330 (1996)
45. Fu, G.: A fuzzy optimization method for multicriteria decision making: an application to reservoir flood control operation. Expert Syst. Appl. **34**(1), 145–149 (2008)
46. Gao, S.P., Liu, S.Y.: Two-phase fuzzy algorithms for multi-objective transportation problem. J. Fuzzy Math. **12**(1), 147–155 (2004)
47. Gass, S.I.: On solving the transportation problem. J. Oper. Res. Soc. **41**, 291–297 (1990)
48. Gen, M., Ida, K., Li, Y.: Solving bicriteria solid transportation problem by genetic algorithm. In: 1994 IEEE International Conference on Systems, Man, and Cybernetics, Humans, Information and Technology, vol. 2, pp. 1200–1207 (1994)
49. Geoffrion, A.M.: Generalised benders decomposition. J. Optim. Theory Appl. **10**(4), 237–260 (1972)
50. Greig, D.M.: Optimization, pp. 100–111. Lonman Group Limited, London (1980)
51. Greenfield, S., John, R.I., Coupland, S.: A novel sampling method for type-2 defuzzification. In: Proceedings of UKCI, London, pp. 120–127 (2005)
52. Grzegorzewski, P.: Nearest interval approximation of a fuzzy number. Fuzzy Sets Syst. **130**, 321–330 (2002)
53. Haley, K.B.: The sold transportation problem. Oper. Res. **10**, 448–463 (1962)
54. Hasuike, T., Ishi, H.: A type-2 fuzzy portfolio selection problem considering possibilistic measure and crisp possibilistic mean value. In: IFSA-EUSFLAT, pp. 1120–1125 (2009)
55. Hatami-Marbini, A., Tavana, M.: An extension of the Electre I method for group decision-making under a fuzzy environment. Omega **39**, 373–386 (2011)

56. Hirano, S., Tsumoto, S.: Rough representation of a region of interest in medical images. Int. J. Approximate Reasoning **40**, 23–34 (2005)
57. Hirch, W.M., Dantzig, G.B.: The fixed charge transportation problem. Nav. Res. Logist. Q. **15**, 413–424 (1968)
58. Hitchcock, F.L.: The distribution of product from several sources to numerous localities. J. Math. Phys. **20**, 224–230 (1941)
59. Hsieh, C.H.: Optimazition of fuzzy inventory model under fuzzy demand and fuzzy lead time. Tamsui Oxf. J. Manage. Sci. **20**, 21–35 (2005)
60. Jiménez, F., Verdegay, J.L.: Uncertain solid transportation problems. Fuzzy Sets Syst. **100**, 45–57 (1998)
61. Jiménez, F., Verdegay, J.L.: Solving fuzzy solid transportation problems by an evolutionary algorithm based parametric approach. Eur. J. Oper. Res. **117**, 485–510 (1999)
62. Kaufmann, A.: Introduction to the Theory of Fuzzy Subsets, vol. I. Academic Press, New York (1975)
63. Karnik, N.N., Mendel, J.M.: Centroid of a type-2 fuzzy set. Inf. Sci. **132**, 195–220 (2001)
64. Kaur, A., Kumar, A.: A new approach for solving fuzzy transportation problems using generalized trapezoidal fuzzy numbers. Appl. Soft Comput. **12**(3), 1201–1213 (2012)
65. Kiesmüller, G.P., de Kok, A.G., Fransoo, J.C.: Transportation mode selection with positive manufacturing lead time. Transp. Res. Part E **41**, 511–530 (2005)
66. Kikuchi, S.: A method to defuzzify the fuzzy number: transportation problem application. Fuzzy Sets Syst. **116**, 3–9 (2000)
67. Kirca, O., Satir, A.: A heuristic for obtaining an initial solution for the transportation problem. J. Oper. Res. Soc. **41**, 865–871 (1990)
68. Klir, G.J., Yuan, B.: Fuzzy Sets and Fuzzy Logic: Theory and Applications. Prentice-Hall Inc., N.J. (1996)
69. Koopmans, T.C.: Optimum utilization of the transportation system. Econmetrica **17**, 3–4 (1949)
70. Kowalski, K., Lev, B.: On step fixed charge transportation problem. Omega **36**(5), 913–917 (2008)
71. Kumru, M., Kumru, P.Y.: Analytic hierarchy process application in selecting the mode of transport for a logistics company. J. Adv. Transp. (2013). doi:10.1002/atr.1240
72. Kundu, P., Kar, S., Maiti, M.: Multi-objective multi-item solid transportation problem in fuzzy environment. Appl. Math. Model. **37**, 2028–2038 (2013)
73. Kundu, P., Kar, S., Maiti, M.: Some solid transportation models with crisp and rough costs. Int. J. Math. Comput. Phys. Quant. Eng. **7**(1), 8–15 (2013)
74. Kundu, P., Kar, S., Maiti, M.: Multi-objective solid transportation problems with budget constraint in uncertain environment. Int. J. Syst. Sci. **45**(8), 1668–1682 (2014)
75. Kundu, P., Kar, S., Maiti, M.: Fixed charge transportation problem with type-2 fuzzy variables. Inf. Sci. **255**, 170–186 (2014)
76. Kundu, P., Kar, S., Maiti, M.: Multi-item solid transportation problem with type-2 fuzzy parameters. Appl. Soft Comput. **31**, 61–80 (2015)
77. Kundu, P.: Some transportation problems under uncertain environments, Ph.D. thesis, National Institute of Technology Durgapur, Durgapur (2014). Supervisors: Dr. Samarjit Kar and Prof. Manoranjan Maiti

78. Lee, H.-S.: A fuzzy multi-criteria decision making model for the selection of the distribution center. In: Wang, L., Chen, K., S. Ong, Y. (eds.) ICNC 2005. LNCS, vol. 3612, pp. 1290–1299. Springer, Heidelberg (2005)

79. Lee, S.M., Moor, L.J.: Optimizing transportation problems with multiple objectives. AIIE Trans. **5**(4), 333–338 (1973)

80. Li, L., Lai, K.K.: A fuzzy approach to the multi-objective transportation problem. Comput. Oper. Res. **27**, 43–57 (2000)

81. Li, Y., Ida, K., Gen, M.: Improved genetic algorithm for solving multi-objective solid transportation problem with fuzzy numbers. Comput. Ind. Eng. **33**(3–4), 589–592 (1997)

82. Liang, T.F.: Applying fuzzy goal programming to production/transportation planning decisions in a supply chain. Int. J. Syst. Sci. **38**(4), 293–304 (2007)

83. Lin, T., Yao, Y., Zadeh, L.: Data Mining, Rough Sets and Granular Computing. Springer, Heidelberg (2002)

84. Liu, B.: Minimax chance constrained programming model for fuzzy decision systems. Inf. Sci. **112**(1–4), 25–38 (1998)

85. Liu, B.: Dependent-chance programming with fuzzy decisions. IEEE Trans. Fuzzy Syst. **7**(3), 354–360 (1999)

86. Liu, B.: Theory and Practice of Uncertain Programming. Physica-Verlag, Heidelberg (2002)

87. Liu, B.: Inequalities and convergence concepts of fuzzy and rough variables. Fuzzy Optim. Decis. Making **2**, 87–100 (2003)

88. Liu, B.: Uncertainty Theory: An Introduction to its Axiomatic Foundations. Springer, Berlin (2004)

89. Liu, B.: A survey of credibility theory. Fuzzy Optim. Decis. Making **5**(4), 387–408 (2006)

90. Liu, B.: Theory and Practice of Uncertain Programming, 3rd edn. UTLAB (2009). http://orsc.edu.cn/liu/up.pdf

91. Liu, F.: An efficient centroid type-reduction strategy for general type-2 fuzzy logic system. Inf. Sci. **178**, 2224–2236 (2008)

92. Liu, B., Iwamura, K.: Chance constrained programming with fuzzy parameters. Fuzzy Sets Syst. **94**(2), 227–237 (1998)

93. Liu, L., Lin, L.: Fuzzy fixed charge solid transportation problem and its algorithm. Fuzzy Syst. Knowl. Discov. **3**, 585–589 (2007)

94. Liu, B., Liu, Y.K.: Expected value of fuzzy variable and fuzzy expected value models. IEEE Trans. Fuzzy Syst. **10**, 445–450 (2002)

95. Liu, F., Mendel, J.M.: Encoding words into interval type-2 fuzzy sets using an interval approach. IEEE Trans. Fuzzy Syst. **16**(6), 1503–1521 (2008)

96. Liu, G., Zhu, W.: The algebraic structures of generalized rough set theory. Inf. Sci. **178**, 4105–4113 (2008)

97. Liu, L., Zhu, Y.: Rough variables with values in measurable spaces. Inf. Sci. **177**, 4678–4685 (2007)

98. Liu, S.T., Kao, C.: Solving fuzzy transportation problems based on extension principle. Eur. J. Oper. Res. **153**(3), 661–674 (2004)

99. Liu, Z.Q., Liu, Y.K.: Type-2 fuzzy variables and their arithmetic. Soft. Comput. **14**, 729–747 (2010)

100. Liou, T.S., Wang, M.J.: Ranking fuzzy numbers with integral value. Fuzzy Sets Syst. **50**(3), 247–255 (1992)

101. Mendel, J.M.: Uncertain Rule-Based Fuzzy Logic Systems: Introduction and New Directions. Prentice-Hall, NJ (2001)

102. Mendel, J.M.: Fuzzy sets for words: a new beginning. In: Proceedings of IEEE International Conference on Fuzzy Systems, St. Louis, MO, pp. 37–42 (2003)
103. Mendel, J.M.: Computing with words: Zadeh, Turing, Popper and Occam. IEEE Comput. Intell. Mag. **2**(4), 10–17 (2007)
104. Mendel, J.M.: Computing with words and its relationships with fuzzistics. Inf. Sci. **177**, 988–1006 (2007)
105. Mendel, J.M., John, R.I.: Type-2 fuzzy sets made simple. IEEE Trans. Fuzzy Syst. **10**(2), 307–315 (2002)
106. Mendel, J.M., John, R.I.: Advances in type-2 fuzzy sets and systems. Inf. Sci. **177**(1), 84–110 (2007)
107. Mendel, J.M., John, R.I., Liu, F.L.: Interval type-2 fuzzy logical systems made simple. IEEE Trans. Fuzzy Syst. **14**(6), 808–821 (2006)
108. Mendel, J.M., Wu, H.: Type-2 fuzzistics for symmetric interval type-2 fuzzy sets: Part 1, forward problems. IEEE Trans. Fuzzy Syst. **14**(6), 781–792 (2006)
109. Miettinen, K.M.: Non-linear Multi-objective Optimization. Kluwers International Series. Kluwer Academic Publishers, London (1999)
110. Mikhailov, L., Sing, M.G.: Fuzzy analytic network process and its application to the development of decision support sustems. IEEE Trans. Syst. Man Cybern. **33**(1), 33–41 (2003)
111. Mikhailov, L., Tsvetinov, P.: Evaluation of services using a fuzzy analytical hierarchy process. Appl. Soft Comput. **5**, 23–33 (2004)
112. Mitchell, H.B.: Pattern recognition using type-2 fuzzy sets. Inf. Sci. **170**, 409–418 (2005)
113. Monahan, J.P., Berger, P.D.: A transportation mode selection model for a consolidation warehouse system. Math. Methods Oper. Res. **21**(5), 211–222 (1977)
114. Mondal, M., Maity, A.K., Maiti, M.K., Maiti, M.: A production-repairing inventory model with fuzzy rough coefficients under inflation and time value of money. Appl. Math. Model. **37**, 3200–3215 (2013)
115. Moore, R.E.: Interval Analysis. Prentice-Hall, Englewood Cliffs (1966)
116. Mula, J., Poler, R., Garcia-Sabater, J.P.: Material requirement planing with fuzzy constraints and fuzzy coefficients. Fuzzy Sets Syst. **158**, 783–793 (2007)
117. Nahmias, S.: Fuzzy variable. Fuzzy Sets Syst. **1**, 97–101 (1978)
118. Ojha, A., Das, B., Mondal, S., Maity, M.: An entropy based solid transportation problem for general fuzzy costs and time with fuzzy equality. Math. Comput. Model. **50**(1–2), 166–178 (2009)
119. Palekar, U.S., Karwan, M.K., Zionts, S.: A branch-and-bound method for the fixed charge transportation problem. Manage. Sci. **36**, 1092–1105 (1990)
120. Pandian, P., Anuradha, D.: A new approach for solving solid transportation problems. Appl. Math. Sci. **4**(72), 3603–3610 (2010)
121. Pawlak, Z.: Rough sets. Int. J. Inf. Comput. Sci. **11**(5), 341–356 (1982)
122. Pawlak, Z.: Rough Sets - Theoretical Aspects of Reasoning About Data. Kluwer Academatic Publishers, Boston (1991)
123. Pawlak, Z., Slowinski, R.: Rough set approach to multi-attribute decision analysis (invited review). Eur. J. Oper. Res. **72**, 443–459 (1994)
124. Pawlak, Z., Skowron, A.: Rough sets: some extensions. Inf. Sci. **177**, 28–40 (2007)
125. Polkowski, L.: Rough Sets, Mathematical Foundations, Advances in Soft Computing. Physica Verlag, A Springer-Verlag Company, Heidelberg (2002)
126. Pramanik, S., Roy, T.K.: Intuitionistic Fuzzy goal programming and its application in solving multi-objective transportation problems. Tamsui Oxf. J. Manage. Sci. **23**(1), 1–17 (2007)

127. Qin, R., Liu, Y.K., Liu, Z.Q.: Methods of critical value reduction for type-2 fuzzy variables and their applications. J. Comput. Appl. Math. **235**, 1454–1481 (2011)
128. Ramakrishnan, C.S.: An improvement to Goyals modiffed VAM for the unbalanced transportation problem. J. Oper. Res. Soc. **39**, 609–610 (1988)
129. Saad Omar, M., Abass Samir, A.: A Parametric study on transportation problem under fuzzy environment. J. Fuzzy Math. **11**(1), 115–124 (2003)
130. Schell, E.D.: Distribution of a product by several properties. In: Proceedings 2nd Symposium in Linear Programming, DCS/comptroller, HQ US Air Force, Washington D.C., pp. 615–642 (1955)
131. Shafiee, M., Shams-e-alam, N.: Supply chain performance evaluation with rough data envelopment analysis. In: 2010 International Conference on Business and Economics Research, vol. 1. IACSIT Press, Kuala (2011)
132. Tao, Z., Xu, J.: A class of rough multiple objective programming and its application to solid transportation problem. Inf. Sci. **188**, 215–225 (2012)
133. Tsumoto, S.: Mining diagnostic rules from clinical databases using rough sets and medical diagnostic model. Inf. Sci. **162**(2), 65–80 (2004)
134. Tuzkaya, U.R., Önüt, S.: A fuzzy analytical network process based approach to transportation-mode selection between Turkey and Germany: a case study. Inf. Sci. **178**, 3133–3146 (2008)
135. El-Wahed, W.F.A.: A multi-objective transportation problem under fuzziness. Fuzzy Sets Syst. **117**, 27–33 (2001)
136. Wang, P.: Fuzzy contactability and fuzzy variables. Fuzzy Sets Syst. **8**, 81–92 (1982)
137. Wang, Y.J.: Fuzzy multi-criteria decision-making based on positive and negetive extreme solutions. Appl. Math. Model. **35**, 1994–2004 (2011)
138. Wang, H., Lee, C.Y.: Production and transport logistics scheduling with two transport mode choices. Naval Res. Logistics **52**, 796–809 (2005)
139. Wang, Y.J., Lee, H.S.: Generalizing TOPSIS for fuzzy multi-criteria group decision-making. Comput. Math. Appl. **53**, 1762–1772 (2007)
140. Wang, Y.J., Lee, H.S., Lin, L.: Fuzzy TOPSIS for multi-criteria decision-making. Int. Math. J. **3**(4), 367–379 (2003)
141. Wang, Y.M., Elhag, T.M.S.: Fuzzy TOPSIS method based on alpha level sets with an application to bridge risk assessment. Expert Syst. Appl. **31**, 309–319 (2006)
142. Wang, Y.M., Parkan, C.: Multiple attribute decision making based on fuzzy preference information on alternatives: ranking and weighting. Fuzzy Sets Syst. **153**(3), 331–346 (2005)
143. Wang, Y.M., Yang, J.B., Xu, D.L., Chin, K.S.: On the centroids of fuzzy numbers. Fuzzy Sets Syst. **157**, 919–926 (2006)
144. Wang, X., Ruan, D., Kerre, E.E.: Mathematics of Fuzziness - Basic Issues. Springer, Heidelberg (2009)
145. Wu, H.C.: The central limit theorems for fuzzy random variables. Inf. Sci. **120**, 239–256 (1999)
146. Wu, D., Mendel, J.M.: Uncertainty measures for interval type-2 fuzzy sets. Inf. Sci. **177**, 5378–5393 (2007)
147. Wu, D., Tan, W.W.: Computationally efficient type-reduction strategies for a type-2 fuzzy logic controller. In: Proceedings of IEEE FUZZ Conference, Reno, NV, pp. 353–358 (2005)
148. Wu, Z., Chen, Y.: The maximizing deviation method for group multiple attribute decision making under linguistic environment. Fuzzy Sets Syst. **158**(14), 1608–1617 (2007)

149. Xiao, S., Lai, E.M.-K.: A Rough programming approach to power-aware VLIW instruction scheduling for digital signal processors. In: ICASSP (2005)
150. Xu, J., Yao, L.: A class of two-person zero-sum matrix games with rough payoffs. Int. J. Math. Math. Sci. (2010). doi:10.1155/2010/404792
151. Xu, J., Li, B., Wu, D.: Rough data envelopment analysis and its application to supply chain performance evaluation. Int. J. Prod. Econ. **122**, 628–638 (2009)
152. Yang, L., Gao, Z., Li, K.: Railway freight transportation planning with mixed uncertainty of randomness and fuzziness. Appl. Soft Comput. **11**, 778–792 (2011)
153. Yang, L., Liu, L.: Fuzzy fixed charge solid transportation problem and algorithm. Appl. Soft Comput. **7**, 879–889 (2007)
154. Yuan, Y.: Criteria for evaluating fuzzy ranking methods. Fuzzy Sets Syst. **44**, 139–157 (1991)
155. Yuste, A.J., Triviño, A., Casilari, E.: Type-2 fuzzy decision support system to optimise MANET integration into infrastructure-based wireless. Expert Syst. Appl. **40**(7), 2552–2567 (2003)
156. Zadeh, L.A.: Fuzzy sets. Inf. Control **8**, 338–353 (1965)
157. Zadeh, L.A.: The concept of a linguistic variable and its application to appromximate resoning -I. Inf. Sci. **8**, 199–249 (1975)
158. Zadeh, L.A.: Fuzzy sets as a basis for a theory of possibility. Fuzzy Sets Syst. **1**, 3–28 (1978)
159. Zimmermann, H.J.: Fuzzy programming and linear programming with several objective functions. Fuzzy Sets Syst. **1**, 45–55 (1978)
160. Zimmermann, H.-J.: Fuzzy Set Theory-and its Applications, 3rd edn. Kluwer Academic Publishers, Boston (1996)
161. Zhang, Q., Fan, Z., Pan, D.: A ranking approach for interval numbers in uncertain multiple attribute decision making problems. Syst. Eng. Theory Pract. **5**, 129–133 (1999)
162. Zhang, D., Wang, Y., Huang, H.: Rough neural network modeling based on fuzzy rough model and its application to texture classification. Neurocomputing **72**(10–12), 2433–2443 (2009)
163. Zhang, Z., Shi, Y., Gao, G.: A rough set-based multiple criteria linear programming approach for the medical diagnosis and prognosis. Expert Syst. Appl. **36**(5), 8932–8937 (2009)

# Author Index

Printed in the United States
By Bookmasters